D0601663

Dictionary of Ceramic Science and Engineering

Dictionary of Ceramic Science and Engineering

Loran S. O'Bannon

Senior Technical Advisor (Retired)
Battelle Columbus Laboratories

Past President
Fellow
Honorary Life Member
The American Ceramic Society

Plenum Press • New York and London

Library of Congress Cataloging in Publication Data

O'Bannon, Loran S., date–
Dictionary of ceramic science and engineering.

Bibliography: p.
Includes index.
1. Ceramics—Dictionaries. I. Title.
TP788.O2 1983 666′03′21 83-13936
ISBN 0-306-41324-8

A Division of Plenum Publishing Corporation
233 Spring Street, New York, N.Y. 10013

Printed in the United States of America

TO MARIE

whose gentle but persevering encouragement,
whose most valuable suggestions and criticisms,
and whose labors have contributed so much to this effort

this work is dedicated.

Preface

Ceramic science and engineering is one of the oldest, yet one of the newest, fastest growing, and most thoroughly exciting of man's endeavors. The products of the modern ceramic industry embrace every facet of modern living.

What are ceramics? The American Ceramic Society, the National Institute of Ceramic Engineers, the Ceramic Educational Council, The American Society for Testing and Materials, and other learned societies and associations define ceramics as inorganic, nonmetallic materials that are subjected to elevated temperatures during their processing and manufacture—temperatures of 540°C (1000°F) and above. Although the origin of ceramics is lost in antiquity, it is believed that the industry began during neolithic times with the production of pottery vessels for the storage of food and water, the vessels being formed by hand and baked in the sun. Later, the vessels probably were baked in bonfires or campfires to obtain a stronger and more durable product. By early biblical days, the art had attained a semblance of industry status; the city and tower of Babel were built of brick, and slime was used as the mortar (Genesis 11:3). Abraham sent Hagar into the wilderness with food and a ''bottle'' of water (Genesis 21:14). Today, through modern research and development, ceramic materials, products, and processes are so extensive, and so diverse, that they stagger the imagination. They encompass virtually every phase of modern life and living. Ceramic products are produced in every country of the world, and the list of these products is endless.

Many of the raw materials used in ceramics occur in nature and have been available since the days of the creation. These include the oxides, silicates, carbonates, nitrates, phosphates, chlorides, sulfates, chromates, fluorides, and other naturally occurring substances. Many of these are now upgraded by the use of beneficiation techniques, resulting in improved, more useful, and even more beautiful products of manufacture. During the past century, particularly since the outbreak of World War II in 1939, scientists, engineers, and production personnel have synthesized a host of new minerals and chemicals that have extended the influence of ceramics into previously undreamed of horizons. Some of these new products are the aluminides, beryllides, borides, carbides, nitrides, silicides, sulfides, and tellurides. These new materials have opened the door to the development of completely new industries and technologies. The development of ceramic dielectric components, nuclear fuel elements and other reactor components, lasers, magnetic ceramics and glasses, cermets, high-temperature lubricants, ceramic–glass and ceramic–metal composites, and similar materials such as the piezoelectric materials have permitted man to enter and to conquer scientific and technical worlds that previously were impossible to overcome, particularly the world of hostile environments.

At the risk of oversimplification, the ceramic industry at the turn of the century was divided into six generalized industrial classifications: whitewares, structural clay products, refractories, portland cement, glass, and porcelain enamels. Whitewares may be hypothesized to be the oldest of these categories on the basis of their production and use during neolithic times. Structural clay products may have followed with the advent of the adobe brick. Then came the refractories which were used to line the crucibles, furnaces, and tanks in which iron and steel, glass, portland cement, and porcelain enamels are made. These relative dates of origin are mentioned only as a matter of possible interest and not as a matter of fact. It has been stated, but unconfirmed, that the first manufacturing industry in what is now the United States was a glass factory established one year after the arrival of the Jamestown colonists in Virginia.

Whitewares

Whitewares are defined as products formed of raw materials which fire to a white color—materials such as clay, feldspar, potter's flint, whiting, and steatite. The term whitewares actually is a misnomer, however, as many products of this classification are produced in a myriad of colors. The more common whitewares are the earthenware, porcelain, and vitreous china products known as dinnerware or tableware used in the home and in commercial dining rooms, ceramic cooking ware such as casseroles and bowls, floor and wall tile, pottery, and a wide variety of artware such as figurines, vases, lamp bases, ash trays, and other decorative products. Whiteware products of industrial importance and use are the electrical porcelains such as spark plugs and insulators, laboratory ware such as porcelain crucibles, combustion boats, combustion tubes, grinding mill linings, grinding balls, and spheres for use in cracking and condensing towers. The list is quite long, and the examples given are presented to serve as illustrations of the industry and its products.

Structural Clay Products

The structural products of the ceramic industry are many and are so well known and commonplace that they are taken for granted. Prominent among such products are building brick, face brick, paving brick, drain tile, roofing tile, hollow building tile, quarry tile, wall coping, stoneware, conduits, and terra cotta. The uses and potential uses of these materials are indicated by the designations assigned to them. All of these products are made of clays native to the area in which they are manufactured, and are characterized by high strength, excellent load-bearing properties, and excellent durability and permanence in terms of weathering and underground corrosion.

It should be mentioned that many ceramic products other than structural clay products, per se, also are used in the construction industry. Concrete, mortar, plaster, and plasterboard are examples. Ceramic wall and floor tile are used where cleanliness and beauty are essential. Wall tile is used extensively in vehicular and pedestrian tunnels and subways. Because they are impervious to water and general wastes, and because they are easily cleaned, ceramic products are used in water-supply systems, waste disposal plants, and general sanitation systems.

Refractories

Refractories virtually laugh at fire. They are structural materials that are resistant to molten metals, glass, slags, and other corrosive substances at elevated temperatures. They are made of a wide variety of raw clays or minerals such as the refractory clays, kaolin, magnesite, chrome ore, olivine, diaspore, and bauxite. For more severe or for special service conditions, refractory products are made from synthetic compositions such as the carbides, borides, beryllides, and silicides, or from upgraded or purified minerals such as titania, zirconia, silica, chromium oxide, and alumina.

Refractory products are formulated and processed into shapes and forms as required to meet specific service requirements. In granular form, they are used as a cover for furnace bottoms, as mortars for the laying or setting of the various refractory shapes, and as castables for placement in furnaces and kilns by manual or mechanical means. The more widely used and better known of the refractory products are the fabricated products such as bricks, cubes, cones, rods, cylinders, pyramids, and many other geometrical forms and sizes used in heat resisting and structural applications.

The shaped refractory products are generally identified as neutral refractories (which are composed of alumina, silicon carbide, carbon, and other neutral ingredients), acid refractories (which are composed principally of silica and which react with lime, alkalies, and other basic materials at elevated temperatures), and basic refractories (which are composed of magnesite, dolomite, chrome ore, and forsterite which react with silica and other acidic materials at elevated temperatures).

Refractory materials are used to line furnaces and kilns. They also are used in the production of crucibles, saggers, and other containers. These products are employed in the production of iron and steel, a number of nonferrous metals, and in the melting of glass and porcelain enamel and glaze frits. They also are used in the production of portland cement and in the firing of pottery, spark plugs, glazes, porcelain enamels, and hundreds of other products as well as in the production of power and steam and in the refining of petroleum.

Special refractories are used in rocket motors, insulating tile, nose cones, and numerous other parts of space vehicles and missiles to resist the high temperatures and erosion conditions during the blast-off and the tremendous heat generated by friction with the atmosphere during re-entry. Refractories also provide resistance to solar and space radiations as well as insulation from heat and cold in order to keep the interior of space vehicles comfortable and safe for the astronauts and for their life-sustaining and other facilities. Even the launching pads for the space vehicles are of refractory compositions to insure thermal and physical stability during the blast-off.

A group of refractory products which might deserve a classification all its own is the group known as abrasives, one of the least publicized of the ceramic products. Although there are many natural abrasives such as the diamond, emery, and sand, the commercially produced synthetic materials are industrially the more important; these include alumina, silicon carbide, boron carbide, boron nitride, tungsten carbide, etc.

The abrasives are used in a wide variety of products from the abrasive cloths, papers, and stones which are used manually, to the abrasive disks and wheels for grinding and polishing machines. These are used in hundreds of applications from fingernail files, wood sanders, and tool sharpeners to the machines used for the grinding of shafts and bearings, the grinding and polishing of glass, and the grinding and polishing of other hard ceramics, many to dimensional tolerances that can be obtained in no other way. A significant development in recent years is the ceramic cutting tool which permits the faster, cooler, and more efficient machining of metals.

Not all abrasives are used in the form of a fabricated tool or wheel. Many are used in compressed-air streams or in liquid suspensions for high-speed cutting and polishing. Others are used as jewels and bearings in watches and a host of scientific instruments.

Cement and Concrete

Portland cement, concrete, mortar, plaster, and other pozzolanic products are ubiquitous. They are, perhaps, the most visible and well-known of all ceramic products. They are used extensively in the building of dams, canals, aqueducts, highways, bridges, airport runways, sidewalks, and other pavements. They are used in all types of building construction—foundations, walls, floors, and roofs of houses, schools, churches, office buildings, factories, hotels, hospitals, condominiums, apartments, and virtually every other type of permanent structure. In the artistic or aesthetic world they are used in the production of statuary, finials, copings, gargoyles, and bird baths. Concrete products are found in every country of the world.

Portland cement is a hydraulic cement made by calcining lime-bearing and other clay minerals to incipient fusion and then grinding the resultant clinker to a fine powder. Pozzolanic cements are made of slags and siliceous and aluminous products which react with slaked lime in the presence of water at room temperature. Concrete is a blended mixture of portland cement with aggregates such as sand and gravel with sufficient water to produce a workable mass that will set to rock-like hardness. Mortars are mixtures of portland or other pozzolanic cements with lime or gypsum plaster and water, usually of a trowelable consistency, used to bond brick, stone, concrete block, and other materials together in masonry construction. Plaster, an easily trowelable mixture of lime, sand, and water which will harden on drying, is used to coat walls, ceilings, and partitions. Plasterboard, a flat, sheet-like manufactured product of hardened gypsum plaster core encased in a paper, felt, or pulpboard envelope, is often used as a backing or substitute for plaster. All of these products are used as fireproofing or fire-resisting, waterproofing, and soundproofing materials of construction.

In addition to the conventional cements which are placed and shaped at the site of use, a number of other cements are of industrial importance. There are the reinforced concrete products which contain bars, rods, meshed wire, or metal or other fibers as the reinforcing medium. These are designed to carry tensile stresses such as are encountered in beams, columns, poles, chimneys, bridges, stairways, and other engineering structures. Under severe conditions, the reinforcing steel bars or rods may be placed under tension while the concrete is hardening to produce a product known as prestressed concrete, an even stronger product which enables the construction of longer bridge spans as well as shallow, light-weight, and more graceful concrete structures.

Precast concrete, which is formed in a factory

and hauled to the site of construction, is used in the production of beams, culverts, columns, bridge sections, and pipe, the latter being employed for the movement of water and sewage. Panels of unusual and attractive appearance are made by the incorporation of aggregates of various colors and textures in the concrete slurry before pouring or casting.

Building blocks of a variety of shapes and sizes, some solid and some with hollow cores, are commonplace.

Lightweight concrete, a concrete made with lightweight aggregates such as perlite, vermiculite, or expanded products such as foamed clay, slate, shale, clinker, or slag, is used in the production of light-weight, unreinforced concrete of high insulating properties.

Packaged concrete mixtures are used in the repair of damaged or deteriorated concrete. In some instances in which the deteriorated areas are relatively large and deep, the voids are filled with aggregate and a portland cement grout is pumped over and around the aggregate. This procedure has proved useful in making underwater repairs. Aluminous cements made of lime and bauxite are rapid setting and frequently are used in making road repairs and for the insulation of furnaces. However, the aluminous cements usually will not set under water.

Terrazzo, a concrete containing colored aggregate, is widely used as a decorative and practical flooring in office buildings, hospitals, and other public buildings. After the concrete has set, it is ground smooth and highly polished on the job.

Glass

Glass, an ancient discovery, also is a modern miracle. It is all about us, serving us in many ways. Because of its tremendous versatility, glass also is a paradox.

Glass, for example, may be hard or soft, flexible or brittle, weak or strong—stronger even than steel. Some glasses are water soluble, and others are chemically inert—resistant even to boiling aqua regia. Some glasses are transparent, some are translucent, and others are completely opaque. They may be made to absorb, disperse, or reflect light, and they may be tinted or made in every color of the rainbow. The most obvious of the glass products are tableware, bottles, jars, mirrors, and the windows in our homes, automobiles, office buildings, factories, and hotels. These are known as soda-lime glasses and are made of sand, sodium carbonate, limestone or lime, and sodium sulfate, and comprise approximately ninety percent of the total glass production in the United States.

Borosilicate glasses are composed of blends of conventional glass-making materials plus a minimum of five percent of boric oxide. They are resistant to both heat and chemicals, properties which make them particularly useful in the production of domestic and commercial kitchenware, cookware, mixing bowls, laboratory beakers and flasks, and vessels in which chemicals are manufactured.

Lead-alkali glasses, which contain substantial amounts of lead oxide, are widely employed in artware, thermometer and barometer tubes, shielding windows for nuclear hot cells, and in applications where high electrical resistance is required.

Aluminum silicate glass, a glass in which aluminum oxide has been incorporated, is employed in instances where high thermal-shock resistance is required. Opal glasses, which are made translucent or opaque by the addition of fluorine compounds, are employed in the production of goblets, bowls, and artware, and are widely used in light fixtures and lamp bases.

Optical glass, a glass of high transparency and free of imperfections, is used in eyeglasses, microscopes, telescopes, and a wide variety of scientific and technical instruments. Light-sensitive glasses, which darken when exposed to bright light, also are used in the production of ophthalmic lens and sunglasses. Photosensitive glasses are employed in the production of permanent patterns in glass such as designs and photographs.

Fibrous glass products which have come into prominence are molded furniture, drapery materials, golf-club shafts, fishing rods, boat hulls, thermal and acoustic insulating materials, and an impressive number of consumer and industrial products. Fine filaments of special glasses have recently been developed to replace metals in telephone and other communication systems which have resulted in improved and more efficient services.

Cellular, or foamed, glass is being used widely as thermal and acoustic insulation in both domestic and industrial products.

Coated glass products, glass on which a coating or thin film of another material such as a metal has been applied, are employed as mirrors, resistors for electrical and electronic circuitry, space heaters, self-defogging windows, and heat-reflecting windows.

Glass ceramics, glass which has been converted from a glassy to a crystalline state, are used as missile nose cones, counter tops, cookware, and other products in which low thermal expansion and high resistance to thermal shock are required. Special compositions may be made magnetic and of high capacitance for use in dielectric components. The magnetic glasses are used as insulators and in some small direct current motors, computer memory cores, television yokes, telecommunication systems, and antennae.

Fused or vitreous silica, sometimes known as silica glass, is composed almost entirely of silica and is made in transparent and translucent forms. These are used in telescope mirrors, envelopes for mercury vapor lamps, and covers for solar cells. Glasses containing ninety percent silica and four percent other materials are employed as view ports in space craft and numerous industrial applications.

Mammoth mirrors for use in tapping the energy of the sun to heat furnaces to temperatures of 2760°C (5000°F) and above are in the process of development. Glass and ceramic-coated pistons may be used to improve the performance of automobile and other internal-combustion engines. Glass-coated materials have been shown to improve the resistance of agricultural, industrial, and other equipment to weathering, corrosion, and wear. Glass-fiber and ceramic-fiber timbers have been developed as lightweight construction materials for use in fire-proof, rot-proof, vermin-proof, storm-proof, and earthquake-proof structures. Ceramic–metal batteries also have been developed which eventually may be improved to the point where they will supply power for space equipment and under-the-ocean cities.

Ceramic–Metal Systems

Porcelain enameling, the application of relatively low-melting glasses and ceramic coatings to iron, steel, and other metals began before the birth of Christ; specimens may be seen in museums throughout the world. The early enamels first were applied to gold, then to silver and bronze, and finally to copper in the manufacture of small items such as jewelry and items of an artistic nature. In the nineteenth century, techniques for the enameling of cast iron were developed and a huge modern industry was born. Today, cast iron, sheet iron, steel, aluminum, copper, and other metals and alloys are coated with a wide variety of enamels and ceramic coatings in the production of hundreds of important household, industrial, and military products.

Available in all colors, porcelain enamels are used on kitchen ranges, refrigerators, laundry equipment, dishwashers, bath tubs, wash basins, kitchen sinks, hot-water heaters, and other household appliances. They are used in the manufacture of farm silos and other structures for the storage of animal and human foods. Since they are easily cleaned and sterilized, they are used in medical and hospital equipment. Enameled iron and aluminum both are used in commercial and residential construction, particularly as curtain walls and murals, because of their resistance to weathering and corrosion. They are used in the manufacture of automobile mufflers, signs, chalkboards, smokestacks, and to line milk tanks, beer tanks, and vessels used in the production of food and chemical products. Glass-coated copper wire increases the permissible operating temperatures of electric motors and similar equipment. Glass-to-metal seals are essential in the electrical and electronic industries, as are the refractory coatings in the aircraft, space, and nuclear industries. The development of new materials and techniques—vapor deposition, the fluidized bed, electrostatic deposition, chemical deposition, electrochemical deposition, plasma spraying, and the like—promises a bright future for the ceramic coatings industry.

Closely related to ceramic coatings for metals are the cermets which are heat-resistant and wear-resistant composite mixtures of ceramics and metals. These are formed into a wide variety of shapes and fired to high temperatures. Cermets currently are being used in brake-shoe linings, jet-engine components, nuclear fuel elements, electrical resistors, and other oxidation-resistant products.

Piezoelectrics

Piezoelectric ceramics are electromechanical transducers which convert mechanical vibrations to electric voltage, or which will convert electric voltage to mechanical force. Examples of such materials are barium titanate, lead zirconate titanate, potassium sodium niobate, and similar polycrystalline materials. They are made in a manner similar to that employed in the manufacture of ceramic insulators and are subjected to high direct-current voltage to develop permanent polarization.

The piezoelectric materials are used in crystal microphones, crystal loudspeakers, and cartridges for phonographic pickups. They are employed in television circuitry, mobile radio equipment, sonar transmitters and receivers, and ultrasonic cleaning equipment. They also are used in gauges to measure the blast pressures in firearms and explosives.

Electronics

Electronics is a branch of physics which deals with the emission, behavior, and effects of electrons, and the application of devices that utilize the flow of electrons as in a semiconductor, a transistor, a vacuum, or a gas contained in an appropriate tube. Although some understanding of the science was developed many generations ago, it has been only in recent years that electronic devices have exploded across the scientific, technical, and industrial world.

Among the more prominent or better known of the electronic ceramic materials are the semiconductors such as silicon, gallium arsenide, lead telluride, copper oxide, magnesium iodide, mercury indium telluride, zinc sulfide, and cadmium selenide; these are materials which exhibit electrical resistivities between those of a metal and those of an insulator. Ferroelectric materials are crystalline substances, such as barium titanate, potassium dihydrogen phosphate, and Rochelle salt, which exhibit spontaneous electric polarization, piezoelectricity, and electric hysteresis, and are used in dielectric amplifiers, acoustic transducers, and ceramic capacitors. The transducers are devices which are activated by power from one system and, in turn, supply power in another form to another system such as microphones, telephone receivers, loud speakers, automobile horns, door bells, barometers, phonographic pickups, and photoelectric cells. Thermisters are electrical resistors whose resistances vary sharply with changes in temperature. Varistors are two-electrode semiconductor devices whose resistance properties are dependent on applied voltage. Sensors are devices which will respond to a physical stimulus such as light, heat, pressure, or sound, and will transmit or convert the stimulus to a useful purpose such as the operation of a control mechanism, a television camera, or an information gathering system. Transistors are three-terminal semiconductor devices, typically containing two rectifying and one nonrectifying (ohmic) junctions, which are used as amplifiers, detectors, or switches.

Electronic devices are employed in a huge variety of domestic, business, commercial, industrial, military, space, medical, and scientific products. In communications, they are used in all facets of the telephone, radio, television, sound recording and reproduction systems, and similar products. High-speed computers, dictating machines, typewriters, duplicating equipment, and the like are commonplace in modern business, commercial, and industrial offices, nearly all containing electronic components. Domestic products employing these devices are the microwave ovens, refrigerators, laundry equipment, dishwashers, burglar alarms, light switches, air conditioners, and clocks. Industrial uses include controls for materials handling equipment, chemical processes, safety devices, and instruments which contribute to factory automation, including process and quality control. Electronic military equipment includes radar, navigational devices, sonar, supersnoopers, rocket instrumentation systems, gunfire control systems, guided missile systems, and many other such devices. The medical profession is heavily dependent on the electronic industry for such instruments and devices as X-ray machines, electrocardiographs, electroencephalographs, electromyographs, and other diagnostic devices, as well as X-ray diathermy, pacemakers, defibrillators, electroshockers, ultrasonic machines, hearing aids, and closed-circuit television.

Nuclear Energy

Nuclear energy is energy or power developed by one of two processes: (1) Nuclear fission, which is the splitting of an atomic nucleous into fragments, usually two fragments of comparable mass; and (2) nuclear fusion, which is the fusion or combining of two light nuclei to form a more massive nucleous (and possibly other reaction products) with a simultaneous release of energy. At the present time, the nuclear fusion process, which is in an advanced stage of development, offers a safe and inexhaustible source of tremendous power for the needs of mankind once it becomes available.

The ceramic materials used in the production of fuels for the fission process are the oxides and carbides of uranium, plutonium, and thorium. Graphite, beryllium, deuterium, and hydrogen are employed as moderators which slow down, control, or moderate the neutrons from the high velocities at which they were created during the fission process.

High-strength concrete, occasionally containing

iron or lead particles of various sizes and shapes as a supplementary aggregate, is widely used in shielding structures around nuclear reactors and other sources of radiation to contain or reduce radiation to a safe level. Special high-lead-bearing glasses which maintain their optical transparencies when exposed to high levels of radiation are used as "hot cell" windows. Whitewares, refractories, structural clay products, electrical and electronic materials and instruments, porcelain enamel and ceramic coated metals, and other ceramic or ceramic-containing products are prominent in all nuclear installations.

Nuclear energy is most widely known as a source of electric power and for its use in military weaponry. From a humanitarian point of view, however, many peace-time uses have emerged as a result of research and development efforts by nuclear scientists and engineers. Nuclear energy has proved to be a safe and efficient source of power in propulsion systems for warships, submarines, and space vehicles. It is being used effectively in mineral explorations and mining. In public works projects, it is being used to create underground storage caverns for the storage of petroleum, for the disposal of wastes, and for the extraction of geothermal power.

Radiation research has provided the medical profession with a host of invaluable diagnostic and therapeutic tools. Radioactive tracers have been employed to detect irregularities in the blood and lymphatic systems, and to follow the progress and effects of medicines as they pass through the body. Radiation is being used for the sterilization of instruments and food. Radioactive autographs are used to detect brain and body tumors, and a radioactive breath analyzer is used to detect diabetes. Plutonium oxide is employed to power pacemakers and artificial hearts.

In the field of agriculture, radioactive treatments have resulted in plant mutations which have produced monosexual beets, fruits, and possibly other foodstuffs that do not require cross-fertilization for continued propagation. Radioactive tracers in fertilizers are being used to determine the availability and effect of trace elements in vegetables and fruits.

The industrial applications of radiation technology are many. Among these are instruments and techniques to determine the internal structures of materials, particularly materials in inaccessible locations, a particularly valuable procedure to locate areas of potential material failures. Instruments also are in use to assess the integrity of welds in pipelines, bridges, boilers, and jet engines. Gauges to measure the thickness of materials and coatings, and to assess the degree and effects of wear on materials in service are now available. The use of radiation in the curing of paints, as a catalyst in expediting chemical reactions, and in the cross polymerization of plastic materials is becoming increasingly important.

Aerospace

The exploration of outer space is a subject which has intrigued man for many centuries, beginning, perhaps, in the second century when Ptolemy of Alexandria proposed his concept of the movement of the sun and the planets about the earth. At about the same time, Samosata wrote of an imaginary flight to the moon. In later years, Copernicus, Sir Isaac Newton, and others presented their more accurate dissertations on the movement of bodies in space. Jules Verne, in 1865, wrote his interesting and amazingly accurate novel, *From the Earth to the Moon*.

The modern aerospace industry actually began in 1903 with the development of the first airplane by the Wright brothers in their Dayton, Ohio, bicycle shop. Today, the aerospace industry is the largest and the most complex industry in the United States. The industry generally is considered to be divided into three parts: Aircraft (55%), missiles (20%), and space vehicles (25%).

Ceramic raw materials, manufactured ceramic products, and products which contain ceramics are employed in virtually every segment of the aerospace industry. Concrete and refractory cements are used in the launching pads, and other refractories are used in rocket nozzles, heat shields, and other heat-resisting components. Special requirements for extreme environments not previously encountered are placing demands on ceramic research, development, and production personnel for new and improved products in virtually every segment of the industry—refractories, glass, coated metals, whitewares, structural clay products, electrical and electronic ceramics, nuclear ceramics, and so on, not to mention the fabrication and production techniques developed by the industry. These products are employed in the production of gauges, meters, and a host of instruments and devices used in airborne, space, and ground-based control systems. They are used in rocket motors and power-plant systems, auxiliary power-plant systems, land-based and airborne computers, life-support systems, guidance systems, tracking stations and communication systems, take-off and re-entry heat shields, and similar sophisticated instrumentation and equipment, to say nothing of simple hardware such as tubing, valves, knobs, actuators, and small electric motors.

The list is without end.

Lasers

Lasers, which came into prominence in 1960, are devices which convert input power into very narrow, very intense beams of light in the ultraviolet, visible, and infrared regions of the spectrum by utilizing the natural oscillations of the atoms. They have contributed much, even revolutionized, scientific and technological fields such as interferometry, spectroscopy, metrology, holography, etc., and may well be the key to opening the door to nuclear fusion.

The original laser employed the ruby, a crystalline aluminum oxide ceramic containing a small amount of chromium, as the light-emitting medium. The yttrium aluminum garnet doped with neodymium is another. Glass doped with neodymium radiates in the infrared region.

Lasers are used in hundreds of scientific, engineering, commercial, industrial, military, and space applications. When focused on solid surfaces such as ceramics or metals, tremendous heat is generated which may be used to cut through the solid. The heat also may be used to weld parts together to form monolithic structures, to drill holes in ruby watch jewels, to drill holes and to cut grooves in electrical and electronic circuit boards, and to resize holes in worn wire-drawing dies.

In the field of medicine, the laser is used to repair torn or detached retinas, to remove blemishes and tatoos in skin surgery, and to excise and treat malignant tumors.

The laser-beam printer is finding wide application as a nonimpact printer for all types of documents. The laser camera is used extensively, particularly in making airborne photographs at night.

In the field of scientific instrumentation, laser anemometers are employed to measure wind velocities, laser ceilometers are used to measure the altitude of clouds, laser extensometers measure the flow of the tides, and laser-equipped transits improve the speed and accuracy of land surveys. The laser also is used in earthquake alarms, in guidance systems for missiles, bombs, projectiles, and aircraft control systems. They are vital components in computer memory cores and in communication systems such as radio, television, office equipment, and data control systems.

In Conclusion

It must be emphasized that the preceding discussions of ceramics, ceramic products, and their uses are brief. The illustrations that have been given are intended to illustrate the character and properties of some of the many and diverse products of the ceramic industry and their importance to mankind; to present a more comprehensive review is an impossible task. For example, no mention was made of many industrial ceramics such as paint and plastic pigments, paper fillers, roofing granules, molds for the casting of a variety of materials, high-temperature lubricants, valve bushings, toothpaste abrasives, pigments used to produce the white sidewalls of automobile tires, and many thousands of other uses. Tonnagewise, the ceramic industry is one of the largest, perhaps even *the* largest industry of the world.

Why, then, the need for a dictionary of ceramics? In this day of the interdisciplinary character of the various fields of science and engineering, the research and development personnel in technical areas other than ceramics are constantly seeking materials, products, and processes which the ceramic industry might supply to meet their needs. Similarly, ceramic scientists and engineers in one segment of this huge and diverse industry are constantly seeking information from the other segments which they might use. This is indicated, certainly, by the large attendance at the many conventions, meetings, symposiums, and conferences sponsored by ceramic societies, associations, colleges, and universities, as well as other technical organizations each year.

On an international scale, a ceramic dictionary will promote the standardization of terms and enhance the communication and understanding of the ceramic terms employed in the scientific and engineering communities throughout the world, particularly since the terms and definitions employed in one country frequently are quite different from those employed in another country for the same subject. These suggestions were first offered by Dr. Nicolae Ciontea, Director of the National Research Institute for Glass and Fine Ceramics of Romania, and have since been repeated by Ministers of Commerce and Industry, corporate executives, directors of national and corporate laboratories, and scientists and engineers in numerous countries of the world in which the author has served as a technical and science advisor.

Finally, a comprehensive dictionary should be a valuable asset in the libraries of colleges and universities, in the libraries of corporate and other laboratories, and even in the technical sections of public libraries to assist students and personnel pursuing scientific and technical careers.

It must be mentioned that this *Dictionary of Ceramic Science and Engineering* is a dictionary—a reference book listing the words, terms, materials, processes, products, and some of the more prominent business terms that are important to the ceramic and related industries. As a dictionary, it is not to be likened to an encyclopedia or to a textbook to be read from cover to cover, although such a reading certainly should be educational. As a dictionary, or reference book, it is intended to present only sufficient information on a particular subject to establish the interest of the reader as to the appropriateness of a more detailed search of the literature—technical journals, encyclopedias, textbooks, the informational literature of manufacturers and suppliers—for the information needed.

The entries for this dictionary were compiled from several hundred sources such as textbooks, glossaries, technical journals, trade journals, and other technical dictionaries. No word has been included which has not been gleaned from other sources.

The intent of the author has been to provide only definitions. No effort was made to include pronunciations, derivations, or syllabication of the entires.

It is the sincere hope of the author that this *Dictionary of Ceramic Science and Engineering* will be a valuable and useful contribution to the technical literature, and will promote a better understanding of ceramic terms, not only by those who are active in the ceramic industry, but by scientists, engineers, and production personnel in other technologies and industries as well.

Loran S. O'Bannon

Contents

abamurus. A buttress-like unit or a second wall, usually of concrete, concrete block, or other masonry, supporting or reinforcing a structure.

Abbe value. A number designating the deviation of light waves by an optical glass, expressed as the reciprocal dispersive power of the glass by the equation:

$$v = (n_D - 1)\ (n_F - n_c)$$

in which v is the Abbe value, n_D is the index of refraction of the sodium line at 589.3n, and n_F and n_c are the indices of the hydrogen lines at 486.1n and 656.3n, respectively. Also known as Abbe number, *nu* value, and constringence.

ablation. The process of wearing or wasting away of the surface of an object by erosion, melting, evaporation, or vaporization.

ablative material. A body or a coating of low-thermal conductivity, such as a ceramic or a glass-reinforced plastic, from which the surface layer is removed by a pyrolytic process, thereby resulting in the absorption or dissipation of heat from a substrate.

abopon. A viscous, liquid sodium borophosphate complex used in porcelain enamels and glazes as a suspension agent and binder.

Abrams' law. The strength of a concrete or mortar, with given concrete materials and conditions of tests, is governed by the quantity of mixing water employed so long as the mix is of workable plasticity; it may be calculated by the equation:

$$S = A/B^r$$

in which S is the strength, and A and B are constants, r being the water to cement ratio of the compacted cement.

abrasion. The wearing, grinding, or rubbing away of the surface of a solid by friction induced by moving solids, liquids, or gases.

abrasion hardness. The relative hardness of a solid substance in terms of its capacity to scratch, abrade, or indent another solid material or itself be scratched, abraded, or indented. See also Brinell test, Knoop hardness, Mohs hardness, Rockwell hardness, scleroscope.

abrasion resistance. A measure of the ability of a material to resist wear by friction; samples may be evaluated on the basis of loss in weight, loss of gloss, or by the degree or permanence of discoloration when a lead pencil, dye, or fine powder of contrasting color is drawn or rubbed across the abraded area.

abrasion tester. A laboratory device, usually provided with a scouring, cascading, or jet-propelled abrasive acting on the surface of a solid, employed in the evaluation of the abrasion-resistant properties of the surface. See Taber abrasion tester, Kessler abrasion tester.

abrasion-wear index. The ratio of the degree of wear on the surface of a solid material as a function of the conditions or the results of abrasion.

abrasive. Any substance which, by virtue of its hardness or other property, is used for grinding, cutting, or polishing, such as diamond, silicon carbide, alumina, sand, ceria, rouge, etc.

abrasive annealing. The heating and cooling of a solid material on a prescribed schedule as a treatment to remove stress, to induce softness, to refine its crystalline, molecular, or other structure, or to alter its physical properties.

abrasive belt. A band or endless loop of cloth, paper, leather, or sheet of other flexible substance to which an abrasive product has been bonded for use in grinding and polishing operations.

abrasive blasting. The cleaning, etching, or finishing of a surface by the impingement of a blast of air or steam in which an abrasive has been entrained.

abrasive cloth. A strong, usually pliable fabric or cloth to which an abrasive has been bonded, and which is used in manual or mechanical grinding and polishing operations.

abrasive cone. A solid, cone-shaped, bonded abrasive

product mounted on a spindle for use in high-speed grinding and machining operations.

abrasive disk, bonded. A disk-shaped, bonded abrasive product mounted on a face plate for use on a grinding or milling machine; work is ground or polished on the side of the abrasive disk opposite the face plate.

abrasive disk, coated. A circular, usually flexible, paper, cloth, fiber, or other sheet material coated on one side with a mixture of abrasive and binder for use in mechanical grinding and polishing operations.

abrasive-jet cleaning. The process of removing dirt and soil from a solid surface by the impingement of an abrasive-bearing stream of liquid or gas on the surface of the solid.

abrasive, levigated. A very fine abrasive powder used as a burnishing medium.

abrasive machining. The technique of forming or shaping a solid item by grinding, drilling, or some similar mechanical process.

abrasive, mild. An abrasive, such as talc, having a hardness (Mohs) of 1–2.

abrasiveness. The ability of a material to wear down or rub away the surface of a solid material by friction.

abrasive paper. A strong paper sheet to which an abrasive has been bonded for use in grinding and polishing operations, such as sandpaper or emery paper.

abrasive sand. A sharp-grained sand, usually graded to a mesh size, used as an abrasive.

abrasive tumbling. A process to improve the surface finish or to deburr solid materials by tumbling in a rotating cylinder containing abrasive particles.

abrasive wheel. A grinding wheel or disk composed of an abrasive grit and an appropriate bonding material used for the grinding, polishing, shaping, or cutting of a solid surface.

absolute density. The weight of a unit volume of a substance under specified conditions of pressure and temperature, excluding its pore volume and interparticle voids.

absolute gravity. The specific gravity or density of a fluid under standard conditions of pressure and temperature.

absolute humidity. The weight of the water vapor contained in a unit volume of air.

absolute temperature. Temperature measured from absolute zero on an accepted scale of temperature measurement, such as the Celsius (Kelvin) or the Fahrenheit (Rankine) scale.

absolute zero. Temperature characterized by the complete absence of heat, or at which all particles whose motions constitute heat cease to move; believed to be equivalent to −273.16°C or −459.69°F.

absorbency. The ability of a material to penetrate into or to be penetrated by another material.

absorption. The process in which fluid molecules are taken up by, and distributed through, a solid or another liquid.

absorption, dye. A test in which the depth or degree of penetration of a dye solution into a nominally nonporous body, frequently under prescribed conditions of temperature and pressure, is taken as a measure of the porosity of the body.

absorption rate. The amount of water absorbed by a brick or other body during partial or complete immersion for a specified period, usually one minute; expressed in ounces or grams per unit of time for a sample of specified size.

absorption ratio. The ratio of the weight of water absorbed by a masonry unit during immersion in cold water to the weight absorbed during immersion in boiling water for an equivalent period of time. See absorption test (2).

absorption test. (1) A test in which a ceramic body is immersed in or is subjected to a spot of dye solution, and the depth of penetration is taken as a measure of the porosity of the body.

(2) A test in which a body is immersed in a selected or specified solution for a designated time and temperature, and the ratio of the weight of solution absorbed to the weight or the volume of the dry specimen is reported as the absorbency of the body.

absorption test, accelerated. An absorption test in which the end point is hastened by testing under conditions more severe than those anticipated in actual service.

absorption, x-ray. The absorption of energy from an x-ray beam by a medium through which the beam is passing.

abutment. The portion of a structure which receives the thrust or pressure of the arch in a furnace or kiln, and which generally consists of a skewback brick and steel support.

a-c. Symbol for alternating current.

acacia gum. A water-soluble gum derived from various acacia plants which is used as a binder in porcelain-enamel and glaze slips; also known as gum arabic, gum Senegal, and gum Kordofan.

accelerated-service life. The elapsed time required to reach the end point in a service test conducted under conditions more severe than those which will be encountered during the normal use of a product.

accelerated test. Any test of a property which is conducted under conditions more severe than will be encountered during the normal life of a product or material.

accelerator. A chemical admixture introduced into a batch of concrete, stucco, mortar, plaster, or similar material as a catalyst to hasten hydration or other reaction, thereby causing the batch to develop strength more rapidly than normally would be attained; examples are the alkali carbonates, potash alum, and powdered gypsum.

acceptability. The quality of a product in terms of its ability to meet minimum standards specified for its use.

acceptance level. The maximum and minimum limits of quality standards between which a product is considered to be acceptable for its intended use.

acceptance limits. The test levels used in the sorting of specimens that establish the rating group into which a material or product under test should be assigned.

acceptance number. The maximum number of defective pieces allowable in a sample of specified size.

acceptance standard. A specimen of a material or product selected to be used as a reference standard to indicate the acceptable measure of quantity, weight, extent, value, or quality of a material or product.

acceptance test. A test to determine the conformance of a product to a purchase order or contract, or to determine the degree of uniformity of the product, as a basis for its acceptance by the purchaser.

accessory. (1) A subordinate material such as a fastener, closure, coupling, gasket, or the like which is necessary for the installation or performance of a product.
(2) A product which in itself is not essential, but which adds to the convenience or effectiveness of another product.

accessory mineral. A mineral found in a subordinate quantity in another mineral, but which is not essential and which does not affect the character or the properties of the parent mineral.

accountability. (1) The section of nuclear materials management which is responsible for the measurement, control, record, and report systems of transfers and inventories to provide current and accurate information relative to the chemical and physical form, location, availability, and source of nuclear materials.
(2) The position of being responsible, answerable, liable, or accountable for an activity.

accuracy. (1) The degree of precision existing between an experimentally determined value and an accepted reference value.
(2) Freedom from error.

A. Cer. S. The official abbreviation for The American Ceramic Society.

achromatic glass. A glass which will transmit light without dispersing it into its constituent colors.

achromatic lens. A combination of two or more lenses of different focal powers which will transmit light free of undesired colors.

acicular. Needlelike in shape.

acid. In the ceramic context, an oxide of the general empirical formula RO_2, in which R represents an element such as silicon, titanium, zirconium, tin, etc., that will react chemically as an acid at elevated temperatures.

acid annealing. A process for preparing metal shapes for porcelain enameling in which the metal is coated with acid followed by annealing to remove oils, rust, and other soil from the surface by scaling, and to relieve stresses in the metal prior to application of the enamel coating.

acid bottom and lining. The exposed bottom and lining of a steel-making furnace composed of materials such as silica brick, sand, siliceous rock, or other refractories, which will react as an acid with the molten metal and slag at operating temperatures. See acid open-hearth furnace, acid refractories, acid slag, acid steel.

acid clay. A clay which will release hydrogen ions on contact with water.

acid embossing. The process in which the surface of glass is obscured by treatment with hydrofluoric acid or its compounds. See frosted.

acid-extractable material. Substances which may be dissolved and removed from a material by treatment with an acid, usually under specified conditions.

acid frosting. The etching of glass, particularly glass tableware, by treatment with hydrofluoric acid or its compounds.

acid gold. A decoration of gold applied to the surface of a glaze which previously was etched with hydrofluoric acid or other fluoride to improve adherence.

acidic oxide. Any oxide which will display acidic properties at elevated temperatures, such as SiO_2, TiO_2, ZrO_2, SnO_2, CeO_2, GeO_2, PrO_2, Sb_2O_3,As_2O_3, and P_2O_5.

acid open-hearth furnace. An open-hearth furnace lined with a highly siliceous refractory brick, the lining sometimes being coated with a fritted layer of silica sand.

acid polishing. The process of polishing glass surfaces by means of an acid treatment to minimize roughness.

acid refractories. Refractories containing substantial amounts of silica which may react with basic refractories, slags, or fluxes at high temperatures.

acid-refractory furnace. A furnace or cupola lined with an acid-type refractory, such as silica brick.

acid resistance. The degree to which porcelain enamels, glazes, glasses, and other ceramic surfaces are resistant to attack by acids.

acid-resisting brick. A fired, clay brick of low-water absorption and high resistance to attack by acids; usually used with acid-resisting mortars.

acid-resisting enamel. A porcelain enamel exhibiting high resistance to attack by acids, particularly household, fruit, and cooking acids.

acid scaling. The process of dipping or spraying raw metal with acid followed by annealing at a red heat as a means of removing oils, rust, and other soils prior to the application of a porcelain enamel to the metal.

acid slag. Slag in which the silica content is greater than the content of basic ingredients, such as lime and magnesia.

acid spar. A fluorspar containing 98% or more of calcium fluoride and 1% or less of silica.

acid steel. A grade of steel produced in furnaces lined with silicate refractories.

acid, white. A mixture of hydrofluoric acid and ammonium bifluoride employed in the etching of glass surfaces.

ACL kiln. A type of traveling-grate preheater employed to preheat a portland-cement batch before it is charged into the rotary cement kiln as a means of minimizing the length of the kiln required for the calcining operation.

acoustic insulation. Specially textured plaster, tile, or other product employed to diminish the intensity of sound; for example, a perforated tile.

acoustic plaster. A plaster having a chemically or mechanically textured or roughened surface which will absorb or prevent the transfer of sound.

acoustic tile. A thin, decorative tile of plaster, ceramic, fiber, or other material having sound-absorbing properties, which is used as a covering for walls, ceilings, and other surfaces.

acrylic polymer. A thermosetting resin used as a binder in laminated products; made by polymerization of acrylic acids, acrylates, etc.

actinic glass. A glass that transmits more of the visible components of light and less of the infrared and ultraviolet components.

activated alumina. A granular, highly porous form of aluminum oxide, Al_2O_3, exhibiting high-absorptive and catalytic properties. See alumina, activated.

activated carbon. A family of highly porous carbonaceous substances of high surface area per unit of volume manufactured in powdered, granular, or pelletized form by processes that develop high-absorptive properties. Also known as activated charcoal.

activated carbon, granular. Activated carbon in particle sizes predominately larger than 80 mesh. See activated carbon.

activated carbon, powdered. Activated carbon in particle sizes predominately smaller than 80 mesh. See activated carbon.

activated clay. A clay, such as bentonite, which is treated with acid to improve its bleaching and adsorptive properties.

activation. Any process, such as chemical treatment, heat, or radiation, which is employed to improve the reactivity or absorptive properties of a material.

activation analysis. A sensitive technique for the identification of trace elements based on the induced radiation characteristics of a specimen exposed to neutrons in a nuclear reactor.

activity. A general term describing the ability or capacity of a material to absorb or to react in a desired manner.

adapter. (1) A type of flange used to mount a grinding wheel on a shaft of smaller diameter than the center hole in the wheel.

(2) A device or attachment designed to connect or attach two dissimilar parts in an apparatus.

addition. A material added in relatively small quantities to a ceramic coating, body, or other composition to influence the manufacturing, working, or performance properties of the composition.

additive. A substance added in relatively small quantities to bring about a change in or to enhance the properties of another.

adherence. (1) In general ceramic usage, the bond or union developed at the interface between two substances by fusion or by chemical or physical reaction during fusion.

(2) The degree to which a porcelain enamel, glaze, or other ceramic coating adheres to its substrate.

(3) A measure of the stress necessary to cause one material to separate from another at their interface.

adherence failure. The separation of a porcelain enamel from its base metal, usually exposing bright metal in the fractured area; the traditional measure of the degree of failure is the ratio of bright metal to adherent enamel fragments remaining in an indented area which was deformed by a plunger in a specified manner to a specified size.

adhesion. The degree or strength of attachment of a material in contact with another.

adhesion-type ceramic veneer. Thin sections of a ceramic held in place by the adhesion of a mortar to the unit and to the backing without the use of metal or other fasteners.

adhesive. A mucilaginous or cementitious substance placed or spread between two solid surfaces to bind the surfaces together.

adhesive strength. The force required to separate two bonded surfaces.

adiabatic. An occurrence which takes place without the loss or gain of heat, such as the expansion or contraction of bodies during drying at constant temperature.

admixture. A material added in small quantities to a batch to alter the working or performance characteristics of the batch in a desired manner.

adobe. (1) A structure made of unfired brick or clay.
(2) A clay from which unfired brick are made.
(3) Large, roughly molded, sun-dried brick of varying dimensions which sometimes are reinforced by the incorporation of straw in the batch.

adsorbate. Any substance which, in molecular, atomic, or ionic form, will condense on, penetrate into, and be retained by another liquid or solid.

adsorbed water. Water that is attracted to and held on the surface of a material in thicknesses of one or more molecules, and adherence being due to molecular forces.

adsorbent. Any solid or liquid, such as charcoal, activated alumina, silica, water, and mercury, having the ability to attract and concentrate significant quantities of another substance on its surface and to be penetrated by this substance.

adsorption. The attraction and adhesion, in extremely thin layers, of molecules, atoms, or ions of gases, liquids, or dissolved substances to the surface of solid or liquid materials in which they come in contact.

adsorption, anion. The adsorption of anions of clay on basal OH surfaces in which structural hydroxyls are replaced by other anions, or on edges where unsatisfied positive bonds occur; exchange of edge hydroxyls also may occur.

adsorption, cation. The adsorption of cations of clay on basal surfaces where negative charges occur, possibly as a result of isomorphous replacements within the lattice, and/or adsorption on prism surfaces where unsatisfied negative bonds may occur; surface adsorption occurs in three-layer clays, while edge adsorption predominates in kaolin-type clays.

adsorption, chemical. The binding of an adsorbate to the surface of a solid by forces exhibiting energy levels approximating those of a chemical bond.

adsorption, countercurrent. An adsorption process in which the flow of fluid is in the direction opposite to the movement of activated carbon.

adsorption, heat of. The heat developed during the adsorption process, as when one mole of substance is adsorbed on another at constant pressure.

adsorption, hydraulic. The adsorption of weakly ionized acids or bases formed by the hydrolysis of some types of salts in aqueous solution.

adsorption, integral heat of. The sum of the differential heats of adsorption from zero to a given level of adsorption.

adsorption, irreversible. Adsorption in which the desorption isotherm is displaced toward higher equilibrium adsorption capacities from the adsorption isotherm.

adsorption isotherm, Freundlich. A logarithmic plot of the quantity of component adsorbed per unit of substance versus the concentration of that component at equilibrium and at constant temperature which approximates a straight line as represented by the equation:

$$\frac{x}{m} = kp^{1/n}$$

in which x is the mass of gas adsorbed, m is the mass of adsorbent, p is the gas pressure, and k and n are constants for the temperature and system.

adsorption isotherm, Langmuir. A modification of the Freundlich adsorption isotherm expressed by the equation:

$$f = ap/(1 - ap)$$

in which f is the fraction of the surface covered, p is the gas pressure, and a is a constant; the equation is useful chiefly for gaseous systems.

adsorption, preferential. The adsorption which occurs when one or more components are adsorbed to a much greater extent than other components.

adsorption, reversible. Adsorption in which the desorption isotherm approximates the adsorption isotherm.

adsorption test, accelerated. A test in which the end point of an adsorption test is hastened by conducting the test under conditions more severe than those that are anticipated in normal service. See adsorption.

adsorption theory, Langmuir. The surface of an adsorbent has only uniform energy sites and, therefore, adsorption is limited to a monomolecular film.

adsorption, van der Waals. The binding or adherence of an adsorbate to the surface of a solid is the result of an attractive force produced between two atoms or nonpolar molecules when a fluctuating dipole moment in one atom or molecule induces a dipole moment in the other, thereby causing the two dipole moments to interact.

adsorption zone. The area on an adsorbent in which the concentration of an adsorbate in a fluid decreases from the influent concentration to the lowest detectable concentration.

adsorptive capacity, dynamic. The quantity of a material adsorbed per unit of a powdered substance from a fluid moving through a bed of the substance at breakpoint for that material.

adsorptive capacity, equilibrium. The quantity of a material adsorbed per unit of a substance from a fluid at equilibrium temperature and concentration or pressure.

AEA. Acronym for air-entraining agent.

aerate. To introduce air into a slurry by stirring or other means of agitation.

aerated concrete. Concrete containing a substantial amount of entrapped air which was introduced into the mass by foaming or other process. See aeration of concrete.

aeration of concrete. (1) The process by which air or other gas is introduced into concrete to produce a product having a density substantially less than that of normal concrete, to improve the workability and frost resistance of the concrete, and to reduce bleeding and segregation in the concrete; the gas-forming ingredients usually are introduced into the cement clinker during grinding or into the concrete batch during mixing.

(2) The exposure of cement to moist air during storage, causing erratic effects in the setting characteristics of the concrete in which it is used. See air-entraining agent.

AFA rammer. A device consisting of a 14-pound weight falling from a height of two inches onto a plunger of a two-inch mold to form test specimens of particulate refractory compositions, foundry sands, and similar materials or products.

after-expansion or after-contraction. The permanent linear change measured on a refractory material reheated to a specified temperature for a prescribed time; reported as a percentage of the original length.

agalmatolite. $Al_2SiO_4O_{16}OH$; sp. gr. 2.8–2.9; hardness (Mohs) 1–2; a natural hydrous aluminum silicate of the pyrophyllite family.

agate. A variegated chalcedony, SiO_2, with its colors arranged in stripes, blended in clouds, or displaying moss-like forms; sp. gr. 2.65; hardness (Mohs) 6.5–7.0; used in the production of agate mortars and pestles, as grinding balls in ball mills, and as a burnisher or polisher of gold in ceramic-ware decorations.

agate glass. A multicolored glass resembling natural agate in appearance; made by blending glasses of two or more colors while in the molten or plastic state, or by rolling a transparent glass into other glasses of various colors.

agate mortar and pestle. A highly polished, blemish-free, abrasion-resistant mortar and pestle made of natural quartz; resistant to all acids and alkalies except HF and NaOH; used to pulverize materials when minimal contamination is required.

agateware. Ceramic and porcelain-enameled ware characterized by veins of color distributed through the body or coating in a pattern resembling the appearance of natural agate.

agglomerate. To gather into a ball, cluster, or other mass, or a material that has been processed into a ball, cluster, or similar mass.

aggregate. An inert material, such as sand, gravel, slag, shell, or broken stone, which is to be mixed with cement to form concrete or mortar.

aggregate, coarse. An aggregate in its natural condition or which has been graded to a diameter greater than ¼ inch (4.75 mm.) for use in the production of concrete.

aggregate, expanded. A bloated clay, shale, or similar material produced by heating the material to a temperature sufficient to cause the surface to fuse, thereby entrapping air and steam within each fused particle which subsequently swells and explodes; used primarily in the production of light-weight concrete and as a thermal and sound insulator.

aggregate, exposed. A concrete finished in a manner to expose the upper surface of the aggregate particles at the concrete surface so as to produce a pleasing architectural effect.

aggregate, fine. An aggregate which has been graded to a diameter less than ¼ inch (4.75 mm) for use in the production of concrete.

aggregate, heavy. Aggregate of barite, magnetite, steel punchings, or other materials of high-specific gravity used in the production of high-density concrete for use in radiation shielding, counterweights, and similar applications where a heavier than normal concrete is required.

aggregate, light-weight. Expanded or bloated clay, shale, slate, or similar material used in the production of light-weight concrete.

aggregate, reactive. An aggregate of siliceous minerals and rocks which reacts with alkalies in cements to produce a destructive internal expansion in concrete after it has hardened.

aggregate, separated. Concrete aggregate classified into fine and coarse components.

aging. (1) The storage of porcelain enamels, glazes, slips, slurries, or frit powders before use.
(2) The change occurring in slips, slurries, or frit powders with the passage of time.
(3) The curing of prepared ceramic materials by storage for a definite period under controlled conditions.
(4) The curing of mortars and cements for periods of sufficient duration to develop necessary strength before exposure to severe conditions of use. Also known as maturing, souring.

agitator. (1) An instrument or apparatus employed to stir, shake, or mix.
(2) A vehicle, employed to transport cement, which mixes or agitates the premixed batch in a drum en route to a job site.

agricultural tile. An unglazed tile of tubular shape designed for burial in the ground to form a piping system to drain excess water from agricultural lands.

air. The normal atmosphere of the earth which consists essentially of 78% of oxygen, 21% of nitrogen, one % of argon, and small amounts of carbon dioxide, neon, helium, krypton, xenon, and radon.

air bells. A defect in optical glass consisting of irregularly shaped bubbles formed during pressing and molding operations.

air brick. A fired brick essentially of standard size in which holes are formed through its length, as opposed to its depth, to permit the circulation of air in structures.

air chain. A chain or string of air bubbles or inclusions contained in glass, glaze, porcelain enamel, a vitreous or near-vitreous body, or similar product, usually as a defect.

air classification. The separation and grading of solid particles of a material by density or size by a technique of progressive suspension or settling as in a rising stream of air of controlled velocity, each grading being reported as a percentage of the original sample.

air, compressed. Air exhibiting a density and pressure greater than the surrounding atmosphere.

air compressor. A device which increases the density and pressure of air by mechanical compression.

air content. The volume of voids in a cement paste, mortar, or concrete, excluding the pore space in the aggregate particles; expressed as a percentage of the total volume of the paste, mortar, or concrete.

air conveyor. See conveyor, air.

air-cooled blast-furnace slag. Molten blast-furnace slag cooled under normal atmospheric conditions or cooled in an accelerated manner by the application of water to the solidified slag surface.

air current. A moving stream of air.

air drying. The removal of moisture from a material, glaze, porcelain enamel, or body by exposure to air.

aired ware. Defective ceramic ware on which the glaze has become partially devitrified or some volatilization of glaze ingredients has occurred.

air elutriator. A device designed to remove impurities from the air, as by washing or filtering.

air-entrained concrete. A concrete containing purposefully introduced air bubbles of minute sizes as a means of improving its durability and other properties. See aeration of concrete.

air-entraining agent. A material or admixture, such as a soap, resin, or grease-like substance, which reduces the surface tension of water in concrete to facilitate the entrapment of minute bubbles of air in the batch as a means of improving the durability or other properties of the concrete; the agent sometimes may be added to the cement during grinding.

air-floated. Clays and other materials which are finely milled and separated or graded by density or size by the use of an air classifier. See air classification.

air-fuel ratio. The ratio of the air supply to the fuel supply during combustion, expressed in terms of volume or weight.

air-hardening refractory cement. A finely ground, refractory cement containing admixtures to promote setting of mortars and cements at temperatures at or above room temperature but below vitrification temperature.

air inclusions. (1) Small bubbles of air or other gas enclosed in glass, glazes, porcelain enamels, or bodies which become evident after firing; usually a defect but sometimes intentional as a form of decoration.
(2) Gaseous inclusions in mica which appear as grayish areas in transmitted light and as silvery areas in reflected light.

air line. (1) A system of pipes and tubing moving compressed air from one point to another.
(2) A fine, elongated cord or bubble having the appearance of a hairline which is considered a fault in glassware, particularly in glass tubing.

air permeability. The measure of the rate of flow or diffusion of air through a porous ceramic; expressed as a unit of volume or pressure gradient per unit of area.

air pocket. A sizable bubble of air found in clay bodies during wedging or throwing.

air, primary. The initial air supplied to fuel during the early stages of combustion during the firing of a furnace or kiln.

air pump. A device to exhaust air from a closed container or tank to produce a partial vacuum; to introduce air into a closed container, such as a spray tank; or to move air from one location to another.

air ramming. The shaping of refractory or other ceramic products by means of pneumatic hammers.

air-relief valve. A small automatic or manually operated valve placed at a high point in a pipe line to exhaust air or other gases from the line.

air sampling. The collection of a specimen of air for analysis.

air, saturated. Air containing the maximum amount of water vapor possible at any particular temperature or barometric pressure.

air seal. A moving curtain of air across the entrance or exit of a furnace or other enclosed area as a means of minimizing heat loss or to minimize the movement of air in or out of the area.

air, secondary. Air introduced into a furnace, kiln, or other combustion chamber to complete the combustion of fuel during firing.

air separator. A device in which a stream of air at a controlled velocity is used as a means of separating particles of solid material as they remain suspended in the stream or settle from the stream.

air set. The property by which a material develops strength during the process of losing moisture by evaporation.

air-setting cement. A cement or mortar which develops high strength in air during the loss of moisture under normal conditions of temperature and atmospheric pressure.

air-setting refractories. Refractory mortars, ramming mixes, gunning mixes, cements, and similar compositions which are tempered with water for placement; the mixtures develop a strong bond and strength on drying which is retained during subsequent service at elevated furnace and kiln temperatures.

air-swept ball mill. A continuous ball mill in which the finely milled particles of the mill charge are removed by a current of air as the coarser particles continue to be ground.

air, tertiary. Preheated air charged into the waste-gas flue of a kiln being fired under reducing conditions as a means of minimizing the emission of smoke from the stack.

air twist. A process in which twisted capillaries are incorporated in the stems of glass tableware to produce a pleasing decorative effect.

air void. An air-filled space of irregular shape sometimes occurring in freshly mixed concrete; the voids are larger in size than intentionally entrained air bubbles, and are considered to be defects.

alabaster. Compact, fine-grained, white or delicately shaded gypsum, $CaSO_4 \cdot 2H_2O$.

alabaster glass. A glass containing inclusions of materials having different indices of refraction, and which shows no color reaction to light; resembles alabaster or onyx in appearance.

Albany slip. A clay of high-flux content and fine particle size found in the vicinity of Albany, New York; the clay fires in the temperature range of cones 6 to 9, and is used as a glaze for electrical porcelain and stoneware bodies, and as a bond in the manufacture of vitrified grinding wheels.

albite. $Na_2O \cdot Al_2O_3 \cdot 6SiO_2$; a soda feldspar used as an ingredient in ceramic bodies and glazes, and as a substitute for Cornish stone. Sometimes known as white feldspar, soda spar, white schorl, sodaclase. See Cornish stone.

albolite. A plastic cementitious material composed essentially of silica and magnesia.

alcove. The narrow channel through which molten glass flows from the fining chamber to the forehearth, or to a revolving pot, for gathering by an Owens machine. See Owens process.

alginates. Hydrophilic, colloidal salts of the alginic acids, chiefly sodium or ammonium alginate; used as binders and suspension agents in ceramic bodies, glazes, porcelain-enamels, and similar slurries, and as a waterproofing agent in concretes.

aliquot. A representative sample of a large quantity of a material.

alite. $3CaO \cdot SiO_2$; a constituent of portland-cement clinker.

alkali. A general term applied to the oxides, hydroxides, and carbonates of sodium and potassium, the alkaline earth metals, and other alkaline metals; used primarily as fluxing agents in ceramic compositions.

alkali-aggregate reaction. A deleterious reaction between the siliceous portions of aggregates and the alkalies contained in portland cement, the reaction usually occurring in concrete after it has hardened.

alkaline earths. The oxides of barium, calcium, magnesium, strontium, radium, and beryllium; the oxides of barium and calcium are used primarily as fluxes in porcelain enamels and glazes, and magnesium oxide is used extensively in refractories.

alkaline glaze. Glazes containing high percentages of alkaline materials, such as Na_2O, K_2O, Li_2O, CaO, MgO, BaO, etc.

alkaline metal. Any metal of the family including sodium, potassium, lithium, cesium, and rubidium which react with water to form alkalies.

alkali resistance. The relative degree to which porcelain enamels, glazes, and other ceramic surfaces will resist attack by aqueous alkaline solutions, the term most frequently referring to the resistance of these products to alkaline materials used in the home.

alkali-resisting enamel. A porcelain enamel exhibiting high resistance to soaps, detergents, cleaning fluids, and other household alkalies.

alkyd. Any of a group of thermoplastic resins prepared by the reaction of some polybasic alcohols, such as glycol or glycerine, with dibasic acids or anhydrides, such as phthalic anhydride; used extensively as an adhesive for glass fibers.

alligator hide. A defect characterized by an extreme roughness of a porcelain-enamel surface which resembles the hide of an alligator in appearance; it is somewhat analogous to a severe case of orange peel.

allophane. $Al_2O_3 \cdot SiO_2 \cdot nH_2O$; sp. gr. 1.8–1.9; hardness (Mohs) 3; a clay mineral sometimes containing appreciable amounts of P_2O_5.

alloy. A fused metallurgical mixture of two or more materials; specific alloys are used as a source of special nuclear ceramics.

alluvial clay. A brickmaking clay deposited in or near river beds by flowing water; more plastic, less refractory, and darker in color than residual clays.

alpha activity. The spontaneous emission of doubly charged helium ions from the nucleous.

alpha particle. A positively charged helium-4 nucleous emitted by several radioactive materials.

alpha phase. Quartz with a trigonal-trapezohedral structure which is stable at temperatures below approximately 573°C (1060°F).

alumina. Al_2O_3; mol. wt. 101.94; m.p. 2030°C (3686°F); sp. gr. 3.4–4.0; hardness (Mohs) 9; an amphoteric material second only to silica in importance to the ceramic industry; acts as a refractory in low-temperature products and as a flux in high-temperature compositions; used extensively in the manufacture of abrasives, refractories, whitewares, refractory coatings, protective surfaces for transistors, glass, and cermets; examples of specific products include thread guides, clutch and brake linings, spark plugs, mill linings, blasting nozzles, nuclear fuel elements, welding-rod coatings, color modifiers, jewel bearings for watches and scientific instruments, electron tubes, infrared windows, resistors, semiconductors, lasers, gas-turbine parts, radomes, rocket equipment, and prosthetics, such as artificial teeth and bones, etc.

alumina, activated. A highly porous, granular form of Al_2O_3 used as a catalyst, catalyst carrier, and absorbent. It is chemically inert to most gases, will not swell, soften, or disintegrate in water, exhibits high resistance to thermal and mechanical shock and abrasion, and will hold moisture without change in form or properties.

alumina balls. (1) High-density, abrasion-resisting balls used as grinding media in ball mills where contamination by iron or other metallic grinding media is to be avoided.

(2) Spheres ranging from ¼ to ¾ inch in diameter which exhibit high heat and chemical resistance when used in reactor or catalytic beds.

alumina brick. Refractory brick containing 50% or more of Al_2O_3; used in high-temperature applications, such as liners for kilns and furnaces, particularly in areas where the service conditions are severe.

alumina bubble brick. A lightweight, insulating refractory product made by pressing the brick or other shapes from Al_2O_3 into which air bubbles have been introduced by passing a stream of air over a molten Al_2O_3 batch.

alumina, calcined. Al_2O_3 subjected to one or more of a variety of thermal treatments above 1093°C (2000°F) to produce products which are more friable, fluffier, and which contain less of the alpha phase than the tabular form. Characterized by high purity, extreme hardness (Mohs 9), high density, good thermal conductivity, good thermal and mechanical shock resistance, and good electrical resistivity at high temperatures. Employed in abrasive products, glass, porcelains, spark plugs, electrical insulators, and similar products.

alumina cement. A hydraulic cementitious product formed by sintering mixtures of bauxite with limestone; the resulting cement will set to maximum strength in about 24 hours. Employed in applications where a cement resistant to elevated temperatures is required.

alumina fiber. A strong, usually short thread or fiber of Al_2O_3 used in the production of plastic-bonded insulating products and dielectrics.

alumina, friable. A medium-pure form of alumina which fractures more readily than normal alumina, but not so readily as white alumina.

alumina, fused. Alumina produced by heating a mixture of calcined bauxite or a relatively pure grade of Al_2O_3 with iron borings to temperatures above 1980°C (3596°F) in an electric arc furnace; grains of increased toughness may be obtained by the addition of TiO_2; used in bushings, spindles, and other applications where excellent abrasion resistance is required.

alumina, hydrated. $Al_2O_3 \cdot 3H_2O$, mol. wt. 156.0; sp. gr. 2.42; used in glass, whitewares, refractories, and as a setter material to prevent ware and kiln furniture from sticking during firing.

alumina, manufactured. Any alumina produced or refined by artificial means, such as by sintering or crystallization, usually for use as an abrasive.

alumina, microcrystalline. A tough, abrasive grade of

Al_2O_3 recrystallized from a molten bath by a process which results in particle or crystal sizes generally smaller than regular alumina; used as a fine abrasive and polishing compound.

alumina, natural abrasive. A naturally occurring, somewhat impure grade of Al_2O_3 such as corundum, emery, sapphire, and ruby.

alumina porcelain. A high-grade, dense, strong porcelain made of bodies in which Al_2O_3 is a major component; used in manufacture of spark plugs and electric insulators.

alumina, regular. A recrystallized grade of Al_2O_3 of relatively large crystal size, the Al_2O_3 content being approximately 95%.

alumina, ruby. An abrasive grade of corundum having the characteristic ruby-red color produced by the presence of chromic oxide.

alumina, semifriable. A hard, abrasive grade of recrystallized alumina having an Al_2O_3 content ranging from 96 to 98%.

alumina-silica refractories. A class of refractories consisting essentially of alumina and silica, including the high-alumina, fireclay, and kaolin refractories.

alumina, single crystal. Al_2O_3; mol. wt. 101.94; mp. 2040°C (3704°F); sp. gr. 3.98; hardness (Mohs) 9; a tough, abrasion-resistant, stable form of alumina produced by recrystallization from a molten bath; employed in watch and instrument bearings, fiber-forming dies, lasers, scientific and military products, and other products where resistance to high temperatures and mechanical damage is required.

alumina, sintered. A refractory product produced by heating natural Al_2O_3 to temperatures just below fusion to produce grains ranging from microcrystalline to coarsely crystalline sizes which are characterized by high density, abrasion resistance, physical strength, and dielectric strength, and having a low-power factor; employed in the production of abrasives, automotive and aircraft sparkplugs, machine tools, thread guides, ceramic-metal seals, high-temperature coatings, and electrical components.

alumina, tabular. An alumina of 99.5% purity prepared by heating Al_2O_3 to temperatures above 1980°C (3596°F) until near 100% conversion to alpha alumina is obtained. Composed of tablet-like crystals exhibiting high-electrical capacity, exceptional strength, and volume stability at elevated temperatures. Used in high-quality refractories for lining industrial furnaces for the ceramic and metallurgical industries.

aluminate. Compounds of various metal oxides with alumina, and having the general formula $M_xO_y \cdot x\text{-}Al_2O_3$; characterized by high-strength and oxidation resistance; melting points ranging from approximately 1400°C (2603°F) to 2140°C (3875°F); employed most widely in structural applications. See appropriate metal aluminates.

aluminate cement. See alumina cement.

alumina, tough. A relatively impure, block-shaped regular alumina in which the Al_2O_3 content ranges from 90 to 96%, the balance being impurities.

alumina, white. A recrystallized alumina abrasive.

alumina whiteware. Any ceramic product with an essentially white body, such as artware, dinnerware, wall tile, sanitary ware, spark plugs, and other products in which the major crystalline phase is Al_2O_3.

aluminides. Aluminides are intermetallic compounds of aluminum and are of importance because of their good strength and oxidation resistance at temperatures up to 1093°C (2000°F). Although data on the properties of the aluminides are scant, the reported melting points range from 1460°C (2660°F) to 2160°C (3920°F), specific gravities from 4.00 to 9.67, linear thermal expansions less than 1.3% at 815°C (1500°F), and microhardnesses below 500 kg/mm^2 (Vickers) or 73 Rockwell A. See aluminides of chromium, cobalt, molybdenum, nickel, niobium, palladium, tantalum, thorium, titanium, tungsten, uranium, vanadium, and zirconium.

aluminous cement. See alumina cement.

aluminum. In the ceramic context, a series of aluminum compounds and alloys, which are designed to receive porcelain enamels and ceramic coatings; available in sheet, cast, extruded, or foil form.

aluminum antimonide. AlSb; mol. wt. 148.7; m.p. 1080°C (1976°F); used in the production of semiconductors, transistors, rectifiers, and similar electronic products.

aluminum borate. (1) $2Al_2O_3 \cdot B_2O_3 \cdot 3H_2O$; mol. wt. 291.5; dissociates at approximately 1035°C (1897°F); employed as an ingredient in glass and other vitreous and semivitreous products. (2) $9Al_2O_3 \cdot 2B_2O_3$; mol. wt. 1056.7; melts at about 1950°C (3542°F); employed in bodies requiring good thermal-shock resistance and refractoriness under load.

aluminum boride. (1) AlB_2; mol. wt. 48.6, m.p. 1654°C (3010°F); dissociates at about 980°C (1796°F); sp. gr. 3.16; hardness 980 Knoop. (2) AlB_{10}; mol. wt. 135.17; m.p. about 2421°C (4390°F); sp. gr. 2.54; hardness 2650 Knoop. (3) AlB_{12}; mol. wt. 156.81; m.p. 2163–2213°C (3925–4015°F); sp. gr. 2.56–2.60; hardness 2400–2600 Knoop.

aluminum carbide. Al_4C_3; mol. wt. 143.88; m.p. about 2704°C (4900°F); sp. gr. 2.99.

aluminum enamel. A relatively low-melting porcelain enamel formulated specifically for application to aluminum and aluminum alloys.

aluminum fluoride. AlF_3; mol. wt. 83.97; sublimes at about 1260°C (2300°F); sp. gr. 2.89. Employed as a source of alumina and as a source of fluorine for its fluxing and opacifying properties.

aluminum fluoride hydrate. $AlF_3{\cdot}3\frac{1}{2}H_2O$; mol. wt. 143.5; sometimes used in the production of white porcelain enamels.

aluminum fluosilicate. $Al_2(SiF_6)_3$; mol. wt. 480.2; sp. gr. 3.58; used in porcelain enamels and glass. Also known as sodium silicofluoride, topaz.

aluminum hydroxide. $Al_2O_3{\cdot}xH_2O$; loses water at 300°C (572°F); sp. gr. about 2.4; a white gelatinous precipitate used in the manufacture of glassware and glazes.

aluminum metaphosphate. $Al(PO_3)_3$; mol. wt. 264.03; m.p. about 1537°C (2799°F); used in porcelain enamels, glazes, and glasses, and as a high-temperature insulating cement.

aluminum monohydrate. $Al_2O_3{\cdot}H_2O$; mol. wt. 118.9; sp. gr. 2.4; used as an inorganic thickener and suspension agent, coating material, binder, high-temperature adhesive, and as a source of alpha alumina or corundum in bodies formed by hot pressing.

aluminum niobate. $Al_2O_3{\cdot}Nb_2O_5$; mol. wt. 368.14; m.p. 1549°C (2820°F).

aluminum nitride. AlN; mol. wt. 40.98; m.p. 2000°C (3632°F); sp. gr. 3.26; hardness (Mohs) 6–7; used as a component in the manufacture of crucibles for the melting of aluminum.

aluminum oxide. Al_2O_3; mol. wt. 101.94; m.p. 2030°C (3686°F); sp. gr. 3.4–4.0; hardness (Mohs) 9; used in the natural form or as a prepared compound as a component in abrasives, refractories, electrical insulators and electronic products, crucibles and laboratory ware, whitewares, and a wide variety of ceramic products in which strength, toughness, thermal durability, chemical resistance, and similar properties are of primary importance. See also alumina, bauxite, corundum.

aluminum oxide, hydrous. $Al_2O_3{\cdot}nH_2O$; sp. gr. about 2.4; a white, gelatinous precipitate used in the manufacture of glass, glazes, and vitreous or near vitreous ware.

aluminum phosphate. (1) $AlPO_4$; mol. wt. 121.99; m.p. 1500°C (2732°F); used as a binder in refractory products and in dental cements. Also known as aluminum orthophosphate. (2) AlP; mol. wt. 57.99. (3) $Al_2O_3{\cdot}P_2O_5$; mol. wt. 243.98; m.p. 1593°C (2900°F); sp. gr. 2.6.

aluminum potassium borate. $(AlO)_2K(BO_2)_3$; mol. wt. 253.5; m.p. below 1800°C (3272°F); sp. gr. 3.4.

aluminum silicate. (1) $Al_2O_3{\cdot}3SiO_2$; mol. wt. 282.15; decomposes at 1810°C (3290°F); sp. gr. 3.15; used as a refractory component in glass and various ceramic compositions. Also known as andalusite, kyanite, sillimanite. (2) $3Al_2O_3{\cdot}2SiO_2$; mol. wt. 425.96; m.p. 1849°C (3360°F); sp. gr. 3.13–3.26.

aluminum silicate, calcined. $3Al_2O_3{\cdot}2SiO_2$ (essentially 95% mullite); mol. wt. 425.96; m.p. 1810°C (3290°F); soft. temp. 1650°C (3002°F); sp. gr. 3.15; used in the manufacture of various refractory products, porcelains, vitreous ware, and laboratory ware.

aluminum silicate refractories. See refractories, aluminum silicate.

aluminum silicofluoride. See aluminum fluosilicate.

aluminum sodium sulfate. $AlNa(SO_4)_2{\cdot}12H_2O$.

aluminum sulfide. Al_2S_3; mol. wt. 150.1; sp. gr. 2.02; m.p. 1100°C (2012°F).

aluminum titanate. $Al_2O_3{\cdot}TiO_2$; mol. wt. 181.84; m.p. 1860°C (3380°F); stable from 1260–1865°C (2300–3389°F); sp. gr. 3.68; used in the production of special ceramics resistant to thermal shock.

alundum. A trade name for fused alumina used as an abrasive or refractory material.

alunite. $KAl_3(SO_4)_3(OH)_6$; mol. wt. 414.2; sp. gr. 2.6–2.8; hardness (Mohs) 3.5–4.0; a calcined material employed in the production of high-alumina refractories.

amber glass. A glass tinted to colors ranging from pale yellow to brown or reddish brown by the addition of iron oxide and sulfur compounds to the batch.

ambetti. A translucent antique glass containing minute opaque specks of crystallized particles from the molten batch.

ambient. Surrounding; a term describing the conditions or character of an encompassing environment, such as the atmosphere or fluid, in terms of its temperature, composition, pressure, etc.

amblygonite. A mineral reported variously as $Li(AlF)PO_4$, $Li(FOH)AlPO_4$,$AlPO_4LiF$, and $2LiF{\cdot}Al_2O_3{\cdot}P_2O_6$; m.p. 1170°C (2138°F); sp. gr. 3.1; hardness (Mohs) 6; used as a flux in low-temperature porcelain enamels and to promote opacity in glass dinnerware.

ambonite. A mineral consisting essentially of cordierite-bearing horneblende, biotite, and andesite.

Amer. Ceram. Soc. Abbreviation for The American Ceramic Society.

American bond. The bond in which a header course of brick is used every fifth, sixth, or seventh course, with stretcher courses being used between the header courses. Also known as common bond.

American hotel china. A heavy, moderately translucent dinnerware of high strength and a water-absorption value of less than 0.3%; the ware is coated with a glaze highly resistant to commercial soaps and detergents, food chemicals, and physical damage.

American household china. A glazed, highly translucent, vitreous dinnerware of the type employed as household dinner service.

American method of shingle application. A technique for laying asbestos-cement shingles of rectangular shape in which double coverage with head lap and no side lap are provided.

ammonia. NH_3; added to iron-oxide bodies of the sgraffito-decorated type to deflocculate and control the segregation of iron oxide and to stabilize the red color over a firing range wider than normal.

ammonium alum. $Al_2(SO_4)_3 \cdot (NH_4)_2SO_4 \cdot 24H_2O$ or $AlNH_4(SO_4)_2 \cdot 12H_2O$; used to increase the set of porcelain-enamel ground coats and acid-resisting cover coats.

ammonium bicarbonate. NH_4HCO_3; used with fluorine compounds in an etching bath to produce frosted surfaces on glass, such as obtained on electric-light bulbs.

ammonium bifluoride. NH_4HF; used in combination with hydrofluoric acid to produce frosted surfaces on glassware.

ammonium metavanadate. NH_4VO_3; mol. wt. 116.99; sp. gr. 2.3; used as a colorant to produce yellow, green, and turquoise glazes and porcelain enamels, frequently in conjunction with the oxides of tungsten, molybdenum, and zirconium.

ammonium molybdate. $(NH_4)_6Mo_7O_{24} \cdot 4H_2O$; mol. wt. 1235.95; decomposes on heating; sp. gr. 2.38–2.95; sometimes used as an adherence-promoting agent in clear and white porcelain-enamel ground coats.

ammonium paratungstate. $5(NH_4)_2 \cdot 12WO_3 \cdot 5H_2O$; mol. wt. 3773.5; sp. gr. 2.3; used in the production of tungsten trioxide by calcination.

ammonium stearate. Employed as a waterproofing additive in hydraulic cements.

ammonium sulfate. Natural $PbSO_4$; used to reduce melting time for glass batches.

amorphous. Having no determinable form or crystalline structure.

ampelite. A carbonaceous schist containing alumina, silica, and sulfur; sometimes used as a refractory.

amperage. The strength of an electric current expressed in amperes, which see.

ampere. The unit of electric current equal to the current produced by an electromotive force of one volt through a resistance of one ohm, and equal to one coulomb per second.

amphibole. Any of a group of minerals containing calcium, magnesium, iron, sodium, and aluminum in combination with silica; for example, asbestos and horneblende.

amphora. A large ceramic jar with a narrow neck and with two handles that rise almost to the level of the mouth.

amphoteric. A substance which is capable of reacting either as an acid or base; for example, Al_2O_3, B_2O_3, Fe_2O_3, and Cr_2O_3.

amplifier. A device which will increase the voltage, current, or power of a system or signal.

ampoule. A small bulbous glass container which may be filled and then sealed by fusion of the neck.

analog. Something similar to something else, particularly in terms of features or properties on which comparisons may be made.

analysis. The separation and measurement of the constituents of a substance, and the interpretation of these results. Also identified as chemical content, mineral content, physical properties.

analysis, gravimetric. A quantitative chemical analysis based on reactions that produce a material to be weighed.

analysis, mechanical. The separation of particles, such as aggregate, by mechanical means as on a vibrating nest of sieves of specified sizes to determine the particle-size distribution in a parent material.

analysis, optical. The study of the chemical composition, particle size, and other characteristics of a material or mixture by means of transmitted light; for example, absorption, polarization, refraction, and scattering.

analysis, proximate. A mineralogical analysis of a substance calculated from its chemical composition.

analysis, qualitative. An analysis in which all constituents in a sample are identified.

analysis, quantitative. An analysis in which the relative or actual amounts of any or all constituents are determined.

analysis, rational. See analysis, proximate.

analysis, screen. A technique to determine the particle size or particle-size distribution of powders and the solid constituents of porcelain enamels, glazes, and other slips and slurries by calculating the percentage of solids retained on each of a graduated series of sieves of various sizes.

analysis, size. The determination of the proportion of particles of a particular size in a granular or powdered sample.

analysis, statistical. The analysis of data by statistical techniques.

analysis, ultimate. The mineralogical composition of a material or body reported in terms of its oxide components as determined by chemical analysis.

analysis, volumetric. A quantitative analysis in which accurately titrated volumes of standardized chemical solutions are used to determine the amount of a particular constituent in a solution.

analysis, x-ray. The determination of the atomic distribution, structure, chemical analysis, and anisotropic behavior of crystalline materials by means of x rays.

analytical-reagent grade. A classification adopted by the American Chemical Society to designate the quality of a chemical or chemical reagent in terms of its composition and degree of purity.

anatase. TiO_2; mol. wt. 79.9; m.p. about 1885°C (3425°F); sp. gr. 3.9–4.2; hardness (Mohs) 5.5–6; the tetragonal form of TiO_2 used as an opacifier and pigment in porcelain enamels, glazes, and glass. See titanium dioxide, rutile, brookite.

anchor. An L-shaped supporting device used to mount glass, masonry, concrete, or other panels or units to a wall or other surface.

anchored-type of ceramic veneer. Any ceramic panel or sheet laid superficially over a permanent backing and then anchored in place.

anchor, storm. A corrosion-resistant metal fastener designed to mount and hold asbestos-cement shingles and other exterior panels in place under severe weather conditions.

andalusite. $Al_2O_3 \cdot SiO_2$; mol. wt. 162.0; dissociates to yield principally mullite on firing at 1350°C (2462°F); sp. gr. 3–3.5; hardness (Mohs) 7–7.5; used as a component in refractory, spark plug, insulator, and whiteware bodies. See also kyanite.

andesine. A soda-lime feldspar in which the principal constituents are $NaAlSi_3O_8$ and $CaAl_2Si_2O_8$.

Andreasen sedimentation pipette. An instrument in which differences in settling rate are employed as a means of determining the particle size distribution in clays and materials of similar character.

Andrews elutriator. A device consisting of a sequence of classifiers and a graduated cylinder for use in making particle-size analyses. See classifier.

aneroid barometer. A barometer in which variations in atmospheric pressure are measured by fluctuations of a thin elastic metal covering a partially evacuated chamber and indicated by a pointer on a calibrated dial.

angle bead. A slender, curved item of ceramic tile designed to finish the internal or external corners of a wall-tile installation.

angle brick. Any brick shaped to fit an angle or corner.

angle of drain. After dipping ware in a porcelain-enamel slip, the angle at which ware is placed on a rack to drain to obtain a desired thickness.

angle of incidence. The angle which a ray of light makes on a surface and a line perpendicular to that surface.

angle of nip. The maximum angle of the jaws, rolls, mantle, or ring of a jaw crusher which will accept and grip a solid mass for crushing.

angle of refraction. The angle made by the refracted part of a light ray with a line perpendicular to the surface of the refracting medium through the point of incidence of the refracted ray.

anglesite. $PbSO_4$; a source of lead oxide in ceramics.

angle tile. A tile designed or cut for placement in an angular space.

angstrom unit. A one-hundred millionth of a centimeter; used primarily to express wavelengths of light or electromagnetic radiation.

anhydrite. $CaSO_4$; mol. wt. 136.1; m.p. 1450°C (2642°F); sp. gr. 2.96; hardness (Mohs) 3–3.5; used as a drying agent and as a substitute for gypsum in cement.

anhydrous. Without water, both free water and water of crystallization.

anion. A negatively charged ion.

anion adsorption. The adsorption of anions of clay on basal OH surfaces in which structural hydroxyls are replaced by other ions, or on edges where unsatisfied bonds occur; exchange of edge hydroxyls also may occur.

anionic exchange. A type of ionic exchange in which the negative ions in a solution are exchanged with the negative ions in a solid, the superficial physical structure of the solid being unaffected.

anionic exchange capacity. A measure of the ability of a solid substance, such as a clay, to exchange or adsorb ions; usually expressed in milliequivalents of ion per 100 grams of solid.

anisotropic. Having a crystal structure other than cubic, or which does not exhibit the same properties in all directions with reference to light, thereby resulting in more than one index of refraction.

anneal. (1) A process of heating and cooling glass on a prescribed schedule to prevent or release stresses which contribute to brittleness.

(2) The heating of metal shapes to a red heat or above as a means of removing scale, rust, and surface contaminants prior to cleaning and pickling the ware for porcelain enameling.

annealing, abrasive. The process of heating and cooling a material on a prescribed schedule as a treatment to remove stress, refine structure, induce softness, or alter its properties.

annealing, acid. The process of coating metal shapes for porcelain enameling with acid followed by annealing to remove oils, rust, scale, and other soil from the metal surface by scaling, and also to remove inherent stresses.

annealing, bright. The heating of an item of enameling iron or steel to a red heat or above in a reducing atmosphere to produce a clean, bright surface on the metal prior to the application of a porcelain-enamel coating.

annealing, fine. The maintenance of a steady temperature at the end of a firing operation to make certain that all parts of an item being fired attain the same temperature throughout.

annealing fire. (1) The heat treatment of glass and metals to remove internal stresses.

(2) The heat treatment of metal shapes prior to cleaning for porcelain enameling to burn off scale, dirt, grease, and other contaminants, and sometimes to temper the metal.

annealing furnace. The furnace or oven in which the temperature, and sometimes the atmosphere, are controlled for the annealing of glass or metal.

annealing, glass. The process of heating and, particularly, cooling of glassware in accordance with a prescribed schedule to reduce residual thermal stresses to a specified level, and in some instances to modify the structure of the glass.

annealing point. The temperature, or the temperature-time relationship, at which internal stresses in a glass are substantially reduced or relieved.

annealing, porcelain-enameling. The heating of shapes of enameling iron or steel to a red heat or above to remove rust, scale, grease and oils, and other types of soil from the metal surfaces prior to the cleaning and pickling operations in preparation for porcelain enameling.

annealing range. The range of temperatures in which the inherent internal stresses in glass can be reduced or relieved, and which generally is at a rate considered feasible for commercial production.

annealing temperature. Any temperature within a temperature range at which internal stress in a glass can be substantially reduced or relieved, usually, for commercially practical purposes, within a matter of minutes.

annular coil. An electromagnetic coil of the encircling type.

annular kiln. A kiln of the type in which ware is placed in stationary compartments, and the firing zone is moved through each compartment in a successive manner by adjustment of the fuel input.

annular nozzle. A nozzle equipped with a ring-shaped orifice.

anodic cleaning. See anodic pickling.

anodic pickling. An electrolytic process for cleaning and pickling metal for porcelain enameling, or any other finishing treatment in which the metal is used as the anode in the cleaning and pickling bath. Also known as anodic cleaning.

anomalous. Deviating from the normal or the expected value.

anorthite. $CaO \cdot Al_2O_3 \cdot 2SiO_2$; mol. wt. 278.2; sp. gr. 2.74–2.76; hardness (Mohs) 6.0–6.5; a calcium feldspar used in concretes, porcelain enamels, glazes, abrasives, abrasive bonds, artificial teeth, glass, insulating compounds, and conventional ceramic bodies. Also known as calcium feldspar.

anorthoclase. A feldspar of a composition between albite and orthoclase. Also known as soda orthoclase, soda microcline, anorthose.

anthracite-coal-base refractory. A refractory composition containing appreciable amounts of calcined anthracite coal as a source of carbon.

anthracite duff. Briquets composed of mixtures of powdered anthracite and bituminous coals sometimes used in chain-grate stokers for cement kilns.

antiflux. A material which acts as a flux at high temperatures, but hinders fusion at lower temperatures, such as CaO, SnO_2, TiO_2, ZrO_2, CeO, ZnO, MgO, and Al_2O_3.

antifogging compound. A chemical compound which will prevent or minimize the condensation of moisture on windshields, lenses, and other glass products, such as a mixture of castor oil, Na_2SiO_3, kaligen, KNO_3, and water.

antimicrobial agent. A chemical which will inhibit the growth of microbial organisms in ceramic slips, glazes, porcelain enamels, and other slurries during storage.

antimonate of lead. $Pb_3(SbO_4)_2$; mol. wt. 993.18; employed as a yellow pigment in glazes, porcelain enamels, and glass. Also known as lead antimonate, Naples yellow, antimony yellow.

antimony. In the ceramic context, antimony oxide, Sb_2O_3.

antimony oxide. Sb_2O_3; mol. wt. 291.5; m.p. 656°C (1213°F); sp. gr. 5.2–5.7; used as an opacifier in porcelain enamels and as a minor adherence-promoting agent in white porcelain-enamel ground coats, as a constituent in Naples yellow pigments, as a decolorizer and fining agent in glass manufacture, and as a component in glass which is transparent to infrared radiation.

antimony sulfide. Sb_2S_3; mol. wt. 339.7; m.p. 546°C (1015°F); sp. gr. 4.6; used as an aid in the production of ruby and amber glasses, to promote opacity in opal glasses, and occasionally as a minor adherence promoting agent in porcelain enamels. Also known as stibnite, antimony trisulfide, antimony orange, antimony black, antimony needles, antimonous sulfide.

antimony yellow. $Pb_3(SbO_4)_2$; mol. wt. 993.18; a yellow pigment used in glass, glazes, and porcelain enamels. Also known as Naples yellow, lead antimonate, antimonates of lead.

antioch process. A technique for the production of plaster casting molds in which an aqueous slurry of plaster of paris is poured over a mold, following which the mold is steam treated, allowed to set in air, oven dried, and then cooled for use.

antique. (1) An item, such as old glass or pottery, which is of particular value because of its age or historic background.

(2) A type of glass similar in appearance and character to the medieval glasses used in stained-glass windows, which usually is produced in the form of hand-blown cylinders that are cut in the soft or plastic state and allowed to sag to flatness on a suitable, smooth or textured surface.

antiscale compound. A preparation applied to alloy burning tools to protect them from oxidizing and scaling during the firing of porcelain enamels.

antiskid finish. A textured or intentionally roughened surface on porcelain-enamel, tile, concrete, or other facing area to prevent or minimize the possibility of accidental slipping or skidding.

antistatic tile. Floor tile containing a material which will dissipate or disperse charges of static electricity, particularly for use in areas where sparking may be hazardous.

Antonoff's rule. The surface tension at the interface between two saturated liquid layers at equilibrium is equal to the difference between the individual surface tensions of similar layers when exposed to air.

anvil. A piece of wood, a pebble, or other hard substance used to prevent the distortion of a pot during forming by pressing the anvil against the inside wall at the point opposite the point where the forming or shaping pressure is applied.

AP. Acronym for annealing point.

apatite. $Ca_5(F, Cl, OH)(PO_4)_3$; sp. gr. 3.1–3.2; hardness (Mohs) 5; a natural calcium phosphate used as an opacifier in the manufacture of opal glass and as a substitute for bone ash in whiteware bodies.

aplastic. (1) Not exhibiting growth or change.

(2) A noncommittal term describing a mineral or similar substance difficult to identify, or appearing to be essentially a fine gravel.

aplite. A granitic mineral consisting mainly of quartz and feldspar; used as a source of alumina in glass, porcelain and whitewares, pottery, and porcelain enamel.

apparent density. The weight per unit volume of a material, including voids inherent in the material. See apparent specific gravity.

apparent initial softening point. The initial or lowest temperature at which softening or plastic flow of a body, such as glass, glaze, porcelain enamel, etc., begins, and the physical rigidity of the body is overcome.

apparent porosity. The ratio of the open pore space of a body to its bulk volume, expressed in percent; calculated by the formula:

$$P = \frac{W_s - W_f}{V} \times 100$$

in which P is the apparent porosity, W_s is the weight of the water-saturated specimen in grams, W_f is the weight of the original fired specimen in grams, and V is the volume of the specimen in cubic centimeters.

apparent solid density. The ratio of the mass of a body to its apparent solid volume.

apparent solid volume. The total volume occupied by a body, including open and sealed pores.

apparent specific gravity. The ratio of the weight of a unit volume of a body to an equal volume of water at the same temperature determined by the formula:

$$G = \frac{W_f}{V - (W_s - W_f)}$$

in which G is the apparent specific gravity, W_f is the weight of the fired specimen in grams, V is the volume of the fired specimen in cubic centimeters, and W_s is the weight of the water-saturated specimen in grams. Also known as apparent density.

apparent volume. The volume of a body, including its sealed pores, as indicated by the equation:

$$V_a = V_t + V_s = \frac{D}{da}$$

in which V_a is the apparent volume, V_t is the true volume, V_s is the volume of the sealed pores, D is the dry weight, and da is the apparent density (or apparent specific gravity).

appliance. An item of household equipment, such as a washer, dryer, range, refrigerator, toaster, etc., used to perform domestic chores.

application weight. The weight of an application of a porcelain-enamel coating per unit of area covered, usually expressed in grams per square foot for cover coats (one side of test panel) or ounces per square foot (both sides of test panel) for ground coats; normally, the term refers to dry weight unless specifically indicated to be wet weight.

applied research. Research to develop specific knowledge for application in the solution of a particular problem.

apron. (1) A protective refractory shielding arrangement designed to protect the undercarriage of kiln cars from hot gases emanating from the firing chamber of a tunnel kiln situated immediately above the cars; the system consists of vertical metal plates attached to the sides of the kiln car which slide through sand contained in troughs along the bottom of the inside walls of the kiln.

(2) A slab of concrete, metal, wood, or other material placed over the opening to a cistern, barrel, drum, or similar vessel.

(3) A platform of concrete, metal, wood, or other material protecting an item of machinery.

apron conveyor. A conveyor consisting of a series of metal or wood plates mounted at right angles on an endless chain to transfer materials or products from one location to another.

apron feeder. A modification of an apron conveyor designed to feed pulverized materials to a process or packaging unit at a controlled rate.

aqua regia. A mixture of three parts by volume of concentrated nitric acid and one part of concentrated hydrochloric acid.

aqueous solution. A solution in which water is employed as the solvent.

arabesque. An ornate type or style of decoration consisting of flowers, foliage, animals, and figures applied to pottery and artware by painting, low-relief carving, etc., so as to produce intricate patterns of interlaced lines.

Arabian luster. A pottery overglaze containing carbonates or sulfides of copper or silver which are reduced during firing to produce a metallic appearance.

arabic, gum. A water-soluble gum from a variety of acacia trees which is used as a binder to improve the green strength of porcelain enamels and glazes. Also known as acacia gum, gum Senegal, gum Kordofan.

aragonite. $CaCO_3$; mol. wt. 100.1; decomposes at 825°C (1517°F); sp. gr. 2.93; hardness (Mohs) 3.5–4.0; used in refractories, whitewares, glass, electronic bodies, and similar products.

arbitration. A means of settling management-labor disagreements in which both sides agree to be bound by the decision of one or more neutral persons by some mutually agreed method.

arbor. A spindle or shaft on which a grinding wheel, cutting tool, or other rotating part is mounted.

arbor hole. The hole in the center of a grinding wheel, cutting tool, or other rotating part by which the part is mounted on the spindle or shaft of a machine.

arc. The discharge of electricity between two electrodes.

arc of contact. The portion of a grinding wheel in contact with the material or object being ground.

arc furnace. A furnace in which the heat is generated by means of an electric arc.

arc furnace, direct. A furnace in which the electric current passes through the furnace charge.

arc furnace, indirect. A furnace in which ware is heated indirectly by an electric arc struck between electrodes.

arch. (1) A curved structure spanning an open space such as the working zone in a furnace or kiln, thereby forming the roof of the furnace or kiln.

(2) To heat a crucible or glass-melting pot in a pot furnace.

arch, bearer. An arch or series of arches which support the checkerwork of a regenerator or heat exchanger which heats air or gas before combustion.

arch brick. (1) A wedge-shaped brick designed for use in an arch.

(2) An extremely hard-fired or overburned brick from an arch of a kiln.

arch, catenary. An arch designed in the form of an inverted catenary so as to exhibit a minimum of stresses. See catenary arch.

arch, chimney. An arch in the base of a chimney or flue to expedite the entry of combustion and other flue gases.

arch, cooling. A stationary lehr or oven in which glass is annealed.

arch, curtain. An arch of refractory brickwork which supports the wall between the upper part of a gas producer and the gas uptake.

arch, drop. An auxiliary arch projecting below the inner surface of the arched roof of a furnace.

arch, flat. An arch in a furnace or kiln in which the inner and outer surfaces are horizontal and parallel; the inner arch may have a large radius. Also known as jack arch.

arch furnace. A furnace or kiln having a roof which spans and which is supported by two walls.

arch, ignition. An arch constructed with a grate in a furnace or kiln which is designed to expedite the ignition of fuel as it moves beneath the hot brickwork of the furnace.

Archimedes principle. A body immersed in a liquid undergoes an apparent loss in weight equal to the weight of the fluid it has displaced.

Archimedes screw. A spiral tube around an inclined axis or an inclined tube containing a tight-fitting, broad-threaded screw designed to raise water from one level to another.

architectural concrete. A concrete of particularly high quality and free from blemishes; used as the exposed surface on the interior or exterior faces of buildings and other structures.

architectural terra cotta. Hard-fired, glazed or unglazed clay building units generally larger than brick or conventional facing tile; the units may be machine extruded or hand molded and they may be plain or ornamental.

archive sample. A sample retained for purposes of record.

arch, jack. A sprung arch having an outer surface which is horizontal and an inner surface that may be horizontal or curved.

archless kiln. An updraft kiln having no permanent parts which is constructed with walls of either burned or unburned brick; after loading, the kiln is covered with brick, earth, or ashes and fired with solid, liquid, or gaseous fuels.

arch, main. The central part or crown of a furnace or glass-melting tank.

arch, pot. A furnace in which a glass pot is preheated or fired.

arch, relieving. A sprung arch in the substance of a wall above an opening in a furnace wall designed to support the wall and reduce the strain on a second arch constructed immediately below.

arch, rider. An arch or series of arches which support the checkerwork of a furnace regenerator.

arch, rise. The vertical distance between the spring line and the highest point of the undersurface of the arch.

arch, rowlock. An arch constructed of wedge-shaped brick arranged in concentric rings.

arch, saddle. One of a series of arches supporting the checkerwork in a furnace.

arch, segmental. A circular arch in which the inner curved surface is less than a semicircle.

arch, sprung. A curved structure spanning the working zone in a furnace, and which is supported by abutments at the sides or at the ends of the furnace.

arch, suspended. A furnace roof consisting of brick shapes suspended from overhead supports.

arc-image furnace. A furnace which produces very high temperatures by focusing the rays of an electric arc into a relatively small area by means of lenses, mirrors, or other technique.

arc, material transfer. The movement of contact material by the action of an electric arc.

arc melting. The melting of a substance in or by means of an electric arc.

arc spraying. The deposition of molten refractory materials, such as oxides, carbides, nitrides, and silicides, on ceramic or metal surfaces by blowing in an atomized state at high speeds as obtained by the use of a plasma jet.

area of contact. The total area of the surface of a grinding wheel in contact with the item being ground.

area, surface. (1) The total exposed area of the surface of a pulverized solid, usually expressed as some unit of area per gram.

(2) The measured extent of an area, excluding thickness.

arenaceous clay. Sandy clay. Sometimes known as arenite or sandstone.

argillaceous. Containing or consisting of clay or clay minerals such as Albany slip. Sometimes known as mudstone.

argon. An inert gas used as a protective atmosphere surrounding materials which are sensitive to atmospheric gases during firing; used in plasma-jet torches during the application of highly refractory materials to metals.

aridized plaster. Plaster treated with calcium chloride during hydration as a means of increasing its strength and the uniformity of its properties.

ark. A large container or vat used for the mixing and storage of clay slips.

arkose. A sedimentary sandstone composed of fragments containing a high ratio of feldspar and quartz. Also known as feldspathic sandstone.

Armco iron. A relatively pure grade of iron made by the open-hearth process; used in porcelain enameling.

armoring. A metal encasement for refractory brick which is used to protect brick exposed to corrosive atmospheres at the top of the stack of a blast furnace.

arris. The short edge or angle at the junction of a building brick and a ridge tile at the hip or ridge of a roof, molding, or raised edge.

arsenic. In the ceramic context, a term for arsenic oxide, As_2O_3, which see.

arsenic acid. $H_3AsO_4 \cdot 1/2H_2O$; sometimes used as a source of arsenic in glass. See arsenic oxide.

arsenic oxide. As_2O_3; mol. wt. 198.0; sublimes at 193°C (379°F); used as a fining agent and decolorizer in glass and as an opacifier in glazes. Also known as arsenious oxide, arsenic trioxide, white arsenic.

artificial discontinuity. Discontinuities such as grooves, notches, or holes which are introduced into bodies intended to be used as reference standards to provide accurately reproducible sensitivity levels for electromagentic test equipment.

artificial weathering. A test, frequently accelerated, to estimate the resistance of a material or product to weathering in which specimens are subjected to infrared radiation, water, salt water, ultraviolet radiation, and other conditions simulating those encountered in nature.

asbestine. A fibrous variety of talc exhibiting properties similar to asbestos.

asbestos. A group of impure minerals which occur in fibrous form, such as chrysotile, amosite, tremolite, actinolite, crocidolite, etc.; used for fireproofing, heat and electrical insulation, building materials, and similar applications.

asbestos board. A fire-resistant board made of mixtures of asbestos and portland cement.

asbestos cement. A mixture of asbestos, portland cement, and water used in the production of fire-resistant flat and corrugated sheets, shingles, tile, piping, siding, wallboard, and similar products.

asbestos-cement pipe. A pipe manufactured from asbestos cement for use in drainage applications and in corrosive environments.

asbestos felt. Asphalt-impregnated asbestos used as a vapor barrier for concrete.

asbestos fiber. Milled and screened asbestos in fiber form.

asbestos insulation. A fibrous asbestos used as thermal insulation at temperatures above 815°C (1499°F); frequently bonded with clay and sodium silicate.

asbestos shingle. A shingle resistant to weather, fire, and general deterioration which is formed by compressing mixtures of asbestos fiber and portland cement; used as roofing, siding, and similar applications in building construction.

asbolite. An impure earthy mixture of cobalt and manganese oxides used in the production of underglaze blue colors when fired under reducing conditions. Also known as asbolane, black cobalt, cobalt ocher, earthy cobalt.

ash. The noncombustible solid residue remaining from the burning of a fuel or other organic material.

ashes. The residue of burned trees, land plants, bones, seaweed, and marsh plants; sometimes used as a flux in high-temperature bodies and glazes.

ash furnace. A fritting furnace used in the production of materials used in the production of glass. See frit.

ashlar brick. A brick produced with a rough-hackled face resembling the appearance of stone.

ashlar masonry. A type of masonry construction of fired-clay block of a size larger than conventional brick, and with the exposed faces of square or rectangular shape, laid in mortar in a uniform pattern; sometimes sawed, dressed, tooled, or quarry-faced stone is used in the place of the ceramic block.

ashlar masonry, random. An ashlar masonry unit composed of fired-clay block of various sizes so as to provide a random pattern.

as-is basis. A material or product offered and accepted in the condition or shape in which it exists at the time without making changes.

asphalt. A bitumen occurring naturally or distilled from petroleum. See bitumen.

asphalt felt. A sheet of felt-like material impregnated with asphalt for use in roofing and waterproofing applications, frequently in conjunction with asbestos-cement products.

asphalt rock. A porous rock such as sandstone or dolomite, which has become impregnated with asphalt in its natural location.

aspirating screen. A sieve through which particles are drawn by a combination of vibration and suction.

assay. A qualitative or quantitative measurement of the components of a material.

assay, chemical. A chemical measurement of the quantity of one or more components of a substance.

assay, physical. An assay of a material made essentially by physical means.

assembly, joint. A dowel assembly and supporting framework to hold the dowels in place during the placement of concrete, especially in pavement construction. Sometimes called a dowel basket.

assembly line. A more or less direct-line, mass-production arrangement of workers, machines, and equipment along which work moves consecutively from one operation to the next until it is completed.

assurance, quality. All testing and inspection activities undertaken by a producer or supplier to assure a customer that products delivered to him are of a quality level acceptable to him in all respects.

ASTM. Acronym for the American Society for Testing and Materials.

astringent clay. A clay containing an astringent salt such as alum.

atmosphere. (1) The gaseous mass surrounding the earth which is composed of 21 parts of oxygen and 78 parts of nitrogen by volume (23 parts of oxygen and 77 parts of nitrogen by weight), 1% of argon, 0.02% of carbon dioxide, and some aqueous vapor.

(2) The gaseous environment existing in a furnace or kiln, particularly in the zone in which ware is being fired.

(3) A unit of pressure equal to 1.013250×10^6 dynes/cm^2 the air pressure at mean sea level.

atmosphere, controlled. A specified concentration of gas at a specified temperature, and sometimes a specified humidity, serving as the environmental medium during the storage, processing, firing, or use of ceramic materials or products.

atmosphere, neutral. A gaseous environment which is neither oxidizing nor reducing, particularly as applied to the storage, processing, firing, or use of ceramic materials or products.

atmosphere, oxidizing. A gaseous environment in which an oxidizing reaction will take place, particularly during the storage, processing, firing, or use of ceramic materials or products.

atmosphere, reducing. A gaseous environment in which a reducing reaction will take place, particularly such as occurs during the incomplete combustion of fuels or when reducing gases are introduced into the firing zone of a furnace or kiln.

atom. The smallest particle of an element that will enter into the composition of a molecule.

atomic absorption spectrometry. The measurement of light adsorbed by the unexcited atoms of an element as a means of identifying the composition and properties of a substance.

atomic number. The number of electrons orbiting around the nucleous of an atom.

atomic physics. The branch of science concerned with the structures of the atoms, the characteristics and properties of the elementary particles of which an atom is composed, the arrangement of the energy state of the atoms, and the processes involved in the interactions of radiant energy with matter.

atomic weight. The mean weight of an atom of an element in relation to one atom of carbon isotope having a standard weight of 12.0.

atomization. The process of converting liquids and solids to a fine spray, minute particles, or a fine dust.

atomized oil. Fuel oil combined with air under pressure to facilitate its combustion.

atomizing air. A stream of fast-moving air employed to convert liquids or solids to fine sprays or dusts.

attapulgite. A fibrous clay mineral of the general composition $(MgAl)_2Si_4O_{10}\cdot 4H_2O$; used as a suspension agent in various ceramic slips. Also known as palygorskite.

attribute sampling. A method of quality-control inspection in which sampled ware is classified only as passable or defective.

attribute testing. A reliability test procedure in which specimens are evaluated and classified on the basis of qualitative properties or characteristics.

attrition. Wear and disintegration of a surface by rubbing or friction. Also known as scouring, scoring.

attrition mill. A machine in which materials are pulverized between toothed metal disks rotating in opposite directions.

at. wt. Abbreviation for atomic weight.

auger. A machine which forces or extrudes moist clay and similar bodies through a die by means of a revolving screw contained in a closed cylinder or barrel.

autoclave. An air-tight vessel in which materials are subjected to and treated under high-steam pressure.

autoclave cure. A means of accelerating the curing reactions of concrete, asbestos cement, and similar products at elevated temperatures and pressures in saturated steam, particularly when siliceous materials have been incorporated in a cementitious matrix such that a hydrothermal reaction takes place between the silica and the cement.

autocombustion. An automatic system designed to improve the efficiency of oil combustion by means of electric or electronically controlled impulses.

autogeneous grinding. Grinding in a rotating cylindrical mill without the use of balls or rods, the grinding media being incoming additions of the coarse material to be ground.

autogeneous healing. (1) A self-healing of cracks in concrete under favorable conditions of temperature, moisture, and lack of movement.

(2) The self-healing of cracks, pinholes, etc., in porcelain enamels and glazes under the influence of heat.

autogenous mill. A closed, rotating cylinder or mill in which the grinding medium is the coarse feed of incoming material to be ground.

automatic control. A system in which regulating and switching operations are controlled automatically by some responsive device which is sensitive to certain specific or prescribed conditions.

automatic data processing. The performance of tasks involving informational data by means of an appropriate mechanical-electronic system.

automatic dryer. A dryer in which the temperature and atmosphere are controlled by means of an appropriate control device.

automatic snagging. The removal of surface defects and excess metal from a product by the use of automatic or semiautomatic grinding machines, where the pressure between the grinding surface and the work, as well as the traverse wheel over the work, is controlled mechanically or hydraulically from a control station apart from the grinding wheel.

available energy. Energy existing in bodies or systems under conditions that work theoretically may be obtained from them.

available heat. The amount of heat per unit mass of a substance that may be transformed into some form of work, such as in an engine or other system, under ideal conditions.

aventurine. A glass or glaze containing colored, opaque spangles of nonglassy materials such as copper, gold, chrome, or hematite which give the glaze a shimmering appearance.

average coefficient of cubical expansion. The average change in the unit volume of a body or substance per unit change in temperature over a prescribed temperature range.

average coefficient of linear expansion. The average change in the unit length of a body per unit change in temperature over a prescribed temperature range.

average particle size. The average of the dimensions of particles of a material or a mixture of materials.

azurite. $Cu_3(OH)_2(CO_3)_2$ or $2CuCO_3 \cdot Cu(OH)_2$; mol. wt. 344.7; sp. gr. 3.77–3.83; hardness (Mohs) 3.5–4.0; a basic carbonate of copper used as a blue pigment. Also known as blue copper, blue malachite, chessylite.

B

bacile. A deep ceramic dish or basin.

backer strip. An asphalt-coated felt strip employed as a water-repellent backing for the vertical joint between asbestos-cement shingles.

background. The natural or developer-coated surface of a test specimen employed in the liquid-penetrant process of inspecting magnetic or electromagnetic materials for discontinuities or other possible defects.

background fluorescence. The fluorescent residues observed on the surface of a test specimen during fluorescent-penetrant inspection.

backing. (1) The portion of a wall or structure installed behind a facing course to attain a particular property in the structure, such as strength, insulation, or economy.

(2) A backing material such as cloth, paper, fiber, etc., used as the backing for coated abrasives.

(3) The flexible carrier for the magnetic oxide coatings employed on magnetic tapes.

backlog. An accumulation of unfilled orders or work.

back-off. To remove a cutting tool or grinding wheel from contact with an item being processed.

back order. An order designating future completion or delivery, or the replacement of an order for a previously ordered material or product which was not available at that time.

back stamp. A mark made on the back or bottom of a product to identify its origin or manufacturer; a hallmark.

back wall. The wall at the charging end of a glass-melting furnace.

backwear. A worn condition on the back of an abrasive belt caused by high speed, high pressure, or both, which results in friction between the belt and its backup at the point of contact with a work piece.

baddeleyite. Naturally occurring ZrO_2; mol. wt. approximately 123.22; m.p. 2300–2950°C (4172–5342°F); sp. gr. 5.5–6.0; used in refractory and corrosion-resistant applications such as furnace linings and muffles and as an ingredient in low-expansion ceramic bodies.

badging. The marking of glassware and other ceramic products to identify the manufacturer, ownership, capacity, composition, or other information.

baffle. (1) A partition consisting of a panel, plate, screen, wall, or other device designed to check, regulate, or deflect the flow of something, such as a shield placed in a position to protect ware from combustion gases in a furnace or kiln during firing.

(2) The part of a glass-forming mold designed to shut off the delivery of molten glass into the mold.

baffle mark. A mark or seam line visible on a bottle or

other glass product caused by the joint between the mold and the baffle.

baffle wall. A wall constructed in a furnace or kiln to protect items being fired from flames and combustion gases.

bag. A flexible container made of paper, cloth, plastic, or similar material for the transport or storage of substances.

bagasse. The crushed fibrous material remaining after the juice is extracted from sugar cane; employed as a reinforcement and filler in plaster products, such as acoustic tile.

bag filter. An apparatus containing porous cloth, paper, or felt bags designed to collect dust from dust-laden gases passed through the apparatus.

baghouse. A chamber containing an arrangement of bag filters for the removal of air-borne particles from air or gas streams emanating from furnaces, drymixers, or other dust-producing equipment or operations.

bag wall. A refractory wall in a furnace or kiln designed and placed to deflect a flame to prevent it from striking ware being fired.

Bailey meter. A flowmeter of helical vane construction used to measure the weight of powdered or granular materials passing through an essentially vertical shaft or other enclosed passage.

bait. A tool dipped into a bath of molten glass to start a drawing operation.

balance. A weighing device consisting essentially of a horizontal beam having a fulcrum at the center with a pan suspended from each end, one holding the object being weighed and the other holding equivalent weights.

balance, dynamic. The condition which permits a grinding wheel or other rotating part to rotate at high speeds with no vibration or whip due to uneven distribution of weight through its mass.

balance, material. The comparison of input and output of material quantities for a particular process; generally, the comparison of inventory plus receipts at the beginning of a process with the inventory plus shipments at the end of the process over a specific time interval.

balance, material area. An area within a factory where the material records are maintained in such a way that a balance may be taken at any time during operations to show the amount of material for which the area is responsible.

balance, static. The condition which permits a grinding wheel or other rotating part centered on a frictionless horizontal shaft to remain at rest in any position.

balancing. Testing for balance by adding or subtracting weight to put a grinding wheel or other rotating part into either static or dynamic balance. See static balance, dynamic balance.

ball clay. A kaolinic type of clay characterized by high plasticity, fine-grained particles, high dry strength, long vitrification range, and a white-to-cream color after firing; employed in ceramic bodies to provide plasticity during forming and to induce vitrification during firing, as a suspension agent in porcelain enamels and glazes, and as a bonding agent in nonplastic refractories.

ball, grinding. A hard, dense, abrasion-resistant sphere used as the crushing body in a ball mill; usually composed of flint, dense porcelain, alumina, steel, or heavy alloy.

balling. The tendency of a material to conglomerate or cluster, particularly during mixing.

ball mill. A closed-end rotating cylinder, usually consisting of a steel jacket with an abrasion-resistant porcelain or porcelain-like lining and containing pebbles or porcelain balls as the grinding media, in which materials are wet or dry ground as a means of mixing or reducing the particle size; the mill and grinding media may be of steel or alloy compositions if contamination is not a factor.

ball mill, air-swept. A continuous ball mill in which finely milled particles are swept from the mill by an air current.

ball milling. The process of grinding and mixing materials, with or without liquid, in a rotating cylinder or conical mill partially filled with grinding media, such as pebbles or porcelain balls.

ball mill, Krupp. A grinding device consisting of chilled iron or steel balls grinding together in a die ring of perforated steel plates, each overlapping the next; the ground material is discharged through a cylindrical screen.

ball mill, vibrating. A ball mill in which conventional milling is combined with a vibrating or bouncing action of the mill to obtain more efficient and rapid grinding.

ball test. (1) A test in which a ball of specified size and weight is dropped or forced onto the surface of a body, glaze, porcelain enamel, or other material under prescribed conditions as a means of evaluating a property such as resistance to impact, degree of adherence, etc.
(2) An on-site test of the consistency of concrete.

bamboo ware. A type of brownish or cane-colored stoneware.

Banbury mixer, A heavy-duty mixer consisting of two rotors, the faces of which turn in opposite directions; used in mixing viscous compositions and pastes.

banding. The application of a decorative line or band of color to the edges, sides, and facial surfaces of chinaware, pottery, and similar products.

bank kiln. A kiln constructed on a slope or bank of earth, the incline serving in the place of a flue for the removal of combustion gases.

bank run. Concrete aggregate in the condition as excavated from banks or pits.

banks. The sloping refractory section of an open-hearth furnace located between the hearth and the front and back walls.

bank sand. A sand of low clay content used in making casting cores.

bannering. The leveling of saggers in a kiln to facilitate stacking.

barite. $BaSO_4$; mol. wt. 233.4; m.p. 1580°C (2876°F); sp. gr. 4.3–4.6; hardness (Mohs) 2.5–3.5; employed in glasses as a flux to reduce seeds, increase toughness, improve brilliance, and reduce annealing time. Also known as barytes.

barium aluminate. (1) $3BaO \cdot Al_2O_3$; mol. wt. 562.0; employed as a source of barium oxide in glass compositions to decrease the solubility and increase the brilliance of the glass; also used in cathode coatings for vacuum tubes. (2) $BaO \cdot Al_2O_3$; mol. wt. 255.30; m.p. 1998°C (3630°F); sp. gr. 3.99. (3) $BaO \cdot 6Al_2O_3$; mol. wt. 765.0; m.p. 1860°C (3380°F); sp. gr. 3.64.

barium aluminum silicate. $BaO \cdot Al_2O_3 \cdot 2SiO_2$; mol. wt. 375.45; m.p. 1716°C (3120°F); sp. gr. 3.21–3.30.

barium arsenate. $3BaO \cdot As_2O_5$; mol. wt. 689.94; m.p. 1604°C (2920°F); sp. gr. 5.10.

barium boride. BaB_6; mol. wt. 202.3; m.p. 2270°C (4118°F); sp. gr. 4.32; hardness (Vickers) approximately 3000.

barium calcium silicate. $BaO \cdot 2CaO \cdot 3SiO_2$; mol. wt. 445.70; m.p. 1320°C (2410°F).

barium carbide. BaC_2; mol. wt. 161.36; m.p. >1760°C (3200°F); sp. gr. 3.57.

barium carbonate. $BaCO_3$; mol. wt. 197.4; m.p. 1360°C (2480°F); sp. gr. 4.4; employed as a flux in porcelain enamels and glazes and to improve elasticity, brilliance, mechanical strength, and acid resistance and to prevent scumming; used as an ingredient in flint glass, pressed tableware, television tubes, and laboratory glassware to lower the melting point, improve workability, improve brilliance and hardness, and to improve dielectric constants and resistivity; used to obtain maximum flux density in hard core permanent magnets; used in structural clay products to prevent scum and efflorescence; and employed in steatite, forsterite, zircon porcelain, and titanate electronic components to reduce dielectric loss.

barium chloride. $BaCl_2 \cdot 2H_2O$; mol. wt. 244.3; m.p. 960°C (1760°F); sp. gr. 3.097; used as a set-up agent and scum preventative in porcelain enamels by precipitating soluble sulfates as insoluble barium sulfate.

barium chromate. $BaCrO_4$; mol. wt. 253.4; sp. gr. 4.5; used in the production of yellow and pale green overglaze colors. Sometimes known as chrome yellow.

barium crown glass. An optical crown glass containing barium oxide as a major component. See crown glass, optical.

barium flint glass. An optical flint glass containing barium oxide as a major component. See crown glass, optical.

barium fluoride. BaF_2; mol. wt. 175.4; m.p. 1280°C (2336°F); sp. gr. 4.83; used as an opacifier and flux in porcelain enamels.

barium fluosilicate. $BaSiF_6$; decomposes at 300°C (572°F); sp. gr. 4.3; used as a flux and an opacifier in porcelain enamels and glazes. Also known as barium silicofluoride.

barium glass. A glass in which part of the calcium oxide is replaced by barium oxide.

barium metaphosphate. $Ba(PO_3)_2$; mol. wt. 295.4; m.p. 849°C (1560°F); used as a precoating treatment for metals to prevent primary boiling in sheet steel enamels, and as an ingredient in glass.

barium molybdate. $BaMoO_4$; mol. wt. 297.4; m.p. >1300°C (2372°F); sp. gr. 4.65; used as an opacifier and adherence-promoting agent in porcelain enamels.

barium monohydrate. $Ba(OH)_2 \cdot H_2O$; mol. wt. 189.4; used in the manufacture of barium ferrite magnets.

barium niobate. $6BaO \cdot Nb_2O_5$; mol. wt. 1186.36; m.p. 1927°C (3500°F); sp. gr. 5.98.

barium nitrate. $Ba(NO_3)_2$; mol. wt. 261.4; m.p. 575°C (1067°F); sp. gr. 3.244; used to improve homogeneity and opacity in porcelain enamels and as an ingredient in optical glasses. Also known as nitrobarite.

barium octahydrate. $Ba(OH)_2 \cdot 8H_2O$; mol. wt. 299.5; loses water of crystallization at 78°C (172°F); m.p. of anhydrous $Ba(OH)_2$ 408°C (770°F); sp. gr. 1.656; used in ceramics as a source of high-purity BaO. Also known as barium hydroxide.

barium oxide. BaO; mol. wt. 153.4; m.p. 1923°C (3490°F); sp. gr. 4.73–5.46; hardness (Mohs) 3.3; used as a fluxing ingredient in glass.

barium peroxide. BaO_2; mol. wt. 169.4; m.p. 450°C (842°F); decomposes at 800°C (1470°F); sp. gr. 4.58; limited use in glass manufacture.

barium phosphate. $3BaO{\cdot}P_2O_5$; mol. wt. 602.12; m.p. 1727°C (3140°F); sp. gr. 4.1.

barium phosphide. Ba_3P; mol. wt. 567.18; sp. gr. 3.18; hardness (Knoop) 3200.

barium selenite. $BaO{\cdot}SeO_2$; mol. wt. 280.6; sp. gr. 4.4.

barium silicate. (1) $BaSiO_3$; mol. wt. 213.4; m.p. 1640°C (2984°F); sp. gr. 4.4. (2) $BaO{\cdot}2SiO_2$; mol. wt. 273.48; m.p. 1419°C (2585°F); sp. gr. 3.73. (3) $2BaO{\cdot}SiO_2$; mol. wt. 366.75; m.p. >1755°C (3190°F); sp. gr. 5.20. (4) $2BaO{\cdot}3SiO_2$; mol. wt. 486.90; m.p. 1449°C (2640°F); sp. gr. 3.93.

barium stannate. $BaSnO_3{\cdot}3H_2O$; mol. wt. 358.1; loses H_2O at 280°C (536°F); used as an additive to barium titanate bodies to decrease the Curie point for use as capacitors of high-dielectric constant; also used in glass enamels to improve alkali resistance.

barium sulfate. $BaSO_4$; mol. wt. 233.4; m.p. 1580°C (2876°F); sp. gr. 4.25–4.5; used in porcelain enamels to improve workability and to reduce tendency to shoreline and dimple. Also known as blanc fixe, barite, barytes.

barium sulfide. BaS; mol. wt. 169.4; m.p. >1660°C (3020°F); may be fired in bodies at 1450°C (2642°F), but will vaporize at 1600°C (2912°F); sp. gr. 4.25; used in the manufacture of crucibles for the melting of cerium and uranium.

barium telluride. BaTe; mol. wt. 264.86; m.p. 1527°C (2780°F).

barium thorate. $BaO{\cdot}ThO_2$; mol. wt. 417.48; m.p. 2299°C (4170°F); sp. gr. 7.66.

barium titanate. (1) $BaTiO_3$; mol. wt. 233.3; m.p. 1618°C (2950°F); widely used in piezoelectric and ferroelectric applications because of its high-dielectric constant. (2) $BaO{\cdot}2TiO_2$; mol. wt. 313.6; m.p. 1320°C (2410°F). (3) $BaO{\cdot}3TiO_2$; mol. wt. 394.0; m.p. 1356°C (2475°F); sp. gr. 4.70. (4) $BaO{\cdot}4TiO_2$; mol. wt. 473.8; m.p. 1420°C (2600°F), sp. gr. 4.60. Barium titanates are employed in guided missiles, sonar, ultrasonic cleaning, accelerometers, filters, measuring instruments, etc.

barium titanium silicate. (1) $BaO{\cdot}TiO_2{\cdot}SiO_2$; mol. wt. 293.32; m.p. 1398°C (2550°F). (2) $BaO{\cdot}TiO_2{\cdot}2SiO_2$; mol. wt. 353.38; m.p. 1248°C (2280°F).

barium tungstate. $BaWO_4$; mol. wt. 385.4; sp. gr. 5.04; used as a white pigment and as a phosphorescent.

barium zirconate. $BaO{\cdot}ZrO_2$; mol. wt. 276.6; m.p. 2620°C (4748°F); sp. gr. 2.63; used as an addition to barium titanate bodies to improve their dielectric properties.

barium zirconium silicate. $BaO{\cdot}ZrO_2{\cdot}SiO_2$; mol. wt. 336.6.

Barker-Truog clay treatment. An alkali treatment for clay to obtain pH values ranging from 7 to 10, depending on the original acidity of the clay; such clays exhibit improved plasticity for use in the shaping of brick.

bar mat. A mat of preassembled steel bars for installation as a reinforcement in a concrete slab, usually a paving slab.

barn. A unit expressing the probability of the occurrence of a specific nuclear reaction; numerically, it is 10^{-24} cm^2.

barometer. An instrument designed to measure the pressure of the atmosphere.

barometer, aneroid. A barometer in which variations in atmospheric pressure are measured by fluctuations of a thin elastic metal covering a partially evacuated chamber and indicated by a pointer on a calibrated dial.

barometer, mercury. A barometer in which variations in atmospheric pressure are measured by the rise and fall of a column of mercury contained in a partially evacuated vertical glass tube sealed at the top, the open end of the tube resting in a reservoir of mercury exposed to the atmosphere.

barrel. A unit of measure of cement equal to 376 pounds or four sacks.

barrel finishing. Improving the surface or removing burrs from the edges of work by tumbling the work in a rotating cylinder containing suitable particles or grains of abrasives.

barrier. A panel, wall, or other structure designed to bar or deflect the passage of something, such as a baffle placed to deflect combustion gases in a furnace from impinging on ware being fired.

barrier, moisture. A material or coating applied to concrete to retard the passage of moisture into a wall.

barrier, vapor. See barrier, moisture.

bar, runner. An iron casting attached to a circular grinding head or runner for the grinding of plate glass.

bars, Holdcroft. Bars of selected compositions designed to soften at different temperatures for use as pyroscopes, which see.

barytes. $BaSO_4$; mol. wt. 233.4; sp. gr. 4.3–4.6; hardness (Mohs) 2.5–3.5; used as a flux in glasses to reduce seeds, increase toughness, improve brilliance, and reduce annealing time; also used in ceramic bodies, glazes, and porcelain enamels to minimize or prevent scumming. Also known as barite.

basalt. A crystalline volcanic rock composed essentially of soda-lime feldspar, pyroxene, magnetite, olivine, magnesite, and ilmenite.

basalt, fusion cast. A hard, crush-resistant and abrasion-resistant product obtained by casting molten basalt into appropriate shapes and cooling in accordance with a prescribed schedule; used as a flooring and lining material in areas of severe abrasion.

basalt ware. A hard, black, unglazed vitreous ware having an appearance similar to that of basalt rock.

base. (1) An alkaline substance, either ionic or molecular, which tends to accept a proton from another substance or which will react with an acidic material.

(2) The bottom of a container, bottle, or other item.

(3) The compacted earth or granular material upon which a paving slab is placed.

(4) The foundation that supports a printed circuit or the pins, leads, or other terminals of a bulb or tube to which an external electrical or electronic connection is to be made.

base charge. The monetary amount per unit of source material, special nuclear material, or other product set forth in the code of Federal regulations.

base coat. A fired coating over which another coating is applied.

base course. The concrete foundation over which a wall, pavement, or other structure is to be erected or placed.

base exchange. A surface property exhibited by colloidal inorganic materials, such as clays, whereby certain ions are replaced by other ions in a surrounding medium.

base, manhole. A concrete-slab foundation or the bottom manhole riser section with an integrally cast concrete floor over which a manhole is constructed.

base metal. The metal to which porcelain enamel is applied.

basic brick. Refractory brick composed essentially of basic ingredients, such as lime, magnesite, chrome ore, or dead-burned magnesite, which will react chemically with the acidic refractories, clays, or fluxes at high temperatures.

basic brick, direct-bonded. A fired refractory brick in which the refractory grains are united predominately by a solid-state diffusion mechanism.

basic brick, pitch-bonded. Unburned basic-refractory shapes bonded with pitch; if the shapes subsequently are heat-treated sufficiently to minimize softening of the bond on reheating, they are designated as a tempered product. See pitch.

basic brick, pitch-impregnated. Burned basic refractory shapes which are impregnated with pitch after they have been fired. See pitch.

basic fiber. Untreated glass fiber as it is obtained from the forming equipment.

basic-lined. A furnace, kiln, converter, or similar structure lined with basic refractory shapes made of materials such as lime, magnesite, chrome ore, etc.

basic open-hearth furnace. An open-hearth furnace constructed of basic refractories covered with magnesite or burned dolomite, and which is employed in the production of basic pig iron.

basic oxide. A metallic oxide, such as sodium oxide or potassium oxide, which will form a hydroxide when combined with water, and which will react chemically with acidic materials.

basic refractory. A refractory composed of basic refractory materials, such as lime, magnesite, chrome magnesite, etc., and which will react with acidic slags and fluxes at elevated temperatures.

basic research. Investigations designed to advance knowledge in which practical application of the knowledge is not an immediate consideration.

basic slag. A slag rich in basic ingredients produced as a by-product in the steel-making process.

basket, pickle. A corrosion-resistant basket or open-meshed container in which metals being prepared for porcelain enameling are pickled. See pickle.

basket-weave checkerwork. An arrangement of corrosion-resistant refractory brick serving as flues in regenerators and other structures in which the ends of each brick are placed at right angles to the center of each adjacent brick to form a pattern resembling the weave of the splints in a basket.

bas-relief. A type of artware in which the figures project slightly above the background surface.

basse taille. A process in which transparent or translucent porcelain enamels are applied and fired over a metal background which has been carved in low relief.

bastard ganister. A mineral of ganister appearance, but differing in properties.

bat. (1) A plaster slab or disk upon which clay is worked, or upon which ware is formed and dried.

(2) A fireclay slab upon which ware is placed and fired in a kiln.

(3) A fragment of hardened clay or brick.

(4) A slab of moist clay.

(5) A brick cut transversely so as to leave one end whole.

(6) A sheet of gelatin used in bat printing.

batch. A quantity of raw materials blended together for subsequent processing, such as a glass batch or furnace charge.

batch blending. Stepwise changes in the composition of a batch to arrive at a desired composition of a final product.

batch charger. A mechanical device employed to introduce a batch into a smelter or melting tank.

batch dryer. A periodic dryer in which the ware being dried remains stationary in a circulating stream of air, usually warm or hot, until dry.

batcher. A type of equipment in which the ingredients of a batch are measured and collected before discharging into a process operation, such as a ball mill or concrete mixer.

batch feeder. A mechanical device, such as an auger, employed to charge a glass or porcelain enamel batch into a melting tank or smelter.

batch furnace. A furnace into which ware is charged, fired, and removed before the introduction of another charge.

batch house. The area in a factory in which materials are received, stored, handled, weighed, and mixed preparatory for movement to a subsequent manufacturing operation.

batching sequence. The process of introducing raw materials into a batch mixer or process in an ordered, stepwise sequence.

batch operation, contact. The adsorption process in which activated carbon is dispersed in a fluid to be treated and then separated when practical equilibrium is reached.

batch process. A manufacturing operation or process which is carried to completion before the same operation or process is repeated; that is, the process is not continuous.

batch, raw. (1) A batch of thoroughly mixed ingredients ready for the next manufacturing operation.

(2) A glass batch containing no cullet.

(3) Any batch of mixed ingredients ready for processing.

batch smelter. A periodic smelter or glass-melting tank into which a charge is introduced, melted, and discharged as a unit process in accordance with a prescribed time and temperature cycle.

batch truck. A dump truck in which the body is partitioned into compartments for the transport of weighed batches of cement and aggregate from the weighing areas to the mixer.

batch-type mixer. A machine into which all ingredients of a batch are weighed, mixed, and discharged as a unit operation before introduction of a subsequent charge.

bath. (1) A liquid preparation, such as water, cleaner, acid, neutralizer, or other solution, in which something is immersed for treatment.

(2) Liquid penetrants into which parts are immersed for inspection.

(3) Penetrants retained in immersion tanks for reuse.

batt. An alternative spelling of bat.

batten. A thin strip of material employed to seal, conceal, or reinforce a joint as, for example, a strip of flat or corrugated asbestos cement used to conceal butt joints of flat or corrugated asbestos-cement sheets.

batter. The upward slope or the angle at which the outer face of a wall slopes from the vertical.

batt printing. A process for printing on ceramic ware in which a design is transferred from an engraving plate to ware by means of a bat of solid glue or gelatin.

bat wash. A slurry of refractory materials applied to kiln setters to prevent the sticking of ware during firing.

Baumé. Either of two calibrated hydrometer scales to estimate the specific gravity of liquids. For liquids lighter than water, the specific gravity equals 140 ÷ (130+°Bé) at 15.6°C (60°F); for liquids heavier than water, the specific gravity equals 145 ÷ (145−°Bé) at 15.6°C (60°F).

bauxite. Rocks consisting largely of hydrates of alumina, together with varying amounts of iron and titanium oxides, silica, and other impurities. Bauxites fuse at 1800°C (3270°F) and above, and have specific gravities varying from 2.45 to 3.25. As a major source of alumina, bauxites are employed extensively in the manufacture of grinding wheels, abrasive stones, abrasive cloth and paper, polishing and grinding powders, refractories for kilns and glass tanks, electroceramics, and quick-setting alumina cements.

bauxite clay. A natural mixture of bauxite and clay containing not less than 47% nor more than 65% of alumina on a calcined basis.

Bayer process of alumina extraction. A process in which alumina ores are digested in hot solutions of caustic soda and removed as soluble aluminates.

Bé. Symbol for Baumé.

bead. (1) An enlarged, rounded edge of a glass tumbler or other glass article.

(2) An excess of porcelain-enamel slip or powder along the edge of a coated ware.

(3) An application of porcelain enamel, usually of a contrasting color, to the edge or rim of a porcelain-enameled article.

(4) A small piece of glass tubing used to enclose a lead wire.

beader. An operator who applies a beading enamel to a porcelain-enameled article.

beader off. An operator who removes a bead of excess porcelain enamel or smoothes the edges of the coating on porcelain-enameled ware.

beading. (1) The process of applying porcelain enamel, usually of a contrasting color, to the edges or rims of porcelain-enameled articles.
(2) The removal of excess slip from the edge of dipped ware.

beading enamel. Any of the special porcelain enamels applied as a beading on ware for purposes of decoration and protection of exposed edges of the ware.

bead test. A test of the softening and flow characteristics of glaze, glass, and porcelain-enamel compositions in which a bead or button-like specimen of specified size and shape is compared with standard compositions at elevated temperatures.

bead thermister. A thermister consisting of two wire leads cemented together by a molten droplet of a semiconducting material.

beaker. A thin-walled, flat-bottomed vessel, usually of glass, but sometimes of metal or plastic, with a wide mouth and pouring spout.

beam, reinforced. A concrete beam placed in tension, compression, or torsion by steel bars, wire mesh, rods, etc., embedded in the concrete.

bearer arch. One of a series of arches that supports the checkerwork in a regenerator or heat exchanger which heats air or gas before combustion.

bed. (1) The layer of mortar upon which brick and stone are laid.
(2) The prepared base or foundation upon which ware is placed for processing, such as the floor of a kiln.

bed depth, critical. The minimum depth of an adsorbent bed required to maintain the mass-transfer zone for activated carbon.

bedder. A plaster-of-paris shape for forming a bed of powdered alumina on which bone china is fired.

bedding. The process of placing ceramic ware on a suitable refractory grain or powder as a support to prevent warpage during firing.

bed, expanded. A bed of granular activated carbon or other substance through which a fluid flows upward at a rate to elevate and separate the particles slightly without changing their positions.

bed, fixed. A bed of granular activated carbon or other substance through which a fluid may flow without causing substantial movement of the bed.

bed, fluidized. A bed of free-flowing granular or finely divided material in which the solids are suspended in a rising current of air or medium, causing them to behave in the manner of a fluid; it is finding extensive and expanding use in chemical and physical technologies involving the application of uniformly controlled heat for drying, calcining, and quenching as required in coatings, food preparation, fuel combustion, fluidized-bed nuclear reactors, and the like.

bed, intermittent-moving. An adsorption process characterized by the upward flow of a fluid through a fixed bed of granular activated carbon with periodic withdrawal of spent carbon from the bottom of the bed and additions of reprocessed or virgin carbon to the top of the bed.

beehive kiln. A circular beehive-shaped kiln characterized by a domed roof and fired through chambers stationed around the circumference.

Belgian kiln. A longitudinal-arch, side-fired kiln in which the fire is directed to grates stationed at regular intervals along the bottom of the structure.

bell. (1) The enlarged end of a concrete or other pipe which overlaps the end of an adjoining pipe.
(2) A refractory funnel placed to receive molten steel from the nozzle of a ladle.

bellarmine. A fat, narrow-necked, salt-glazed bottle or jug usually having a bearded face stamped or engraved on the neck as a decoration.

bell damper. A bell-shaped, sand-seal type of damper frequently used in annular kilns.

bell dresser. A tool consisting of rotating metal cutters employed in the truing, shaping, and dressing of grinding wheels.

Belleek china. A thin, highly translucent chinaware having zero water absorption which is composed of a body containing substantial amounts of frit, and which normally is coated with a soft luster glaze.

belly. (1) The side of a clay pot.
(2) The section of a converter in which steel is collected before it is poured.
(3) The widest section of a blast furnace.

belshazzar. A wine bottle of approximately 16-quart capacity.

belt. An endless flexible band passing around two or more pulleys; used to convey materials or objects, or to transmit motion from one pulley to one or more other pulleys.

belt, abrasive. A coated abrasive product in the form of a belt for use with a powered grinding or polishing machine.

belt conveyer. An endless belt running between head and tail pulleys used to transport loose materials or products from one point to another.

belt drive. A mechanism actuating a ball mill or other item of equipment by means of a friction belt rotating around a pulley mounted on a rotating shaft.

belt feeder. A mechanical device which delivers raw materials from one point to a processing station by means of a moving belt.

belt grinding. Grinding the surface of a material or product by means of a continuous abrasive-coated belt.

belting. A finishing operation for concrete pavement in which a wide belt is dragged back and forth across a fresh slab of concrete and advanced along the slab.

belt kiln. A kiln through which ware being fired is transported by means of an endless, high-temperature-resistant alloy belt.

belt marks. Marks made on the bottom of glass articles as they ride through the lehr on a slightly overheated chain belt.

belt, segmented. A coated abrasive belt made of segments spliced together, the segments being necessary to obtain belts wider than 50 inches, the widest coating width generally available. See abrasive belt.

bench. The floor of a pot furnace, often called a siege.

bench grinder. An offhand grinding machine supported on a bench, the grinding mechanism consisting of one or two grinding wheels mounted on a horizontal spindle.

bench scale. A process, test, or other procedure carried out on a small scale as on a laboratory bench or work table.

bend. A pane of glass which has been bent to fit an opening. See bending.

bending. The manipulation of glass, particularly flat glass, in a kiln to form curved shapes or bends.

bend test. (1) A measure of the transverse or cross-bending strength.

(2) A test in which bisque or fired porcelain-enameled panels are distorted by bending to determine the resistance of the coating to cracking or fracture.

beneficiation. Any process of upgrading or improving the physical or chemical properties of a material to enhance its use such as washing, flotation, etc.

bent glass. Flat glass that has been shaped into cylindrical, curved, or other shapes while hot.

bentonite. A clay derived from volcanic ash and characterized by an extremely fine grain size. Its chief constituent is montmorillonite ($Al_2O_3 \cdot 5SiO_2 \cdot 7H_2O$) plus 5 to 10% of alkalies or alkaline-earth oxides. One type, which will absorb large quantities of water, will swell enormously; used to increase dry and fired strengths and reduce absorption in whiteware bodies; also used as suspension agent in porcelain-enamel slips.

beryl. $Be_3Al_2(SiO_3)_6$; mol. wt. 537.36; m.p. 1410°C (2570°F); sp. gr. 2.63–2.80; hardness (Mohs) 7.5–8; inert to most reagents except hydrofluoric acid; employed as a dielectric, to reduce firing shrinkage, and to improve transverse strength, resistance to thermal shock, and impact resistance in spark-plug bodies. Also used in mat glazes for talc bodies, as a green colorant in other glazes, and in the production of glass windows for Roentgen-ray tubes.

beryllia. BeO; mol. wt. 25.02; m.p. 2570°C (4658°F); sp. gr. 3.016; exhibits high-thermal conductivity, excellent dielectric characteristics, good resistance to wetting, relatively good physical strength; employed in microwave parts, solid-state devices, and gyroscopes, and as a moderator, reflector material, and matrix for fuel elements in nuclear applications. Also known as beryllium oxide.

beryllides. Intermetallic compounds in which one element is beryllium, the general formula being Me_xBe_y; characterized by high melting temperatures ranging from approximately 1427°C (2600°F) to 2080°C (3775°F), excellent resistances to oxidation up to 1260°C (2300°F) and some to as high as 1540°C (2804°F), high strength and strength retention at elevated temperatures, and excellent thermal-shock resistance; reported specific heats range from 0.20 to 0.40 Btu/lb./°F; thermal conductivities range from 20 to 50 Btu/hr./ft./°F between 371°C (700°F) and 1483°C (2700°F); linear thermal expansions of about 2% at 1371°C (2500°F); bend strengths of about 30×10^3 psi. between 21°C (70°F) and 1231°C (2250°F); Vickers hardness values between 500 and 1300 kg./mm.^{2} [2.5 kg. load at 21°C (70°F)]; and Young's moduli below 50 $\times 10^6$ psi at 21°C (70°F); potential materials for use in structural applications and spark-resistant tools. See beryllides of chromium, cobalt, hafnium, iron, manganese, molybdenum, nickel, niobium, palladium, platinum, plutonium, tantalum, thorium, titanium, tungsten, uranium, vanadium, yttrium, and zirconium.

beryllium aluminate. $BeO \cdot Al_2O_3$; mol. wt. 126.96; m.p. 1870°C (3398°F); sp. gr. 3.50–3.84; hardness (Mohs) 8.5. Also known as chrysoberyl.

beryllium boride. (1) BeB_2; mol. wt. 30.66. (2) Be_2B; mol. wt. 28.86. (3) BeB_6; mol. wt. 73.94. Other properties not reported. See borides.

beryllium carbide. Be_2C; mol. wt. 30.04; decomposes above 2950°C (5342°F); unstable in oxygen above 982°C (1800°F); sp. gr. 1.90; hardness (Mohs) approximately 9; modulus of rupture 16,000 psi; compressive strength 105,000 psi; employed as a neutron moderator in nuclear applications and in applications where hardness, toughness, elasticity, and corrosion resistance at moderately high temperatures are important.

beryllium nitride. Be_3N_2; mol. wt. 55.22; m.p. 2200°C (4028°F); sp. gr. 2.71; oxidizes in air above 600°C (1112°F); used in incandescent mantles and in applications where hardness, elasticity, corrosion resistance, and toughness at moderately high temperatures are required.

beryllium oxide. BeO; mol. wt. 25.02; m.p. 2570°C (4658°F); sp. gr. 3.016; hardness (Mohs) 9; exhibits excellent dielectric properties, good physical strength and resistance to wetting by metals and nonmetals, and high-thermal conductivity; employed in rocket nozzles, crucibles, insulators, radomes, thermocouple protection tubes, microwave parts, solid-state devices, and gyroscopes, and as a moderator, reflector material, and matrix for fuel elements in nuclear applications. Also known as beryllia.

beryllium phosphide. Be_3P_2; mol. wt. 89.10; sp. gr. 2.06.

beryllium silicate. $2BeO \cdot SiO_2$; mol. wt. 110.10; m.p. 1560°C (2840°F); sp. gr. 2.99.

beryllium sulfide. BeS; mol. wt. 41.08; sp. gr. 2.47.

Bessemer converter. A refractory-lined vessel in which steel is produced by the Bessemer process, which see.

Bessemer process. A process for making steel by blowing air through molten pig iron, whereby most of the carbon and impurities are removed by oxidation.

beta activity. The spontaneous emission of electrons from a nucleus.

beta particle. An electron, of either positive or negative charge, which has been emitted by an atomic nucleus or neutron in the process of transformation.

beta phase. See quartz inversion; cristobalite; tridymite.

betatron. An apparatus in which high velocities are imparted to electrons.

bevel brick. A brick with one edge or surface sloping to another surface at an angle which is not a right angle.

beveling. The process of edge-finishing flat glass to a desired bevel angle.

bias. A constant or systematic error as opposed to a random error, manifested as a persistent positive or negative deviation of the method average from the accepted reference value.

bias, statistical. A constant or systematic error in test results which can exist between the true value and a test result obtained from one method, between test results obtained from two methods, or between two test results obtained from a single method as, for example, between operators or between laboratories.

Bicheroux process. An intermittent process employed in the fabrication of plate glass of high quality in which molten glass is cast between rolls onto driven conveyor rolls or a flat moving table which delivers the strip to a lehr where the glass is slowly cooled while passing between a series of asbestos-covered rollers. The sheet then is cut to specified lengths, ground, and polished as individual plates.

bichromate of potash. $K_2Cr_2O_7$; mol. wt. 294.22; m. p. 396°C (745°F); decomposes at 500°C (932°F); sp. gr. 2.692; employed with whiting and zinc oxide to make carnation pink or red ceramic colors.

bid. To make an offer to supply a specified material, product, or service at a specified price.

bidet. A low, basin-like item of ceramic sanitaryware designed for personal hygiene.

Bierbaum scratch hardness. A measure of the hardness of a solid material based on the width of a scratch made by drawing a diamond point across the surface under preset pressure conditions, the measurement being made by use of a microscope.

bimetal. A bonded laminate of two dissimilar metals having different expansion properties; employed in thermocouples to measure differences in temperature.

bin. A relatively large enclosed area in which raw materials are stored prior to use.

binder. A cementing medium, or a substance added to a powder or granular material, to give formed items workability and green or dry strength sufficient for handling and machining in all stages prior to firing, and which usually is expelled during sintering or firing; normally a material of relatively low-melting point added to a powder mixture for the specific purpose of cementing together powder particles which alone could not be handled without danger of breakage or which would not sinter or fire into a strong body.

binder course. A bituminous layer serving as a bonding course between the foundation layer and the wearing layer of a concrete installation.

binder tape. A paper or other material employed to wrap groups of insulated wire into cable configuration prior to sheathing.

Bingham plastometer. An instrument designed to assess the deformation and flow of materials in which a slurry is forced through a capillary under various pressures.

biotite. A common mineral of the mica family having the general composition $K(Mg, Fe)_3AlSi_3O_{10}(OH)_2$; sp. gr. 2.8–3.2; hardness (Mohs) 2.5–3; a frequent impurity in feldspar and nepheline syenite. Usually dark in color.

Biot number. A numerical evaluation to estimate the thermal-shock resistance of a material by its heat-transfer properties by the formula hr/k, in which h is the heat-transfer coefficient, r is the distance between a specific plane and the surface of a specimen, and k is the thermal conductivity of the material.

bipolar field. The longitudinal magnetic field within a part or object having two magnetic poles.

birefringence. The double bending of light rays as observed in an anisotropic crystal viewed under cross nichols when characteristic and measurable colors are produced to indicate the difference in the minimum and maximum indices of refraction of the crystal.

biscuit. (1) A term employed in some industries having the same meaning as bisque.

(2) A small setter composed of refractory clays on which pots are placed for firing.

bismuth chromate. $Bi_2O_3 \cdot Cr_2O_3$; mol. wt. 547.95; used as an orange-to-yellow pigment in porcelain enamels and glazes.

bismuth oxide. Bi_2O_3; mol. wt. 466.0; m.p. 820–860°C (1508–1580°F); sp. gr. 8.2–8.9; employed as a fluxing component in optical glasses, as a flux and bonding agent for metallic components in ceramic glazes, as a flux in cast-iron porcelain enamels, and in ceramic colors; its ceramic properties are similar to those of lead oxide, but it is more fusible.

bismuth selenide. Bi_2Se_3; mol. wt. 654.9; m.p. 706°C (1300°F); sp. gr. 6.82; used in some thermoelectric applications.

bismuth stannate. $Bi_2(SnO_3)_3 \cdot 5H_2O$; mol. wt. 1008.2; dehydrates at 200°C (392°F) and above to form $Bi_2(SnO_3)_3$, mol. wt. 918.1; used as an additive in barium titanate capacitors to produce bodies of intermediate dielectric constant.

bismuth subcarbonate. $(BiO)_2CO_3$; mol. wt. 309.97; sp. gr. 6.86; used as a flux and opacifier in glass and porcelain enamels.

bismuth subnitrate. $4BiNO_3(OH)_2 \cdot BiO(OH)$; mol. wt. 1360.06; decomposes at 260°C (392°F); sp. gr. 4.928; used to give pearly luster to glasses and glazes, as a constituent in high-refractive glass, and in low-temperature porcelain enamels and colorants.

bismuth telluride. Bi_2Te_3; mol. wt. 800.98; m.p. 585°C (1085°F); sp. gr. 7.3; hardness (Mohs) 1.5–2; thermoelectric material employed in cooling devices.

bismuth trioxide. Bi_2O_3; mol. wt. 495.96; m.p. 820°C (1508°F); sp. gr. 8.76; used as a yellow pigment in porcelain enamels, glazes, and other ceramics.

bisque, bisque ware. (1) Unglazed ceramic ware which has been subjected to a single fire.

(2) A coating of wet-process porcelain enamel which has been dried but not fired.

bisque fire. The kiln firing of ceramic ware before application of a glaze.

bit gatherer. An operator who gathers small quantities of glass on an appropriate tool for use in the decoration of hand-blown glassware.

bit stone. Refractory particles, such as flint fragments or sand, placed in saggers to prevent ware from sticking to the sagger bottoms during firing.

bitumen. Any of a variety of hydrocarbons obtained naturally, such as asphalt or tar, or by distillation from coal and petroleum.

bituminous. See bitumen.

bituminous concrete. Concrete in which a bituminous material has been incorporated as a binder.

blackboard enamel. A special, slightly roughened porcelain enamel which will provide a suitable writing surface for blackboard chalk.

black body. A body which will absorb all radiation and which will emit radiant energy at a maximum rate for a given temperature for use in determining the temperature of a closed furnace when viewed through a relatively small hole with an optical pyrometer.

black core, black heart. A defect occurring in fireclay and other refractory brick when vitrification of the surface areas takes place before oxidation of carbonaceous matter in the interior is complete.

black edge, black edging. A black porcelain enamel applied and fired over the ground coat at the exposed edges of ware for both protective and decorative purposes; subsequent coatings of cover-coat enamels are brushed from the areas prior to firing.

blacking. Graphite applied to the working surface of molds as a parting material to prevent a casting from sticking, and to improve the surface of ware cast in the molds.

black iron oxide. FeO; mol. wt. 71.8; m.p. 1420°C (2588°F); sp. gr. 5.7.

black light. Light in the near ultraviolet range of wavelengths just below the visible range, from 3200 to 4000 Å.

black-light filter. A filter which will suppress transmission of visible light but will permit passage of ultraviolet radiation having wavelengths in the range of 3200 to 4000 Å.

black raku. A rough, thick-walled, very soft, and porous earthenware coated with a lead-borate glaze; used in the tea ceremony in Japan.

black shape. Fabricated ware or shapes prior to porcelain enameling.

black silicon carbide. A black, tough silicon carbide manufactured from coke and silica in an electric furnace, and employed as an abrasive.

black speck. A defect in fired porcelain enamels appearing as visible black specks, usually caused by dirt or scale, but which also may be glass-eye blisters or boiling from the ground coat.

blanc-de-chine. A white, glazed Chinese porcelain.

blank. (1) A parison or preliminary shape from which a finished article is further formed, or a mold for producing such a shape.

(2) Any article of glass on which subsequent forming or finishing is required.

(3) A piece cut from a metal sheet from which a finished article for porcelain enameling is to be fabricated.

blanket feed. A technique for charging a glass batch into a furnace to produce a broad, thin layer of even distribution across the width of the furnace.

blanking. The process of cutting and forming metal shapes for porcelain enameling by means of a mechanically operated die and plunger press.

blank mold. A metal mold employed in the manufacture of glass holloware to give the item its initial shape or form.

blank, optical. Optical glass formed to the approximate dimensions required, and from which final lenses are made.

blank, pressing. Optical glass, formed by pressing into a specified rough size and shape, from which a finished article is produced.

blast. Air blown into a furnace or kiln under pressure.

blast furnace. A large, vertical, refractory-lined furnace employed to smelt or extract iron from its ore.

blast-furnace slag. The nonmetallic product, consisting essentially of silicates and aluminosilicates of calcium and other base materials, that is developed in a molten condition simultaneously with iron in a blast furnace.

blast-furnace slag, air cooled. The material resulting from the solidification of molten blast-furnace slag under atmospheric conditions; subsequent cooling may be accelerated by application of water to the solidified surface.

blast-furnace slag, expanded. The lightweight, cellular material obtained by the controlled processing of molten blast-furnace slag with water or water and other agents such as steam or compressed air, or both; used in lightweight concrete and as both a thermal and acoustic barrier.

blast-furnace slag, granulated. The glassy granular material formed when molten blast-furnace slag is rapidly chilled by immersion or fritting in water.

blasting. The process of cleaning metal, especially cast iron, for porcelain enameling in which the surface of the metal is subjected to the abrasive action of sharp abrasive particles carried in a fast-moving stream of air.

bleb. A blister or bubble defect on the surface of pottery.

bleed back. The ability of a penetrant to bleed out of a discontinuity after it has been cleaned from the surface of a specimen.

bleeding. The autogenous flow of mixing water within, or its emergence from, newly placed concrete or mortar, caused by the settlement of the solid materials or drainage of the mixing water.

bleedout. The action of an entrapped penetrant in emerging from surface discontinuities.

blemish. (1) A defect or flaw in a product consisting of a stain, disfigurement, or strained area attributable to the normal composition, forming, or extraneous factors encountered in the production of the item.

(2) An insignificant imperfection in a dry-process porcelain enamel.

blend. A combination of materials which are thoroughly mixed.

blending. (1) The process of mixing materials.

(2) The process of evening the rougher part of a surface with the smoother part so that the entire surface is of the same plane or surface, or both.

blending, batch. Stepwise changes in a batch composition to arrive at the final composition in the finished glass or other product.

blending sand. Sand that is added to the normal available sand in concrete to improve gradation.

blibe. A defect in glass in the form of a gas-filled cavity, between a seed and blister in size.

blinding. (1) A surface defect in glazes due to devitrification, resulting in a dull or crystalline appearance.

(2) The clogging of a sieve.

blister. A bubble or gaseous inclusion of relatively large size in a body or at the surface of a glaze or porcelain enamel after firing.

blister copper. A partially refined form of copper having a blistered surface after smelting due to the gases generated during solidification.

blistering. (1) The development of enclosed or broken macroscopic bubbles or vesicles in a body, glaze, porcelain enamel, or other coating during firing.

(2) Nonadherence of color in firing.

blister, metal. In porcelain enameling, the bloating of the metal sheet; a source of enamel defects, particularly blisters.

blister, pipe. A blister or bubble occurring in handblown glassware resulting from scale or other impurities on the blowpipe.

blister, weld. Blisters which occur along the line of a weld during porcelain enameling, and which originate in the weld; a defect.

bloach. An imperfection resulting from the incomplete grinding of plate glass caused by a low point in the glass which retains a part of the original rough surface.

bloat. To cause solid particles, such as clays and slags, to puff or swell due to sudden expansion of air or moisture contained in the material when subjected to a blast of a super-heated air, hot flame, or other source of heat of high temperature.

bloating. The permanent expansion or swelling of a ceramic material or body during heating which produces a vesicular structure in the substance being heated.

block. (1) A master mold made from an original pattern from which case molds are produced.

(2) Hollow translucent glass units having various patterns molded on their interior and exterior surfaces, or both, and usually made in two halves which are sealed together.

block brick. A brick, larger than standard or jumbo in size, used to bond adjoining or intersecting walls.

block, comparative test. An intentially cracked metal block having two separate but adjacent areas for the application of different penetrants to compare the relative effectiveness of penetrants; the effectiveness of the testing techniques, or the test conditions as an inspection procedure for electromagnetic or magnetic particles or materials.

block density. The weight of a unit volume of a substance, including its pore volume but excluding interparticle voids; determined under specified conditions.

block filter. A hollow, rectangular, vitrified clay masonry unit, sometimes salt glazed, used in trickle-type floors in sewage disposal plants. The block is designed with apertures connecting with drainage channels through the upper surface, and are arranged to form aeration and drainage grilles to pass air into, and liquids from, overlying filter media; the drainage channels convey liquid away from the filter bed.

block, handle. A particular type or style of handle attached to a cup, vase, or other item by means of a clay bar.

blocking. (1) The process of shaping a gather of glass in a metal or wood cavity.

(2) The process of stirring a glass batch by immersing a wooden block or other source of gaseous bubbles in the molten mass.

(3) The process of reprocessing glass to remove surface imperfections.

(4) The mounting of optical glass blanks in a holder for grinding and polishing operations.

(5) The process in which a furnace is idled at a reduced temperature.

(6) The process of setting refractory blocks in a furnace.

block mold. A one-piece mold used in glass making.

blockout. An opening or cavity formed in concrete to facilitate subsequent construction operations, such as an opening in a wall for the installation of a pipe or other item; the opening frequently is sealed with mortar or concrete when the installation has been completed.

block, quarl. A refractory shape employed as a burner or burner segment for the injection of gaseous or liquid fuel into a glass-melting tank.

block reek, block rake. A scratch or cullet-cut imperfection in glass caused by a particle of cullet lodged in the polishing felt during the polishing operation on flat glass.

block, rotary kiln. A modified circle brick, usually with a 9-inch outside chord and a smaller inside chord, 6 or 9 inches in radius length and 4 inches thick; used to line circular and rotary kilns.

block, scotch. A rammed refractory port in an open-hearth furnace.

block, scouring. A chemically bonded abrasive block composed of Al_2O_3, SiC, or similar abrasive used in the grinding and polishing of metal and ceramic surfaces.

block, skimmer. A refractory block placed as a wall in a glass tank or a porcelain-enamel frit smelter to prevent slag and impurities from flowing into the feeder channel or fining chamber.

block, sleeper. The refractory block used in forming the sides of the throat of the submerged passage between the melting and working ends of a glass tank.

block, soldier. Refractory block, installed on end, which extends below the depth of the molten glass in a glass tank; also used in some types of ladles and furnaces.

block, spreader. A refractory block of triangular cross section employed to divide and distribute coal being charged into a coke oven.

block, tank. A refractory block used in a tank compartment of a glass-melting tank.

block, trimmed. Dressed or crude mica that has been split into prescribed thicknesses and has been side trimmed to remove irregularities, imperfections, and contaminants.

block, tweel or tuille. A refractory block employed in the construction of a counterweighted door of a glass furnace to protect a newly set pot, or to control the flow of molten glass in a furnace.

bloom. (1) A nonreflecting coating on glass.

(2) A surface film on glass resulting from attack by constituents in the atmosphere, or by the deposition of smoke or other vapors.

blotter. A disk of compressive material, usually of blotting paper stock, used between an abrasive grinding or polishing wheel and its mounting flange.

blotting. In liquid penetrant inspections, particularly of electromagnetic and magnetic particles and products, the action of a developer in soaking up a penetrant from the surface of a fault for increased contrast.

blow-and-blow process. The process of forming hollow glassware in which the preliminary and final shapes are formed by air pressure.

blower. (1) An operator who forms glass by blowing.
(2) A machine employed to move or supply air to a particular area for a particular use.

blow head. Part of a glass-forming machine serving to introduce air under pressure to blow a hollow glass article.

blowhole. (1) A large blister such as is formed when contaminants are vaporized along a weld seam during the firing of porcelain enamels.
(2) A device placed in the top of a kiln to facilitate the escape of steam and other gases, particularly during the early stages of the firing operation.

blowing. (1) The shaping of hot glass by air pressure, either by machine or by mouth.
(2) The bursting of pots and crucibles when heated too rapidly.

blowing iron. The pipe used by a glassmaker for gathering and blowing glassware by mouth.

blow mold. The metal mold in which a blown glass article is finally shaped.

blow molding. The shaping of glass in the plastic or molten state by placing a parison in a mold and completing the shaping operation by blowing air into the parison.

blown away. A fault in the neck of a glass bottle which occurs when an insufficient quantity of molten glass is employed during fabrication.

blown enamel. Ridges produced on the surface of ware during the spraying of wet porcelain enamels, usually the result of the coating being too thick or too fluid, or to the use of excessive atomizing air pressure at the spray gun.

blown glass. Glassware formed by air pressure, as by mouth blowing or by the use of compressed air.

blow off. The removal of dust and dirt from the surface of dry, or bisque, porcelain enamels just prior to firing.

blowout. The displacement and lengthening of an electrical arc to facilitate its extinction, as by an air blast, magnetic field, or other method.

blow-over. The thin-walled bubble of glass formed above a blow mold in a handshop operation to facilitate bursting off.

blow pipe. (1) An apparatus employed to produce a hot localized flame by using a mixture of compressed air and coal gas.
(2) A long metal pipe used for the working and forming of glass at the bench.

blowup. The buckling and cracking of a concrete paving slab due to abnormal expansion.

blue enamel. (1) A wet-process porcelain enamel applied too thinly to hide the substrate, particularly a previously fired blue ground coat or other dark coating.
(2) A dry-process porcelain enamel applied so thin that it appears bluish in color.

blue ground coat. A porcelain-enamel composition usually containing additions of cobalt, manganese, and nickel oxides as adherence-promoting agents; the coating, which fires to a dark blue color, is used as a ground coat on sheet iron and steel.

blue, mazarine. An underglaze or overglaze containing approximately 50% of cobalt oxide which produces a rich dark-blue color when fired.

blunge. The agitation or blending of ceramic materials in a mechanical or hand-operated mixer, usually to suspend the materials in water or other liquid.

blunger. A mixer with revolving paddles or other mixing device employed to produce slurries or slips.

blunging. The process of mixing clays and other ceramic materials in a liquid, usually water, to form a slurry or slip.

blurring highlight test. A test, usually visual, to evaluate the resistance or the degree to which porcelain enamels are attacked by acids.

blushing. The discoloration or clouding of a glaze or porcelain enamel during firing.

board. A slab of wood upon which ware is placed for transport from one operation or station to another.

bobbin coil. A coil or coil assembly used for electromagnetic testing by insertion into a test specimen as, for example, an inside probe for tubing.

Boccaro ware. A red, unglazed stoneware with relief decorations.

body. (1) A mixture of clays and nonplastic materials that is workable and has suitable firing properties from which ceramic products are made.
(2) The structural portion of a ceramic article, as distinct from the glaze, or the material or mixture from which the item is made.
(3) The attribute of molten glass associated with homogeneity and viscosity which contributes to its workability.

body mold. The portion of a glass mold which shapes the outer surface of ware during pressing.

boehmite. $Al_2O_3 \cdot H_2O$; mol. wt. 119.96; decomposes at 360°C (680°F); sp. gr. 3.014; a natural hydrated aluminum oxide occurring as a major constituent in bauxite and bauxitic clays.

BOF. Acronym for basic-oxygen furnace used in steelmaking.

bogie kiln. An intermittent box-type kiln in which ware, placed on a truck or kiln car, is charged, fired, and discharged before a subsequent charge is placed in the kiln.

Bohemian glass. A hard, brilliant glass employed in table and chemical ware, usually a lime-potash glass with a high silica content.

boil, boiling. (1) A defect occurring in fired porcelain enamels which consists of bubbles, pinholes, black specks, dimples, or spongy surfaces.

(2) An imperfection in glass which consists of gaseous inclusions or small bubbles, bubbles larger than seeds.

(3) The turbulence caused by the evolution of gases from melting glass, porcelain enamels, or other batches.

boiling through. A term sometimes used to describe the boiling of porcelain enamels, particularly in instances of severity when defects occur in cover coats.

boil, primary. The evolution of gas during the initial firing of porcelain enamel, sometimes resulting in blister-type defects.

bole. Any of a variety of soft unctuous clays used to produce color, or a reddish brown body made from such clays.

bolt-hole. A hole made in a component during the manufacture of an item to facilitate final assembly of the item by means of inserted bolts, screws, or other fasteners.

bolt-hole brush. A special round brush, usually equipped with a centered metallic guide pin, employed to remove bisque porcelain enamel from the inside and edges of small openings in the ware, particularly to prevent chipping during subsequent assembly of the porcelain-enameled product.

bolus alba. Kaolin, which see.

bond. (1) The degree of adhesion of a porcelain enamel or other coating to the metal to which it is applied and fired.

(2) The stress required to cause one material to separate from another at the interface.

(3) The material in a grinding wheel that holds the grains together and supports them while in use.

(4) The intergranular material which provides strength in ceramic bodies.

(5) The adhesion of cement paste to aggregate particles, or of concrete or mortar to reinforcing steel, or of concrete to previously hardened concrete on a construction joint or in a patch.

bond, American. The bond in which a header course in a brick structure is used in every fifth, sixth, or seventh course, with stretcher courses being placed between the header courses. See English bond, Flemish bond.

bond clay. A plastic clay of high dry strength employed as a binder in ceramic bodies containing substantial amounts of nonplastic components.

bonded abrasive disk. A disk-shaped bonded abrasive product mounted on a face plate for use on grinding and milling machines; work is ground or polished on the side of the abrasive disk opposite the face plate.

bonded brickwork. Any regular arrangement of bricks in a structure designed to increase the strength and to enhance the appearance of the structure.

bonded products. Products in which an abrasive and a bonding agent have been intermixed and processed to produce a relatively inflexible abrasive product, such as a grinding wheel or rubbing stone.

bonded roof. The roof of a furnace or kiln in which the transverse joints are staggered.

bonder. A brick of special size and shape employed to begin or finish a course of bonded brickwork.

bond failure. Insufficient adherence of a porcelain enamel to a base metal as indicated by bright metal in the fractured area.

bond, fireclay. A fireclay exhibiting sufficient natural plasticity to bond nonplastic materials in the manufacture of refractory products.

bond, in-and-out. A type of masonry construction consisting of alternate courses of headers and stretchers.

bonding agent. (1) An admixture for improving the bond of mortar and concrete in a patch.

(2) A paint or coating applied to hardened concrete to facilitate the bonding of a new application of concrete or mortar.

bonding materials. Organic materials employed in conjunction with glass and ceramic fibers, sheets, molded shapes, and other products to impart strength, adherence, chemical resistance, weather resistance, electrical properties, and similar properties for use in the production of cloth, laminates, electrical and electronic components, insulating materials, and the like.

Bondley process. A metallizing process in which titanium or zirconium is bonded to the surface of a ceramic body to facilitate soldering or joining of components in the production of electrical and electronic products.

bond, organic. An organic plastic, rubber, resin, shellac, or other product used to bond ceramic materials in the manufacture of a product.

bond, shellac. A bond for ceramic materials, such as grinding wheels, in which shellac or a shellac-base adhesive is a major component.

Bond's hypothesis. The grinding rate of a solid material is proportional to the rate at which a crack will progress through the material.

bond strength. (1) The degree of adherence of a porcelain enamel to the metal to which it is applied and fired.

(2) The strength of a mortar joint or wall in construction applications.

bond vitrified. The strength developed by the fusion of ceramic materials.

Bond and Wang crushing theory. The energy required to pulverize or crush a solid material may be calculated by the formula:

$$h = \left[\frac{0.001748\ C^2}{SE}\right]\left[\frac{(n+2)(n-1)}{n}\right]$$

in which h is the energy required, C is the compressive strength, S is the specific gravity, E is the modulus of elasticity, and n is the approximate reduction ratio.

bone ash. Calcined bones consisting of 67 to 85% of basic calcium phosphate; employed in porcelains, pottery, milk glass, and porcelain enamels as an opacifier and fluxing ingredient.

bone china. A soft, highly translucent chinaware of relatively low-firing temperature made from a whiteware body containing a minimum of 25% bone ash as a fluxing agent, and having a water absorption ranging from 0.3 to 2%; a typical composition is 50% bone ash, 25% china clay, and 25% Cornish stone.

bone dry. Thoroughly dried and free of uncombined water.

bonnet hip. A roofing tile of special angular shape employed as a junction between two faces of a roof.

bookform splittings. Consecutive splittings of mica from the same block, each usually dusted with mica powder to reduce cohesion, arranged in individual books or bunches for use as an electrical insulating material.

boost melting. An auxiliary method of adding heat to molten glass in a fuel-fired tank by passing an electric current through the glass.

boot. A suspended or floating refractory shape in the nose of a glass-melting tank to protect the glass from fuel gases and floating scum, and to serve as an opening for the gathering of the glass.

BOP. Acryonym for basic oxygen process for steelmaking.

boracic acid. H_3BO_3; usually identified as boric acid in ceramic usage.

borate glass. A glass in which boric oxide in combination with silica is employed as the major glass-forming ingredient.

borax. $Na_2B_4O_7 \cdot 10H_2O$; mol. wt. 381.4; m.p. (anhydrous) 741°C (1366°F), sp. gr. 1.7; hardness (Mohs) 2–2.5; employed as a powerful flux and glass-forming agent in glass, glazes, porcelain enamels, etc.

borax glass. (1) Glass in which boric oxide is used as the major glass-forming ingredient in combination with silica.

(2) Vitreous, anhydrous borax used as a glass former and flux in glass, glazes, and porcelain enamels.

boraxon. See boron nitride.

boric acid. H_3BO_3; mol. wt. 61.8; m.p. 184°C (365°F); sp. gr. 1.7; hardness (Mohs) 2–2.5; employed in glazes, porcelain enamels, glass pastes, special glasses, and cements, primarily as a flux.

boric oxide. B_2O_3; mol. wt. 69.6, m.p. above 1500°C (2732°F); sp. gr. 1.83–1.88; used principally in the manufacture of cements, glass, and porcelain enamels as a flux, and in nuclear applications as a thermal-neutron absorber.

borides. Intermetallic compounds in which one element is boron and the other a metal, and having compositions ranging from Me_3B to MeB_{12}; they are harder, higher melting, chemically less reactive, and electrically more resistive than the constituent metallic elements; characterized by high-oxidation resistance and strength retention at elevated temperatures; melting points to as high as 3260°C (5900°F); densities ranging from 2.5 to 16.7 gm/cm^3; specific heats of less than 0.35 Btu/lb./°F to 2205°C (4000°F); linear thermal expansions of 2% or less between 21°C (70°F) and 1649°C (3000°F); elastic moduli ranging between 30×10^6 to 60×10^6 psi. at room temperature; microhardness values ranging between 1300 and 3300 kg./mm^2; potential materials for use as structural materials, particularly in aerospace applications. See borides of aluminum, barium, calcium, cerium, chromium, cobalt, dysprosium, erbium, europium, gadolinium, hafnium, holmium, iron, lanthanum, lutetium, magnesium, molybdenum, neodymium, nickel, niobium, osmium, plutonium, praseodymium, rhenium, rhodium, ruthenium, samarium, scandium, strontium, tantalum, terbium, thorium, thulium, titanium, tungsten, uranium, vanadium, ytterbium, yttrium, zirconium, and zirconium-boron composites.

boron. B; atomic weight 10.811; m.p. 2300°C (4172°F); sp. gr. 2.45.

boron carbide. B_4C; mol. wt. 52.28; m.p. 2350°C (4262°F); sp. gr. 2.6; second only to diamond in hardness; produced by reduction of boric oxide by carbon in an electric furnace; employed as an abrasive in grinding wheels, belts, papers, and powders; in articles of high resistance to abrasion, in nozzles for high-temperature applications, in control rods for nuclear reactors, and in electrical-resistance heating elements for high-temperature furnaces.

boron content, equivalent. A concentration of natural

boron that would provide a thermal neutron cross-section equivalent to that of a specific impurity element.

boron content, equivalent factor. The factor used to convert the concentration of an impurity element to a neutron cross-section equivalent amount of natural boron.

boron content, total equivalent. The sum of the equivalent boron content values.

boron equivalent. The absorptive capacity for thermal neutrons of weights of various elements expressed in terms of the weight of natural boron.

boron nitride. BN; mol. wt. 24.83; m.p. about 3000°C (5432°F); sp. gr. 2.25; used as a refractory in crucibles, parts for chemical equipment and pumps, vacuum-tube separators, seals and gaskets, rocket nozzles, furnace insulation, lubricant in glass molds, machine tools, abrasive in special grinding operations, and as a neutron absorber in nuclear applications. Also known as borazon.

boron oxide. B_2O_3; mol. wt. 69.6; m.p. above 1500°C (2732°F); sp. gr. 1.83–1.88; used in the production of glass, glazes, porcelain enamels, and special cements, primarily as a fluxing agent.

boron phosphate. BPO_4; mol. wt. 105.84; vaporizes at 1400°C (2552°F); sp. gr. 1.873; used in ceramic bodies and special glasses.

boron phosphide. BP; mol. wt. 41.8; m.p. above 2000°C (3632°F); sp. gr. 2.97; electroluminescent material.

boron silicide. (1) B_6Si; mol. wt. 92.98; m.p. 1946°C (3540°F); sp. gr. 2.43. (2) B_4Si; mol. wt. 71.34; decomposes at 1093°C (2000°F); sp. gr. 2.46. (3) B_3Si; mol. wt. 60.52; m.p. 1927°C (3500°F); sp. gr. 2.64. Also known as silicon boride.

boron, soluble (in boron carbide). The boron that dissolves from boron carbide by separate reflux digestions with two different acids, 0.1 *M* hydrochloric acid (the boron assumed to be boric acid) and 1.6 *M* nitric acid (the boron assumed to be boric acid plus free boron).

boron value. See boron equivalent.

borosilicate crown glass. An optical crown glass containing substantial quantities of silica and boric oxide. See crown glass, optical.

borosilicate glass. A silicate glass containing not less than 5% of boric oxide.

bort. An imperfect diamond or diamond fragments employed principally as an abrasive or as a bonded tip on a cutting tool.

bosh. The section of a blast furnace between the hearth and stack in which iron ore is reduced to metallic iron.

bossing. The removal of brush marks from painted pottery by patting or striking the design with a silk bag stuffed with soft cotton or wool, particularly designs which first are painted in oil and then dusted with powdered pigments.

Bottger ware. A dark red stoneware.

botting clay. A refractory clay of high plasticity used to plug the tapping spouts of cupolas and furnaces containing molten materials.

bottle kiln, bottle oven. An updraft kiln in the shape of a tapered bottle, the tapered neck serving as the flue.

bottle, vacuum. A bottle or other container equipped with an evacuated liner to prevent the influx of heat or cold into the contents of the container from the surrounding environment.

bottom pouring. Discharging the contents of a smelter, melting tank, ladle, or other container from the bottom.

bottom, slugged. An imperfection in the bottom of a bottle or container in which the glass is thick on one side and very thin on the other.

bottom tap. A hole for the drainage of molten compositions and slags from the bottom of a furnace, smelter, or melting tank.

bottom teeming. The filling of ingots or molds in which the molten batch enters the molds from the bottom.

boule. A pure crystal such as silicon or sapphire, frequently a pear-shaped mass having the atomic structure of a single crystal, formed in a special furnace by rotating a small seed crystal while slowly pulling it out of the molten bath; used as bearings, thread guides, phonograph needles, etc.

bowing. The tendency of a length of coated abrasive or other material to curve or bend; caused by excess moisture (expansion) or lack of moisture (shrinkage) on one side of the abrasive strip.

bowl. The portion of a feeder which delivers molten glass to the forming unit, and which consists of the orifice, revolving tube, needle, etc.

boxcar roof. The roof of an open-hearth furnace in which the transverse and horizontal ribs form boxlike shapes along the top.

box furnace, box kiln. An intermittent box-shaped furnace in which ware is placed, fired, and removed on a scheduled basis before the introduction of a subsequent charge.

boxing. The manner of arranging cups rim-to-rim to prevent distortion during firing.

box section. A concrete pipe of rectangular cross section.

boy, mechanical. A mechanism to manipulate the mold in the hand-forming of glass.

b.p. Abbreviation for boiling point.

braid. (1) A shield for insulated cables and conductors consisting of woven metallic wire.
(2) A woven, fibrous, protective covering over an insulated conductor or cable.

brake lining. A covering of asbestos, cermet or other ceramic material molded to the brake shoe or brake band which presses against the rotating drum to apply resistance to the motion of a body.

brass wire. Wire of selected diameters employed to cut clay and unfired ceramic products.

braze. To join or solder two or more metal components with a material having a melting point lower than the metals being joined.

brazing. The process of joining two or more metal parts by fusing a solder between the adjoining surfaces to form a vacuum-tight bond; in ceramic technology, the braze is made between a metallized ceramic and a mating metal.

breadboard. An experimental model of an item being considered for production, particularly a proposed electronic product, to establish the feasibility of the item and to detect areas for its improvement.

breakdown voltage. The difference in the potential at which electrical failure occurs in an electrical insulating material located between two electrodes under specified conditions. Also termed dielectric and electric breakdown voltage.

breaking strength. The resistance of a material to breaking, usually under tension.

breaking stress. The stress required to fracture a material, by tension, compression, or shear.

breakout. A defect in dry-process porcelain enamels characterized by an area of blisters with well-defined boundaries.

break point. The first appearance in the effluent of an absorbate on activated carbon under prescribed conditions.

breasts. The sloping refractory components below the ports and adjoining brickwork of an open-hearth furnace which serves to join the hearth with the furnace ends.

breast wall. (1) The entire side wall of a furnace between the flux block and crown, excluding the ends.
(2) The refractory wall between pillars of a pot furnace and in front of or surrounding the front of a pot.

breeze coal. The residue from coke and charcoal making; used in concrete and bricks.

breezing. A thin layer of buckwheat anthracite coal or coarse sand spread on the refractory floor of a glass furnace before the setting of pots.

Brenner gauge. A device calibrated to estimate the thickness of porcelain enamels as a function of the force required to lift a metal pin from contact with the coating surface against a known magnetic force acting beneath the undersurface of the base metal.

Brewster. A unit of photoelasticity equivalent to a retardation of 10^{-13} cm^2*dyn*$^{-1}$.

Brewster's window. A glass window of special composition used in each end of some gas lasers to transmit one polarization of the laser output beam without loss.

brianchone luster. A luster in which a reducing agent is incorporated as a component of a ceramic glaze.

brick. A block of clay or shale formed into a rectangular prism while in a plastic condition, and hardened by firing in a kiln or by sun baking (adobe) for use as a masonry unit in building and other construction.

brick, acid resisting. Brick of suitable composition and treatment exhibiting low-water absorption and a high degree of resistance to chemicals; usually employed in conjunction with acid-resisting mortars.

brick, air. A brick of essentially standard size containing holes along the lateral axis to permit the circulation of air in structures.

brick, alumina. See alumina brick.

brick, angle. Any brick shaped to an oblique angle to fit a corner.

brick, arch. (1) A wedge-shaped brick for use in the construction of an arch.
(2) An extremely hard-fired brick from an arch in a scove kiln.

brick, basic. Refractory brick composed essentially of basic materials, such as lime, magnesia, chrome ore, or dead-burned magnesite, which react chemically with acid refractories, acid slags, or acid fluxes at high temperatures.

brick, bauxite. See alumina brick.

brick, block. A brick, larger than standard or jumbo brick in size, used to bond adjoining or intersecting walls.

brick, brindled. A brick having high-crushing strength made from iron-bearing sedimentary clays which are partially reduced at peak-firing temperatures.

brick, building. Common brick which are not produced for texture, color, or other decorative effects; used in building construction.

brick, center. A hollow refractory brick with a hole in its upper face connected by the hollow center with holes in the side faces for use in the bottom pouring of molten steel.

brick, checker. Refractory brick of special design to permit the passage of hot gases through a regenerator.

brick, chemically bonded. Brick manufactured by processes in which mechanical strength is developed by use of chemical bonding agents rather than by firing.

brick, chrome. A refractory brick produced substantially or entirely from chrome ore. See chrome brick.

brick, chrome-magnesite. A refractory brick, which may be burned or unburned, manufactured substantially from a mixture of refractory chrome ore and dead-burned magnesite in which the chrome ore is, by weight, the predominating ingredient.

brick, chuff. A relatively soft, underfired brick of salmon color.

brick, circle. A curved-face brick used to form cyclindrical structures.

brick clays. Clays possessing properties suitable for the production of brick. Such clays, which usually fire to a red color, are somewhat impure, containing considerable amounts of fluxing ingredients, will mold readily, fire to an appropriate degree of hardness at a relatively low temperature, and will be resistant to warping and cracking during firing. Grades which contain lesser amounts of impurities and soluble salts, and which fire to greater hardness, lower porosity, greater strength, and more uniform colors are used in the manufacture of face brick.

brick, clinker. A very hard-fired brick of distorted or bloated shape due to nearly complete vitrification as a result of overfiring.

brick, concrete. A building brick made to specification from portland cement and appropriate aggregate materials.

brick, cored. A brick that is 75% solid in any plane parallel to the bearing surface.

brick, crown. A brick tapered for use in closing an arch or crown.

brick, de-aired. Brick formed from a batch which has been subjected to a partial vacuum.

brick, dolomite. A refractory brick which has been manufactured substantially or entirely from dead-burned dolomite.

brick, dolomite-magnesite. A refractory brick manufactured substantially from a mixture of dead-burned dolomite and dead-burned magnesite, in which the dead-burned dolomite is the predominating ingredient.

brick, dome. A brick of tapered structure for use in the construction of a dome.

brick, double. A brick 5⅓×4×8 inches (13.5×10.2×20.3 centimeters) in size.

brick, drop-machine silica. Brick formed by dropping a prepared quantity of a plastic silica-brick mix from a considerable height so as to fill and compact the mix in a mold before pressing.

brick, dry-pressed. Brick formed in molds under high pressure from a relatively dry clay of 5 to 7% moisture content.

brick earth. A loamy, relatively impure clay used in making some types of common brick.

brick, economy. A brick having nominal dimensions of 4×4×8 inches (10.2×10.2×20.3 centimeters).

brick, electrocast. Refractory shapes made by fusing refractory oxides in an electric furnace and casting the molten material into molds to form the finished products.

brick, end-cut. Extruded brick, the ends of which are cut by wire.

brick, engineered. A brick having nominal dimensions of 3⅕×4×8 inches (8.1×10.2×20.3 centimeters) which are harder and more dense than ordinary brick; used in bridge construction, some buildings, and other applications, where high strength and frost resistance are required.

brick, facing. Brick, often having a particular face texture, made for facing purposes, as well as for structural use; nominally 2⅔×4×8 inches (6.8×10.2×20.3 centimeters), the size varying with the geographical location, the manufacturer, and the architectural requirements.

brick, feather. A brick modified so that one of the larger faces is inclined from one side to the opposite side where the thickness is reduced to approximately ⅛ inch.

brick, fire. Any refractory brick, but more specifically a brick made of fire-clay and containing less than 50% of alumina; used to line kilns, furnaces, chimneys, etc.

brick, flashed. Brick subjected to reducing conditions near the end of the firing cycle to develop a desired color.

brick, floor. Smooth, dense, abrasion-resistant brick used as a finished surface of floors.

brick, furring. A type of hollow brick which has been grooved to receive and retain a coating of plaster in the construction of walls.

brick, glass. A hollow glass block with plain or patterned surfaces used in the construction of walls, partitions, windows, etc.

brick, graphite. A brick produced from mixtures of coke and pitch which has been heat-treated to develop a graphitic crystal structure; used as a refractory.

brick, green. Unfired brick.

brick, hand-made. Brick formed by hand or shaped in molds by hand, working from a body of suitable consistency; such brick may or may not be pressed in a hand-operated press after partial drying.

brick, hard-burned. Any brick, usually a refractory brick, fired at a higher than normal temperature.

brick, high-alumina. See alumina brick.

brick, high-duty fireclay. Fireclay brick having a pyrometric cone equivalent between cones 31½ and 33.

brick, industrial floor. A brick having extremely high resistance to wear, mechanical damage, chemicals, and temperature conditions such as may be encountered in industrial and commercial installations.

brick, in-wall. Fireclay brick used to line the in-wall section of a blast furnace.

brick, ipre. An I-shaped paving brick.

brick, jack. A type of refractory brick employed as a base on which melting pots are placed, and which are designed with openings or holes to accommodate the fork of a fork-lift truck for easy movement of the pots from one location to another.

brick, jamb. A brick with the corner of one side rounded to a radius approximately equal to the width of the brick; used to construct curved walls and other curved structures.

brick, jumbo. A brick larger than standard size, usually 4×4×12 inches (10.2×10.2×30.5 centimeters); sometimes produced to dimensional specifications.

brick, key. A tapered brick used to close and tighten an arch.

brick, ladle. A refractory brick of low porosity and permanent expansion used to line ladles for the containment of molten metal.

brick, lattice. A hollow or perforated type of building brick employed for heat insulation.

brick, low-duty firebrick. A fireclay brick having a pyrometric cone equivalent between cones 15 and 29.

brick, lug. A brick formed with lugs to facilitate spacing with adjoining brick in construction.

brick, magnesia. See brick, magnesite.

brick, magnesite. A brick produced substantially or entirely of dead-burned magnesite, sometimes with additions of as much as 15% of other oxides; used in linings for furnaces where corrosion by basic slags may be severe.

brick, magnesite-chrome. A refractory burned or unburned brick made substantially of a mixture of dead-burned magnesite and refractory chrome ore, and in which the dead-burned magnesite is the predominant ingredient.

brick, magnesite-dolomite. A refractory brick made substantially of a mixture of dead-burned magnesite and dead-burned dolomite, and in which dead-burned magnesite is the predominant ingredient.

brick, medium-duty fireclay. A fireclay brick having a pyrometric cone equivalent between cones 29 and 31½.

brick, merch. A discolored, off-size, or distorted building brick.

brick, metalkase. Basic refractory brick enclosed in thin steel casings or box-like enclosures as protection against heat and hostile environments.

brick, modular. A brick of a size which will fill a 4-inch modular unit when laid, including the mortar joint.

brick, mold. A brick of the insulating type shaped to fit the top of an ingot mold.

brick, neck. A brick shaped so that one large face is tapered toward one end.

brick, nine-inch. A standard brick size, 9×4⁷⁄₁₆ ×2½ inches (22.9×11.3×6.4 centimeters) used in the refractories industry.

brick, Norman. A building brick, 2⅔×4×12 inches (6.8×10.2×30.5 centimeters), in size.

brick, nozzle. A tubular refractory shape used in a ladle through which steel is teemed at the bottom of the ladle, the upper end of the shape serving as the seat for the stopper.

brick, packaged. One or more brick enclosed in a packaged unit for protection and ease in handling.

brick, panel. A long silica refractory laid as a stretcher in a coke oven. See stretcher.

brick, paving. A vitrified brick of high-strength and low-water absorption, frequently furnished with spacing lugs, made with smooth or wire-cut surfaces; used to construct roads, driveways, and other pavements.

brick, perforated. A building brick with symmetrically arranged holes parallel with the face of the brick to reduce weight.

brick, pitch-bonded basic. Unburned basic refractory shapes bonded with pitch; when heat-treated to minimize softening of the bond on reheating, they are identified as tempered.

brick, pitch-impregnated. Burned basic-refractory shapes which subsequently are impregnated with pitch. See pitch.

brick, place. A relatively soft, low-quality brick of salmon color used in temporary or noncritical construction.

brick, pressed. Brick pressed in molds under pressure from clay of relatively low moisture content (5 to 7%).

brick, radial. Brick with curved faces for use in the forming of concentric cylinders and other circular construction.

brick, refractory. Any refractory brick which will be subjected to high temperatures during use.

brick, repressed. Bricks formed by subjecting blanks of a clay mixture to pressure in a mechanical press.

brick, Roman. A brick 2×4×12 inches (5.1×10.2×30.5 centimeters) in size.

brick, rubbing. A block of bonded abrasive used for rubbing down castings, scouring chilled-iron rolls, polishing marble, and similar applications.

brick, runner. A refractory shape having a hole or holes through which molten metal is conveyed during the teeming of bottom-poured ingots. See bottom teeming.

brick, salmon. A relatively soft, underfired salmon-colored brick, which is used in applications where strength is not a critical or major requirement.

brick, sand-lime. A brick made of a mixture of silica sand and lime which is cured under the influence of high-pressure steam.

brick, sand-struck. A brick produced by molding relatively wet clay of approximately 20 to 30% moisture content, frequently by hand, in a sanded mold to prevent sticking of the brick to the mold.

brick saw. A mechanically operated abrasive disk used to cut brick.

brick, scove. An unfired refractory brick used in the construction of scove kilns, which see.

brick, SCR. A brick 2⅔×6×12 inches (6.8×15.2×30.5 centimeters) in size.

brick scratchers. A wire comb employed to texture the surface of brick following the extrusion operation.

brick, semi-silica fireclay. A fireclay brick containing 72% or more of silica.

brick, sewer. A low-absorption, abrasion-resistant brick used in drainage systems and structures.

brick, side-arch. Wedge-shaped brickwork with walls inclined toward each other.

brick, side-cut. A brick wire-cut along the sides instead of at the ends.

brick, silica. A highly refractory brick containing 90% or more of silica which has been bonded with lime. See silica brick.

brick, siliceous fireclay. Fireclay brick containing substantial quantities of uncombined silica and relatively low quantities of fluxing ingredients.

brick, sleeve. A tube-shaped firebrick used to line slag vents.

brick, soap. A brick of one-half the usual width, approximately 2 inches.

brick, soft-mud. Brick produced by molding, frequently by hand, a highly plastic body containing from 20 to 30% of water.

brick, standard. A brick 2⅔×4×8 inches (6.8×10.2×20.3 centimeters) in size.

brick, stiff mud. An extruded brick formed from a stiff, but plastic, clay body containing roughly 12 to 15% of moisture.

brick, straight. A rectangular brick, 13½ inches or less in length, in which the thickness is less than the width.

brick, sun-dried. A brick of earth or clay, sometimes containing straw, which is roughly molded and sun-dried instead of fired. Also known as adobe.

brick, superduty fireclay. A fireclay refractory brick having a pyrometric cone equivalent to not less than cone 33, a linear shrinkage of 1% or less in the 1598°C (2910°F) reheating test, and 4% or less failure in the panel spalling test [preheated at 1649°C (3000°F)].

brick, superduty silica. A term applied to silica brick in which the total alumina, titania, and alkali content is significantly lower than normal.

brick, tapestry. A brick having a rough, unscored, textured surface.

brick, textured. A brick with its surface treated to change its appearance from that produced by the die, such as by scratching or scoring.

brick, triple. A brick 5⅓×4×12 inches (13.5×10.2×30.5 centimeters) in size.

brick, tuyere. A refractory shape containing one or more holes through which air and fuel are introduced into a furnace.

brick, unburned. Brick produced by processes which do not involve a firing process, such as adobe, chemically bonded refractories, etc.

brick, unfired. See brick, unburned.

brick, water-struck. Soft-mud brick formed in molds which are dampened in advance to prevent the brick from sticking to the molds.

brick, wedge. A brick having its largest faces sloping toward each other at an acute angle.

brick, wire-cut. A brick cut from an extruded column of clay with a wire.

brickwork. Any masonry structure or pavement made of brick.

brickwork, reinforced. Brickwork or masonry units strengthened by metal bars, rods, wire, etc., embedded in the bed or mortar joints.

brick, zirconia. See zirconia brick.

bridge. The structure formed by the end walls of the adjacent melter and refiner compartments of a glass tank, the covers spanning the gap between the end walls.

bridge cover. A refractory block spanning the space between the bridge walls of a glass-melting tank.

bridge-material transfer. Material transfer that occurs without the presence of a gaseous electric discharge. The filament of molten contact material that connects the two separating electrical contacts does not rupture in the middle; thus, there is a gain of material on one contact and a loss of material from the other.

bridge wall. The part of a glass-melting tank which forms a bridge or separation between the melting and the refining sections.

bridging oxygen. An atom of oxygen situated between and bonded to two silicon atoms in a glass structure.

bright annealing. The heating of steel to a red heat or above in an inert or reducing atmosphere which inhibits or prevents oxidation, the surface of the metal remaining bright.

bright glaze. A white, colored, or clear ceramic glaze having a high gloss.

bright gold. An inexpensive luster of gold resinate combined with other metal resinates and a flux; used as a decoration when fired on glass, porcelain enamel, glaze, or other surfaces. See luster.

brights. Any portion of decorated glass forming a part of a design, but which has not been acid treated.

brilliance. The property of being very bright in appearance; in glasses or glassy compositions, the property is influenced by the index of refraction, the transparency, and the surface polish of the item being observed.

brilliant cutting. A process of decorating flat glass in which designs are cut in the glass by abrasives and polishing wheels.

brindled brick. A brick of high-crushing strength made of iron-bearing sedimentary clays which are partially reduced during firing.

Brinell test. A measurement of the hardness of a material obtained by pressing a steel ball one centimeter in diameter into the material being tested under a prescribed load; the spherical-surface arc area of the resulting indentation is measured and divided by the applied load; the results are reported as the Brinell number, kilograms per square millimeter.

briquetting. The process of forming powdered or granular materials into cubes, blocks, or other shapes in dies under pressure.

Bristol glaze. An unfritted zinc-bearing glaze for stoneware, terra cotta, and similar bodies.

British thermal unit (Btu). The unit of heat required to raise the temperature of water at maximum density (air-free) 1°F under a constant pressure of 1 atmosphere, the equivalent of 252 calories.

brittle. The property of being easily broken or fractured without prior deformation.

brittle fracture. A fracture occurring in a metallic or ceramic body due to its brittle nature.

brittleness. The tendency of a material to fracture without appreciable deformation under stress.

brittle-ring test. A tensile test in which maximum stress is applied to the inner periphery of a ring specimen by application of a load to the outer periphery of the ring.

broken-joint tile. A roofing tile laid over the center of the head of a tile immediately below.

broken seed. A fractured bubble on the surface of plate glass after polishing.

Brongniart's formula. A formula to calculate the solid content of a suspension in which

$$W = (P-20)S/(S-1)$$

in which W is the weight of solid in 1 pint of the slurry in ounces, P is the weight of 1 pint of the slurry, and S is the specific gravity of the dry solid material.

Brookfield viscometer. An instrument to measure the viscosity of a porcelain-enamel or glaze slip in which the resistance of an electrically operated cylinder to rotation in the slip is determined.

brookite. A black, brown, or reddish orthorhombic form of titania, which is trimorphous with anatase and rutile, having a specific gravity of 3.87–4.08 and a hardness (Mohs) of 5.5–6.

brown coat. A mortar or plaster which has been strengthened by the addition of hair or other fibrous material, and over which a finish coat is applied.

brownies. A synonym for copperheads in porcelain enamels.

brucite. $Mg(OH)_2$; mol. wt. 67.43; sp. gr. 2.38–2.40; hardness (Mohs) 2.5; used in refractories as a source of dead-burned magnesite, and as a component in welding-rod coatings.

bruise. An area of small cracks in glassware resulting from impact.

Brunauer-Emmett-Teller equation. A technique for determination of the surface area of a powder or porous solid by computing the monolayer area from the volume of a gas adsorbed on the surface of a sample of known weight; an extension of Langmuir's isotherm equation.

brush. A conductor arranged to make electrical contact between a stationary and one or more sliding components.

brush, bolt-hole. A round, stiff-bristled brush, equipped with a metallic guide pin, used to remove bisque porcelain enamel from the inside and edges of small openings in the ware to prevent subsequent chipping of the fired coating during the assembly of the finished product.

brush, edging. A stiff-bristled brush, equipped with a straight metal guide, used to remove bisque porcelain enamel from the edges of ware to prevent chipping and to enhance the appearance of the subsequently fired product.

brush force. The force required to close, maintain, and open electrical contacts.

brushing. (1) The removal of bisque porcelain enamel from ware before firing by brushing through a stencil or along an edge to produce a design or edging.

(2) The removal of bedding material from ceramic ware after the bisque fire.

brush marks. A defect or blemish in glassware consisting of fine lines having the appearance of brush marks.

Btu. An acronym for British thermal unit, which see.

bubble cap. A ceramic cap, serrated along the bottom to permit the passage of vapors, for use in distillation and de-acidifying towers in chemical processes.

bubble glass. A decorative product containing bubbles of prescribed size and arrangement.

bubble-pressure pore-size determination. A method of estimating the maximum pore size of a material by calculating the pressure required to force a bubble of air through the material wetted by a liquid of known surface tension.

bubble structure. The size and distribution of voids in a fired porcelain-enamel coating.

bubbly clay. A clay containing organic impurities which cause bubbles in porcelain enamels and glazes during firing.

buck. A special support employed in the firing of heavy porcelain-enameled ware.

bucket conveyor. A conveyor of bulk material consisting of a series of scoops or bucket-like containers mounted on an endless belt or chain.

bucking coils. A coil connected and positioned in such a way that its electric or magnetic field opposes the electric or magnetic field of one or more other coils so that an unbalance is produced in the system to yield an indication.

buckstave, buckstay. A steel bracing employed to take the thrust of the refractory structure, such as the roof, in the construction of a furnace.

buffing wheel. A flexible disk coated with a very fine abrasive which is used in the buffing or polishing of surfaces.

bugholes. Small pits, bubbles, or voids in the surface of formed concrete.

buhr mill. A pulverizing machine in which materials are ground between a siliceous rock rotating against a stationary surface of the same material.

builder. A scrap refractory used as a filler in the construction of kiln bottoms and similar items.

building block. Hollow concrete or fired-clay blocks used in the construction of walls which usually are to be covered with a finishing material such as stone.

building brick. A brick formed and fired to a stable unit from clay, but not especially produced for color or texture, for use in the general construction industry.

building clay. A clay suitable for the production of brick for use in the construction industry.

bulb edge. The heavy rounded edge or bead on sheet-drawn glass.

bulb trailer. An instrument for squeezing out the flow lines of slip on a clay surface.

bulged finish. A distended top section of a glass bottle.

bulk density. The ratio of the weight of an object or material to its total volume, including pore space, expressed as grams per milliliter or as pounds per cubic foot.

bulkhead. A panel of brick built into a wall for easy replacement.

bulking. The tendency of fine particles of a material to occupy a greater volume when moist.

bulk modulus of elasticity. The ratio of the compressive or tensile forces applied to a material per unit of surface area to the change in the volume of the material per unit of volume.

bulk specific gravity. The ratio of the mass of a material to that of a quantity of water which has a volume equal to the bulk volume of the material at the temperature of measurement. Calculated by the formula:

$$G_b = \frac{W_f}{V_f}$$

in which G_b is the bulk specific gravity, W_f is the weight of the fired test specimen in grams, and V_f is the volume of the fired test specimen in cubic centimeters.

bulk volume. The volume of a solid material, including the volume of open and sealed pores. Calculated by the formula:

$$V_b = P_o + P_s + V_t = \frac{D_w}{d_b}$$

in which V_b is the bulk volume, P_o is the volume of open pores, P_s is the volume of sealed pores, V_t is the true volume of the solid, D_w is the dry weight of the specimen, and d_b is the bulk density of the specimen.

Buller rings. Unfired ceramic rings, 2½ inches in diameter with a hole ⅞ inch in diameter in the center, of prescribed compositions which by their respective shrinkages are used as an indication of the thermal history to which accompanying ware has been exposed during firing.

bullet-resisting glass. A special laminated safety glass from ¾ to 3 inches thick.

bull float. A finishing tool with a handle several feet long which will permit a worker, standing from a distance, to finish a slab of concrete from the interior to the edge.

bull header. A bull-nosed or jamb brick laid on its face so that the normal bedding area is visible in the wall face.

bullion. The central portion of a disk of crown glass to which the blowing iron was attached.

bullnose. A brick having the corner of one end and side rounded to a radius approximately equal to the width of the brick.

bull's eye. A circular window.

Bull's kiln. A clamp kiln in which brick are placed and fired in trenches.

bundle, fiber. A package of parallel, long, thin, flexible glass fibers used to transmit images from one end of the bundle to the other in fiber optics.

bung. (1) A group of saggers or pots stacked in a kiln.
(2) A removable roof section built in a kiln.

bunker C fuel oil. A special grade, No. 6, fuel oil; used by industry and for large-scale heating operations.

bunker fuel oil. A heavy fuel oil formed by the stabilization of the residual oil remaining after the cracking of crude petroleum, and used in large-scale heating and power-production applications.

burley clay, burley flint clay. A rock containing nodules of aluminous or ferruginous materials, or both, bonded by fireclay.

burn. (1) The controlled heat treatment of ceramic ware and coatings in a furnace or kiln.
(2) Synonym for firing.

burnable poison. A neutron absorber, such as boron, purposely included in a reactor to help control long-term reactivity changes by its progressive burnup.

burner. (1) The mechanism by which air and fuel are mixed and directed into a combustion chamber.
(2) The operator whose duty is to tend a ceramic kiln.

burner block. A refractory block with one or more orifices through which fuel is introduced into a furnace or kiln.

burner, premix. An oil or gas burner in which the fuel and air are premixed before ignition in a combustion chamber.

burning. (1) The process of firing ceramic bodies, glazes, porcelain enamels, and other coatings and products in a furnace or kiln for the purpose of developing bond or other necessary or desired physical and chemical properties.
(2) The heat treatment, vitrification, or curing of a grinding wheel to produce desired bond properties.
(3) The change in a material being ground or polished caused by heat generation during the grinding operation, frequently accompanied by discoloration of the material.
(4) Overpickling of metal for porcelain enameling, often producing pits in the metal surface.

burning bars, points, or tools. A heat-resistant metal alloy used to support porcelain-enameled ware during the firing operation.

burning, draw. The removal of a load of porcelain-enameled ware from the furnace for a short period of time prior to completion of the firing operation in order to equalize heating of all areas of the ware, particularly when the underlying metal is of varying thicknesses.

burning off. Over-firing of porcelain enamels resulting in a rough, dark surface saturated with undissolved iron oxide.

burning shrinkage. See firing shrinkage.

burning-tool marks. A defect in porcelain enamels occurring on the sheet-metal surface opposite the point of contact with the supporting burning tool.

burning zone. The area in a continuous furnace where the major amount of heat is supplied to ware during the firing operation.

burnished gold. A durable type of gold applied to glazed ware as a suspension in oil, fired, and rubbed with agate or other polishing material to a bright finish.

burnishing. Polishing of an overglaze gold, leather-hard clay, or other material with agate, stone, sand, or steel tool to produce a bright surface.

burnishing, pattern. The process of obtaining special effects on the surfaces of clay ware by burnishing.

burn-off. (1) The process of severing an unwanted portion of a glass article by fusing the glass.
(2) Slag-like area resulting from an insufficient coating of porcelain enamel which occurs during firing.

burn-out. The removal of organic additives from unfired nuclear-fuel shapes by the application of heat.

burnt lime. Calcined dolomitic limestone or calcite, or a mixture of these.

burnup. Nuclear transformations induced during nuclear operations. The term may be applied to fuel or to other materials or to the amount of depletion due to nuclear transformation.

burr. (1) A thin, ragged edge of metal resulting from punching, cutting, or grinding of a metal sheet.
(2) A fragment of excess material, or of a foreign material, adhering to the surface of a body.

burring. The removal of sharp edges or fins from punched, cut, or ground metal items.

burr mill. A mill consisting of two ribbed disks of stone or metal rotating against each other; used in the grinding of solid materials and in homogenizing mixtures of pigments in a suitable liquid medium to produce pastes for the decoration of ware.

bursting. The disintegration of refractories containing chrome ore when exposed to iron oxide at high temperatures; characterized by having the exposed face swell and grow until it breaks away from the brick mass following a permanent increase in volume.

bursting expansion. A term sometimes used as a synonym for bursting.

bursting off. The breaking of the thin-walled bubble of glass formed above a blow mold.

bursting strength. The ability of a material to withstand pressure without rupture.

burst pressure. The maximum inside pressure a material or object can withstand without rupture.

bushing. (1) The liner of an orifice that delivers molten glass to a forming machine, or the liner of the unit through which molten glass is drawn in the production of glass fibers.
(2) A bearing that lines the supporting structure for a rotating shaft.

bushing, reducing. An insert of any suitable material used to reduce the arbor hole of a rotating shaft, such as the shaft of a grinding wheel, so as to accommodate a shaft or spindle of smaller size.

bustle pipe. A large refractory-lined pipe which encircles and delivers a hot-air blast to a blast furnace.

butting contacts. Electrical contacts in which the motion of the moving contact is perpendicular to the contact faces, and which open and close with no appreciable sliding or rolling action.

button. A section in pressed glassware so designed that it may be knocked out to form a hole of specified dimensions in the parent glass.

button test. A test in which button-like specimens of prescribed form, and sometimes density, are employed to evaluate the fusion and flow characteristics of frits, glasses, and powders.

buttress. A projection designed to increase the resistance of a wall in a structure to lateral forces.

butt seal. A ceramic-metal or glass-metal seal which will withstand high temperatures and a high vacuum without leakage, such as is obtained when a flat or washer-like metal is brazed to a metallized ceramic for use in electric and electronic components or products.

buyer. An individual or organization issuing a purchase order or making an offer to purchase something.

bytownite. A soda-lime feldspar.

C

°C. Symbol for degrees Centigrade or Celsius.

cadmium acetate. $Cd(CH_3COO)_2 \cdot 3H_2O$; mol. wt. 283.53; loses water at 130°C (266°F); m.p. of hydrate 256°C (493°F); sp. gr. 2.01–2.34; used in the production of iridescent glazes.

cadmium antimonide. CdSb; mol. wt. 234.2; m.p. 452°C (847°F); a semiconductor.

cadmium carbonate. $CdCO_3$; mol. wt. 172.42; decomposes below 500°C (932°F); sp. gr. 4.258; used to improve the stability of cadmium-selenium red colors.

cadmium fluoride. CdF_2; mol. wt. 150.41; m.p. 1000°C (1832°F); sp. gr. 6.64; used in electronic and optical applications and as a starting material for laser crystals.

cadmium niobate. $Cd_2Nb_2O_7$; mol. wt. 523.02; an antiferroelectric having low-loss properties at high frequencies.

cadmium nitrate. $Cd(NO_3)_2 \cdot 4H_2O$; mol. wt. 304.07; m.p. 59.5°C (139.1°F); b.p. 132°C (270°F); sp. gr. 2.455; used as a reddish-yellow colorant in porcelain enamels and glass.

cadmium orange. An impure form of cadmium sulfide used as a ceramic colorant.

cadmium orthophosphate. $Cd_3(PO_4)_2$; mol. wt. 527.27; m.p. 1500°C (2732°F).

cadmium selenide. CdSe; mol. wt. 191.61; m.p. above 1350°C (2462°F); sp. gr. 5.81; used in the production of red ceramic colors.

cadmium silicate. $CdSiO_3$; mol. wt. 188.47; m.p. 1242°C (2268°F); sp. gr. 4.93.

cadmium sulfide. CdS; mol. wt. 144.47; sublimes at 980°C (1796°F); sp. gr. 3.9–4.8; employed as a component in the production of red, orange, and yellow ceramic colors. Also known as cadmium yellow.

cadmium telluride. CdTe; mol. wt. 239.91; m.p. 1090°C (1994°F); sp. gr. 6.2; employed in rectifiers, solar batteries, and optical systems.

cadmium titanate. $CdTiO_3$; mol. wt. 208.31; a ferroelectric.

cadmium tungstate. $CdWoO_4$; mol. wt. 360.41.

cadmium yellow. A series of colors, usually cadmium sulfide co-precipitated with barium sulfate, ranging from golden yellow to a greenish or reddish yellow.

cadmium zirconate. $CdO \cdot ZrO_2$; mol. wt. 251.4; used as a depressant of the dielectric constant of barium titanate capacitors at the Curie temperature.

cage. A preassembled unit of reinforcements for concrete pipe or piling consisting of circumferential and longitudinal bars or wire mesh.

cage-mill disintegrator. A machine consisting of high-speed rotating vanes employed to disintegrate soft particles in the beneficiation of coarse concrete aggregate.

cake. A slab of damp clay or ceramic body as removed from a filter press.

calcareous clay. Clay containing calcium-bearing minerals, usually sulfate or carbonate.

calcine. A material or mixture of materials which has been heated to a high temperature, but without fusion, to eliminate volatile constituents and to produce desired physical changes.

calcined alumina. Al_2O_3; mol. wt. 101.94; m.p. about 2040°C (3704°F); sp. gr. 3.4–4.0; index of refraction 1.765; available in several grades based on heat treatment; contains traces of residual water; contains less of the alpha phase than the tabular grades, and is easier to mill. See alumina, calcined.

calcined clay. Ball or china clay which has been heated until the combined water is removed, and the plastic character is destroyed.

calcined gypsum. $CaSO_4 \cdot xH_2O$; known commercially as plaster of paris.

calcined kaolin. Nominally $3Al_2O_3 \cdot 2SiO_2$ plus amorphous siliceous materials; mol. wt. approximately 425.9; melting point 1770°C (3218°F); deformation temperature 1750° to 1770°C (3182° to 3218°F); used in refractories, kiln furniture, castables, investment molds, low-expansion and insulating bodies, and other high-temperature products to improve refractoriness, mechanical strength, thermal-shock resistance, load-bearing properties, and resistance to corrosion by molten glasses, fritted glazes, porcelain-enamel frits, and slags.

calcined limestone. Limestone which has been heat-treated for the removal of CO_2.

calcined refractory dolomite. Refractory dolomite which has been heated for a sufficient time and temperature to remove volatile matter and to decompose the carbonate structure.

calcined soda. The commercial grade of sodium carbonate used in the manufacture of glass.

calcining refractory materials. The process of heating raw refractory materials to remove volatile constituents and to produce changes in volume.

calcite. $CaCO_3$; mol. wt. 172.42; sp. gr. 2.72; hardness (Mohs) 2; the principal ingredient in limestone; employed as a major component in portland cement, in soda-lime glassware, in pottery bodies, and for insulating coatings for capacitors and printed circuits. Also known as calcspar.

calcite dolomite. A carbonate rock consisting of 10 to 50% calcite and the balance dolomite.

calcium acrylate. $(CH_2CHCOO)_2Ca$; used as a binder for clay products and foundry molds.

calcium aluminate. (1) $CaO \cdot Al_2O_3$; mol. wt. 158.02; m.p. 1600°C (2912°F); sp. gr. 3.67. (2) $CaO \cdot 2Al_2O_3$; mol. wt. 259.9; m.p. 1760°C (3200°F), but incongruent; sp. gr. 2.90. (3) $3CaO \cdot 5Al_2O_3$; mol. wt. 677.8; m.p. 2230°C (3130°F). (4) $3CaO \cdot Al_2O_3$; mol. wt. 270.2; m.p. 1538°C (2800°F); sp. gr. 3.0.

calcium aluminum silicate. (1) A slag-like product

consisting essentially of CaO, Al_2O_3, MgO, and SiO_2; used in amber, green, and other glasses. (2) $CaO·Al_2O_3·2SiO_2$; mol. wt. 278.2; m.p. 1549°C (2820°F); sp. gr. 2.77. (3) $2CaO·Al_2O_3·SiO_2$; mol. wt. 274.2; m.p. 1596°C (2895°F); sp. gr. 3.04.

calcium antimonate. $CaO·Sb_2O_3$; mol. wt. 347.60; limited use as an opacifier in porcelain enamels and glazes.

calcium boride. CaB_6; mol. wt. 105; m.p. 2235°C (4055°F); sp. gr. 2.45; hardness 2740 Vickers.

calcium carbide. CaC_2; mol. wt. 64.07; m.p. 2160°C (3920°F); sp. gr. 2.04.

calcium carbonate. $CaCO_3$; mol. wt. 100.1; decomposes at 825°C (1517°F); sp. gr. 2.7–2.95; used as a component in portland cement, in soda-lime glassware and pottery bodies, and as an insulating coating for printed circuits and capacitors. Also known as calcite, aragonite.

calcium chloride. $CaCl_2$; mol. wt. 111; m.p. 772°C (1420°F); sp. gr. 2.15; used as a mill addition in porcelain-enamel slips, as a flocculant in glazes, as an accelerator in portland cement, and as a water proofer in concrete.

calcium chromate. $CaCrO_4·2H_2O$; mol. wt. 192.13; loses water at 200°C (392°F); used as a yellow colorant.

calcium chromite. $CaO·Cr_2O_3$; mol. wt. 208.12; m.p. 2161°C (3940°F); sp. gr. 4.80.

calcium cyanamide. $CaCN_2$; mol. wt. 90.1; m.p. 1200°C (2192°F); sp. gr. 1.083.

calcium dialuminate. $CaO·2Al_2O_3$; mol. wt. 259.96; m.p. 1705°C (3101°F).

calcium ferrite. (1) $2CaO·Fe_2O_3$; mol. wt. 271.9; m.p. 1438°C (2620°F); sp. gr. 3.98. (2) $CaO·Fe_2O_3$; mol. wt. 215.8; m.p. 1215°C (2220°F); sp. gr. 5.08.

calcium fluophosphate. $Ca_5(F,Cl,OH)(PO_4)_3$; the mineral apatite; sp. gr. 3.1–3.2; hardness (Mohs) 5; used as a substitute for bone ash in translucent whiteware bodies and in the manufacture of opal glass.

calcium fluoride. CaF_2; mol. wt. 78.1; m.p. 1360°C (2480°F); sp. gr. 3.18; hardness (Mohs) 4; used as an opacifier and flux in porcelain enamels, glass, and glazes, as a flux in whiteware bodies, and as a glass etchant; also used as a component in crucibles for the melting of uranium.

calcium fluosilicate. $CaSiF_6$; mol. wt. 182.14; sp. gr. 2.662.

calcium hafnate. $CaHfO_3$; mol. wt. 266.68; m.p. 2470°C (4478°F); sp. gr. 5.73; coefficient of thermal expansion 7×10^{-6}.

calcium hexaluminate. $CaO·6Al_2O_3$; mol. wt. 667.72; forms corundum and liquid phase at 1850°C (3362°F).

calcium hydroxide. $Ca(OH)_2$; mol. wt. 58.1; loses water at 580°C (1076°F); sp. gr. 2.34; used in mortars, plasters, and cements. Also known as hydrated lime.

calcium magnesium silicate. (1) $CaO·MgO·SiO_2$; mol. wt. 156.5; m.p. 1499°C (2730°F), but incongruent; sp. gr. 3.2. (2) $CaO·MgO·2SiO_2$; mol. wt. 216.6; m.p. 1390°C (2535°F); sp. gr. 3.28. (3) $2CaO·MgO·2SiO_2$; mol. wt. 272.7; m.p. 1460°C (2660°F); sp. gr. 2.94. (4) $3CaO·MgO·2SiO_2$; mol. wt. 328.8; m.p. 1574°C (2865°F); sp. gr. 3.15.

calcium metaborate. $Ca(BO_2)_2$; mol. wt. 125.72; m.p. 1100°C (2192°F).

calcium metasilicate. $CaSiO_3$; mol. wt. 116.14; m.p. 1544°C (2812°F); sp. gr. 2.8–2.9; hardness (Mohs) 4.5–5; used in pottery bodies, wall tile, cements, wallboard, mineral wool, and special low-loss electroceramics.

calcium molybdate. $CaMoO_4$; mol. wt. 200.08; used as an adherence-promoting agent in some antimony-bearing porcelain-enamel ground coats.

calcium monoaluminate. $CaAl_2O_4$; mol. wt. 158.02; m.p. 1605°C (2921°F); sp. gr. 2.98; employed in high-alumina cements.

calcium niobate. (1) $3CaO·Nb_2O_5$; mol. wt. 434.5; m.p. 1560°C (2840°F), but incongruent; sp. gr. 4.23. (2) $2CaO·Nb_2O_5$; mol. wt. 378.4; m.p. 1565°C (2850°F); sp. gr. 4.39. (3) $CaO·Nb_2O_5$; mol. wt. 322.3; m.p. 1560°C (2840°F); sp. gr. 4.72.

calcium nitrate. $Ca(NO_3)_2·4H_2O$; mol. wt. 236.16; m.p. 42°C (107.6°F); sp. gr. 1.82; used as an oxidizing agent in zirconia and titania opacified porcelain enamels.

calcium nitride. Ca_3N_2; mol. wt. 196.3; sp. gr. 2.06.

calcium orthosilicate. $2CaO·SiO_2$; mol. wt. 172.22; m.p. 2130°C (3866°F); sp. gr. 3.27; a constituent in portland cement and some dolomite refractories.

calcium oxide. CaO; mol. wt. 56.08; m.p. 2570°C (4658°F); sp. gr. 3.40; a fluxing ingredient used extensively as whiting, dolomite, limestone, burned lime, fluorspar, and wollastonite in a wide variety of ceramic products: glass, pottery, glazes, porcelain enamels, portland cement, mortar, and plaster. Also known as lime, calcia.

calcium phosphate, dibasic. $CaHPO_4·2H_2O$; mol. wt. 172.15; decomposes at 25°C (77°F); sp. gr. 2.306; used in the manufacture of glass, principally as a fluxing and glass-forming ingredient.

calcium phosphate, tribasic. $Ca_3(PO_4)_2$; mol. wt. 310.28; m.p. 1670°C (3038°F); sp. gr. 3.18; used in porcelain, pottery, porcelain enamels, and milk glass both as an opacifier and as a glass former.

calcium plumbate. Ca_2PbO_4; mol. wt. 367.38; sp. gr. 5.71; used in glass manufacture as a flux.

calcium potassium silicate. $CaO \cdot K_2O \cdot SiO_2$; mol. wt. 210.4; m.p. 1631°C (2970°F).

calcium pyrophosphate. $Ca_2P_2O_7$; mol. wt. 254.2; m.p. 1230°C (2246°F); sp. gr. 3.09.

calcium scandate. $CaO \cdot Sc_2O_3$; mol. wt. 194.3; sp. gr. 3.89.

calcium silicate. (1) $CaSiO_3$; mol. wt. 116.14; m.p. 1544°C (2812°F); sp. gr. 2.8–2.9; hardness (Mohs) 4.5–5. Also known as pseudowollastonite. (2) $3CaO \cdot 2SiO_2$; mol. wt. 288.5; decomposes at 1899°C (3450°F). (3) $2CaO \cdot SiO_2$; mol. wt. 172.3; m.p. 2130°C (3865°F); sp. gr. 3.28. (4) $3CaO \cdot SiO_2$; mol. wt. 228.4; decomposes at 1465°C (2670°F).

calcium soap. Calcium resinate used as a binder in ceramic inks and pastes.

calcium stannate. $CaSnO_3$; mol. wt. 206.78; m.p. $>$1200°C (2192°F); employed in barium titanate bodies to lower Curie temperature, and as a base for phosphors.

calcium sulfate. $CaSO_4$; mol. wt. 136.14; m.p. 1450°C (2642°F); sp. gr. 2.964; as plaster of paris ($CaSO_4 \cdot 2H_2O$) is used extensively in models and molds, as a bedding agent in the grinding and polishing of plate and optical glasses, as an occasional batch ingredient in glass and glazes, as a binder in low-density insulation, and as a flocculant in glazes and other slips to prevent settling.

calcium sulfide. CaS; mol. wt. 88.16; sp. gr. 2.61.

calcium titanate. $CaO \cdot TiO_2$; mol. wt. 135.98; m.p. 1915°C (3479°F); sp. gr. 3.17–4.02; a high dielectric material used in barium and other rare-earth titanates and zirconates for piezoelectric applications.

calcium titanium silicate. $CaO \cdot TiO_2 \cdot SiO_2$; mol. wt. 196.3; m.p. 1382°C (2520°F); sp. gr. 3.5.

calcium tungstate. $CaWO_4$; mol. wt. 288.08; m.p. 1535°C (2795°F); sp. gr. 5.9–6.1; hardness (Mohs) 4.5–5; index of refraction 1.93 (approx); good mechanical strength and chemical stability.

calcium uranate. $CaO \cdot UO_3$; mol. wt. 312.27; m.p. 1799°C (3270°F); sp. gr. 7.45.

calcium zinc silicate. $2CaO \cdot ZnO \cdot 2SiO_2$; mol. wt. 313.8; m.p. 1427°C (2600°F).

calcium zirconate. $CaZrO_3$; mol. wt. 179.3; m.p. 2350°C (4262°F); sp. gr. 4.74; low firing shrinkage; used in titanate dielectrics.

calcium zirconium silicate. $CaO \cdot ZrO_2 \cdot SiO_2$; mol. wt. 239.2; m.p. 1582°C (2880°F).

calcrete. A mixture of gravel and sand cemented by calcium carbonate.

calcspar. A synonym for calcite or limestone.

calibration. (1) Determination of the values of the significant parameters by comparison with values indicated by a reference instrument or by a set of reference standards.

(2) The process of fixing, checking, or correcting an arbitrary or inaccurate scale of a measuring instrument to absolute values.

calibration, chemical. Calibration of an instrument or method based on chemical standards.

calibration curve. The graphical representation of a relationship between a measured parameter and a concentration or mass of the standard for the substance under consideration.

calibration factor. The slope of the calibration curve, or its inverse, usually in terms of the measured unit per concentration or mass of the element.

calibration result. The result obtained in an analysis of a calibration standard, equal to the known content of the calibration standard, if the apparatus is properly calibrated.

calibration standard. Any of the standards of various types having known parameters which may be used to adjust the sensitivity setting of test instruments at some predetermined level for the periodic adjustment of the sensitivity.

calipers. An instrument consisting of a pair of hinged legs which may be used to measure internal and external dimensions.

calorie, large. The quantity of heat required to raise the temperature of one kilogram of water by 1°C, from 3.5° to 4.5°C.

calorie, small. The quantity of heat required to raise the temperature of one gram of water by 1°C, from 3.5° to 4.5°C.

calorific value. A measure of the quality of fuels; usually expressed as available Btu per unit of weight or volume for complete combustion.

camber. A surface imperfection consisting of a single arch of curvature as opposed to waviness.

camber arch. An arch with a horizontal exterior and a slightly curved interior.

came. Lead strips used for setting glass panes, medallions, mobiles, etc.

campaign. The working life of a furnace, glass tank, or other melting unit between major cold repairs.

Canada balsam. An exudate of the balsam fir tree having an index of refraction similar to that of glass; used in cementing optical lenses and other optical elements.

canal. The section of a glass tank through which molten glass flows from the relatively wide fining area to the drawing chamber or machine.

cane. Solid glass rods of small to medium diameter.

cane clay. A fireclay, sometimes sandy, but less refractory than normal fireclay.

cannon pot. A small glass-melting pot or crucible.

cantilever arch. An arch supported by flat projections on opposite walls.

cant strip. A strip placed under the edge of the lowest row of tiles on a roof to give them the same slope as the other tiles.

cap. (1) A type of bottle closure.
(2) To cut off the ends of a glass cylinder.
(3) The act of preparing a strength specimen for testing in which a fluid or mastic material is applied to the ends of the specimen which will be in contact with the testing machine.
(4) Synonym for crown.

capacitance. The property of a system of conductors and dielectrics which permits the storage of electrically separated charges when potential differences exist between the conductors.

capacitance unbalance. The difference in capacitance of two insulated conductors to the shield, expressed as a percentage of the capacitance between the conductors, or in percent unbalance.

capacitor. A device consisting of conductive plates separated by a dielectric, and which gives capacitance, which see.

capacity. (1) The cubic content or volume which can be contained by a receptacle or a porous substance.
(2) The ability of a material to yield, withstand, or perform.

capacity, dynamic adsorptive. The quantity of a given component adsorbed per unit of a substance, such as activated carbon, from a fluid or fluid mixture moving through a fixed bed at the breakpoint for that substance.

capacity, equilibrium adsorptive. The quantity of a given component adsorbed per unit of substance, such as activated carbon, from a fluid or fluid mixture at equilibrium temperature and concentration or pressure.

capacity insulation. The ability of masonry to store heat as a result of its mass, density, and specific heat.

capillarity. The ability of a brick or other fired ceramic product to conduct liquids through its pore structure by force of surface tension.

capillary. A tube having a very small internal diameter.

capillary drying. The progressive removal of moisture from a porous solid by surface evaporation followed by the capillary movement of more moisture to the drying surface until the core and surface of the solid are of the same moisture concentration.

capillary viscometer. A long narrow tube used to measure the laminar flow of liquids.

capital expenditure. Money spent for long-term improvements, additions, or equipment, and charged to a capital assets account.

cap seat. The ledge inside the mouth of a milk bottle.

carbide. (1) A binary compound of carbon with a more electropositive element such as aluminum, beryllium, chromium, hafnium, lanthanum, niobium, silicon, tantalum, thorium, titanium, uranium, vanadium, tungsten, and zirconium.
(2) A cemented or compacted mixture of carbides used for metal-cutting and machining tools. Carbides are characterized by high-melting points, Mohs hardness values of 8–9, low-impact strength, low-electrical and thermal conductivities, and high moduli of elasticity. They are used in grinding wheels, grinding belts and papers, electrical-resistance heating elements for kilns and furnaces, drill bits, saw teeth, wire-drawing dies, balls for the tips of ball-point pens, and similar applications where thermal and wear resistance are important.

carbide fuel. An oxidation-resistant, high-strength composition prepared from a mixture of a nuclear fuel, metal, and a carbon compound.

carbide tool. A high-heat and wear-resistant cutting and machining tool made of the carbides of tantalum, titanium, or tungsten.

carbofrax. A refractory silicon carbide used in refractory cements, refractory brick, and shapes for furnace walls, domes, checkers, radiant tubes, hearths, etc., where temperatures are severe; a proprietary product.

carbon. C; at. wt. 12.11; sublimes above 3500°C (6332°F); sp. gr. (amorphous) about 2, (graphite) 2.25, (diamond) 3.5; properties vary widely with form and generally exhibit low coefficient of thermal expansion and high resistance to thermal shock, high strength which increases with temperature, high electrical and thermal conductivities, and excellent abrasion, erosion, and corrosion resistance.

carbonaceous deposits. Particles of carbon or a material of substantial carbon content usually occurring as a contaminant in or on the surface of a body or other substance.

carbon, activated. A group of carbonaceous substances having high adsorptive properties. See activated carbon.

carbonate. (1) A salt or ester of carbonic acid.
(2) In ceramic usage, a salt consisting of a metallic element in combination with a CO_3 radical such as, for example, $BaCO_3$, $CaCO_3$, K_2CO_3, and Na_2CO_3; used as convenient source of metal oxides in ceramic bodies.

carbon black. Any of various colloidal black substances consisting essentially of elemental carbon prepared by partial combustion or thermal decomposition of hydrocarbons.

carbon black, microstructure. The arrangement of the carbon atoms within a carbon black particle.

carbon black, structure. The degree or state of agglomeration of particles in carbon black.

carbon-ceramic refractory. A refractory product composed of a mixture of carbon or graphite and one or more refractory ceramic materials, such as fireclay or silicon carbide.

carbon deposition. The deposition of amorphous carbon, resulting from the decomposition of carbon monoxide into carbon dioxide and carbon within a critical temperature range. When deposited within the pores of a refractory brick, the carbon may build up sufficient pressure to destroy the bond and cause the brick to disintegrate.

carbon dioxide. CO_2; a heavy, colorless, odorless gas; a source of defects when formed by decomposition of carbonaceous impurities in bodies and coatings.

carbon dioxide bonding. A bonding process for foundry sands and cores in which formed mixtures of a refractory and sodium silicate are subjected to carbon dioxide.

carbon-film resistor. A resistor consisting of a film of carbon deposited on a ceramic form.

carbon, granular activated. Activated carbon in particle sizes predominately larger than 80 mesh. See activated carbon.

carbon, graphitic. Tiny flakes of pure carbon formed in pig iron during cooling which tend to weaken the metal; causes blisters in porcelain enamels.

carbon, green. A shaped but unfired carbon body.

carbon, impervious. A dense, bitumen-bonded carbon body formed by pressing and then repressing to an essentially pore-free brick; used to line chemical process and storage vessels.

carbon monoxide disintegration of refractories. The disintegration of refractories due to carbon contamination resulting from the dissociation of carbon monoxide in furnace atmospheres.

carbon, porous. An item fabricated from carbon pressed without the use of a binder; it is less resistant to oxidation at elevated temperatures than porous graphite.

carbon, powdered activated. Activated carbon in particle sizes predominately smaller than 80 mesh.

carbon refractory. A refractory product composed substantially or entirely of carbon or graphite, or both. Used in crucibles, stopper nozzles in steel-making furnaces, etc.

carbon refractory, anthracite-coal base. A commercial refractory product made substantially from calcined anthracite coal.

carbon refractory, graphite base. A commercial refractory product composed essentially of graphite.

carbon refractory, metallurgical-coke base. A commercial refractory product made substantially from metallurgical coke.

carbon refractory, petroleum-coke base. A commercial refractory product made substantially from calcined petroleum coke.

carbon, retort. (1) A dense carbon or graphite; formed in the upper parts of the retorts used in coal-gas manufacture.

(2) Carbon or graphite added to a glaze to produce localized reduction during firing.

carbon tetrachloride. CCl_4; a liquid employed as a solvent and cleaner in the degreasing of metals.

carborundum. A trade name for abrasives, refractories, and similar products of silicon carbide, fused alumina, and other materials; employed as abrasive grains and powders for cutting, grinding, and polishing, grinding wheels and stones, rubbing bricks, coated abrasives, tiles, antislip tiles and treads, refractory grains, and as a semiconductor.

carborundum stone. A silicon carbide whetstone used to remove pinpoints, and other imperfections from ware.

carboxymethylcellulose. CMC; employed as a binder, thickener, and suspension agent in porcelain-enamel and glaze slips.

carboy. A large, specially cushioned glass container of 5 to 15 gallon capacity for liquids, especially acids.

carburetor. A refractory-lined apparatus or chamber in which oils are vaporized, cracked, and enriched in the manufacture of carbureted water gas.

car dryer. A dryer in or through which ware is transported on cars.

car, kiln. A movable truck or carriage with one or more platforms on which ware is stacked for transport through a kiln.

carnegieite. $NaAlSiO_4$; an artificial mineral similar to feldspar; mol. wt. 142.03; m.p. 1526°C (2779°F).

carrageen. An Irish moss from which a syrup is made for use as a siccative or suspension agent for glazes and other slips.

carrier. A substance to which a trace element has been

added and which will carry the trace element through a desired chemical or physical process for a particular purpose.

carrier distillation, spectrochemical. An emission spectrographic technique in which a carrier material is added to a sample to facilitate the vaporization of a sample or fractional distillation of a sample.

carrier fluid. The fluid in which fluorescent and nonfluorescent magnetic particles or other active materials are suspended to facilitate their application for testing purposes.

carrier gas. (1) An inert gas that is used to sweep gaseous products through an analysis system, but not be included in the analysis.

(2) The gas which transmits powder from one point to another, as from a spray gun.

carry-in. Manual loading of a lehr.

cartoon. A drawing or sketch used as a model for a product.

car top. The refractory surface of a tunnel-kiln car.

car tunnel kiln. A long kiln, with the firing zone located near the center, through which ware is transported by means of kiln cars.

cascade. The downward flow of particles over one another in a manner resembling a waterfall.

cascade pulverizer. An apparatus in which crushing and grinding is accomplished by the tumbling action of large lumps of a material on other particles of the same material.

case. The outer layer of a substance which is substantially harder than its core.

cased glass. (1) Glassware having a surface composition different from the glass body.

(2) Glass composed of two or more layers of different colors.

case hardening. A process of hardening a substance so that the surface layer or "case" is made substantially harder than the interior or "core."

casement wall. (1) The entire side wall of a furnace between the flux block and the crown, excluding the ends.

(2) A refractory wall between pillars of a pot furnace situated in front of or surrounding the front of a pot.

case mold. A mold replica of an original model used to make a working mold.

casserole. A lidded cooking dish of glass, pottery, etc.

cassiterite. SnO_2; mol. wt. 150.7; sp. gr. 6.8–7.1; hardness (Mohs) 6–7; the mineral from which tin oxide is derived.

cassius purple. A precipitated pigment, obtained by mixing the chlorides of gold and tin; used in glazes at low and medium firing temperatures.

cast. To form a liquid or plastic mass into a specific shape by setting or by cooling in a mold; or an object so formed.

castable. A combination of refractory grain and a suitable bonding agent which, after the addition of a proper liquid, is usually poured or sprayed into place to form a refractory shape or structure which becomes rigid by chemical action; used in the construction and repair of furnaces, cupolas, and similar applications.

castable refractory. A hydraulic-setting refractory suitable for casting into shapes and usually bonded with aluminous cement.

casting. (1) A process of shaping glass by pouring the molten material into or onto molds, tables, or rolls.

(2) The forming of ceramic ware by pouring a body slip into a porous mold which absorbs sufficient water from the slip to produce a semirigid article.

(3) The process of pouring a molten substance into a suitable mold and allowing it to solidify.

(4) An item produced by a casting process.

casting, drain. The forming of ceramic ware by pouring a body slip into an open porous mold, and then draining off the remaining slip when the cast has attained the desired thickness.

casting, fusion (electrocasting). The forming of electrically fused refractory compositions by casting the molten batch into suitable molds, followed by carefully controlled cooling, to produce blocks and shapes having a high degree of crystallinity and density.

casting, hollow. A synonym for drain casting, which see.

casting, investment. The process of forming relatively small items to close tolerances in a fired refractory mold formed from a wax pattern, the wax subsequently being removed by melting. See investment casting.

casting ladle. A refractory-lined steel ladle used to transport molten steel from one location to another, and from which molten steel is poured into molds.

casting plaster. A white gypsum product used in making castings and carvings.

casting, pressure. The process in which slip-cast bodies are subjected to relatively high pressure to minimize drying shrinkage and to speed the rate of production.

casting, refractory. The process in which molten materials are cast at elevated temperatures in fireclay or other refractory molds.

casting slip. A slurry of properly formulated ceramic bodies which are shaped by pouring into appropriate molds.

casting, slip. The process in which a slip is poured into a porous mold which rapidly absorbs water, leaving a body having the inside shape of the mold. See slip casting.

casting solid. The process of forming ceramic ware by introducing a body slip into a two-section porous mold, the outer section forming the outside contour and the inner section forming the inner contour of the ware, allowing a solid cast to form between the two mold faces.

casting spot. A surface defect appearing as a discolored, vitrified spot on the surface of cast pottery, the defect frequently being formed when improperly deflocculated clay makes contact with the mold.

casting strain. Strains which are developed in a cast body during cooling.

casting stress. Stresses which develop in a casting as a result of shrinkage.

casting, vacuum. See vacuum casting.

casting, wet-ground hollow. A synonym for drain casting, which see.

cast iron. Any iron-carbon alloy that contains more than 1.7% carbon, usually between 2.0 and 4.0%.

cast iron, alloy. A cast iron of improved strength, resistance to scaling, and resistance to wear corrosion; usually contains additions of 0.1 to 5.0% of chromium, copper, molybdenum, nickel, and other alloying elements.

cast-iron enamel. A porcelain enamel compounded specifically for use on cast iron.

cast-iron enameling. See dry process enameling; wet-process porcelain enameling.

cast iron, gray. Cast iron in which the carbon is present essentially in an uncombined state.

cast iron, malleable. Cast iron in which the carbon content is reduced by annealing after solidification.

cast iron, white. A silvery cast iron of low-silicon content in which the carbon is chemically combined completely with the iron by sudden chilling while in the molten state.

cast stone. A molded concrete building block shaped to resemble natural stone.

catalysis. The change in the rate of a chemical reaction brought about by the presence of a substance which itself is unchanged at the completion of the reaction.

catalyst. A substance which, by its presence, will change the rate of a chemical reaction but which itself will be unchanged in composition or quantity after the reaction is completed.

catalytic. Thermal decomposition.

cataphoresis. The movement of suspended particles through a fluid by an electromotive force.

catch basin. A reservoir in which water from a process is drained to permit solids to settle for subsequent recovery or disposal.

catenary arch. A sprung-type arch in the form of an inverted catenary, the curve formed by a chain suspended from two points of equal height, the resultant arch exhibiting minimal stresses.

cat eye. An imperfection in glass consisting of an elongated bubble containing a particle of foreign matter.

cathedral glass. An unpolished, translucent sheet glass, usually formed by rolling, with one surface sometimes textured.

cathode. (1) The negative terminal of an electrical system.
(2) The negative terminal of a diode biased in the forward direction.
(3) The primary source of electrons in an electron tube.
(4) The positively charged pole of a storage battery or primary cell.

cathode arc. An arc occurring when the contact spacing exceeds a certain critical value, depending on the contact material and current. Material transfer is from cathode to anode.

cathode drop, cathode fall. The potential difference between the cathode and the electric discharge plasma.

cathode material transfer. The movement of contact metal from the cathode by means of a cathode arc.

cation. A positively charged ion; the ion in an electrolyte that migrates to the cathode.

cation adsorption. The adsorption of cations either on basal surfaces where negative charges, possibly as a result of isomorphous replacements within the lattice, or adsorption on prism surfaces where unsatisfied negative bonds may occur, or both; basal surface adsorption predominates in three-layer clays, while edge adsorption predominates in kaolin clays.

cation exchange. A surface property exhibited by colloidal inorganic materials, such as clays, whereby surface ions are replaced by other ions present in the surrounding medium.

cation exchange capacity. A measure of the ability of a substance, such as clay, to adsorb or exchange cations, usually expressed in terms of milliequivalents of cations per 100 grams of dry substance.

cationic. Having a positive charge which moves toward a cathode in an electrolized solution.

cat scratch. A surface imperfection on glassware consisting of marks resembling a scratch by the claws of a cat.

Cauchy light-dispersion formula. The index of refraction of a medium (n) as a function of wavelength (λ) is expressed by the equation $n = A + (B/\lambda^2)$ in which A and B are constants.

cauliflower ware. Cream-colored ware molded to resemble the appearance and surface configuration of a cauliflower.

caulking. A material used to make a seam or point airtight, watertight, or steamtight by forcing a suitable material or compound into the area.

caustic lime. Calcium hydroxide, $Ca(OH)_2$; mol. wt. 58.1; loses water at 580°C (1076°F); sp. gr. 2.34; used in mortars, plasters, and cements.

caustic potash. KOH; mol. wt. 56.1; m.p. 360°C (680°F); sp. gr. 2.044.

caustic soda. NaOH; mol. wt. 40.0; m.p. 318°C (604°F); sp. gr. 2.13.

cave. A pit under a glass furnace where the fire is located.

cavitation. Pitting or erosion of concrete, as when exposed to high-velocity turbulent flow of water.

cavity wall. A wall constructed in two adjacent sections with an air space between to provide thermal insulation.

C/B ratio. The ratio of the weight of water absorbed by a masonry unit during immersion in cold water to the weight absorbed during immersion in boiling water; an indication of the probable resistance of brick and similar fireclay products to freezing and thawing.

celadon, celadon glaze. A grayish-green, semiopaque glaze fired in a reducing atmosphere in which reduced iron is the colorant.

celeste blue. (1) Any of a number of iron-blue pigments, usually containing a considerable quantity of extender, such as barytes.

(2) A cobalt-blue pigment softened by additions of zinc oxide.

celestite. $SrSO_4$; mol. wt. 183.7; decomposes at 1580°C (2876°F); sp. gr. 3.95; hardness (Mohs) 3–3.5; used to impart iridescence on pottery glazes and glass, and as a fining agent in crystal glass. Also known as celestine.

celite. (1) Diatomaceous earth and products of similar composition composed essentially of silica 92.7, alumina 3.8, ferric oxide 1.4, lime and magnesia 1.0, and potash and soda 0.9%, and which is used as an ingredient in cements and as an abrasive in glass and metal polishing.

(2) A solid-solution constituent in portland cement clinker composed of $4CaO{\cdot}Al_2O_3{\cdot}Fe_2O_3$ and $6CaO{\cdot}2Al_2O_3{\cdot}Fe_2O_3$.

cell. A hollow space enclosed in a hollow-clay building block or similar structure having a minimum dimension of not less than ½ inch and a cross-section area of not less than 1 inch^2.

cell furnace. A glass-tank furnace in which the glass in the melting and auxiliary zones is heated electrically.

cellular concrete. A concrete of reduced density and increased insulating properties prepared by the addition of substances which, by chemical reaction, cause the concrete to foam, entrapping gases in the concrete mass.

cellular glass. A foamed glass block or sheet made from a mixture of powdered glass and a gas-forming material heated to the flow temperature of the glass. Also known as foamed glass.

cellulose gum. Sodium carboxymethylcellulose (CMC); a synthetic gum used in whiteware bodies and glazes as a thickener and binder to improve the green strength.

cellulose nitrate. $C_6H_7O_5(NO_2)_3$: sometimes employed as a binder in conductive and other coatings.

celsian. $BaO{\cdot}Al_2O_3{\cdot}2SiO_2$; mol. wt. 375.42; m.p. 1780°C (3236°F); a barium feldspar sometimes used in refractories for electric furnaces and kilns.

Celsius. A temperature scale in which 0° is the freezing point and 100° is the boiling point of water; a synonym for the Centigrade scale of temperature measurement.

cement. (1) A generic term for plastic materials having adhesive and cohesive properties and which will harden in place.

(2) A powder produced from a calcined mixture of clay and limestone which, when mixed with water, forms a paste that hardens into a stone-like mass, and which is the bonding medium in mortar and concrete.

cement aeration. The effect of atmosphere, particularly moist air and carbon dioxide, on the storage characteristics and subsequent setting properties of portland cement.

cement, air-setting. Any plastic cementitious material which will set in air under normal conditions of temperature and pressure.

cement, aluminous. A slow-setting, rapid-hardening cement consisting of alumina (40%), lime (40%), silica (10%), and impurities (10%); sp. gr. 3.0; will set in 24 hours, and is used as a heat-resisting cement.

cement brick. A molded brick of cement and sand formed under pressure and steam cured at 93°C (200°F); used as backing brick.

cement content. The amount of cement in one cubic yard of concrete, expressed in pounds, sacks, or barrels.

cemented carbide. A cemented or compacted mixture of the carbides of the heavy metals, such as tantalum and tungsten, bound together by a low-melting metal, such as cobalt; used in abrasive products, machining and cutting tools, drills, sandblast nozzles, wear-resistant machine parts, tire studs, hard-facing welding rods, etc., because of toughness, shock resistance, compressive strength, and thermal conductivity.

cement factor. The cement content of concrete.

cement, fireclay. A mixture of dry fireclay and water glass used in the repair of saggers, kiln cracks, and similar damage.

cement, gaize. A cement composed of finely ground mixtures of a pozzolanic material and hydrated lime or portland cement.

cement, grappier. A cement made of finely ground lime which has been overburned or underburned.

cement, gravel. Gravel consolidated by clay, calcite, silica, or other material.

cement gun. (1) A mechanical device employed to place mortar or cement in selected areas.

(2) A machine designed to mix, wet, and apply refractory mortars in the walls of hot furnaces and kilns.

cement, high-alumina. (1) A hydraulic cement produced by calcining a mixture of bauxite and limestone.

(2) A hydraulic, refractory cement of high alumina content.

cement, high-early-strength. A variety of portland cement having a high lime to silica ratio which hardens more quickly with the evolution of more heat than normal portland cement. See high-early-strength cement.

cement, high-temperature. A refractory cement which will not soften, fuse, or spall at elevated temperatures.

cement, hydraulic. A cement that sets and hardens by chemical interaction with water; some will set under water.

cement, insulating. A cement in which a substantial quantity of insulating material such as asbestos has been incorporated, or a lightweight concrete of relatively low density, for use as thermal insulation and fire protection in structural applications.

cement, iron ore. A cement in which iron ore, Fe_2O_3, is employed as a replacement for clay, shale, or alumina; sp. gr. about 3.31; more resistant to some corrosive environments, particularly seawater, than portland cement.

cementite. A hard, brittle iron carbide, Fe_3C, which will scratch feldspar and glass, but not quartz; found in certain steels, cast iron, and iron-carbon alloys.

cementitious material. Any material to which a liquid may be added to form a paste having adhesive and cohesive properties and which subsequently will harden into a solid mass.

cement, Keene's cement. An anhydrous calcined or dead-burned gypsum containing an accelerator, such as potassium sulfate; employed as a hard-finish plaster.

cement kiln. A rotary kiln in which limestone and other ingredients are calcined to produce portland cement.

cement-kiln head. The head of the burner and the discharge end of a rotary cement kiln.

cement, Kühl. A hydraulic cement somewhat similar to portland cement but containing 7% or more each of ferric oxide and alumina and substantially less silica.

cement, lap. A cementitious material employed to seal the side and end laps of corrugated roofing.

cement, lime-slag. A cement produced from a mixture of lime and granulated blast-furnace slag.

cement, low-heat. A cement which sets with the evolution of much less heat, and which contains a higher percentage of dicalcium silicate and aluminoferrite and a lower percentage of tricalcium silicate and tricalcium aluminate than conventional portland cement.

cement, magnesia. Magnesium oxychloride cement produced by adding a magnesium chloride solution to magnesium oxide.

cement, masonry. A hydraulic mortar composed of one or more pozzolanic materials such as portland cement, natural cement, slag cement, hydraulic lime, portland-pozzolan cement, or portland blast-furnace slag cement plus selected additions of specially prepared materials, such as hydrated lime, limestone, chalk, talc, slag, clay, or calcareous shell; used in masonry construction.

cement mill. A mill in which rock is pulverized to powder form for use primarily in the production of cement.

cement mortar. A plastic mixture consisting of one part of portland cement, four parts of sand, and a small amount of lime, all blended in water.

cement, natural. A hydraulic cement produced by calcining a naturally occurring argillaceous limestone at a temperature below its sintering point, followed by grinding to a fine powder.

cement, neat. A plastic mixture of portland cement and water, but without aggregate.

cement paint. A mixture of portland cement, filler, accelerator, water repellant, and water employed as a waterproof coating for concrete, brickwork, and other masonry surfaces.

cement, parian. A gypsum cement to which borax is added to produce a hard finish.

cement paste. A plastic mixture of portland cement and water.

cement, patching. A finely ground mixture of portland cement and other ingredients, sometimes containing additions of organic binders, which become plastic but trowable when tempered with water; used in the repair of damaged concrete and mortars. See patching cement (2).

cement plaster. A gypsum plaster used in mortar for plastering interior surfaces.

cement, plastic. A pliant cement used in the patching of holes, cracks, etc.

cement, portland. A hydraulic cement prepared by calcining powdered mixtures of hydraulic calcium silicates, followed by pulverizing the resultant clinker, together with additions of gypsum or other forms of calcium sulfate, to a fine powder.

cement, portland blast-furnace slag. A pozzolanic cement consisting of an intimately mixed blend of granulated blast-furnace slag and portland cement in various specified proportions.

cement, portland-pozzolan. A cement consisting of an intimate and uniform blend of portland cement or portland-cement clinker, pozzolan, and portland blast-furnace slag cement in various proportions.

cement, Potter's red. A cement consisting of an intimate mixture of portland cement and a powdered or granulated fired red clay; used in decorative applications.

cement pump. A device designed to move plastic concrete from one location to another.

cement, refractory. A finely ground mixture of refractory ingredients which becomes plastic and trowable when tempered with water, and which is used in laying and bonding refractory brick.

cement, refractory patching. A finely ground mixture of refractory ingredients which becomes plastic and trowable when tempered with water, and which is used in the repair of damaged areas in furnaces, kilns, glass tanks, and refractory molds.

cement rock. An argillaceous limestone containing lime, silica, alumina, and magnesia used in the manufacture of portland cement.

cement, silica. A refractory mortar composed of a finely ground mixture of quartzite, silica brick, and fireclay in various proportions.

cement silo. A large structure or silo in which dry, bulk, powdered cement is stored for subsequent use.

cement slag. A hydraulic cement composed of an intimate and uniform blend of granulated blast-furnace slag and hydrated lime, the slag content being greater than a specified minimum.

cement, slaters. A water-resistant caulking compound, usually gray in color, used to cover bolt heads, side and end laps of corrugated roofing, and other exposed areas.

cement, soil. A mixture of soil, cement, and water employed to adjust the engineering properties of the soil.

cement, Sorel. A magnesium oxychloride cement consisting of a mixture of magnesium chloride and magnesium oxide, which presumably reacts with water to form magnesium oxychloride, and may include fillers such as sawdust, wood flour, talc, sand, powdered stone, cork, or powdered metals.

cement, sulfoaluminate. A hydraulic cement consisting of a uniform mixture of gypsum and high-alumina cement.

cement, tarras. A volcanic tuff having pozzolanic properties; used as a hydraulic cement. Also known as trass.

cement, waterproof. A cement containing a hydrophobic ingredient to repel water, or to which an impervious coating has been applied to reduce its permeability.

cement, white. An essentially iron-free cement containing opacifying fillers such as china clay, chalk, and the like.

center. To force a ball of clay into a centered position on a potter's wheel.

center brick. A special, hollow, refractory shape with an opening at the top and along the sides through which bottom-poured molten steel is directed from guide tubes to ingot molds.

center-hole lapping. The cleaning and finishing of center holes by lapping with abrasive grains.

centering. The operation on lens elements wherein the element is optically lined up with the axis of rotation, and the edges are ground concentric with the optical axis.

centerless grinding. Grinding the inside or outside diameter of a cylindrical piece which is supported on a work blade instead of being held between centers and which is rotated by a so-called regulating wheel.

center-reinforced grinding wheel. A grinding wheel in which steel rings have been incorporated near the center to provide additional strength.

centers. Conical steel pins of a grinding machine on which a work piece is centered and rotated during grinding.

Centigrade. A temperature scale in which 0° is the freezing point and 100° is the boiling point of water; a synonym for the Celsius scale of temperature measurement.

central-mixed concrete. Concrete that is mixed in a stationary mixer in a central plant and then delivered in agitators to the job site.

centrifugal casting. The process of casting bodies in rapidly spinning molds as a means of producing bodies and shapes of high density.

centrifugal pipe. Concrete pipe manufactured by spinning the concrete mix in a horizontal form, forcing the concrete to the interior rim of the form by centrifugal force.

centrifuge. A mechanical device rotating at very high speed employed to separate particles of varying densities.

cera magnet. A ferrimagnet composed of $BaO \cdot 6Fe_2O_3$.

ceramel. A term sometimes used to identify mixtures of ceramics and metals.

ceramic. (1) Any of a class of inorganic, nonmetallic products which are subjected to a temperature of 540°C (1000°F) and above during manufacture or use, including metallic oxides, borides, carbides, or nitrides, and mixtures or compounds of such materials.

(2) Pertaining to ceramics.

(3) Pertaining to the manufacture or use of ceramic processes, articles, materials, technology, and science.

ceramic aggregate. Concrete containing porous clay or lumps of ceramic materials.

ceramic amplifier. An amplifier using piezoelectric semiconductors of ceramic compositions.

ceramic article. An article having a glazed or unglazed body of crystalline or partly crystalline structure, or of glass, which body is produced essentially from inorganic, nonmetallic materials and either is formed from a molten mass which solidifies on cooling, or is formed and simultaneously or subsequently matured by the action of heat.

ceramic bond. The mechanical and physical strength developed in a ceramic body by a heat treatment which causes the adhesion and cohesion of adjacent particles.

ceramic capacitor. A capacitor whose dielectric is a ceramic material.

ceramic-carbon refractory. A manufactured refractory composed of carbon, including graphite, and one or more ceramic materials, such as fireclay and silicon carbide.

ceramic cartridge. A piezoelectric ceramic used in microphones, phonograph pickups, and similar elements.

ceramic coating. An inorganic, nonmetallic coating bonded to a substrate.

ceramic coating on metal. An inorganic, nonmetallic, protective coating bonded to a metallic substrate and suitable for use at or above a red heat.

ceramic colorant. An inorganic material employed to impart color to a porcelain enamel, glaze, glass, or ceramic body.

ceramic color glaze. An opaque colored glass of satiny or glossy finish obtained by spraying a clay body with a compound or mixture of metallic oxides, chemicals, and clays which is fired at a sufficiently high temperature to form a fused coating inseparable from the body.

ceramic fiber. A filament formed of a ceramic material for use in light weight units for electrical, thermal, and sound insulation, high-temperature filtration, reinforcement, and packing.

ceramic filter. (1) A fired ceramic of selected porosity through which a fluid is passed to separate out matter in suspension.

(2) A ceramic or glass composition employed to suppress waves or oscillations of certain frequencies.

ceramic fuel elements. Uranium oxide, plutonium oxide, boron compounds, rare earth oxides, etc., employed to form fuel rods for nuclear reactors.

ceramic, glass. See glass ceramic.

ceramic glaze. A ceramic coating, glossy or matte, matured to a glassy state on a formed ceramic article, or the material or composition from which the coating is made.

ceramic ink. An ink consisting of a ceramic pigment suspended in a liquid medium, the pigment developing its color on firing.

ceramic magnet. A permanent magnet made from pressed and sintered mixtures of ceramic and magnetic powders such as barium ferrite, lead ferrite, strontium ferrite, magnesium ferrite, etc. See magnetic ceramics.

ceramic-metal coating. A mixture of one or more ceramic materials in combination with a metallic phase which may be applied to a metal or nonmetallic substrate; the coating may or may not require heat treatment prior to service.

ceramic, metallized. A ceramic to which a thin metallic coating has been deposited, usually to facilitate the making of ceramic-to-metal seals.

ceramic microphone. A microphone in which a piezoelectric cartridge is employed.

ceramic-mold casting. A precision casting process in which carbon, low-alloy, and stainless steels are formed in a ceramic mold and fired at a high temperature.

ceramic mosaic tile. An unglazed tile formed either by dust pressing or by plastic forming to ¼ to ⅜ inch in thickness having a facial area of less than 6 square inches; the tiles usually are mounted on paper sheets approximately 2 × 1 ft. to facilitate setting. They may be of clay or porcelain.

ceramic nuclear fuel. A fissionable material mixed with a ceramic and a high-temperature metal or alloy to obtain refractory and damage-resistant properties for use in nuclear reactors.

ceramic paste. A sometimes used synonym for ceramic body.

ceramic pickup. A phonograph pick-up employing a ceramic cartridge.

ceramic, polarized. A substance, such as barium titanate, having high electrochemical efficiency; used as a transducer in ultrasonic systems.

ceramic process. The production of articles or coatings from ceramic materials, the article or coating being made permanent and suitable for its intended use by the action of heat at temperatures sufficient to cause sintering, solid-state reactions, bonding, or by whole or partial conversion to the glassy state.

ceramic reactor. A nuclear reactor in which the fuel and moderator assemblies are formed from ceramic materials, such as metal oxides, carbides, or nitrides.

ceramic-rod flame spraying. A process in which a ceramic coating is applied to a surface by means of a high-temperature gun which atomizes a ceramic rod, delivering and bonding the ceramic to the substrate by an air blast.

ceramics. A general term applied to the art or technique of producing articles by a ceramic process, or to articles so produced.

ceramics, oxide. Ceramics made by dry-pressing or slip-casting essentially pure oxides, such as alumina, beryllia, magnesia, thoria, and zirconia, followed by sintering at high temperatures.

ceramics, solution. A ceramic coating consisting of a decomposable metal salt, porcelain enamel, or thermoplastic resin which is applied and matured on a hot surface, the resulting coating exhibiting high resistance to thermal shock.

ceramic tile. A ceramic surfacing unit, usually relatively thin, composed of a clay body or a body composed of a mixture of ceramic materials, and having a glazed or unglazed face, which is fired at a temperature sufficiently high to produce desired physical properties and other characteristics; used principally for decorative effects and sanitary purposes as well.

ceramic-to-metal seal. An air-tight seal between a ceramic composition and a metal which provides lead-through contacts for electrical and electronic components for use in high-temperature and nuclear environments.

ceramic tools. Cutting tools made from sintered or hot-pressed metal oxides often containing additives to promote sintering.

ceramic transducer. A transducer of ceramic composition which depends on the production of an elastic strain in certain symmetric crystals when an electric field is applied, or which produces a voltage when the crystal is deformed.

ceramic tube. An electron tube having a ceramic envelope capable of withstanding operating temperatures in excess of 500°C (932°F).

ceramic veneer. Thin sections of a ceramic veneer held in place by the adhesion of a mortar to the unit and backing, or thick sections of veneer held in place by grout and wire anchors connected to the backing wall.

ceramic whiteware. A fired ware consisting of a glazed or unglazed body which is usually white and of fine texture, such as china, porcelain, semivitreous ware, earthenware, spark plugs, sanitary ware, and the like.

ceramoplastic. A ceramic insulating material made by bonding synthetic mica and glass.

cereal binder, cereal flour. A finely milled flour used as a binder for core mixtures in a casting process.

cerium aluminate. $CeO \cdot Al_2O_3$; mol. wt. 258.15; m.p. 2704°C (3765°F); sp. gr. 6.17.

cerium bismuthide. (1) CeBi; mol. wt. 349.25; m.p. 1527°C (2780°F). (2) Ce_4Bi_3; mol. wt. 1188.0; m.p. 1604°C (2970°F).

cerium boride. (1) CeB_4; mol. wt. 183.53; sp. gr. 5.74. (2) CeB_6; mol. wt. 205.17; m.p. 2190°C (3975°F); sp. gr. 4.82; hardness (Vickers) 3140.

cerium carbide. (1) Ce_2C_3; mol. wt. 316.50; sp. gr. 6.97. (2) CeC_2; mol. wt. 164.25; m.p. 2538°C (4600°F); sp. gr. 5.56.

cerium chromite. CeO_2; mol. wt. 324.27; m.p. 2438°C (4420°F).

cerium fluoride. CeF_3; mol. wt. 197.13; m.p. 1460°C (2660°F); sp. gr. 6.16; used in arc carbons to increase brilliance.

cerium hydroxide. $Ce(OH)_3$; mol. wt. 191.13; used as an opacifier in porcelain enamels and glazes, and as a yellow colorant in glass.

cerium nitrate. $Ce(NO_3)_3 \cdot 6H_2O$; mol. wt. 434.28; loses $3H_2O$ at 150°C (302°F); decomposes at 200°C (392°F); used in gas mantles.

cerium nitride. CeN; mol. wt. 154.26; sp. gr. 8.09.

cerium oxide. (1) CeO_2; mol. wt. 172.13; m.p. 1950°C (3542°F); sp. gr. 7.65; used as an opacifier in porcelain enamels, as a decolorizer and brightener in glass, and as a polishing agent for glass, marble, and optical surfaces; produces yellow color in glass when used with titania. (2) Ce_2O_3; mol. wt. 328.5; m.p. 2040°C (3705°F).

cerium phosphide. CeP; mol. wt. 171.28; sp. gr. 5.56.

cerium ruthenium. $CeRu_2$; mol. wt. 343.65; m.p. 1538°–1571°C (2800°–2860°F).

cerium selenide. (1) Ce_2Se_3; mol. wt. 518.10; m.p. 1593°–2051°C (2900°–3720°F). (2) CeSe; mol. wt. 219.45; m.p. 1816°C (3300°F).

cerium silicide. $CeSi_2$; mol. wt. 196.37; sp. gr. 5.41.

cerium sulfide. CeS, Ce_2S_3, and Ce_3S_4; mol. wts. 172.19, 376.44, and 557.63, respectively; m.p. 2450°C (4442°F), 1890°C (3434°F); and 2050°C (3722°F), respectively; used for metallurgical melting crucibles for their high chemical and thermal resistance properties.

cerium telluride. (1) Ce_2Te_3; mol. wt. 663.0; m.p. 1666°–1800°C (3030°–3070°F). (2) CeTe; mol. wt. 267.75; m.p. 1587°–1888°C (2890°–3430°F).

cerium vanadate. $2CeO_2 \cdot V_2O_4$; mol. wt. 510.42; m.p. 1832°C (3330°F).

cermet. A heterogeneous body composed of two or more intimately mixed but separable phases, of which at least one is ceramic and the other metallic, combining strength and toughness of metal with the thermal resistance of the ceramic; formed by mixing, pressing, and sintering; used in rocket motors, gas turbines, turbojet engines, nuclear reactors, brake linings, etc., and other products requiring high-oxidation resistance at elevated temperatures.

cermet coating. A mixture of one or more ceramic materials with a metallic phase applied to a metallic or nonmetallic (such as graphite) substrate, and which may or may not require heat treatment prior to service.

cermet resistor. A resistor consisting of a metal and insulating materials fired on a ceramic substrate.

certificate of test. A written, printed or signed document attesting to the validity of a performed test.

cerulean blue. A light blue pigment composed of cobalt stannate, $CoO \cdot n(SnO_2)$.

cesium carbonate. Cs_2CO_3; mol. wt. 325.6; decomposes at 610°C (1130°F); used in specialty glasses.

cesium perchlorate. $CsClO_4$; mol. wt. 232.3; m.p. 250°C (482°F); sp. gr. 3.33; used in optics and specialty glasses.

chain, air. A defect consisting of a chain-like string of air or other gaseous inclusions in a glass or other vitrified ceramic.

chain conveyor. A conveyor consisting of one or two endless chains, equipped with appropriate hooks or crossbars for the movement of materials from one location to another.

chain marks. Marks made on the bottom of glass articles as they ride through a lehr on a slightly overheated chain belt.

chalbite (siderite). $FeCO_3$; mol. wt. 115.8; sp. gr. 3.7–3.9; hardness (Mohs) 3.5–4.5; used as a yellow to red colorant in ceramic bodies and glazes.

chalcedony. SiO_2; mol. wt. 60.1; sp. gr. 2.6–2.65; hardness (Mohs) 6.5–7.0; a cryptocrystalline form of silica.

chalcogenide glass. A glass containing sulfur, selenium, polonium, or tellurium, and which is used in glass switches.

chalk. $CaCO_3$; mol. wt. 100.1; decomposes at 825°C (1517°F); sp. gr. 2.7–2.95; employed as a source of lime or substitute for limestone in ceramic bodies, glazes, porcelain enamels, glass, cements, and polishing powders, and as a medium- and high-temperature flux.

chalkboard enamel. A porcelain enamel having a matte, slightly roughened surface on which writing with chalk may be done; sometimes called blackboard enamel.

chalked, chalky. A condition occurring on porcelain-enameled surfaces and glazes wherein the coating has lost its natural gloss and has become powdery; the powder may or may not be tightly adherent.

chamber, combustion. The area in a furnace or kiln in which the burning of fuel takes place.

chamber, drawing. The area in a glass tank from which sheets of flat glass are drawn.

chamber dryer. A dryer of one or more compartments into which freshly formed ware is placed and dried under reasonably controlled conditions of time, temperature, humidity, and air flow.

chamber kiln. A kiln consisting of one or more compartments into which ware is set on appropriate refractory shapes and fired.

chamber oven. A refractory-lined structure in which gas is produced primarily from coal.

chamotte. A grog produced by firing refractory clay for use as a nonplastic component in refractory compositions.

champlevé. A characteristic form of porcelain-enameled ware in which a design is engraved or carved into the surface of the base metal, frequently copper, gold, or other soft metal; thin, raised divider strips are carved or tacked in place to outline various features of the design; porcelain enamels of selected colors are placed in the various outlined sections and fired to produce pleasing artistic effects.

channel. The section of a forehearth which carries molten glass from the tank to the flow spout, and in which adjustments in temperature are made.

channel, feeder. See channel.

channeling. The greater flow of fluid through passages of lower resistance as may occur in fixed beds or columns of activated carbon or other granular materials due to nonuniform packing, irregular sizes and shapes of the particles, gas pockets, wall effects, or other causes.

charcoal. A porous solid produced by burning carbonaceous materials such as wood, peat, coal, and cellulosic materials in an absence of air to produce a product containing 85 to 90% of carbon.

charge. (1) The glass-forming mixture or batch ready for injection into a smelter or glass-melting tank.

(2) A load of ware placed in a furnace or kiln to be fired.

charger, batch. A mechanical device for introducing batch into a smelter or glass-melting tank.

charging. (1) The process of placing ware in a furnace or kiln.

(2) The process of introducing batch in a smelter or glass-melting tank.

Charlton photoceramic process. A photographic process in which a photosensitive emulsion is applied to a ceramic surface and exposed to a negative in such a manner as to produce a positive image which subsequently is fired on the item.

Charpy impact test. An impact test in which a freely swinging pendulum is permitted to strike and break a notched specimen laid loosely on a support; the position of the pendulum before release is compared with the position to which it swings after breaking the specimen.

chatter. An undesirable repetitive pattern created on the surface of a workpiece, usually at regularly spaced intervals, due to an out-of-round or out-of-balance condition in the abrasive machine.

chatter marks. Surface imperfections on work being ground, usually caused by vibrations transferred from the wheel-work interface during grinding.

check. An imperfection consisting of a crack in the surface of a glass article.

checkerboard. An open brickwork in a checkerboard regenerator allowing passage of hot spent gases. See checkerboard regenerator.

checkerboard regenerator, checkerwork. An open checkerboard arrangement of firebrick in a high-temperature chamber that absorbs heat during a processing cycle, and releases it to preheat fresh combustion air during the down cycle.

checker brick, checkers. Refractory shapes used in checkerboard regenerators, checking, which see.

checkers. (1) The firebrick, alternating with openings, in the chambers of a regenerative furnace to permit the flow of hot air to the combustion chamber.

(2) Regenerators constructed in such a fashion.

(3) The refractory pieces used in such a manner. Sometimes spelled chequers.

checkerwork, basketweave. An arrangement of checker brick such that the ends of each brick are placed at right angles to the center of the adjacent brick to form continuous vertical flues, the plane view resembling the weave of splints in a basket.

checking. (1) Cracking or crazing of ceramic bodies or glazes.

(2) Crazing or cracking of cast-iron porcelain-enamels resulting from cracks in the ground coat.

check, pressure. An imperfection consisting of a check or crack in a glass article as a result of the application of too much pressure during the forming operation.

cheeks. The refractory side walls of the ports of a fuel-fired furnace.

cheese hard. The degree of hardness of a freshly formed ceramic body at which the plastic shape may be handled without deformation.

chemical adsorption. The process by which an adsorbate is bound to the surface of a solid by forces having energy levels approximating those of a chemical bond. Also called chemisorption.

chemical assay. A chemical measurement of the quantity of one or more components in a material.

chemical brick. See chemical stoneware.

chemical calibration. The calibration of an instrument by a method based on chemical standards.

chemical durability. The physical and chemical lasting quality of a product in terms of chemical and physical changes in the product surface or changes in the contents of a vessel.

chemical etching. The formation of a characteristic surface texture when a polished glass surface is etched by suitable reagents.

chemical glass. A chemically durable glass suitable for use in laboratory and production equipment subjected to hostile materials or environments.

chemically bonded brick. Brick manufactured by processes in which mechanical strength is developed by chemical bonding agents instead of by firing.

chemically combined water. Water which is chemically a part of a clay mineral and can be released only upon dissociation of the clay at or about red heat.

chemically strengthened glass. Glass treated by an ion-exchange process to produce a surface layer of high-compressive stress.

chemical porcelain. Vitreous ceramic whiteware containers of high chemical and physical durability in which chemicals are contained, reacted, or transported.

chemical reprocessing. (1) The separation and recovery of the source and special nuclear material contained in irradiated reactor material, usually in the form of purified nitrate solution or oxides of uranium and plutonium and, in certain cases, thorium.

(2) The recovery of valuable components from used materials, wastes, and materials of low concentration by chemical processing.

chemical resistance. The ability of a product to resist chemical attack, decomposition, solution, or other chemical changes when in contact with gaseous, liquid, or solid substances encountered in service environments.

chemical-resistant concrete. A type of portland cement of high tetracalcium aluminoferrite and low tricalcium aluminate plus additions such as calcium soaps, water glass, and other materials which render the product resistant to chemicals.

chemical separation. The removal, isolation, or separation of a desired substance from the remainder of a sample by chemical techniques as opposed to physical or mechanical separations.

chemical stoneware. A ceramic product highly resistant to acids, alkalies, and other chemicals made essentially from lime- and iron-free clays, and of relatively low sand content; such bodies exhibit low-firing shrinkage, low-water absorption (0.4%), sp. gr. 2.2, ultimate tensile strength of 2000 psi, ultimate compressive strength of 80,000 psi, and a modulus of rupture of 5000 psi, the values being approximate.

chemisorption. The binding of an adsorbate to the surface of a solid by forces exhibiting energy levels approximating those of a chemical bond.

chequer. An alternative spelling of checker, which see.

chert. A fine-grained variety of silica or quartz.

chest knife. A tool for removing the moil from hand-blown glassware.

chilling. The rapid removal of heat from a body or product after firing by means of a cold-air blast, water spray, immersion in water, or other liquid, etc.

chill mark. A wrinkled surface on glassware as a result of uneven cooling during the forming operation.

chimney arch. An arch in the base of a chimney used to admit a flue.

china. A vitreous ceramic whiteware, glazed or unglazed, such as dinnerware, sanitary ware, artware, and other products of nontechnical use.

china, Belleek. A thin highly translucent whiteware of zero water absorption composed of a body containing a significant amount of frit and normally having a soft glaze of high luster.

china, bone. A soft translucent dinnerware made from a whiteware body containing a minimum of 25% of bone ash; an approximate composition is 50% bone ash, 25% china clay, and 25% Cornish stone.

china clay. A refractory clay consisting of minerals of the kaolin family which fires to a white or nearly white color. See kaolin.

china, frit. A thin translucent whiteware made from a body of zero water absorption containing a substantial quantity of frit of high fluxing properties; usually coated with a soft glaze.

china, ironstone. A durable dinnerware made from a fine, hard vitrified earthenware body of high strength.

china process. A process of manufacturing glazed dinnerware, sanitaryware, artware, and the like by which a ceramic body is fired to maturity, following which a glaze usually is applied and fired at a lower temperature.

china sanitaryware. Glazed, vitrified whiteware designed for sanitary functions.

china, semivitreous. A dinnerware or other ceramic product exhibiting a moderate degree of water absorption.

china stone. A weathered granitic-type stone sometimes used as a flux in pottery and earthenware bodies.

china, vitreous. A completely matured whiteware product characterized by low water absorption, usually of the order of 0.3%.

Chinese blue, Chinese cobalt. A black mineral aggregate containing hydrated oxides of manganese and cobalt used for underglaze porcelain blues.

Chinese red. Various red and orange colors produced by mixtures of lead chromate, $PbCrO_4$, and lead oxide, PbO, in different proportions. Also known as chrome red.

chip. (1) Breaking of a fragment from an otherwise regular surface, particularly along an edge or corner.

(2) A tiny semiconductor mounted on an appropriate substrate to form a diode, transistor, or similar device.

chipped glass. An intentionally chipped surface on a glass article.

chipping. (1) The process of removing thin sections or fins of extra glass from glass articles prior to grinding.

(2) A defect in porcelain-enameled ware in which fragments of the fired coating are broken away from the surface.

(3) Fragments unintentionally broken from a body, glaze, or glass.

chipping, spontaneous. A defect in porcelain-enameled and other surfaces characterized by chipping, frequently on a radius or edge, that occurs without apparent external causes. Also called spontaneous spalling.

chittering. Small ruptures occurring along the edges or rims of ceramic ware as a result of improper fettling.

chloroplatinic acid. $H_2PtCl_6 \cdot 6H_2O$; employed to produce pleasing gray color effects in the decoration of high-quality porcelains.

choke. An imperfection consisting of an insufficient opening in the neck of a glass container.

choke crushing. The grinding of materials in a roll crusher with the space between the rolls being completely filled with the material to gain the added effect of the particles acting on each other.

chroma. The purity of color determined by its degree of freedom from white or gray; color intensity.

chromate red. See chrome red.

chromatic. Of or relating to color.

chromatic value system, Adams. A system of color measurement based on lightness, the amount of red or green, and the amount of yellow or blue in the color being measured.

chromatography. The separation of complex solutions into chemically distinct layers by seepage or by percolating through a selectively adsorbing medium.

chrome. A general term for chrome-bearing pigments.

chrome alum. See chromium potassium sulfate.

chrome-alumina pink. A family of pink ceramic colors consisting of combinations of Cr_2O_3, Al_2O_3, and ZnO.

chrome brick. A refractory brick produced substantially or entirely of chrome ore. Frequently used as a substitute or replacement for magnesia brick in furnaces and kilns because of lower cost. Also, because of their high resistance to chemical reaction with both basic and acidic oxides at elevated temperatures; also used as a spacer between the silica-brick roofs and magnesia-brick walls of open-hearth and similar furnaces.

chrome cake. A green form of salt cake, Na_2SO_4, containing small amounts of chromium.

chrome glue. A glass cement or a waterproofing agent made by mixing glue with ammonium or potassium dichromate or with chrome alum.

chrome green. Any of various brilliant green ceramic colorants containing or consisting of chromium compounds.

chromel. Trade name for a series of nickel-chromium alloys, sometimes with additions of iron, used as thermocouples and load-bearing accessories in kilns and furnaces.

chrome-magnesite brick. A burned or unburned refractory brick consisting substantially of refractory chrome ore and dead-burned magnesite in which the chrome ore, by weight, is the predominant ingredient.

chrome orange. See chrome red.

chrome ore, refractory chrome ore. A refractory ore consisting essentially of chrome-bearing spinels with only minor amounts of accessory minerals, and with properties suitable for making refractory products.

chrome oxide green. A pigment consisting essentially of chromic oxide; made by burning sodium dichromate with a reducing agent; used in finishes for concrete surfaces; not to be confused with chrome green.

chrome red. Pigments containing varying proportions of $PbCrO_4$ and PbO to produce colors ranging from light orange to red.

chrome refractory. A refractory product made entirely of chrome ore, which see.

chrome spinel. A natural or synthetic oxide of magnesium, aluminum, and chromium used as a refractory.

chrome-tin pink. A glaze colorant consisting of chromic oxide and tin oxide in the presence of lime.

chrome yellow. A series of yellow pigments composed essentially of lead chromate or other lead compounds.

chrome-zircon pink. A glaze colorant similar to chrome-tin pink but with a substantial portion of the tin oxide replaced by zircon.

chromic acid, chromic oxide. Cr_2O_3; mol. wt. 152.02; m.p. 1990°C (3614°F); sp. gr. 5.04; used as a blue, brown, green, yellow, etc. colorant in glazes, porcelain enamels, and glass.

chromite. A mineral composed of the oxides of chromium, iron, aluminum, and magnesium; used in refractories and pigments.

chromium. Cr; at. wt. 52.01; m.p. 1900°C (3452°F); b.p. 2200°C (3992°F); sp. gr. 7.1; used in some ferrite compositions.

chromium aluminide. CrAl; mol. wt. 79.0; m.p. 2160°C (3920°F); coefficient of thermal expansion 10×10^{-6}; excellent resistance to oxidation.

chromium beryllide. CrBe; mol. wt. 70.50; m.p. 1838°C (3340°F); sp. gr. 4.34; hardness (Vickers) 1290.

chromium boride. (1) CrB_2; mol. wt. 72.6; m.p. 2760°C (5000°F); sp. gr. 5.6; tensile strength 106,000 psi; thermal expansion 4.6×10^{-6}; poor resistance to oxidation and thermal shock at high temperatures, (2) Cr_4B; mol. wt. 218.86; m.p. 1649°C (3000°F); sp. gr. 6.24. (3) Cr_2B; mol. wt. 114.84; m.p. 1832°C (3300°F); sp. gr. 6.53. (4) Cr_5B_3; mol. wt. 292.5; m.p. 1899°C (3450°F); sp. gr. 6.12. (5) CrB; mol. wt. 62.83; m.p. 1999°C (3630°F); sp. gr. 6.11. (6) Cr_3B_4; mol. wt. 198.41; m.p. 1927°C (3500°F); sp. gr. 5.76. (7) Cr_2B_5; mol. wt. 158.12; m.p. 1999°C (3630°F).

chromium carbide. (1) Cr_3C_2; mol. wt. 180.03; m.p. 1890°C (3434°F); sp. gr. 6.88; employed in bearings, seals, valve seats, orifices, and chemical equipment. (2) Cr_4C; mol. wt. 220.04; m.p. 1521°C (2770°F); sp. gr. 6.99. (3) Cr_7C_3; mol. wt. 400.07; m.p. 1779°C (3235°F).

chromium hafnium. Cr_2Hf; mol. wt. 386.64; m.p. 1483°C (2700°F).

chromium niobium. Cr_2Nb; mol. wt. 301.14; m.p. 1483°–1710°C (2700°–3110°F).

chromium nitride. CrN; mol. wt. 66.02; decomposes at 1500°C (2732°F); sp. gr. 6.1; hardness approx. 1100 Vickers.

chromium oxide. Cr_2O_3; mol. wt. 152.02; m.p. 1990°C (3614°F); sp. gr. 5.04; hardness (Mohs) 9.0; used primarily as a green, pink, and red colorant in glass, glazes, and porcelain enamels, and as an ingredient in some refractory bricks.

chromium phosphide. (1) CrP_2; mol. wt. 114.06; sp. gr. 4.5. (2) CrP; mol. wt. 83.04; m.p. 1359°C (2480°F); sp. gr. 5.49; hardness 632 Vickers. (3) Cr_2P; mol. wt. 239.06. (4) Cr_3P; mol. wt. 187.06; sp. gr. 6.51.

chromium platinum. Cr_3Pt; mol. wt. 351.26; m.p. 1499°C (2730°F).

chromium potassium sulfate. $CrK(SO_4)_2 \cdot 12H_2O$; mol. wt. 499.23; employed as a red or green ceramic colorant.

chromium ruthenium. $CrRu_2$; mol. wt. 255.41; m.p. 1538°–1571°C (2800°–2860°F).

chromium silicide. (1) Cr_3Si; mol. wt. 184.09; m.p. 1710°C (3110°F); sp. gr. 6.45; hardness (Vickers) 1000. (2) Cr_3Si_2; mol. wt. 212.15; m.p. 1560°C (2840°F); sp. gr. 5.60; hardness (Vickers) 1280. (3) CrSi; mol. wt. 80.07; m.p. 1543°C (2810°F); sp. gr. 5.43; hardness (Vickers) 1000. (4) Cr_2Si_3; mol. wt. 292.22. (5) $CrSi_2$; mol. wt. 108.13; m.p. 1538°C (2800°F); sp. gr. 5.00; hardness (Vickers) 1000–1600. (6) Cr_2Si_7; mol. wt. 404.46. Bodies exhibit moderate strength, excellent oxidation resistance, good resistance to thermal shock, poor resistance to impact loading; used in wear-resistant components for high-temperature applications.

chromium sulfate. $Cr_2(SO_4)_3$; mol. wt. 392.2; sp. gr. 3.01; used in ceramic glazes as a green colorant.

chromium sulfide. (1) CrS; mol. wt. 84.07; m.p. 1570°C (2840°F); sp. gr. 4.09. (2) Cr_5S_6; mol. wt. 452.41; sp. gr. 4.26. (3) Cr_3S_4; mol. wt. 284.29; sp. gr. 4.16. (4) Cr_2S_3; mol. wt. 200.21; sp. gr. 3.92–3.97.

chromium tantalum. Cr_2Ta; mol. wt. 181.50; m.p. approx. 1977°C (3590°F).

chromium zirconium. Cr_2Zr; mol. wt. 195.24; m.p. 1677°C (3050°F); sp. gr. 6.8.

chrysoberyl. $BeO \cdot Al_2O_3$; mol. wt. 126.9; sp. gr. 3.5–3.8; hardness (Mohs) 8.5; non-gem quality material used as source of BeO and Al_2O_3 in bodies.

chrysolite. $3MgO \cdot 2SiO_2 \cdot 2H_2O$; the principal mineral in asbestos.

chuck. A device for holding grinding wheels or special shapes, or the work being ground or shaped.

chuff brick. A relatively soft, underfired brick of salmon color.

chün glaze. A thick, high-temperature opalescent glaze often decorated with a splash of red or purple.

chunk glass. Optical glass obtained by breaking open the pot in which it has been melted and cooled.

chunks. Random sizes of glass sheets which are smaller than standard sizes of work sheets.

chute. A passage or conduit, often inclined, through which objects and free-flowing substances may be conveyed at high velocity.

chute conveyor. A channel or series of channels through which the movement of materials is expedited by vigorous vibration.

chute, grizzly. A chute equipped with grizzlies of decreasing size, each grizzly separating coarse lumps from smaller lumps in decreasing size groups.

ciment fondu. A slow-setting, rapid-hardening aluminous cement composed of 40% of lime, 40% of alumina, 10% of silica, and 10% of impurities.

cinder. See slag.

cinder block. A hollow concrete block made of a mixture of cement and cinders.

cinder concrete. A concrete in which cinders are employed as the aggregate.

cinder notch. An opening in the bottom of the wall of a

blast furnace to permit the flow of slag from the furnace.

circle brick. A brick formed as a segment of a circle used in the construction of cylindrical structures.

circular kiln. A tunnel kiln constructed in the form of a circle with loading and unloading stations side by side; that is, the entrance and exit of the kiln are in the same location.

circular magnetic field. A magnetic field surrounding any electrical conductor or part as a current passes through the conductor or part from one end to the other.

circular reinforcement. A circular-shaped line of reinforcement for concrete pipe.

circulating pump. A pump employed to move slurries and liquids, which have been processed, back into the process system.

circumferential coil. An encircling coil used in electromagnetic testing.

circumferential reinforcement. Reinforcement that is approximately perpendicular to the longitudinal axis of a concrete pipe.

clad. (1) To coat, encapsulate, or contain source and special nuclear materials.
(2) To enclose or encapsulate a substance or item as a protection against a hostile condition or environment.

cladding glasses. Special glasses used for curtain walls, either colored or with a colored coating fused on the surface during manufacture. See curtain wall.

clam, clamming. A mixture of clay, sand, and water, or similar composition used to seal the door of a kiln to prevent heat loss during firing.

clamp kiln. A periodic, updraft, open-top kiln of semipermanent construction; similar to a scove kiln except that it has walls containing fire arches which are laid up with scove brick.

classification. The process of improving or changing the gradation of aggregate or other substance by screening or other sorting equipment.

classifier. A device for separating mixtures into the constituents according to particle size, density, or other property.

clay. Any of a group of natural mineral aggregates consisting essentially of hydrous aluminum silicates which become plastic when sufficiently wetted, rigid when dried en masse, and vitrified when heated to a sufficiently high temperature. Used in many ceramics, including whiteware, pottery, brick, tile, stoneware, drain tile, mortars, molds, firebrick, cement, etc.

clay adsorption, anion. The adsorption of anions either on basal surfaces (that is, the replacement of structural hydroxyls by other anions) or on edges where unsatisfied positive bonds may occur; exchange of edge hydroxyls also may take place.

clay adsorption, cation. The adsorption of cations either on basal surfaces where negative charges occur (possibly as a result of isomorphous replacements within the lattice) or adsorption on prism surfaces where unsatisfied negative bonds may exist, or both; basal surface adsorption predominates in three-layer clays while edge adsorption predominates in the kaolin-type clays.

clay, alluvial. A sedimentary clay transported and deposited by streams and rivers; a brickmaking clay which is more plastic, but less refractory and darker in color than residual clays.

clay, arenaceous. Sandy clay.

clay, ball. Sedimentary, lignite-bearing aluminum silicates that are plastic, fine-grained, easily slaked in water, and which fire to a clean, white to cream color, and which cannot be classified as kaolins or fireclays. See ball clay.

clay, bauxitic. A natural mixture of bauxite and clay containing not less than 47% nor more than 65% of alumina on a calcined basis.

clay, bottling. A plastic refractory clay consisting essentially of fireclay and sand used as a stopper in tap holes of cupolas and other melting furnaces.

clay, brick. See brick clays.

clay, burley or burley flint. A rock containing aluminous or ferruginous nodules, or both, bonded by fireclay.

clay, calcareous. Lime-bearing clay.

clay, china. A refractory clay consisting of minerals of the kaolin family which fires to a white or nearly white color.

clay, clear. A kaolinic clay free of organic and other deleterious impurities.

clay, diaspore. A rock consisting essentially of diaspore, an aluminum monohydrate, bonded by fireclay.

clay, enamel. A fatty clay having ball clay characteristics; employed as a suspension agent for milled porcelain-enamel slips. See ball clay.

clay, expanded. A product made from common brick clays by grinding, screening, and then subjecting to sudden heat, 1482°C (2700°F), changing the ferric oxide to ferrous oxide and causing the formation of bubbles which, in turn, cause the clay to bloat.

clay, fat. A highly plastic clay of high-green strength.

clay, fire. A soft embedded clay rich in hydrated aluminum silicates or silica, low in alkalies and iron, and which can withstand high temperatures without fusion; used in refractory brick, kiln and furnace linings, glassmaking pots and tanks, crucibles, etc.

clay, flint. A nonplastic, kaolin-type refractory clay.

clay, foamed. A cellular clay formed by the mechanical or chemical generation of bubbles in a slurry, and which are retained after the slurry is dried; used as thermal and acoustic insulation.

clay, fusible. A clay which will vitrify and deform when heated to approximately 1200°C (2192°F).

clay, glaze. A fine-grained ball clay of high purity, but containing appreciable quantities of colloidal organic matter; employed as a suspension agent in glaze slips and as a binder in the dried, but unfired, glaze; becomes an intrinsic part of the glaze on firing.

clay inclusions. (1) Unreacted clay or other solid material remaining in a porcelain enamel or glaze after firing; a defect.

(2) Earthy inclusions in mica which appear in various colors when observed in any type of light.

clay, lamellar. A clay which contains disk or sheet formations characteristic of the plastic clays.

clay, lean. A nonplastic clay having poor-green strength.

clay, long. A plastic or fat clay having high-green strength.

clay, marl. A smooth-textured, white, chalky clay; used in brickmaking and as an anticrazing ingredient in stoneware.

clay-mortar-mix. A finely ground clay used as a plasticizer in masonry mortars.

clay, open. Porous, sandy-textured clay.

clay, paving brick. Impure refractory clays or shale which are used to form paving brick of high strength and physical durability.

clay, plastic. A clay which will form a moldable mass when blended with water.

clay, pipe. A fine-grained plastic clay, marl, or fireclay containing little or no iron.

clay, plug. A mass of damp plastic clay used to seal tap holes of smelters.

clay, pneumatolytic. Clay that has been subjected to hot gases and liquids during its natural process of formation.

clay, pot. Refractory clay used to make glass-melting pots and crucibles.

clay, potters. Any ball clay used in the production of pottery.

clay press. A device which removes water from clay-water slurries by filtering under pressure.

clay, primary. A derivative of mother-rock feldspar such as china clay; a residual clay.

clay, pure. Aluminosilicic acid compositions which theoretically consist of 39.45% of alumina, 46.64% of silica, and 13.91% of water.

clay, red. Any of the ferruginous clays producing a red color; used in production of roofing tile, brick, and some types of pottery.

clay, refractory. A clay having a melting point above 1600°C (2912°F) used in refractory products such as firebrick, furnaces, kilns, reactors, etc.

clay, residual. A clay which geologically remains at the site of its formation.

clay, rich. A long or plastic clay having good workability and green strength.

clay, saddler. A clay of fine particle size and high-flux content which fuses at a low temperature, and which may be used as a natural stoneware or electrical porcelain glaze; for example, the Albany slip, which see.

clay, sagger. An open-firing refractory clay of suitable uniformity, having good resistance to repeated heating and cooling, and which is used in the production of saggers.

clay, secondary. A type of kaolin which has been mixed by natural processes with impurities such as alkalies, iron oxides, lime, and magnesia; usually more plastic than primary clay.

clay, sedimentary. A clay that geologically has been transported from its place of formation to another.

clay, short. Nonplastic clay characterized by poor-green strength.

clay shredder. An apparatus designed to chop and fragment plastic clays preparatory for further processing.

clay, slip. A clay containing a high percentage of fluxing ingredients, and which will fuse at a relatively low temperature to form a natural glaze or glass; characterized by a fine-grained structure and low-firing shrinkage.

clay, stove. See fireclay.

clay, surface. An unconsolidated, unstratified clay occurring on the surface of the earth.

clay, swelling. A clay which will absorb large quantities of water.

clay, tap-hole. A damp, plastic, refractory clay formed into a wad used to seal the tap hole of a smelter or melting furnace.

clay tile, natural. A tile made either by the dust-pressed method or by the plastic method from clays that produce a dense body; such tile have a distinctive, slightly textured appearance.

clay tile, structural. Hollow, burned-clay masonry building units with parallel cells or cores, or both.

clay, treading. Clay kneaded under pressure by the bare human heel; a process used in primitive-type potteries.

clay, vacuumed. Clay which has been subjected to a vacuum treatment to remove air bubbles as a means of increasing its density and improving its green strength in ceramic bodies.

clay, varved. A clay deposited in distinct layers, some layers being more silty than others.

clay, vitrification. A clay having a tendency to vitrify when heated to elevated temperatures, usually without deformation until its vitrification temperature is reached.

clay, ware. A synonym for ball clay, which see.

clay, washed. Purified clay of low silica and grit content obtained by stirring the clay into water to form a thin slurry and then allowing the impurities to be removed by settling.

clay-water pastes, yield point. The minimum stress at which continuous flow takes place, the yield point being evident in tensile, compression, torsion, and shear tests, but which (1) will vary with the rate of stressing and (2) will depend on the manner of specimen preparation.

clay, white. A kaolin of high quality which burns to a white color.

clay winning. The mining and processing of clay raw materials to make them suitable for subsequent use.

clean. Free of interfering contamination.

cleanability. The relative ease that soils can be removed from a material, particularly from the surface of the material.

cleaner. A solution, usually alkaline, but sometimes an organic solvent, used to remove oil, grease, drawing compounds, dirt, etc., from a metal surface being prepared for porcelain enameling.

cleaning, immersion. The removal of surface contamination from a surface by immersion in a cleaning liquid.

cleaning, post. The removal of penetrant inspection materials from a test piece after the completion of a penetrant inspection test.

cleanup. The act of preparing a construction joint or rock foundation to receive concrete in which the surface is scrubbed or sandblasted to remove dirt, laitance oil, and other foreign matter.

clear ceramic glaze. An inseparable, fire-bonded, translucent or tinted glaze having a lustrous finish.

clear ceramic glazed tile. Facing tile having facing surfaces covered by a tinted or translucent glaze with a glossy finish.

clear clay. A kaolinic clay free of organic and other deleterious impurities.

clear frit. A frit that remains essentially transparent when processed into a porcelain enamel.

clear glaze. A colorless or tinted transparent ceramic glaze.

clinker. A fused or partly fused by-product of the combustion of coal, but including lava and portland cement clinker and partially vitrified slag and brick.

clinker brick. A very hard-fired brick whose shape is distorted or bloated due to overfiring; that is, to nearly complete vitrification.

clinkering zone. The high-temperature section of a cement kiln where the clinker is formed.

clinoclore. $5MgO \cdot (Al,Cr)_2O_3 \cdot 3SiO_2 \cdot 4H_2O$; sp. gr. 2.6–3.1; hardness (Mohs) 2.0–2.5; possible use in porcelain enamels, glazes, and welding-rod coatings.

clip. The portion of a brick cut to a desired or specified length.

clip tile. Tile designed as a base fitting around the flanges of an I-beam.

clobbering. The process of decorating ware of another artist or producer without permission.

cloisonné. An art form of porcelain enamel, pottery, and tile in which differently colored enamels or glazes are separated by fillets applied along the outlines of a design; for porcelain enamel, the fillets are wire or thin strips of metal secured to the base metal, while for pottery and tile the fillets consist of a ceramic paste squeezed on the base-body surface through a small-diameter orifice.

closed chip. A fractured area on the surface, edge, or corner of an item where the material or coating has not actually broken away from the item.

closed-circuit grinding. A continuous grinding or milling process in which particles of acceptable fineness are removed from the grinding system by a screen or cyclone classifier, while oversized particles are returned to the pulverizer for further processing.

closed pore volume. See sealed pores.

closed pot. A glass-melting pot having a crown to protect the batch from the combustion gases.

closer, king. A brick cut diagonally so that one end is cut to a 2-inch width while the opposite end remains at full width.

closure strip. A preformed filler strip of asphalt or rubber having the same shape and pitch as the corrugated asbestos-cement product, and which is used to close openings or joints in the corrugated sheets at window beads, eaves, lower edges of siding, and similar places.

cloth. A firm-textured woven fabric of cotton or linen which is dampened and placed over filter-press cakes, pugged clays, and the like to prevent loss of moisture before use.

cloth, wire. Wire mesh or screen woven or crimped in a pattern of squares or rectangles; used in sieves and screens.

coadsorption. The adsorption of two or more components on activated carbon or similar substance, each component affecting the adsorbability of the other.

coal, breeze. The residue remaining after the making of charcoal or coke; used in brickmaking and as a filler in concrete.

coalescence. The physical attraction and merging of particles to form larger particles.

coal gas. A gaseous mixture produced by the destructive distillation of coal; used as a commercial fuel.

coarse aggregate. (1) The mineral materials, such as sand and stone, in their natural conditions.
(2) The portion of concrete aggregate which is retained on a ¼ inch (4.75 millimeter) sieve.

coated abrasive. An abrasive product in which the abrasive particles are bonded to paper, cloth, fiber, or other backing material by a resin or glue.

coated abrasive disk. A paper, cloth, fiber, or other disk coated on one side with a mixture of abrasive and binder; used in mechanical grinding.

coating. A film of a substance applied over the surface of another solid.

coating, ceramic. An inorganic, nonmetallic coating applied over a surface and which is bonded in place by firing, such as a porcelain enamel or glaze.

coating, flow. A ceramic coating in slip form which is poured or flowed over a shape and allowed to drain.

coating immersion. The process in which an object is submerged in a ceramic coating in slip form, removed, and then allowed to drain to a relatively uniform thickness; a coating so applied.

coating, pyrolytic. A coating formed on the surface of an article by thermal decomposition of a volatile material, such as a coating of silica resulting from the decomposition of silicon tetrachloride, deposited on an article in a vacuum.

coating, refractory. A coating composed of refractory ingredients used for the protection of brickware, metals, and other materials which are subjected to elevated temperatures.

coating, roller. The transfer of designs from patterned surfaces to the surfaces of ware by means of a roller.

coating, slip. Any ceramic coating applied to a body or shape in the form of a slurry or slip and subsequently fired to maturity.

coating, vacuum. The deposition of a vaporized coating of a material on the surface of another material in a vacuum. Also known as vapor deposition.

cobalt aluminate. $CoO{\cdot}Al_2O_3$; mol. wt. 176.8; m.p. 1960°C (3560°F); sp. gr. 4.37; a ceramic colorant ranging from blue to blue-green.

cobalt aluminide. CoAl; mol. wt. 85.8; m.p. 1630°C (2966°F); sp. gr. 6.04; hardness (Vickers) ~440. See aluminides.

cobalt arsenate. $Co_3(AsO_4)_2{\cdot}8H_2O$; mol. wt. 598.8; sp. gr. 2.95; used as a blue colorant in glass and ceramic inks.

cobalt beryllide. CoBe; mol. wt. 67.96; m.p. 1505°C (2740°F).

cobalt blue. A blue to blue-green pigment composed of cobalt and aluminum oxides.

cobalt boride. (1) Co_3B; mol. wt. 187.64; m.p. 1093°C (2000°F). (2) Co_2B; mol. wt. 128.70; m.p. 1259°C (2300°F). (3) CoB; mol. wt. 69.76. (4) CoB_2; mol. wt. 80.58. See borides.

cobalt carbonate. $CoCO_3$; mol. wt. 118.94; decomposes on heating; sp. gr. 4.13; used in the production of blue and black ceramic colorants.

cobalt chloride. (1) $CoCl_2$; mol. wt. 129.9; sublimes when heated; sp. gr. 3.348. (2) $CoCl_2{\cdot}6H_2O$; mol. wt. 238.02; m.p. 86.75°C (187.9°F) sp. gr. 1.924; employed as a decolorizer in iron-tinted glass.

cobalt chromate. $CoCrO_4$; mol. wt. 174.95; used with aluminum and zinc oxides to produce light blue and light green colors in porcelain enamel and glazes.

cobalt ferrate. $CoO{\cdot}Fe_2O_3$; mol. wt. 234.6; m.p. 1571°C (2860°F); sp. gr. 5.30.

cobalt hafnium. Co_2Hf; mol. wt. 296.48; m.p. 1571°C (2860°F).

cobalt molybdenum. Co_7Mo; mol. wt. 508.58; m.p. 1494°C (2720°F).

cobalt niobium. Co_2Nb; mol. wt. 210.98; m.p. 1571°C (2860°F).

cobalt nitrate. $Co(NO_3)_2 \cdot 6H_2O$; mol. wt. 291.1; m.p. 56°C (132.8°F); sp. gr. 1.88; used as a metal treatment to promote adherence of porcelain enamels to iron and steel.

cobaltous ammonium sulfate. $CoSO_4 \cdot (NH_4)_2SO_4 \cdot 6H_2O$; mol. wt. 395.3; sp. gr. 1.902.

cobalt oxide. (1) CoO; mol. wt. 74.9; decomposes at 1800°C (3290°F); sp. gr. 5.7–6.7. (2) Co_2O_3; mol. wt. 165.9; decomposes at red heat; sp. gr. 4.81–5.60. (3) Co_3O_4; mol. wt. 240.8; sp. gr. 6.07. Employed as a colorant, and sometimes as a decolorizer or masking agent, in glass, underglazes, overglazes, porcelain enamels, decals, and similar decorative applications, and as an adherence promoting ingredient in porcelain-enamel ground coats.

cobalt phosphide. (1) CoP_3; mol. wt. 152.03; sp. gr. 4.26. (2) CoP; mol. wt. 89.96; sp. gr. 6.24. (3) Co_2P; mol. wt. 148.87; m.p. 1385°C (2525°F); sp. gr. 7.55.

cobalt silicate. $2CoO \cdot SiO_2$; mol. wt. 283.87; m.p. 1253°C (3590°F); sp. gr. 4.68; hardness (Mohs) 5–7.

cobalt silicide. $CoSi$; mol. wt. 87.0.

cobalt sulfate. (1) $CoSO_4$; mol. wt. 155.0; m.p. 989°C (1812°F); sp. gr. 3.47; (2) $CoSO_4 \cdot 7H_2O$; mol. wt. 281.1; m.p. 96.8°C (206°F); sp. gr. 1.92; used to impart blue and blue-white colors in whiteware bodies.

cobalt sulfide. CoS; mol. wt. 91.0; m.p. 1100°C (2012°F); sp. gr. 5.45.

cobalt tantalum. Co_2Ta; mol. wt. 299.38; m.p. 1604°C (2920°F).

cobalt tungsten. Co_2W; mol. wt. 301.88; m.p. 1749°C (3180°F).

cobalt zirconium. Co_2Zr; mol. wt. 209.10; m.p. 1560°C (2840°F); sp. gr. 8.46.

cobble mix. Concrete containing aggregate up to 6 inches in diameter.

cock spur. A triangular item of kiln furniture with a single sharp point on which plates and similar ware are placed for firing.

coefficient of confidence. A stated proportion of the times the confidence interval is expected to include the population parameter.

coefficient of friction. The ratio of frictional force between two bodies in parallel contact to the force with which the bodies press against each other.

coefficient of saturation. The ratio of the weight of water absorbed by a masonry or other unit during immersion in cold water to weight absorbed during immersion in boiling water, the ratio to be taken as an indication of the resistance of brick to freezing or thawing.

coefficient of scatter. The ratio of the increase in reflectance with thickness of a porcelain enamel or other coating applied over an ideally black backing.

coefficient of thermal expansion. The fractional change in the length or volume of a body per degree of temperature change.

coercive force. An opposite magnetic force required to return certain ferromagnetic materials to their original nonmagnetic orientation.

coherence. The property of substances being held physically together by mutual attraction of the particles of the substances.

cohesion, cohesiveness. The tendency of substances of like composition to hold together as a result of intermolecular attractive forces.

coil. The process of making specialty or art items by forming the object from ropes or coils of plastic clay.

coil, annular. An electromagnetic testing coil of the encircling type.

coil, bobbin. An electromagnetic testing coil better known as the ID type. See bobbin coil.

coil, bucking. An electromagnetic testing coil of the differential type. See bucking coil.

coil, circumferential. An encircling coil used in electromagnetic testing.

coil, comparator. A system of electromagnetic testing in which two or more coils are connected in series opposition but arranged so that there is no mutual induction or coupling between them, such that any electric or magnetic condition, or both, that is not common to the test standard and the test specimen will produce an unbalance in the system and yield an indication.

coil, differential. A system of electromagnetic testing in which two or more coils are electrically connected in series opposition, such that any electric or magnetic condition, or both, that is not common to the areas of the specimen being tested will produce an unbalance in the system and yield an indication.

coil, encircling. A coil or coil assembly that surrounds the part of a specimen being electromagnetically tested.

coil, feed-through. An electromagnetic testing coil of the encircling type.

coil, ID. A coil or coil assembly used for electromagnetic testing in which the probe is inserted inside the test specimen.

coil, inserted. A coil of the ID type used in electromagnetic testing.

coil, inside. A coil of the ID type used in electromagnetic testing.

coil method of magnetization. A method of magnetization in which part or all of a component is encircled by a current-carrying coil.

coil, probe. A small coil or coil assembly which is used in electromagnetic testing which is placed on or near the surface of a test specimen.

coil, reference. The section of a coil assembly that excites or detects, or both, the electromagnetic field in the reference standard in a comparative system.

coil size. The geometry or dimensions of a coil such as length or diameter.

coil spacing. The axial distance between two encircling coils in a differential system of electromagnetic testing.

coil, test. The section of a coil assembly that excites or detects, or both, the magnetic field in a material being tested in a comparative system.

coke. The solid product resulting from the incomplete combustion of coal, consisting principally of carbon; used chiefly as a fuel in metallurgy to reduce metallic oxides to metal.

coke oven. A refractory-lined oven in which coal is fired in an essentially oxygen-free atmosphere to produce coke.

Colburn process. A method of forming flat glass in which a ribbon of molten glass is drawn upward from the glass tank, rolled flat, annealed, and then cut into desired sizes and shapes.

colcather. Red iron oxide, Fe_2O_3, used as a pigment and as a polishing agent.

cold, joint. The surface between two successive pourings of concrete in which the first pouring has set and can no longer be blended into the second pouring.

cold-rolled steel. A low-carbon, cold-reduced sheet steel used in porcelain enameling.

colemanite. $2CaO{\cdot}3B_2O_3{\cdot}5H_2O$; mol. wt. 411.2; sp. gr. 2.42; hardness (Mohs) 4.0–4.5; used as source of CaO and B_2O_3 in pink and maroon raw-lead glazes, and as a flux in glazes.

collar in, collaring. To reduce the diameter of a pot, particularly the opening, by pressure from the outside while turning it on a wheel.

collet. (1) A split sleeve used to hold work or tools during machining or grinding.

(2) The neck of a glass bottle after removal from the blowing iron.

colloid. A substance in the form of submicroscopic particles which do not settle out when in solution or suspension; such a substance together with gaseous, liquid, or solid substance in which it is dispersed.

colloidal clay. A very fine natural clay which usually swells when it takes up water, and which is used as a binder for nonplastic materials.

colloidal formation. A high-speed grinding device capable of making very fine dispersions of liquids or solids by breaking down particles in an emulsion or paste.

colloid mill. A high-speed grinding device capable of making very fine dispersions of liquids or solids by breaking down particles in an emulsion or paste.

color. The wavelength composition of light, particularly with reference to its visual appearance; a color other than white, black, or gray.

colorant, ceramic. An inorganic material used to impart color to porcelain enamel, glazes, glass, and ceramic bodies.

colored frit. A frit containing a colorant to produce a strong color in porcelain enamel or other ceramic coating.

color filter. A transparent material, such as glass, with selective or nonselective properties with respect to the absorption of light waves according to wavelength.

colorimeter. An instrument that measures color by determining the intensities of the three primary colors that comprise a particular color.

coloring agent. Any substance which will impart color to another substance or product.

color, metallic. Metal particles of gold, silver, and platinum suspended in a suitable medium or oil used to decorate dinnerware and other ceramic products.

color, metameric. A color which will appear the same under one condition or type of light, but will appear as a different color or shade under a different condition or type of light.

color oxide. An oxide of a metal which is used to color glass, glazes, porcelain enamels, ceramic bodies, and other products.

color stability. The resistance of a product to a change in color.

color variations. The property of nonuniform color exhibited by a product during some stage of the manufacturing operation or before or after some condition of service, such as weathering.

column, reinforced. A concrete column in which longitudinal metal bars, often ties or circular materials, are incorporated as reinforcing agents.

combed finish. Articles, such as tile or brick, having face surfaces intentionally altered by scratches or scarves during manufacture to give increased bond with mortar, plaster, stucco, or other mastic used in installations.

combed ware. Ware which has been finished either by combing or by flowing several wet slips or glazes together.

combined sewer. A pipeline intended to convey sewage and storm water.

combined water. Water that is combined chemically with clays and minerals and which can be expelled only by heating to relatively high temperatures. Also known as water of crystallization, water of hydration, water of constitution.

combing, feather. A decorative technique in which a tool containing many sharp points is drawn across superimposed layers of damp slips of various colors for an artistic effect.

comb rack. (1) A comb-shaped burning tool used to support ware during firing.

(2) A comb-like tool used to support metal ware during the cleaning and pickling operation.

combustible. Easily ignited and burned.

combustion. The process of burning.

combustion air. Air introduced into a firing chamber or zone to support the combustion of fuel.

combustion chamber. The area in a furnace or kiln in which fuel is burned.

combustion efficiency. The ratio of the heat actually developed during combustion to the heat theoretically possible under ideal conditions.

combustion, incomplete. A burning process in which oxidation of the fuel is incomplete, sometimes resulting in a reducing atmosphere in direct-fired furnaces and kilns.

combustion, surface. The impingement of fuel gases on an incandescent surface as a means of obtaining more uniform and complete combustion.

comeback. The time required for a porcelain enameling or other furnace to return to temperature after introduction of a load of ware.

comminution. Any process for reducing the particle size of a material.

common brick. A block of clay material usually fired to form a stable mass and used for general building purposes.

compact. (1) To treat glass in a manner, such as by heat treatment, to approach maximum density.

(2) To densify by any means.

compaction. (1) A technique for reducing space requirements for a material.

(2) Increasing the dry density of a material.

(3) The preparation of a compact or object produced by the compression of a powder, generally while confined in a die, with or without the inclusion of lubricants, binders, etc., and with or without the concurrent application of heat.

comparative measurements. Experiments conducted to determine if one product, procedure, or system is better than another.

comparative standard. A reference material used as a basis for comparison or calibration to detect any property or condition that is not common to the test subject and the standard.

compatability. Capable of existing in a homogeneous mixture with another substance without separation or chemical reaction.

complete fusion. Complete liquification under the influence of heat.

component. (1) A constituent part of a mixture.

(2) The smallest number of independently variable substances able to form all of the constituents of a system in whatever proportion they may be present, and from which the composition of each phase can be quantitatively expressed.

composite. A material composed of a mixture of distinct parts, such as a mixture of ceramic materials and a metallic phase, intended to produce a material of specific properties; for example, glass-fiber-reinforced plastics and metals used in the production of boats, cars, radomes, nose cones, aircraft parts, etc.

composite coating. A mixture of one or more ceramic materials in combination with a metal phase applied to a metallic substrate, or a nonmetallic substrate such as graphite, which may or may not require heat treatment prior to service.

composite coating, refractory. A combination of heat-resistant ceramic materials applied to a metallic substrate, or to a nonmetallic substrate such as graphite, and which may or may not require heat treatment prior to placement in service.

composite column. A concrete column reinforced with a metal core, usually steel.

composite compact. A compact composed of one or more layers of different substances, with each substance retaining its own identity.

composite lot sample. A single sample prepared from several containers or lots by combining them in the same ratio as the net weight of the materials sampled.

composite wheel. A bonded abrasive product in which two or more specifications are bonded together into one wheel.

composition. The combination of elements or compounds comprising the whole of a material or product.

compound. A material resulting from the union of two or more elements or ingredients.

compound rolls. A pulverizing system consisting of two or more pairs of rolls arranged vertically, one pair above the other, with the spacing between the rolls being decreased in descending order so that the particle size of a material is reduced as it passes from the upper set of rolls to the next.

compressed air. Air under pressure greater than the surrounding atmosphere.

compression. Reduction in volume of a substance under pressure.

compression failure. The breaking or disintegration of a solid under some form of pressure.

compression test. A test made on a specimen of a material placed under load to determine its compressive strength.

compressive strength. The maximum resistance of a material to compressive loading, or the specified resistance used in design calculations, based on the original area of the specimen cross section.

compressive stress. A stress developing in a solid under the influence of some form of pressure.

concentrate. To increase the amount of a substance in a mixture, solution, or ore.

concentration, threshold. The minimum concentration at which a substance can be detected by a taste or odor test.

concentric wheel. A bonded abrasive product containing two or more concentric sections of different abrasive specifications.

concrete. A homogeneous mixture of portland cement, aggregates, and water; also may contain selected admixtures.

concrete, aerated. Cement containing a high proportion of entrained air introduced by mechanical or chemical means.

concrete aggregate. Sand, gravel, crushed rock, slag, and similar materials blended with portland cement to form concrete.

concrete, air entrained. Concrete containing bubbles or spheroids of air purposefully introduced by the use of an air-entraining agent.

concrete, architectural. Concrete of high quality and free from blemishes which is exposed on the exterior or interior of a building.

concrete beam. A structural beam of reinforced concrete designed for loadbearing functions.

concrete block. Concrete fashioned in the form of hollow and solid blocks of various sizes (frequently 8×8×16 inches) for use in construction and other applications.

concrete brick. Concrete formed in the sizes and shapes of conventional brick, and having high compressive strength and resistance to the conditions of weathering, for use in construction and other applications.

concrete bucket. A cylindrical container in which concrete is transported, usually by crane or a similar mechanical device.

concrete buggy. A cart designed to carry concrete from a mixer or hopper to pouring forms.

concrete, cellular. An air-entrained concrete.

concrete, centrally mixed. Concrete that is mixed in a stationary mixer and hauled to the job site in agitators.

concrete chute. A round-bottomed trough to convey concrete to a lower level.

concrete column. A vertical structure of reinforced concrete designed to carry loads.

concrete finish. The surface texture or smoothness of hardened concrete.

concrete, foamed. Concrete containing a high proportion of air and gas bubbles induced by mechanical or chemical means.

concrete form oil. An oil which is employed to coat the forms into which concrete is cast to facilitate the removal of the concrete from the forms after it has set.

concrete, fresh. Concrete that has not attained its initial set.

concrete, green. Concrete that has attained its final set, but has not yet developed appreciable strength.

concrete hardener. An additive to a concrete mix, such as sodium hydroxide, sodium chloride, or calcium chloride to hasten the set of concrete.

concrete, heavy. High-density concrete, approximately double that of normal concrete, used for counterweights and radiation shielding. See heavy concrete.

concrete, insulating. Lightweight concrete, usually with a density of 90 pounds per cubic foot, used for thermal insulation and fire protection in building construction.

concrete, lightweight. Any concrete made with low-density aggregate.

concrete masonry. Any form of construction composed essentially of concrete block, brick, or tile laid by masons.

concrete, mass. A concrete containing large aggregate and a pozzolan placed in large masses such as in a dam or large footing.

concrete mixer. A rotating cylinder or drum in which concrete is mixed.

concrete, nailing. A lightweight, sawdust concrete capable of receiving and holding nails in certain types of construction. See sawdust concrete.

concrete, no fines. Concrete containing no aggregate less than ⅜ inch in cross section.

concrete pile. A pile or column of reinforced concrete, either cast in place or precast, which is driven into the ground as a support for subsequent construction.

concrete pipe. A porous pipe or conduit made of concrete which generally is used in some type of drainage application.

concrete, plain. Unreinforced concrete, but sometimes containing a fibrous admixture to minimize temperature cracking and shrinkage.

concrete, post-tensioned. A prestressed concrete to which a compressive stress is applied by releasing the prestressed tension in the reinforcing tendons after the concrete has attained sufficient or maximum set.

concrete, precast. Any concrete that is cast in some shape in molds or forms at a location other than where the concrete will be used.

concrete, prestressed. Concrete in which reinforcing steel rods, strands, or wires under tension are embedded; this tensile load becomes a compressive load in the concrete by means of the bond between the concrete and the reinforcing components.

concrete products. Precast concrete such as brick, block, pipe, sills, garden objects, and similar items produced at a central manufacturing plant.

concrete pump. A machine that drives or forces concrete into placing position.

concrete, ready-mixed. Concrete that is mixed before delivery to a job site.

concrete, reinforced. A concrete into which reinforcing rods, bars, mesh, or strands have been embedded while in the plastic state.

concrete retarder. A material added in small quantities to a concrete mix to increase or lengthen the setting time and decrease the rate at which strength is developed; the retarder should have no effect on the concrete after it has set.

concrete, sawdust. A low-strength concrete in which sawdust is employed as an aggregate; used as a lightweight nailing concrete.

concrete, shrink-mixed. A ready-mixed concrete which first is blended in a stationary mixer and then transferred to a truck mixer where mixing is completed in transit to a job site.

concrete slab. (1) A flat, relatively thick plate of concrete of various shapes used as stepping stones, well and pit covers, floors, roofing section, bridge decks and the like.
(2) A concrete pavement.

concrete, transit-mixed. Concrete that is proportioned in a truck mixer and mixed in transit to a job site.

concrete, truck-mixed. Transit-mixed concrete.

concrete, vacuum. Concrete that has been subjected to a vacuum to remove entrapped air and water from the surface to improve the durability, strength, and hardness of the surface.

concrete vibrator. A vibrating device used to consolidate concrete.

concrete, waterproof. A concrete containing a waterproofing admixture or to which a waterproof coating has been applied to its surface.

concrete workability. The ease with which the ingredients of a concrete batch can be mixed and subsequently can be handled, transported, and placed without loss of homogeneity.

concurrent processing. One or more operations taking place at the same time.

condensate. The liquid product from a condenser.

condensation. The process of reducing a gas or vapor to a liquid or solid form.

condenser. Any enclosed vessel in which a vapor is condensed to its liquid state.

conditioning. The process of preparing a material for a subsequent process or use.

conditioning zone. The sections of a glass-melting tank in which temperatures of the molten batch are adjusted for subsequent operations.

conductance. The property of transmitting electricity.

conducting material. Any material through which heat, electricity, or sound will flow.

conductive coating. A porcelain enamel, glaze, metallic, or other coating capable of conducting electricity.

conductivity. The property and rate of conducting heat, electricity, and sound.

conductor. Any substance which will conduct heat, electricity, and sound.

conductor, insulated. A conductor which is coated or surrounded by a nonconducting material to prevent or retard the passage of electric current, heat, sound, or other phenomena of concern.

conduit. (1) A pipe for the conveyance of water or other fluid.

(2) A pipe for the containment of electric wiring as a protection against damage from external causes.

cone classifier. A device consisting of an inverted cone in which solid particles are separated according to size or density by settling in a rising stream of air or water.

cone crusher. A machine for crushing which consists of a cone gyrating within a conical cavity with tapered clearances such that a material is reduced several times during passage.

cone, pyrometric. A trigonal prism of standard shape, position, and composition which is used to indicate the degree of vitrification in process control; the indications are determined by the degree of distortion observed in a series of cones of different compositions which deform at different temperatures.

cone screen test. A technique for measuring the fineness of porcelain enamels in which a cone-shaped sieve is used.

cone wheel. A relatively small abrasive grinding wheel in the shape of a cone which may be mounted in a stationary or portable tool.

confidence. The degree of assurance that a specified rate of failure is not exceeded.

confidence interval. The frequency that a sample or product will meet or exceed specified requirements.

confidence level (coefficient). The stated proportion of the times the confidence interval is expected to be attained.

configuration. The shape or structure of a body or product.

conglomerate. A heterogeneous mixture of solids, usually with no or only minor chemical reaction.

congruent melting. The change of a substance, when heated, from the solid form to a liquid of the same composition, such as, for example, ice.

conical roll. A crushing device in which clay or other substances pass vertically between a set of inverted cone-shaped rolls.

connection, feeder. An opening in a furnace wall through which a feeder channel conveys molten glass from the melting tank to the feeder.

consistency. (1) The properties of a slip that influence its draining, flowing, and spraying behavior.

(2) A measure of the fluidity, softness, or wetness of fresh concrete, determined by the number of inches a sample slumps or subsides when a conical form is removed from the sample; the greater the subsidence, the higher the slump and the wetter or softer the concrete.

consistometer. Any of a variety of instruments designed to measure the fluidity, including the draining, flowing, spraying, and slumping properties, of slips and slurries.

console. A panel consisting of meters, dials, switches, and other instruments by which a manufacturing operation is controlled.

consolidate. To form into a compact mass or to unite as a whole, such as concrete.

constant. A fixed value which does not change during a particular test or process.

constant-weight feeder. A mechanical device for the delivery of a designated weight of raw material from one process to another per unit of time.

constituent. An essential component of a substance or product.

constitution, water of. See combined water.

constriction. The reduction or narrowing of a channel or opening.

constringence. The reciprocal of the dispersive power of a medium such as glass. See nu value.

construction joint. A plane surface between two pourings of concrete, the second pouring being placed on or against the first after the first has hardened.

contact arc. That portion of the circumference of a grinding wheel in contact with the work being ground.

contact area. The total area of the surface of a grinding wheel in contact with the work being ground.

contact pressure. The force of contact between two surfaces per unit of area.

container. Any receptacle used to hold something.

container, glass. A general term applied to glass bottles and jars.

container sample. Samples obtained from individual

containers by use of a sample thief or other approved means.

contaminate. To soil or render unfit by introduction of unwanted or foreign material.

continuity of coating. The degree to which a porcelain enamel or other ceramic coating is impervious; that is, free from pin holes, blisters, bare spots, boiling, copperheads, or other defects which would reduce its protective properties.

continuous cleaning. A term describing a particular type of porcelain enamel which will oxidize and remove food soils accumulated on the interior surfaces of cooking ovens at normal temperatures and conditions of use.

continuous control. An automatic system designed to control a manufacturing process or operation.

continuous dryer. A dryer in which the ware moves through the drying cycle in an uninterrupted flow pattern as opposed to a batch-type dryer.

continuous filament. A glass fiber of great and indefinite length.

continuous furnace. A furnace or kiln into which ware is fed continuously without interruption and through which the ware progresses until the firing operation is complete.

continuous glass tank. A glass furnace in which the molten glass is maintained at a constant level by continuously charging new batch into the furnace in an amount equal to the amount of molten glass withdrawn.

continuous kiln. See continuous furnace.

continuous mixer. A mixer in which materials are charged, mixed, and discharged in a continuous pattern of flow.

continuous moving bed. An adsorption process characterized by the flow of a fluid through a continuously moving bed of granular material, such as activated carbon, with the continuous withdrawal of the spent granular material from the bottom of the bed being replaced by new or reprocessed material at the top.

continuous production. A sequence of production operations involving the continuous flow of materials from one station to the next without interruption.

continuous retort. A refractory or glass lined vessel in which substances are distilled or disintegrated by heat on a continuous basis.

contraction. The process of diminishing in size; for example, the reduction in the size of concrete during setting or the shrinkage of a ceramic body during drying or firing.

contraction crack. A crack developing in a body due to the stresses induced by excessive shrinkage.

contraction joint. An intentionally placed crack or groove in concrete or a masonry unit to create a plane of weakness so that the unit will crack at the weakened groove and minimize the development of random cracks during setting and during the service life of the unit.

contrast. To compare materials and products in such a way to show differences.

contrast ratio. The ratio of the reflectance of a coating over a black backing to its reflectance over a backing having a reflectance value of 80%.

control. The process of directing, checking, testing, and verifying the performance of a process or the quality of a product during a manufacturing operation.

control board. See console.

control, criticality. A mechanism by which it is insured that criticality cannot occur. See criticality.

controlled atmosphere. A specified concentration of gas or mixture of gases at a specified temperature, and sometimes at a specified humidity, in which selected processes take place.

controlled cooling. The cooling of an object from an elevated temperature in a predetermined manner or under specified conditions.

controlled fission. Fission under conditions of continuous adjustment of control rods and of other control devices in a reactor which compensates for the changes in excess reactivity which result from high-power operation and from nuclear reactor-temperature fluctuations.

controlled fusion. The generation of power under controlled thermonuclear fusion reactions.

control panel. See console.

control, process. Controlling the conditions of an operation or process in order to bring about a desired result or product.

control, quality. Activities designed to achieve satisfactory quality of products, particularly process control.

control rod. A device, usually a neutron-absorbing material, used to control chain reactions, particularly in nuclear reactors.

control standard. Any of the standards of various types having known parameters which are used for the evaluation of materials and products, or which may be used to adjust the sensitivity setting of test instruments, or for periodic adjustment sensitivity.

control tests, quality. Tests performed within a system for verifying and maintaining a desired level of quality in a product or process.

convection. The transfer of heat by the circulatory motions in air of fluids due to warmer portions rising and cooler portions sinking.

convergence. To approach a common center or point.

conversion. The change of a compound from one isometric form to another as in the high-temperature conversion of quartz to cristobalite and tridymite.

conversion factor. The numerical factor by which a quantity must be multiplied or divided in order to convert the quantity from one unit of terminology to another.

converter. A refractory lined furnace in which air is blown through or across molten metal to remove impurities by oxidation.

converter, Bessemer. A basic-line, pear-shaped, cylindrical vessel in which steel is produced; air is introduced through openings in the vessel bottom.

conveyor. A machine designed for the continuous transport of items from one location to another.

conveyor, air. A conveyor which transports powdered or granular material through a pipe or duct by means of high-velocity air or by vacuum.

conveyor, apron. A series of overlapping metal plates or aprons mounted on an endless chain for transferring material from one location to another, such as to feed raw material from a bin to a processing station.

conveyor, belt. An endless belt, running between head and tail pulleys, used to transfer loose material or objects from one location to another.

conveyor, chain. A device consisting of two endless linked chains connected by crossbars, or a single endless chain with suspended hooks, platforms, or buckets on which ware is placed; used for the transport of material or objects from one place to another.

conveyor, roller. A series of rollers mounted in a frame in such a manner that flat objects may be moved from one place to another; the rolls may be free-moving or mechanically driven.

conveyor, screw. A conveyor consisting of a rotating screw in a trough or cylinder used to transfer material on a horizontal or inclined plane.

conveyor, slide. A trough or chute for the downward movement of materials under gravitational pull.

conveyor, spiral. See conveyor, screw.

coolant. A liquid applied to the work or grinding wheel during grinding to keep the work from overheating and oxidizing, which keeps the tool cool to prevent reduction in hardness and resistance to abrasion, and which washes away chips and grits, and aids in obtaining a finer finish.

cooler. An auxilliary section in a cement kiln in which the clinker is cooled before grinding.

cooler nail. A cement-coated nail.

cooling arch. A stationary lehr in which glass is annealed.

cooling curve. A time-temperature curve denoting the rate at which a fired or heated product is cooled, usually to room temperature.

cooling-down period. (1) The elapsed time between the opening of a covered glass-melting pot and the time the glass is sufficiently cool to work.

(2) The period between the fining stage and the removal of glass from a furnace.

cooling process. The removal of heat from a substance.

cooling rate. The time required for a glass or fired ceramic to cool between the limits of the working range.

cooling stress. Stress resulting from uneven contraction during the cooling period because of uneven temperature distribution in a body.

cooling zone. The section in a continuous furnace or kiln in which ware is permitted to cool following the firing operation.

cope. The upper portion of a flask, mold, or pattern.

coping. The shaping of stone or other hard nonmetallic by the use of a grinding wheel.

copperas, green copperas. Ferrous sulfate, $FeSO_4 \cdot 7H_2O$, which see.

copper, blister. A partially refined form of copper having a blistered surface after smelting due to gases generated or entrapped during solidification.

copper brazing. Brazing in which copper is employed as the filler metal.

copper carbonate. $CuCO_3$; mol. wt. 221.2; decomposes at 200°C (392°F); sp. gr. 3.7–4.0; used as red, blue, and green colorant in glazes.

copper enamels. Porcelain enamels formulated specifically for use as a decorative and protective coating on copper; usually of high thermal expansion.

copper fluoride. $CuF_2 \cdot 2H_2O$; mol. wt. 137.6; m.p. 785°C (1445°F); sp. gr. 4.23; used in porcelain enamels and glazes, both as a flux and colorant.

copperhead. A defect occurring in porcelain-enamel ground coats that appear as small freckle-like, reddish-brown spots consisting essentially of iron oxide.

copper metaborate. $Cu(BO_2)_2$; mol. wt. 149.2; sp. gr. 3.86; used as pigment in ink for painting on porcelain and other ceramics.

copper oxide, black. CuO; mol. wt. 79.6; m.p. 1064°C (1947°F); sp. gr. 6.32; used to produce blue and green colors on glass, faïance, porcelain, stoneware, and other ceramics when fired in an oxidizing atmosphere, and red colors when fired in a reducing atmosphere.

copper oxide, red. Cu_2O; mol. wt. 143.1; m.p. 1210°C (2210°F); b.p. 1800°C (3272°F); sp. gr. 5.75–6.09; employed in glass, glazes, porcelain enamels, and other ceramics primarily in the production of red colors.

copper sulfate. $CuSO_4 \cdot 5H_2O$; mol. wt. 223.3; m.p. 200°C (392°F); sp. gr. 2.284; used as the colorant in production of copper-ruby glass.

copper sulfide. Cu_2S; mol. wt. 159.2; m.p. 1100°C (2012°F); sp. gr. 3.9–4.6.

copper titanate. $CuO \cdot TiO_2$; mol. wt. 159.4; promotes high-fired density in body.

coral red. Low-temperature color produced in porcelain enamel and glazes by lead chromate.

corbel, corbelling. A supporting projection on the face of a wall; an arrangement of brick in a wall in which each course projects beyond the one immediately below it to form a support, shelf, or baffle.

cord. An attenuated glassy inclusion possessing optical and other properties differing from those of the surrounding glass, such as a glassy dripping from the ceiling of the furnace.

cordierite. $2MgO \cdot 2Al_2O_3 \cdot 5SiO_2$; mol. wt. 584.8; sp. gr. 2.60–2.66; hardness (Mohs) 7.0–7.5; formed or used in electronic-ceramic, stoneware, porcelain, and vitreous-china bodies to improve the thermal-shock resistance of the bodies.

cordierite porcelain. A vitreous ceramic whiteware for technical applications in which cordierite is the essential crystalline phase.

cordierite whiteware. Any ceramic whiteware in which cordierite is the essential crystalline phase.

core. (1) One or more members supported within an extrusion die to form holes in extruded brick or tile.

(2) A cylinder of concrete taken from concrete by means of a core drill for testing or archival purposes.

(3) The central part of a sand mold used in foundries.

(4) The central part of a plaster mold used in solid casting.

(5) A one-piece, heat-insulating shape used at the top of ingot molds.

cored brick. A brick that is at least 75% solid in any plane parallel to the load-bearing surface.

core, strainer. A porous refractory shape used in foundries to control the flow of metal and to prevent slag inclusions and sand from entering a casting.

coring, black coring. A black or gray course in the interior of a brick, usually associated with carbonaceous clays and other organic matter which have had insufficient oxidation before vitrification of the surface.

corner joint. An L-shaped joint formed by two members perpendicular to each other as used in construction.

corner rolls. Half-round units of asbestos cement used to trim and flash corners in asbestos-cement installations.

corner wear. The wear of abrasive wheels on the edges of the outer rims.

Cornish stone. Partially decomposed granite in which quartz, feldspar, and fluorine minerals are the major constituents; used as a flux in the production of ceramic whitewares. Also known as china stone, Cornish clay.

corona. A glow observed around a high-field electrode.

corrosion. The destruction or wearing away of a material by chemical action.

corrosion of refractories. The destruction and wearing away of refractories by the chemical action of external agents such as fluxes.

corrugated. Sheets of materials formed into alternating ridges and grooves.

corrugated asbestos board. Sheets of asbestos cement formed to produce a wavy or corrugated contour.

corrugated glass. Sheets of glass rolled into a wavy, furrowed, or corrugated form.

corundum. Natural Al_2O_3; mol. wt. 101.9; m.p. 2030°C (3686°F); sp. gr. 4.02; hardness (Mohs) 9; used in various abrasive and polishing operations.

cost unit. The total cost of producing a unit of a product.

cottle. The frame placed around a model to hold a plaster slurry until the plaster has set to form a mold.

counter blow. The act of blowing the parison from blown glassware after the initial shaping operation.

countercurrent adsorption. An adsorption process in which the fluid flow is in the direct opposite to the movement of the adsorbent.

coupling. A device or substance for linking together two parts or things.

coupling agents. Molecules of a substance oriented so that selected ions will react and bond with silicon ions on the surface of glass fibers, while the remainder of the molecule will react with resin during the curing operation, thereby coupling or bonding the glass fiber and resin together.

course. A horizontal layer or row of brick, block, or other substance in a structure.

course, rowlock. A course of brick laid on edge with the longest dimension perpendicular to the face of the wall.

coursing joint. A mortar joint between two masonry courses.

cove. A concaved tile or other molding forming the junction between the floor or ceiling of a room and the wall.

cover. (1) A refractory slab placed over a pot or other container to protect the contents from contamination, heat loss, etc.

(2) An item of kiln furniture supporting the posts and top of a firing assembly, and protecting the ware being fired from damage from ware placed above.

cover coat. (1) A coating of porcelain enamel applied and fused over a previously fired ground coat.

(2) A finish-coat porcelain enamel applied and fired on metal without benefit of a ground or intermediate coat.

covered pot. A refractory crucible or glass-melting pot covered with a refractory roof or slab during firing of its contents.

covering power. The degree to which a porcelain enamel, glaze, or other coating obscures the underlying surface.

cove tile. Flanged tile used to complete floor and corner joints in walls.

crack, cracking. (1) A fracture in a wet-process porcelain enamel coating that has been dried but not fired.

(2) A break in a ceramic body or glaze.

(3) The initial opening of a kiln after firing.

crack, grinding. A shallow crack(s) formed in the surface of relatively hard materials due to excessive grinding, heat, or the high sensitivity to fracture of the material during the grinding operation.

cracking, map. The random distribution of cracks on the surface of concrete due to surface shrinkage, internal expansion, or similar stresses.

cracking, pattern. See cracking, map.

cracking, random. See cracking, map.

crackle. (1) A textured effect obtained in wet-process porcelain enamels characterized by a mottled or wrinkled finish.

(2) Glassware, the surface of which has been cracked intentionally by immersion in water and then partially healed by reheating before the final shaping operation.

(3) Decorative, intentional fissures netting the surface of a glaze.

crack-off. The process of separating a glass article from the moil by breaking, as by scratching and by sharp heating.

crack settlement. A crack in the soffit of a beam or the top of a column or wall at its juncture with a slab.

cracks, green. See crack, shrinkage.

crack, shrinkage. Any crack resulting from the shrinkage of a body.

cracks, plastic. Cracks which form in concrete while it is still in the green state.

crank. (1) A refractory support for the firing of glazed flatware.

(2) A low sagger holding one porcelain plate.

craquelé. An alternate spelling of crackle, which see.

crawling. (1) A porcelain-enameling defect in which the fired coating has pulled away or rolled up at the edges of a panel or over dirt or grease, giving the ridged appearance of agglomerates or of irregularly shaped islands.

(2) A parting and contraction of glaze on the surface of ceramic ware during drying or firing, resulting in unglazed areas by the coalesced glaze.

craze, crazing. (1) The cracking which occurs in fired glazes, porcelain enamels, and other ceramic coatings due to critical tensile stresses in the coatings.

(2) Hairline cracks in concrete caused by tensile stresses created when the surface shrinks more rapidly than the interior.

crazing resistance. The resistance of glazes, porcelain enamels, and other ceramic coatings to cracking. See craze.

creased, sand. A type of texture produced on the surface of facing brick by sprinkling or rolling the brick in sand before molding or by texturing the face of the brick during molding.

creep. Plastic flow or deformation of a body under a sustained stress or load, sometimes perceptible at elevated temperatures.

crimp. To cause to become wrinkled, wavy, or bent as a means of strengthening the edges of metal shapes prior to porcelain enameling.

crinkled. A textured porcelain-enameled surface characterized by a fine wrinkled or rippled appearance.

cristobalite. A crystalline allotropic form of silica formed by the inversion of quartz at 1470°C (2678°F); m.p. 1713°C (3115°F); a major component of silica refractories; also used in investment casting of metals; sometimes present in siliceous ceramic bodies.

critical bed depth. The minimum depth of an adsorbent bed required to maintain the mass-transfer zone.

critical humidity. The humidity value above which a solid salt will always become damp and below which it will always remain dry.

criticality. The condition whereby a chain reaction is allowed or accidentally occurs; sustaining a chain reaction is not necessary to the definition of criticality.

criticality controls. Mechanisms by which it is insured that criticality cannot occur. See criticality.

criticality incident. An accident caused by the accumulation of a critical mass of material.

critical mass. The minimum mass of fissile material which can be made critical with a specified geometrical arrangement and material composition.

critical speed. The speed of rotation beyond which the vibration of a spindle carrying an abrasive wheel or point would be hazardous.

crizzle. An imperfection in glass consisting of a multitude of fine surface fractures.

crockery. A thick form of porous opaque pottery often fired at low heat.

crocus martis. A purple or brownish red iron oxide used as a pigment in decalcomanias and glazes.

Crooke's glass. A glass of low ultraviolet transmission containing cerium and other rare earths.

cross-bend test. A test in which bisque and fired porcelain-enameled panels are progressively distorted by bending to determine the resistance of the coating to cracking.

cross-breaking strength. A measure of the resistance of a material to breakage under transverse stress.

cross-feed grinding. The controlled movement of a grinding wheel over a horizontal work piece resting on a work table, the grinding being done at a prescribed rate or depth.

cross-fired furnace. A furnace in which fuel is supplied from side ports.

cross grains. Tangled laminations in a body causing irregular or imperfect cleavage patterns.

cross section. A cut through a substance, especially at right angles to a dimension.

crown. (1) The top or dome of a furnace or kiln.

(2) The top or highest point of the internal surface of the transverse cross section of a concrete pipe.

crown blast. A stream of air introduced at the top of the exit end of a tunnel kiln.

crown brick. A wedge-shaped brick at the crown of an arch that locks other brick in place.

crown flint glass. An optical crown glass containing a substantial addition of lead oxide to produce a higher dispersion of light than the usual optical crown glass. See crown glass, optical.

crown glass. (1) A hard, easily polished, highly transparent optical glass with high refraction and low dispersion, typically containing 72% of SiO_2,15% of Na_2O, and 13% of CaO. See optical crown glass.

(2) A type of window glass shaped by whirling a glass bubble to form a flat circular disk with a lump in the center formed by the glass blower's rod.

crown glass, barium. An optical glass containing a substantial amount of barium oxide as a replacement for calcium oxide as a flux. See crown glass, optical.

crown glass, borosilicate. An optical crown glass containing silica and boric oxide as the major ingredients. See crown glass, optical.

crown glass, lead. Crown flint glass containing a substantial amount of lead oxide as a flux. See crown flint glass.

crown glass, lead-barium. An optical glass containing substantial amounts of lead and barium oxides as fluxing ingredients. See crown glass, optical.

crown glass, optical. An optical glass characterized by low dispersion and low index of refraction, usually forming the diverging elements of an optical system; any optical glass possessing a nu value less than 50.0, or any optical glass having a nu value between 50 and 55 with an index of refraction less than 1.60.

crucible. A refractory vessel or pot in which a material may be melted or calcined at a high temperature.

crucible clays. Refractory ball clays used in the production of high-temperature crucibles or pots. See ball clay.

crude-dressed mica. Crude mica from which dirt, rock, and other contaminants have been removed.

crude mica. Mica in the state as mined, with dirt, rock, and other contaminants still present.

crush. (1) A lightly pitted, dull-gray area on flat glass sheets; a defect.

(2) To grind or break solid substances into small bits or fragments.

crush dressing. The use of steel rolls to form or dress the face of grinding wheels to a desired contour.

crushed gravel. The product resulting from the artificial crushing of gravel with substantially all fragments having at least one face resulting from fracture; used as an aggregate in concrete.

crushed stone. The product resulting from the artificial crushing of rocks, boulders, or cobblestones, substantially all faces of which have resulted from the crushing operation.

crusher. A device which breaks or grinds substances into smaller particles.

crusher cone. A crushing device consisting of a gyrating cone in a conical cavity with tapered clearances such that material is reduced in size several times during passage.

crusher, conical. A crusher in which clay material passes between a moving set of conical rolls.

crusher, disintegrator. An apparatus consisting of a large smooth roll operated at a slow speed and a parallel small-toothed roll operated at a high speed for crushing soft materials.

crusher, grizzly. A type of crusher consisting of moving rods or bars which crush and separate ground material according to size.

crusher, gyratory. A large primary crusher consisting of a rounded crushing head on a vertical shaft within a conical outer shell.

crusher, impact. Any crushing machine that breaks down material by means of a shattering blow such as is imposed by ball mills, rotating hammers, roller bars, etc.

crusher, intermediate. Any crushing device that will reduce the particle size of materials within an intermediate range; that is, from a particle size of approximately 8 mesh to a size of about 20 mesh.

crusher, jar. A small rotating ceramic or ceramic-lined cylinder in which cascading pebbles or porcelain balls are employed as the grinding media.

crusher, jaw. A crushing machine consisting of a moving hinged jaw which swings toward and away from a stationary jaw in an alternating fashion.

crusher, muller. A grinding machine that makes use of mullers riding on the bottom of a revolving pan which is perforated to allow passage of the ground material.

crusher, primary. The first of a series of crushers in which rock, shale, and other minerals are pulverized.

crusher, ring. A type of hammer mill consisting of steel rings held outwardly by the centrifugal force of a horizontal shaft rotating at high speed, the solid material being crushed between the rings and the outer shell of the mill.

crusher, rod. A pulverizer consisting of a rotating tube partially filled with horizontal parallel rods which grind solid materials by rolling over each other.

crusher, roll. A crushing apparatus consisting of two horizontal cylinders which rotate toward each other about their respective axes.

crusher, rotary. A pulverizer in which a cone, rotating at a high speed on a vertical axis, forces solid material against a metal encasement or shell.

crusher, sawtooth. A pulverizer in which material is crushed during passage between sawtooth shafts rotating at different speeds.

crusher, secondary. Crushing and pulverizing machines set next in line after the primary crusher to reduce still further the particle size of rock, shale, and other minerals.

crusher, single roll. A crushing machine consisting of a rotating cylinder with a corrugated or toothed outer surface which crushes material between the teeth of the cylinder and stationary breaking bars.

crusher, vibratory. A ball mill placed on a vibrating mechanism to increase the shattering impact of the grinding media by the bouncing action.

crushing. The process of reducing the size of lump material by mechanical means.

crushing, choke. The grinding of materials in a roll crusher with the space between the rolls being completely filled with particles of the material being ground to gain the added effect of the particles acting on each other.

crushing strength. The property of a material to resist breakdown under externally applied compressive loads, calculated as the load in pounds per square inch or kilograms per square centimeter required to fracture the specimen, based on the cross-sectional area of the specimen.

cryogenics. The study of the production of extremely low temperatures and their effects on materials.

cryolite. Na_3AlF_6; mol. wt. 210.0; m.p. 1000°C (1832°F); sp. gr. 2.95–3.0; hardness (Mohs) 2.5; used in opal glass and porcelain enamels as a flux and opacifier, as a filler in grinding wheels, as a flux in whiteware bodies, and as a constituent in dental cements, light bulbs, and welding-rod fluxes.

cryptocrystalline. A crystalline structure in which the individual crystals are so small that they are not visible under a petrographic microscope.

crystal. A chemically homogeneous solid body having a definite internal molecular structure and, if developed under favorable conditions, having a characteristic external form bounded by plane surfaces.

crystal glass. A colorless, highly transparent glass used for art and tableware.

crystal habit. The size and shape of a crystal, usually contained in a solid matrix.

crystal laser. A solid laser of high-purity crystalline or doped crystalline material, such as pure or doped ruby; used for generating a coherent beam of output light.

crystalline. Composed essentially of crystals. See crystal.

crystalline discoloration. Discoloration appearing as lighter or darker shades of the basic color of mica.

crystalline glaze. A glaze containing macroscopic crystals which have grown during the cooling period following a firing operation.

crystallization, water of. See combined water.

crystallographic x-ray. The study of the structure, identity, texture, and behavior of crystals by x-ray techniques.

crystal, polar. A ferroelectric crystal.

crystal, quartz. A natural or artificial crystal of quartz, SiO_2, having piezoelectric properties; also known as rock crystal.

crystal, rock. (1) A transparent, colorless form of quartz used for lenses and prism components in optical instruments.
(2) Highly polished, handcut or engraved, blown glassware.

crystal, semiconducting. A crystal, such as silicon or germanium, which exhibits an electrical conductivity between that of a metal and an insulator.

crystal structure. The arrangement of atoms or molecules in a crystal.

crystolon. A trade name for an abrasive silicon carbide.

cubical expansion. The change in volume of a material with changes in temperature and pressure.

cull. Material rejected as being below standard and therefore unacceptable.

cullet. Waste or broken glass suitable as an addition to raw batch in the manufacture of glass.

cullet cut. A scratch imperfection in glass caused by a particle of cullet lodged in the polishing felt during the polishing operation.

cullet, raw. A glass charge consisting entirely of cullet. See cullet.

culvert. A covered channel or pipeline under a highway, railroad, canal, or similar construction for the conveyance of water.

cumulative weighing. The weighing of materials successively on the same scales, the weights being added to the previous weights of the batch.

cupel. A small crucible of bone ash used in assay work.

cup gun. A spray gun with a fluid container or cup attached as an integral part so as to feed the fluid into an atomizing nozzle or air stream.

cupola. A circular, vertical furnace, for the melting of iron.

cupping. (1) The pouring of porcelain-enamel slip over an item or part during draining to obtain a smoother and more uniform coating.
(2) A concave or convex arcing of a coated abrasive caused either by an excess or lack of moisture in the backing and the bond.

cup wheel. A cup-shaped or dish-shaped grinding wheel.

cure. The reaction mechanism in which the physical, chemical, and mechanical properties of a hydraulic cement change through the phases of slurry-paste-solid with time, with or without heat, in the presence of water.

cure, autoclave. A means of accelerating the cure reaction of cement at an elevated temperature in saturated steam where reactive siliceous material has been incorporated in the cementitious matrix such that a hydrothermal reaction takes place between the cement and the silica.

cure, normal. A condition of curing cement at atmospheric pressure with incidental external heat.

cure, normal cure. The method of setting or hardening asbestos-cement products wherein the portland cement is allowed to hydrate at atmospheric conditions of pressure, preferably under conditions to inhibit water loss.

curie. A unit of radioactivity defined as the quantity of any radioactive nuclide which has 3700×10^{10} disintegrations per second.

Curie point, Curie temperature. The temperature marking the transition between ferromagnetism and paramagnetism or between the ferroelectric phase and the paraelectric phase.

curing. Protection of concrete for a specified period of time after placement to provide moisture for hydration of the cement, to provide the proper temperature, and to protect the concrete from damage by loading or mechanical disturbance.

curing agent. An additive to cement and asbestos-cement products to increase the chemical activity between the cementitious components with an increase or decrease in the rate of cure.

curing blanket, curing mat. (1) A dampened mat laid over fresh concrete to provide curing moisture.
(2) A dry mat laid over green cement as insulation during cold weather.

curing compound. A liquid sealant sprayed on the surface of fresh concrete as protection against loss of moisture.

curing, membrane. The method of curing concrete by the application of a curing compound to the surface.

curling. A defect in porcelain enamels similar to crawling, which see.

current, eddy. Electrical currents caused to flow in a conductor by the time or space variation, or both, of an applied magnetic field.

current-flow magnetization. A method of magnetizing by passing a current through a component by means of prods or contact heads; the current may be alternating, rectified alternating, or direct.

current-induction magnetization. A method of magnetizing in which a circulating current is induced in a ring component by the influence of a fluctuating magnetic field.

current, inrush. A transient current which exists at the instant an electrical contact is closed and which continues briefly.

current, magnetizing. The flow of either alternating or direct current used to induce magnetism into a part being inspected.

current, steady-state. The current in a circuit after it has reached equilibrium.

current, voltaic. The electric current produced by chemical action as in a battery composed of a primary cell or group of cells.

curtain arch. An arch of refractory brickwork supporting the wall and the upper part of a gas producer and the gas uptake.

curtains. A defect, which may occur in porcelain-enamel ground and cover coats, characterized by a sagged or draped appearance.

curtain wall. An exterior or interior wall section of a building which is neither an integral part of the structure nor load bearing.

curve, calibration. The graphical representation of a relationship between a measured parameter and a concentration or mass of the standard for the substance under consideration.

cut glass. Glassware which has been decorated by grinding figures or patterns on the surface of the ware by means of an abrasive, followed by polishing.

cut glaze. A glazed area in which the coating is of insufficient thickness for good coverage.

cutlery mark. A metallic line or smear on a dinnerware glaze caused by the abrasion of a knife or other instrument on the surface.

cut-off level. The value established above or below which a product is rejectable or distinguished from other items of the same origin.

cut-off scar. A machine-made scar on the base of a glass bottle.

cut-off wheel. A thin, usually organic-bonded, abrasive wheel used for cutting, slicing, or slotting a material.

cut sizes. Flat glass sheets cut to specific dimensions.

cutter. (1) A workman engaged in grinding figures or designs on glass.
(2) A workman who cuts flat glass.
(3) A tool used in cutting glass.

cutter, guillotine. A mechanically or manually operated heavy steel knife used to cut through and trim sheets of materials.

cutter, reel. A device consisting of a series of wires spaced evenly and stretched tightly across the diameter of a circular frame surrounding an extruding clay column in such a position that, when the frame is rotated, the wires will cut the clay column into prescribed lengths.

cutting. (1) Scoring a glass sheet with a diamond or steel wheel and then breaking it along the scratch.
(2) Producing cut glass.

cutting, brilliant. A decorative process by which designs are produced on flat glass by means of abrasive and polishing wheels.

cutting fluid. See coolant.

cutting off. Removing a pot from the potter's wheel by cutting with a wire or string.

cutting rate. The amount of material removed by a grinding wheel per unit of time.

cutting table. A mechanical or stationary table upon which a clay column is severed or sliced.

cutting tool. The portion of the grinding or machining device which contacts and removes material from a workpiece.

cyanide metal treatment. A cleaning and neutralizing treatment of metals in a dilute aqueous bath of sodium cyanide preparatory for porcelain enameling.

cyanite. An alternative spelling of kyanite, which see.

cycle. (1) A complete set of operations that is repeated as a unit.
(2) The time between the first fill of batch and the casting of glass in open-pot practice.

cyclone separator. A device for removing particles from air, water, or other fluids, or for separating substances according to size or density, by centrifugal means.

cyclopean. Mass concrete, such as used in dams and thick structures, containing aggregate larger than 6 inches (15 centimeters).

cylinder. A large steel pipe filled with concrete and used as a pile foundation.

cylinder process. A process for the manufacture of window glass in which molten glass is blown and drawn into the form of a cylinder which subsequently is split or cracked open, reheated, and flattened.

cylinder, test. A concrete test specimen cast in the form of a cylinder.

cylinder wheel. A grinding wheel with a comparatively large hole, typically several inches in diameter, used in surface grinding where the work is done by the side rather than by the peripheral surface of the wheel.

cylindrical grinding. Grinding the outer surface of a part which either is rotated on centers or is centered in a chuck.

cylindrical-screen feeder. An apparatus in which a plastic clay is forced through a cylindrical screen by a bladed shaft rotating on the same axis as the center of the screen; the shredded clay then is delivered to a subsequent processing unit.

Czochralski process. A technique for growing single crystals by pulling a rotating seed crystal from a bath of molten material of the same composition.

D

damage. Harm to a product, facility, equipment, or other item of value, usually short of complete destruction.

damp air. Air having a high relative humidity.

damp course. A layer or sheet of any impervious material, such as a plastic, placed over or around an area such as a wall to prevent the seepage of water into the area.

damper. A movable panel or valve designed and placed to regulate the flow or draft of air into a furnace or kiln.

damping capacity. The ability of a material to absorb vibrations.

damp-proofer. A substance, such as sodium silicate or a fluosilicate of aluminum or zinc, which is added to a batch of concrete or applied as a coating to the surface of hardened concrete to decrease the capillarity of the concrete.

Danner process. A continuous process of producing glass rod or tubing in which molten glass is drawn from a tank and formed by means of a rotating mandrel.

darby. A flat-surfaced metal or wood tool used for smoothing freshly applied plaster.

Darcy's law. The permeability of a substance is the rate at which a fluid will flow through the substance times the pressure drop per unit of length of flow divided by the viscosity of the fluid.

dark adaptation. The adjustment of the eyes in passing from a bright to a darkened area, and vice versa.

dark plaster. A plaster made from calcined, but unground, gypsum.

data. Information obtained through experimentation which is organized for analysis or used as a basis for a decision.

data analysis. The evaluation and interpretation of data.

data, raw. Data which have been collected and are now ready for analysis.

datolite. $CaBSiO_4(OH)$; sp. gr. 2.9–3.0; hardness (Mohs) 5–5.5; a mineral sometimes used as a flux in glazes.

daughter. The product of radioactive decay of a nuclide; the product may or may not be radioactive.

daylight glass. A glass that absorbs red light so that the transmitted light resembles daylight; used in incandescent light bulbs and similar products.

day tank. A periodic glass-melting tank consisting of a single compartment designed to be charged, fired, and emptied during each day of hand gathering.

d-c. Symbol for direct current.

dead-air space. Sealed air space, such as between the inside and outside panels in a wall.

dead burn. To heat-treat a material, such as a basic refractory, to produce a dense refractory product resistant to atmospheric hydration or recombination with carbon dioxide; usually the treatment is at a higher temperature for a longer period of time than a normal calcine.

dead-burned magnesia. See dead-burned magnesite.

dead-burned magnesite. Magnesite, $MgCO_3$, or other magnesium-bearing substance convertible to magnesia, MgO, which has been heat-treated to temperatures above 1450°C (2642°F) to produce a stable material suitable for use as an ingredient in refractory products.

dead-burned refractory dolomite. Raw refractory dolomite, $CaMg(CO_3)_2$, which has been heat-treated to form calcium oxide, CaO, and periclase, MgO, in a matrix resistant to hydration and recombination with carbon dioxide.

dead plaster. Plaster of paris which has been overfired during manufacture.

dead plate. A stationary plate in a glass-making machine on which a glass article rests to await transfer to a subsequent operation during an automatic production process.

deaired brick. A densified brick in which air has been removed during the forming process by the application of a vacuum.

deairing. The process of removing entrapped or absorbed air from a mass or slurry, usually by application of a vacuum.

debiteuse. A vertically slotted, floating clay block on the surface of molten glass through which glass is drawn upward in the Fourcault process. See Fourcault process.

debug. To locate and remove sources of defects and causes of failure in a process or system.

deburr. To remove burrs, sharp edges, and fins from a product such as by grinding or tumbling in a drum containing loose abrasive particles.

decal, decalcomania. Colored designs printed on specially prepared paper for transfer as decorations on glass, glazed and unglazed ceramic ware, porcelain enamels, and other surfaces. See transfer, slide-off.

decarburized steel. Steel of extremely low-carbon content suitable for porcelain enameling, particularly in the production of one-coat (no ground coat) ware.

decay. The spontaneous transformation of a nuclide into one or more nuclides having measurable lifetimes.

deck. The refractory top of a kiln car.

decking. The loading of ware in multiple layers on kiln cars preparatory for firing.

decolorizer. A material, such as a selenium compound, which is added to glass batches to remove or mask color in finished products.

decolorizing. The process of producing a colorless appearance in glass.

decompose. To separate into constituent parts.

decorated. An item made attractive and pleasing to the eye by the use of designs and colors.

decorating fire. The firing process in which decorations are affixed to glazed and porcelain-enameled surfaces.

decoration, impressed. A decoration pressed into a plastic clay body.

decoration, inglaze. A decoration applied to an unfired glaze on an item and fired concurrently with the glaze.

decoration, monochrome. A decoration consisting of a single color.

decoration, overglaze. A ceramic or metallic decoration applied and fired over a previously fired glaze.

decoration, underglaze. A decoration applied directly to the surface of a ceramic item and subsequently covered with a transparent glaze, the decoration and glaze then being fired concurrently.

de-enameling. The chemical or mechanical removal of a porcelain-enamel coating from its base metal; for example, by immersion of the enameled item in a hot bath of sodium hydroxide or by sandblasting.

deep drawing. The die pressing of sheet-metal shapes to relatively large depth-diameter ratios, the shapes subsequently to be porcelain enameled.

defect. An imperfection or discontinuity in a product that interferes with the usefulness or the aesthetic value of the product.

deflecting wedge. A wedge-shaped refractory placed so as to split and distribute a cascading stream of material such as, for example, a stream of coal onto the floor of a coke oven.

deflection. (1) To alter the direction of flow of a stream of gas or liquid by means of a baffle or other designed obstruction.

(2) The linear measurement of bend when a specimen or beam is loaded at mid-span.

deflocculant. A substance, such as water glass or sodium carbonate, which will disperse the agglomerates in a slurry to form a colloidal or near-colloidal suspension which will result in a more fluid slurry or slip.

deflocculate. To disintegrate and disperse agglomerates in a slurry to form a colloidal or near-colloidal suspension of greater fluidity.

deflocculating. Reducing the viscosity of a slip or slurry by the addition of a deflocculant such as water glass or sodium silicate.

defoamer. Any substance, such as the sulfonated oils and silicones, which will reduce or eliminate foam from glaze or porcelain-enamel slips, cleaning and pickling solutions, etc.

deformation. An alteration in the shape or dimensions of a solid when subjected to conditions of pressure or stress.

deformation, elastic. The deformation of a solid, such as concrete, under load which disappears when the load is removed; that is, the solid will return to its original shape when the pressure is relieved.

deformation eutectic. The minimum temperature, under specified conditions, at which a specimen composed of a mixture of solids will soften/or melt sufficiently to deform.

deformation, plastic. Permanent change in the size or shape of a body subjected to stress without fracture.

deformation point. The temperature at which viscous flow of a glass exactly counteracts the thermal expansion of the glass.

deformation, pyroplastic. The permanent or irreversible deformation of a solid at an elevated temperature.

deformation temperature. The minimum temperature at which a solid substance begins to deform.

deformed bar. A steel rod or bar covered with ribs or indentations to improve or enhance its bond with concrete.

degassing. The removal of gases from liquids and solids, such as by heating or by the application of a vacuum.

degreasing fluid. A solvent or detergent solution employed to remove oil and grease from a surface.

degree of freedom. The number of variables (temperature, pressure, volume, composition, concentration, etc.) which must be specified to define the state of a material or system.

dehydration. The removal of free or combined water from a substance or compound, usually by heating or by evaporation in a vacuum.

deionized water. Water that has been purified of salts and is equivalent to distilled water.

delamination. The separation of a laminate into its constituent parts.

delayed fishscale. Half-moon or fishscale shaped fractures occurring spontaneously in porcelain-enamel coatings at some time after the completion of the porcelain-enameling process.

Delft ware. A soft, buff-colored majolica body covered with an opaque white glaze; decorations are painted over the unfired glaze, often using a characteristic blue, and fired with the glaze.

delivery. (1) The process or the equipment delivering charges or gobs of glass to a forming machine.
(2) The final act of any glass-forming unit or process, including the removal of an article from its mold.
(3) The act of delivering or conveying.

Della Robbia ware. A hard, durable item of terra-cotta artware covered with white and brilliantly colored glazes.

demagnetization. The reduction or removal of magnetism from a ferromagnetic material such as by striking a bar magnet with a hammer to disorient the previously oriented molecules.

demijohn. A narrow-necked glass or stoneware bottle of one- to ten-gallon capacity; the bottle usually may have one or two handles and is enclosed in a wickerwork basket.

dendrite. A branching, tree-shaped growth of one mineral or metal in another mineral or metal, the mineral growth usually being crystalline in nature.

dense. (1) Compressed closely together into a compact mass.
(2) A subclass of optical glass having a higher than normal index of refraction.

density. (1) The mass of a substance per unit of volume, expressed as grams per cubic centimeter or other measure of metric volume or as pounds per cubic inch, cubic foot, or gallon.
(2) The degree of opacity of a translucent material.

density, absolute. The weight of a unit volume of a substance measured under specified or standard conditons, excluding the pore volume and interparticle voids contained in the specimen.

density, apparent. The weight of a unit volume of a substance, measured under specified or standard conditions, including the pore volume and interparticle voids contained in the specimen. Also called apparent specific gravity.

density, block. The weight of a unit volume of a substance, measured under specified or standard conditons, including the pore volume but excluding the interparticle voids. Also called particle density.

density, bulk. The total mass of a body or substance per unit volume.

density, magnetic flux. The strength of a magnetic field expressed in flux lines per unit area.

density, packing. The density of an aggregate packed in a container under controlled conditions; reported as pounds per cubic foot or grams per milliliter.

density, particle. The weight per unit volume of a substance, measured under specified or standard conditions, including the pore volume but excluding the interparticle volume. Also called block density.

density, pour. The weight of a powdered or granulated material poured into a graduated container divided by its volume.

density, powder. The density of a powdered material, including all pores.

density, pressure. The density of a compacted substance before sintering or firing.

density, tap. The apparent density of a powdered or

granulated material resulting when the receptacle containing the material is vibrated or tapped manually under standard or specified conditions.

density, theoretical. The density of a material calculated from the total number of atoms per unit cell and the measurement of the lattice parameters.

density, true. The weight per unit volume of a material, measured under standard or specified conditions, excluding both its pore volume and interparticle voids.

dental porcelain. A bubble-free porcelain of exceptionally high strength and density which is shaped and tinted for oral prosthetic use.

deoxidizer. A material which will reduce the oxygen content of another material.

depleted uranium. Uranium containing less than 0.711% by weight, of the isotope uranium^{235}.

deposition, vacuum. The deposition or condensation of a vaporized coating of a material on the surface of another in a vacuum.

deposits, carbonaceous. Particles of carbon or carbon-bearing materials resulting from the decomposition of organic matter or vapors which are laid down on the surfaces of other materials.

depth of fusion. The distance to which fusion extends into a body from its original surface following exposure of the body to its fusion temperature.

depth of penetration. (1) The distance a penetrant has entered into a solid material as measured from the surface of the material.

(2) The maximum depth at which a magnetic or ultrasonic indication can be measured in a test specimen.

descaling. The removal of scale from iron and steel surfaces which are to be porcelain enameled by treatment in an acid bath or by heating the metals in a furnace to a red heat, or both.

desiccant. A drying agent or substance which will absorb moisture from the atmosphere; for example, calcium chloride.

design, standard. A proven and accepted design for a component or product.

desorption. The removal of adsorbate from a surface; the reverse of absorption or adsorption.

detector efficiency. The fraction of particles or photons striking a detector which give rise to a detected response in the measurement of radioactivity.

detector geometry. The fraction of emissions from a source (particles or photons) which impinge on a detector in the measurement of radioactivity.

detector, germanium. A germanium-semiconductor detector used in high resolution gamma-ray spectrometry.

detergent. A substance or mixture having a cleaning action due to a combination of factors such as the reduction of surface tension and improved wetting, emulsification, dispersion, and foam-forming properties of washing solutions; a detergent may be anionic (such as the alkylaryl sulfonates), cationic (such as the quaternary ammonium halides), or nonionic (such as the alkylomides) in their cleansing actions.

detergent remover. An aqueous solution of a detergent employed to remove penetrants from test specimens.

determination. The ascertainment of the quantity or concentration of a substance in a sample.

developer. A material applied to the surface of a test specimen, after the removal of a penetrant solution, to intensify the marking of a discontinuity in the specimen surface.

developer, soluble. A developer employed in liquid penetrant inspection which is completely soluble in its carrier, but not a suspension of a powder in a liquid, which dries to an absorptive coating.

deviation, relative standard. The standard deviation of a value expressed as a percentage of the mean value.

devitrification. The formation of crystalline structures in a glassy matrix, such as may occur in a glass, glaze, or porcelain enamel during the cooling of a vitreous mass.

devitrified glass. A glassy product containing a crystalline phase produced by incorporating a nucleating agent in the molten glass batch, followed by a predetermined heat-treatment process; such products exhibit high resistance to breakage, thermal shock, and chemical attack, and are particularly useful in high-temperature applications.

dewatering. The removal of water from a slurry or slip by filter pressing, centrifuging, evaporation, or other process.

dextrin. A polymer of glucose having a composition between that of starch and maltose; used as a binder in glazes and as a carrier or binder for ceramic inks and decorating colors.

diagonal bond. A type of masonry construction in which the header brick are laid in a diagonal pattern.

diagram, equilibrium. A phase diagram showing the relationship between the composition, temperature, and pressure of a compositional system.

diagram, phase. A graphical representation showing the temperatures at which transitions will occur between different phases of a compositional system as a function of the system.

diamagnetic material. A material repelled by a magnet and which will position itself at right angles to the magnetic lines of force.

diamond. A mineral or synthetic product consisting essentially of carbon crystallized in the isometric system, usually in octahedral shape; used in polishing powders, abrasive wheels, glass cutters, drill bits, and similar products; sp. gr. 3.51–3.53; hardness (Mohs) 10.

diamond indenter. An instrument equipped with a diamond point which is pressed into the surface of a solid, the depth of penetration under a given load being taken as a measure of the hardness of the material being tested.

diamond paste. Diamond dust dispersed in a paste or slurry for use as a grinding or polishing compound.

diamond point. A cutting tool equipped with a diamond point.

diamond powder. See diamond paste.

diamond-pyramid test. A measurement of the hardness of a solid material in which a diamond point having an angle of 136° between opposite faces is pressed into the surface of the material under variable loads, the depth or width of the indentation being taken as a measurement of the hardness of the material. See Knoop hardness, Vickers hardness.

diamond saw. A saw with diamonds or diamond dust inset on the cutting edge of the saw blade.

diamond, synthetic. A manufactured diamond formed by heating a carbonaceous substance to an extremely high temperature under an extremely high pressure.

diamond tool. Any tool in which the working area is inset with diamonds or diamond dust.

diamond wheel. A bonded grinding wheel in which the abrasive grains are crushed and sized natural or synthetic diamonds.

diaspore. $Al_2O_3 \cdot H_2O$; mol. wt. 120.0; sp. gr. 3.35–3.45; hardness (Mohs) 6.5–7; a mineral used in refractories or as an abrasive.

diaspore clay. A mineral consisting essentially of diaspore bonded by clay.

diatom. Any of a class of minute planktonic unicellular organisms with silicified skeletons.

diatomaceous earth. A light friable highly siliceous material derived from the skeletons of diatoms; m.p. 1715°C (3119°F); sp. gr. 1.9–2.35; used as a thermal insulator in the form of aggregate, brick, blocks, and cement, and sometimes as a mild abrasive.

diatomite. Dense chert-like diatomaceous earth. See diatomaceous earth.

dibase. A basic igneous rock used as aggregate in concrete.

dice. Cubical fragments of tempered glass.

dice block. The refractory shapes which line the submerged passage between the melting and the refining zones of a glass tank.

dichroic glass. A glass which will transmit some colors and reflect other colors, or which will display certain colors when viewed from one angle and different colors when viewed from a different angle.

dickite. $Al_2O_3 \cdot 2SiO_2 \cdot 2H_2O$; a mineral of the kaolin family.

didymium salts. A mixture of rare earth salts sometimes used in substantial amounts as a glass colorant, the color varying with the particular rare earths present; sometimes used in small amounts as a glass decolorizer; also used as a component in temperature–compensating capacitors.

die. (1) A mold in which ware is shaped by pressing or by casting.

(2) A perforated plate through which ware is shaped by extrusion.

die drawing. The pulling of filaments or tubing of molten glass through a die to obtain a desired cross-sectional shape and dimension.

dielectric. An electric insulator in which an electric field can be sustained with a minimum dissipation of power.

dielectric breakdown voltage. The potential difference at which electrical failure occurs, under prescribed conditions, in an electrical insulating material between two electrodes.

dielectric constant. The ratio of the capacitance of a capacitor filled with a given dielectric to that of the same capacitor with a vacuum as the dielectric.

dielectric strength. The maximum electrical gradient a dielectric can withstand without rupture when tested under specified conditions.

die lubricant. A material applied to the work surface of a die or added to the substance or product being formed to facilitate movement of the material to minimize die wear and to ease the removal of the formed product from the die; examples of such lubricants are graphite and molybdenum disulfide.

die pressing. The forming or shaping of an item in a die or mold under pressure.

differential coil. See coil, differential.

differential heat of adsorption. The measure of the heat evolved during the adsorption of an incremental quantity of an adsorbate at a given level of adsorption.

differential heating. The thermal gradient occurring in a body during heating, causing stress to develop in a body.

differential measurements. The measurement of any imbalance in a body or system.

differential pressure. The difference in pressure occurring in a system.

differential thermal analysis. The determination of the temperature at which thermal reactions will occur in a material during the heating of the material to elevated temperatures.

differential thermogravimetric analysis. The measurement of the weight change taking place in a material during heating.

diffraction, x-ray. The scattering of an x-ray beam into many beams at many angles to the original beam in an orderly pattern characteristic of different atoms; used to analyze crystal structures.

diffuse indication. The detection of the presence of some imperfection in a specimen which has not been clearly defined.

diffusion, vacuum. The diffusion of selected impurities into a semiconducting material in a vacuum.

diffusivity, thermal. The measurement of heat flowing through a unit area of a substance per unit of time divided by the product of the specific heat, density, and temperature gradient in the material.

diglycol stearate. $(C_{17}H_{35}COOC_2H_4)_2O$; a white wax-like solid used as a temporary binder in the manufacture of grinding wheels and other abrasive products.

digs. Deep, short scratches on the surface of glass.

dilatancy. The property of a material to thicken or become solid under pressure and to become fluid when the pressure is removed.

dilatometer. Any of several instruments used to measure the thermal expansion of a material.

dilution. The reduction of the concentration of a substance by the addition of another substance.

dilution factor. The ratio of the volume of a diluted substance to the volume of the original substance before dilution.

dimensional coordination. The selection of materials by size and shape in a relationship with other units to facilitate assembly or construction.

dimension, nominal. A dimension that may be greater than a specified masonry dimension by the thickness of a mortar joint.

dimple. A shallow conical depression in a fired porcelain-enamel or glaze surface.

dinnerware. Ceramic and glass articles employed in table service.

diode. (1) A rectifier consisting of a semiconducting crystal with two terminals.

(2) An electron tube containing an anode and a cathode.

diopside. $CaO{\cdot}MgO{\cdot}2SiO_2$; mol. wt. 220.5; m.p. 1392°C (2538°F); sp. gr. 3.28; used as a component in whiteware bodies, glazes, and glass, and as a refractory in welding-rod coatings.

dioxide. A compound containing two atoms of oxygen in combination with another element to form a molecule.

dip coatings. Coatings applied to ceramic bodies or to steel to be porcelain enameled by dipping the item in a solution, slip, or other bath and then allowing it to drain to the desired thickness before firing.

dip encapsulation. The process of enclosing or encasing an item by immersion in an insulating material.

dip mold. A glass-forming mold constructed in one piece having an opening at the top for the entry of the molten glass and for the removal of the finished piece.

dipped joint. A masonry joint in which the masonry first is wetted by a mortar slurry before it is set or placed in the mortar joint, or by pouring the slurry over a course of masonry before laying the next course.

dipper. An operator who applies porcelain-enamel or glaze slips to ware by dipping.

dipping. (1) The process of applying porcelain enamel or glaze to an item by dipping in a slip or slurry; a smooth uniform coating is obtained by allowing the item to drain naturally or by swinging, shaking, and gently spinning the article.

(2) Immersion of a heated cast-iron article in dry powdered frit to obtain an adherent coating which may or may not be subjected to further heating to obtain complete fusion.

dipping weight. The weight of a coating retained on dipped ware per unit of area, reported either as wet or dry weight, the dry weight being the more accurate.

dip rinse. The removal of excess penetrant from an inspection specimen by dipping the specimen in water or a penetrant remover.

dip tank. A receptacle or tank containing a solution or slurry in which ware is dipped.

dip tank, recirculating. A dip tank equipped with a means, usually a pump, by which a slip or slurry may be circulated continuously to keep solids in suspension in a slip or slurry.

direct-arc furnace. A melting furnace in which an electric arc extends directly from electrodes to the batch contained in the furnace.

direct-bonded basic brick. A fired refractory in which the grains are bonded by solid-state diffusion.

direct current. An electric current flowing in one direction.

direct fire. The firing of ware in direct contact with the products of combustion in the furnace or kiln.

direct-fired furnace. A furnace having neither recuperator nor regenerator; that is, the furnace is fired without preheating the fuel or air of the fuel mixture. See recuperator, regenerator.

direct-on enamel. A porcelain-enamel finish coat applied directly to a steel base without benefit of a ground coat.

direct teeming. The pouring of molten iron directly from the ladle into ingot molds.

dirt. Undesirable foreign matter in a body or coating; a cause for rejection.

disappearing filament pyrometer. An instrument for measuring high temperatures in which a heated filament of calibrated temperature, enclosed in a telescope, disappears when focused on an incandescent background or surface at the same temperature.

disappearing highlight test. A test to evaluate the deterioration of a glaze, porcelain enamel, or glass surface, or damage done by chemical or physical action, by observing the loss of gloss or change in surface texture.

discontinuity. A pinhole, fracture, or other break in a coating which impairs or destroys the usefulness, purpose, or value of a coating.

discontinuity, artificial. A discontinuity in a surface or coating which has been made for purposes of comparison or reference.

discontinuity, subsurface. A discontinuity, such as a bubble or blister, which does not extend through the surface of an item or coating.

dish grinder. A grinding machine equipped with a dish-shaped abrasive wheel as a grinding mechanism.

dish wheel. A dish-shaped abrasive grinding wheel.

disintegration. To come apart or separate into components.

disintegration per minute. The number of spontaneous nuclear transformations occurring in a material per minute.

disintegrator. A device for grinding and pulverizing materials.

disintegrator crusher. A two-roll crusher consisting of a low-speed smooth and a high-speed serrated roll between which solids are crushed and passed.

disk feeder. A rotating disk beneath the opening of a bin which delivers material from the bin at a specified rate by controlling the rate of rotation of the disk and the size of the gate opening of the bin.

disk grinder. A grinding machine equipped with a large abrasive disk as the work mechanism.

disk sander. A machine which employs an abrasive-coated disk as the grinding and polishing surface.

disk, strain. An internally stressed glass disk of calibrated birefringence which is used as a comparative standard in the measurement of the degree of stress or the degree of annealing of glassware.

disk wheel. A bonded abrasive wheel mounted on a plate so that grinding may be done on the side of the wheel.

dispersion. (1) A change in the refractive index of a substance which occurs with change in wavelength or frequency.

(2) Widely distributed or scattered particles in a medium.

dissociation. The breakdown of a compound or substance due to a change in physical conditions, such as pressure or temperature.

distinguishing stain. An organic colorant added to a body, glaze, or porcelain-enamel slip as a means of identification of the slip before use, particularly when slips of different compositions are of the same color; the colorant or dye burns out during firing.

distortion. (1) A change in the shape of an item due to improper processing such as uneven pressures, uneven or too rapid heating, etc.

(2) An optical effect due to variations in the thickness of plate glass.

distribution. (1) The degree of dispersion of a substance in another substance.

(2) The range of wall thicknesses in a glass article.

D-line cracks. Weathering cracks in concrete occurring as fine, closely spaced cracks near the edge of the unit; the cracks frequently are filled with calcium carbonate and dirt.

D-load. The supporting strength of a concrete pipe loaded under three-edge bearing test conditions; expressed in pounds per linear foot per foot of inside diameter or horizontal span, or expressed in newtons per linear meter per millimeter of inside diameter or horizontal span.

D-load, 0.001-inch or 0.025 millimeter crack. The maximum three-edge bearing test load supported by a concrete pipe before a crack 0.001 inch (0.025 millimeter wide) occurs throughout a length of at least one foot (310 millimeters) of the pipe.

D-load ultimate. The maximum three-edge bearing load which will be supported by a concrete pipe.

dobbin. A turntable type of dryer upon which ceramic tableware is dried in the mold in which it was formed.

dobie. A hand-shaped crudely formed building or refractory brick, either fired or unfired.

docking. The removal of lime deposits from the surface of building brick and roofing tile by immersion in or washing with water.

doctor blade. A flat metal knife or blade mounted in a device so as to spread a uniform thickness of a material on a surface and to remove excess material from the surface such as, for example, the scraping of excess coloring pastes from roller coaters, etc.

doctor roll. A type of roller device employed to remove filter cake from rotary filter drums.

document glass. An ultraviolet-absorbing glass used as a cover to protect documents and valuable papers against deterioration from strong light.

Dodge crusher. A type of jaw crusher with a stationary jaw and a movable jaw hinged at the bottom of the crushing unit.

dog. (1) A device for holding a workpiece so as to permit the piece to be rotated during machining.
(2) A type of drag for a wheel or traversing table.

dog ears, dog teeth. A torn surface on a column of clay emerging from a pug mill; due to insufficient plasticity of the clay column or to damage or dirty extrusion nozzles.

doghouse. A small boxlike vestibule on a glass furnace into which batch is fed or which facilitates the introduction and removal of floaters.

dolly. (1) A hand-operated, low-platform truck mounted on casters used for the movement of materials and ware.
(2) A type of tool used for mixing glazes and other slips.
(3) A refractory-tipped, glass-gathering iron used in semiautomatic forming machines.

dolomite. $CaCO_3 \cdot MgCO_3$; mol. wt. 188.4; CO_2 expelled at about 900°C (1650°F); sp. gr. 2.9; hardness (Mohs) 3.5–4.0; used in refractories, glass, tile, and pottery bodies and also in glazes, primarily as a fluxing ingredient.

dolomite brick. A refractory brick made substantially or entirely of dead-burned dolomite.

dolomite, calcined refractory. Dolomite from which carbonates and volatile constituents have been removed by calcination.

dolomite, dead-burned refractory. Dolomite, with or without additives, which has been calcined to remove carbonates and other volatile constituents to form calcium oxide and periclase, and which is resistant to hydration and recombination with carbon dioxide.

dolomite, double burned. Dolomite which has been once calcined with additions of iron oxide.

dolomite-magnesite brick. A refractory brick made of dead-burned dolomite and dead-burned magnesite and in which the dead-burned dolomite predominates.

dolomite matte. A matte glaze finish produced by the formation of calcium and magnesium silicates in the glaze during firing.

dolomite, raw refractory. Natural limestone containing essentially equal amounts of calcium carbonate, $CaCO_3$, and magnesium carbonate, $MgCO_3$; a material which is suitable for use as a refractory or as an ingredient in refractories.

dolomitic limestone. A mineral composed of more than 80% of calcium or magnesium carbonate; used as a source of calcium in glazes, as a component in cement, and as a refractory and refractory ingredient.

domain. An area in a ferroelectric crystal in which the electric or magnetic moments are uniformly polarized.

dome brick. A brick of a curved pattern or shape suitable for use in the construction of a dome.

dope. A lubricant, such as graphite, which is applied to glass molds to reduce friction and prevent sticking during the forming of glass articles.

doping agent. A measured element or impurity added to a semiconductor composition to promote the development of a desired property or characteristic. Sometimes called dope, dopant.

dose. The amount of radiation delivered to an area or to which the body of a person has been exposed.

dosimeter. An instrument for measuring doses of x-rays or radioactivity.

dot. A refractory spacer used with kiln furniture.

dottling. The horizontal placement of flatware on refractory pins in kilns preparatory for firing.

double burned. A refractory or brick which has been subjected to two separate firings.

double-burned dolomite. Mixture of dolomite and iron oxide which has been subjected to a single calcination treatment.

double-cavity mold. A two-compartment mold for the concurrent fabrication of two articles of glass.

double-cavity process. Any glass-forming process in which two items of glass are formed at the same time in a single double-cavity mold.

double dipping. The process of applying a glaze to a pottery item by dipping the item into a glaze slip twice before firing; the purpose is to obtain contrasting colors on ware which may be fired at the same time.

double drain. The undesired draining of a once-dipped porcelain-enamel coating a second time after the initial drain has appeared to be complete, resulting in a coating of nonuniform thickness.

double embossing. The treatment of a glass surface with acid to produce a design, followed by two additional acid treatments, so that three different shades are produced on the glass surface.

double-faced ware. Porcelain-enameled ware with a finish coat applied to both surfaces of the metal base.

double-frit glaze. A glaze in which frits of two different compositions are incorporated to obtain a coating with a longer firing range and having improved physical and chemical properties.

double glazing. (1) The application of a glaze over a previously applied and dried glaze on ceramic ware, both coatings being fired concurrently.

(2) The placement of two parallel panes of glass in a window, the panes being separated by a thickness, or cell, of stationary air as a means of sound and thermal insulation.

double-glazing unit. A window assembly consisting of two panes of glass separated by a permanently sealed cavity, compartment, or cell.

double-gob process. The process of forming two glass items simultaneously.

double-roll crusher. A pulverizing machine for minerals consisting of two toothed rolls rotating in opposite directions on parallel axes.

double-screened ground refractory material. A once-graded refractory material which has been screened to remove particles that are both coarser and finer than specified sizes.

double-strength glass. Sheet glass of thicknesses between 0.115 and 0.143 inch.

double-wing auger. An auger equipped with two screws at the discharge end.

dovetail. A joint designed to interlock two or more parts.

dowel. A metal bar extending across a concrete joint to aid in vertical alignment and to equalize the transfer of applied loads.

dowel assembly, dowel basket. A reinforcing network or alignment of dowels around which concrete is poured in construction projects.

down-draft kiln. A kiln in which the hot gases from the firebox are passed to the crown, then directed through the ware being fired, and finally are exhausted into a flue or stack.

down time. The production time lost when an item of equipment is not operating due to malfunction, maintenance, power failure, or other cause.

draft. (1) The difference in pressure which causes air and combustion gases to flow from one area to another, such as from a furnace, kiln, or dryer to a flue.

(2) The taper given to a die or mold so that work can easily be removed or withdrawn.

draft gauge. A manometer or instrument employed to measure pressure differences between two areas, such as between a furnace, kiln, or dryer and a flue.

draft, induced. A draft produced or abetted by mechanical or other means, such as by the use of fans, jets, etc., in the flues of kilns and furnaces.

drag. The resistance of the foot or base of a ceramic article to shrinkage during firing due to friction with the slab or sagger on which it rests.

dragade, drag ladle. Cullet produced by ladling molten glass from the melting chamber and quenching in water.

drag-out. The solution removed from a bath by the ware and equipment, as in the cleaning and pickling of metal for porcelain enameling.

drain. The flow of a porcelain enamel or glaze slip on the surface of a piece to form a smooth, even coating.

drain angle. The angle at which an item to be porcelain enameled or glazed is positioned after dipping to permit the excess slip to drain from the item, and that portion of the coating retained on the item to flow to a smooth, uniform thickness.

drain casting. The forming of a ceramic body by pouring slip into a porous mold and then draining the slip from the mold after the cast body has attained the desired thickness.

draining. The process of removing excess slip from dipped items by gravity flow.

drain line. A line or streak appearing in dipped or flow-coated ware as the result of uneven coating thickness.

drain, storm. A pipeline to carry storm or rain water from an area.

drain tile. Tile of circular cross section designed to collect and convey surface and subsurface water away from an area.

drain time. The time required for a porcelain enamel or glaze slip applied by dipping, slushing, or flow coating to cover the ware uniformly and for drainage to cease.

draw. (1) The quantity of glass delivered by a glass-melting tank per unit of time, usually 24 hours.

(2) To remove a charge of fired ware from a kiln.

(3) The draft in a flue.

drawability. The ease with which a metal can be shaped by deep-drawing.

draw bar. A submerged, refractory block in a glass tank defining the point at which sheet glass is drawn.

draw firing. Removal of a load from a porcelain-enameling furnace prior to completion of the firing operation to permit equalization of the heat in the ware, particularly in areas of greater thickness; the load may or may not be returned to the furnace, depending on the degree of maturity of the coating.

draw gang. A group of workmen employed to cut and handle glass coming from the lehr.

drawing. The unloading of a kiln.

drawing chamber. The section of a glass-melting tank from which molten glass is drawn.

drawing compound. A composition, such as graphite, talc, greases and oils, applied to the surface of metal to serve as a lubricant to prevent draw marks and tearing during drawing and stamping operations.

drawing die. A die in which sheet metal is shaped by drawing and stamping.

draw mark. An imperfection in metal shapes caused by friction with, or a defect in, a die.

drawn glass. Glass made automatically and continuously by drawing from the melting tank, and then rolling or shaping.

drawn stem. Glass tableware in which the stem or base is pulled or drawn from the bowl while in the plastic state.

dredging. The application of powdered, porcelain-enamel frit to a hot metal shape, usually cast iron, by sifting the powder over the surface of the metal.

dress. To shape or return a tool to its original shape and sharpness.

dressed crude mica. Mica from which dirt and rock have been removed.

dresser. An apparatus employed to shape, true, and dress grinding wheels by the use of rotating cutters.

dresser, Huntington. Star-shaped metal cutters rotating on a spindle for use in truing and dressing abrasive grinding wheels.

dresser, star. See dresser, Huntington.

dressing. (1) The process of restoring the efficiency of an abrasive grinding wheel by removal of dulled grains.
(2) Reshaping the faces of grinding wheels to special contours.

dressing, crush. The use of steel rolls to form or dress the face of an abrasive grinding wheel to a desired contour.

drip feed. A technique for supplying oil or paraffin as lubricants for the moving parts in furnaces.

drippings, smelter. Drippings of molten glassy material from an accumulation of the material on the crown of a smelter.

drop arch. An auxiliary arch projecting below the inner surface of the arched roof of a furnace.

drop chute. Flexible sheet-metal tubes forming downspouts to control the flow of concrete during a vertical or downward fall as it is being placed.

drop-machine silica brick. Silica brick formed by dropping the prepared mix into a mold from varying heights, the force of the drop being sufficient to force the mix into all corners to fill the mold.

dropping. The shaping of a glass article by sagging heat-softened glass into a mold without the application of mechanical pressure.

drop throat. The throat of a glass tank situated below the level of the bottom or floor of the melting tank.

dross. Waste and impurities collected on the surface of a molten glass bath.

drum dryer. A heated, rotating drum in which tumbling or cascading raw materials are dried.

dry. Free from or deficient in water or moisture.

dry basis. The weight or volume of a substance exclusive of any moisture which may be present.

dry body. (1) An unglazed body, usually of the stoneware type.
(2) A body from which all moisture has been removed.

dry-bulb temperature. Actual atmospheric temperature as measured by an ordinary dry-bulb thermometer.

dry disk. An apparatus for finishing the face of an abrasive grinding wheel.

dry edging. An imperfection consisting of rough edges and corners on glazed ceramic ware due to insufficient application of glaze to the area.

dryer. A heated chamber, frequently with circulating air, in which ware is placed for the removal of water or moisture by evaporation.

dryer, automatic. A dryer in which heat, air movement, and other drying conditions are controlled by appropriate devices.

dryer, batch. A periodic dryer in which a charge is placed, dried, and removed.

dryer car. A dryer through which ware is transported by means of cars.

dryer, chamber. A periodic dryer consisting of one or

more chambers in which ware is placed, dried, and removed as needed for further processing.

dryer, continuous. A dryer in which ware is charged, dried, and removed in a continuous or uninterrupted flow.

dryer, hot floor. A dryer in which heat is introduced through a steam or gas-heated floor.

dryer, humidity. A dryer in which the humidity is controlled.

dryer, infrared. A dryer in which heat is supplied by infrared electric lamps or incandescent gas burners.

dryer, jet. A dryer in which ware is dried by jets of warm air, steam, or both, injected into the drying chamber.

dryer, pallet. A periodic dryer in which ware, stacked on pallets, is charged, dried, and removed.

dryer, periodic. A dryer in which ware is placed, dried, and removed before recharging.

dryer, pipe-rack. A room in which palleted ware is placed on steam pipes for drying.

dryer, Proctor. A type of steam-heated dryer.

dryer, room. A heated room in which ware is placed, dried, and sometimes stored prior to firing.

dryer, rotary. A heated, rotating tube or drum through which materials are tumbled and cascaded during drying.

dryer, spray. A structure in which an atomized suspension of solids in a liquid is dried by hot gases or by impingement onto a hot surface.

dryer, steam-rack. A room in which palleted ware is stacked on steam-heated racks for drying.

dryer, string. An intermittent, tunnel-type dryer of high humidity used in the treatment of building brick.

dryer, tunnel. A tunnel-shaped dryer through which ware, stacked on cars, is passed.

dryer, waste-heat. A dryer in which heat is supplied from otherwise wasted hot flue gases, heat from the cooling of fired ware, etc.

dryer white. Discoloration of clayware due to the presence of soluble salts at the surface, usually due to the migration of the salts dissolved in the water moving from the interior of the body to the surface during the drying cycle.

dry-film lubricant. See die lubricant.

dry foot. An unglazed base or foot on the underside of fired ceramic ware.

dry gauge. Cullet produced by ladling molten glass from a melting unit into water.

dry grinding. Milling of materials without a liquid medium.

drying. The removal of water and moisture from a body.

drying crack. A fissure in an unfired body, glaze, or porcelain enamel due to stresses incurred during handling or drying.

drying oven. A closed unit in which specimens are dried by heating.

drying rate. The speed at which a moisture-bearing material, body, or coating will dry under specific heating or atmospheric conditions, or both.

drying shrinkage. The contraction of a moist body during the drying process, expressed as linear percent of the original length or volume percent of the original volume.

drying shrinkage, linear. The linear contraction of a moist body during drying; calculated by the formula:

$$\text{Linear drying shrinkage, \%} = \frac{L_p - L_d}{L_p}$$

in which L_p is the length of the specimen in its original wet or plastic state, and L_d is the length of the specimen after drying.

drying shrinkage, volume. The volume contraction of a moist body during drying; calculated by the formula:

$$\text{Volume drying shrinkage, \%} = \frac{V_p - V_d}{V_p}$$

in which V_p is the volume of the specimen in its original wet or plastic state, and V_d is the volume of the specimen after drying.

drying time. The time required for a moist body, material, or coating to dry under particular heating and atmospheric conditions.

drying, vacuum. The technique of expediting the removal of moisture from a material or body by the use of a vacuum in conjunction with a conventional drying system.

dry kiln. A kiln designed to dry ceramic greenware at the lowest possible heat before it enters the firing zone.

dry milling. Reducing the particle size of a substance by milling without the use of a liquid medium.

dry mix. The blending of batch ingredients in the dry state, liquids being added at the time when subsequent processing is required.

dry modulus of rupture. The transverse strength of a

standard specimen in the dry, but unfired, state. See modulus of rupture.

dry pack. A moist mixture of cement and sand used in the repair of deep cracks and cavities in concrete.

dry pan. A muller-type mixer in which materials are ground or blended with a minimum amount of moisture. See muller.

dry powder. A finely pulverized substance or mixture free from or deficient in moisture.

dry press. A mechanically or hydraulically actuated press used in the shaping of moistened ceramic bodies in a mold under pressure.

dry-pressed brick. Brick formed in a mold under high pressure from a body containing 5 to 7% of moisture.

dry pressing. The process of forming or shaping ceramic bodies of low-moisture content (5 to 10% of water) by compression in molds.

dry process. (1) A process for manufacturing portland cement in which the batch is charged into the cement kiln in the dry state.

(2) To process concrete aggregate without the use of water.

dry-process enameling. A process of porcelain enameling in which the base metal, usually cast iron, is heated to a temperature slightly above the fusion temperature of the enamel, followed by sifting finely powdered enamel frit on the metal surface or by dipping the hot metal into a dry batch of powdered frit, and then firing the coating to maturity; the process may be repeated with minimal cooling of the ware until the desired coating thickness is attained.

dry-rubbing test. (1) A test to evaluate the resistance of a glaze, porcelain enamel, or other surface to abrasion by rubbing the surface with a dry abrasive powder under standardized test conditions; in some instances a slurry of an abrasive powder may be used.

(2) A test to evaluate the degree to which a chemical attacks a glaze, porcelain enamel, or other surface by rubbing a finely divided powder of a contrasting color across the test area and observing the degree of color retained by the chemically treated surface.

dry screening. The process of separating small sizes of granular or powdered solids from coarser particles by passing them through a screen of desired mesh size while in the dry state. See screen analysis.

dry shake. A dry mixture of cement and special fine aggregate broadcast over a concrete floor before final finishing to provide a wear-resistant surface.

dry spray. An imperfection having the appearance of a rough sandy surface on porcelain-enameled ware due to improper spraying.

dry strength. The resistance of a dried but unfired ceramic body to physical or mechanical damage.

dry weight. The weight of a porcelain enamel or other coating applied per unit of area to an item after the wet coating has been thoroughly dried.

dual-drum mixer. A mixer consisting of a long drum containing two compartments separated by a bulkhead with a swinging chute extending through the unit.

ductile. Capable of being deformed by elongation without fracture.

dulling. The wearing of the sharp edges of cutting tools or abrasive grains resulting in inefficient or ineffective performance of the tools or grains.

dullness. Lacking in brilliance or luster, as evidenced in porcelain enamel and glaze surfaces.

dummy. A foot-operated device employed for the wetting, raising, opening, and closing of paste molds used in blowing glassware by mouth.

dummy joint. A preformed contraction joint in concrete designed to form a line along which a crack can form in the slab with minimum damage to the adjacent sections of the slab.

Dumont blue. A sintered mixture of cobalt oxide, sand, and potash employed as a colorant in glass, glazes, and porcelain enamels.

dumortierite. $8Al_2O_3 \cdot 6SiO_2 \cdot B_2O_3 \cdot H_2O$; sp. gr. 3.2–3.3; hardness (Mohs) 7; a mineral employed in the manufacture of high-grade porcelains to improve their resistance to thermal shock and physical damage.

dump hopper. A large hopper which can be tipped mechanically to remove its contents.

dunite. An ultrabasic mineral consisting of magnesium-rich olivine; used in the manufacture of forsterite refractories as a source of chromium.

dunk. To plunge hot glazed ware into a cold liquid, usually water, to produce decorative crazing in the glaze.

dunting. The cracking of fired ware which has been cooled too rapidly.

dunting point. The temperature at which the inversion of silica from the alpha crystalline form to the beta form occurs, and vice versa.

durability. The property of an article of being resistant to physical and chemical damage under the usual conditions of service, and of being useful over extended periods of time and use.

dust. Fine dry particles of matter which essentially are larger than colloidal in size, less than 1/16 millimeter in maximum cross section, and which are capable of being suspended in air or other gases.

dust coat. (1) A thin dusty-appearing coating of porcelain enamel or glaze applied by spraying.

(2) A mixture of concrete and fine aggregate distributed over a concrete slab before finishing, and while the concrete is still plastic to improve the resistance of the concrete to wear.

dusting. (1) The sifting of finely powdered porcelain-enamel frit over preheated metal articles, the powdered coating subsequently being fired.
(2) The removal of dust and loose dirt from dried porcelain enamel and glaze surfaces before firing.
(3) The application of a thin dust-like coating on an item of ware by spraying.
(4) An imperfection on glaze and porcelain-enamel surfaces consisting of an inordinate buildup of dry slip during spraying.
(5) The disintegration of refractories by the inversion of one crystal form to another during cooling.
(6) The erosion of concrete surfaces under traffic.

dust pressing. The process of forming ceramic bodies of 1.5% or less water content by pressing in a mold.

dusty spray. The application of porcelain enamel and glaze slips to ware in a manner that a wet film is not produced.

dwell time. The time a penetrant is in contact with a surface during an absorption or penetration test, including the application and drain times.

dye-absorption test. A test of the porosity of a fired specimen in which the specimen is immersed in a dye solution under specified conditions of time, pressure, and temperature; the depth of penetration of the dye into the specimen then is observed or measured.

dyes. (1) Soluble, combustible organic colorants added to glazes, porcelain enamels, and other coatings to assist sprayers in controlling the uniformity and thicknesses of coatings that are difficult to see.
(2) Soluble, combustible organic colorants added to bulk porcelain enamel and glaze slips in storage to assist workers in identifying materials of different composition but of similar outward appearance.
(3) Colorants in solution used to aid in the detection of cracks, pinholes, and other surface and body imperfections in fired ware; the solutions are brushed over and then wiped from areas of potential defects to reveal the imperfections.

dynamic balance. The condition under which a grinding wheel or other rotating part, rotating at a high speed, will exhibit no vibration or whip due to uneven distribution of weight in its mass.

dysprosium aluminate. $Dy_2O_3 \cdot 2Al_2O_3$; mol. wt. 576.92; m.p. 1816°C (3300°F): sp. gr. 6.05.

dysprosium boride. (1) DyB_4; mol. wt. 205.8; sp. gr. 6.74; (2) DyB_6; mol. wt. 227.24; sp. gr. 5.49.

dysprosium carbide. (1) Dy_3C; mol. wt. 499.56; sp. gr. 9.21. (2) Dy_2C_3; mol. wt. 361.04. (3) DyC_2; mol. wt. 186.52; sp. gr. 7.45.

dysprosium niobate. $Dy_2O_3 \cdot Nb_2O_5$; mol. wt. 639.24; m.p. ~ 1949°C (3540°F); sp. gr. 5.49.

dysprosium nitride. DyN; mol. wt. 176.52; sp. gr. 9.93.

dysprosium oxide. Dy_2O_3; mol. wt. 372.9; sp. gr. 7.81; used as a component in control rods for nuclear reactors, as a phosphor activator, and in dielectric compositions.

dysprosium silicate. (1) $Dy_2O_3 \cdot SiO_2$; mol. wt. 433.14; m.p. 1929°C (3505°F); hardness (Mohs) 5–7. (2) $Dy_2O_3 \cdot 3SiO_2$; mol. wt. 926.38; m.p. 1921°C (3490°F); hardness (Mohs) 5–7. (3) $Dy_2O_3 \cdot 2SiO_2$; mol. wt. 493.24; m.p. 1721°C (3130°F); hardness (Mohs) 5–7.

dysprosium silicide. $DySi_2$; mol. wt. 218.64; m.p. 1248°C (2820°F); sp. gr. 6.8. (2) Dy_3Si_5; mol. wt. 788.06.

dysprosium sulfide. (1) Dy_5S_7; mol. wt. 1037.02; m.p. 1540°C (2805°F); sp. gr. 6.35. (2) Dy_2S_3; mol. wt. 421.22; m.p. 1479°C (2695°F); sp. gr. 6.54; (3) DyS_2; mol. wt. 226.64; sp. gr. 6.48.

E

earthenware. A glazed or unglazed, nonvitreous, opaque ceramic whiteware having a water absorption greater than 3%.

eccentric axis. An axis located elsewhere than at the geometrical center of a body.

economic mineral. A mineral of commercial interest or value.

economy brick. Brick nominally 4 × 4 × 8 in. in size.

eddy current. An electric current induced in a conductor moving through a nonmagnetic field or in an area where there is a change in magnetic flux.

eddy-current testing. A nondestructive test in which eddy-current flow is induced in a specimen, and changes in flow are measured.

edge bowl. The hollow bowl-like protrusion containing a slot through which sheet glass is drawn in the Pittsburgh sheet-glass process, which see.

edge effect. (1) An outward-curving distortion of the lines of electrical force near the edges of two parallel plates forming a capacitor.

(2) The disturbance of a magnetic field and eddy current due to the proximity of an abrupt change in geometry, such as an edge.

edge lining. The application of a decorative line around the rim of ceramic ware such as plates, saucers, dishes, and other shapes of dinnerware.

edge polishing. The polishing of the edges of plate glass after it has been cut.

edge, rough and burred. The frayed or serrated edge of a ceramic, metal, or other material after it has been sheared, cut, or fractured.

edge-runner mill. A pulverizing or crushing mill equipped with vertical rollers rotating in a circular enclosure of metal, ceramic, or stone.

edge skew. A brick with one side sloped at an angle, the ends and faces each being parallel.

edge, thick. A mica splitting with an edge or end 1½ times greater than the thickness at any other point, or greater than the maximum average thickness for its particular grade.

edgework. The grinding, smoothing, or polishing of the edges of a glass.

edging. (1) The removal of unfired porcelain enamel from the edges of a piece of ware prior to firing.

(2) The spraying of porcelain enamel over the edge of ware as a decoration or reinforcement.

(3) The grinding of the edge of a piece of glassware to a prescribed size or shape.

edging brush. A stiff-bristled brush with a metal guide used to remove dry, but unfired, porcelain enamel from the edge of ware before firing to prevent chipping and improve the appearance of the final product.

effective depth of penetration, electromagnetic. The minimum depth beyond which a further increase in specimen thickness no longer can be detected.

effective modulus of rupture. See modulus of rupture, effective.

effective permeability, electromagnetic. A hypothetical quantity describing the permeability experienced by a specimen under a given set of physical conditions, such as a cylindrical specimen in an encircling coil at a specific test frequency.

effective porosity. The porosity of a material containing connecting pores, expressed as a percent of the bulk volume occupied by the pores.

efficiency, detector. In reactivity measurements, the fraction of particles or photons striking the detector which result in a detected response.

efficiency, relative. The rating of the absorptive capacity of an activated carbon compared with that of a reference carbon in a defined test.

efficiency, relative detector. The product of the detector efficiency and the detector geometry.

efficiency, thermal. The ratio of input versus output temperature to the maximum temperature range possible.

effloresce. To become powdery due to the loss of water of crystallization, or to become encrusted with crystals of salt from a solution due to evaporation or chemical change.

efflorescence. (1) A formation of powdery or crystalline salt on the surface of concrete, asbestos concrete, masonry, or other surface due to the migration and precipitation of soluble salts from the interior of the body.

(2) Also known as bloom, which see.

efflorwick test. A test to estimate the tendency of a building brick to effloresce in which a fired cylinder of brick clay is immersed in a solution of soluble salt in distilled water, dried, crushed, and analyzed for the presence of the selected salt. See efflorescence.

effluent. A liquid discharged or flowing from a process or place.

eggshell. (1) A fired porcelain enamel or glaze having a semimatte, eggshell-like texture.

(2) A type of very thin, highly translucent porcelain.

E-glass. A glass fiber of low alkali content.

Egyptian blue. A blue frit or a powdered pigment of the general composition $CuO \cdot CaO \cdot 4SiO_2$ contained in a glassy matrix.

Egyptianized clay. A clay to which tannin has been added to make it more plastic.

elastic after-effect. The ratio of deformation remaining in a specimen, which has been subjected to tensile stress over a prescribed period of time, to the degree of deformation immediately after the stress has been relaxed.

elastic deformation. The degree of deformation of concrete or other body under load which disappears when the load is removed or relaxed.

elastic fractionation. A process in which soft aggregate particles are separated from harder particles by hurling the composite aggregates against a steel plate; the harder particles will rebound farther on impact with a hard inelastic surface than particles which are soft and friable.

elastic modulus. The ratio of stress to strain within the elastic range of a substance.

electret. A dielectric material which possesses a permanent or semipermanent polarity in a manner analogous to a permanent magnet.

electrical conductivity. The ability or the measure of the ability of a material to conduct electric current.

electrical conductivity of a particulate substance. A measure of the current flowing through a unit cross section of a particle for an imposed unit gradient under specified conditions of packing.

electrical erosion. The erosion of an electrical insulating material due to the influence of an electrical discharge.

electrical-erosion resistance. A quantitative measure of the amount of erosion of an electrical-insulating material by an electrical discharge under specific test conditions.

electrical porcelain. A porcelain body formulated and designed for use as an electrical insulator.

electric boosting. An auxiliary method of adding heat to a molten glass batch in a conventionally fired glass-melting tank by passing electric current through the molten batch.

electric breakdown voltage. The difference in the potential at which electrical failure occurs in an electrical insulator located between two electrodes.

electric capacitor. An electric nonconductor that permits the storage of energy as the result of electric displacement when opposite surfaces of the nonconductor are maintained at a different potential.

electric conductor. A body which will transmit electricity.

electric contact. Any physical contact between two or more parts which will permit the flow of electricity between the parts.

electric furnace. A furnace or kiln in which the main source of heat is provided by electrical means.

electric, seignette. Any crystalline substance, such as Rochelle salt or potassium sodium tartrate ($KNaC_4O_6{\cdot}4H_2O$), displaying ferroelectric properties which can be used in the production of ceramic capacitors, transducers, dielectric amplifiers, etc.

electric strength. The voltage gradient at which dielectric breakdown occurs in an insulating material.

electrocast refractory. A dense refractory brick or other shape which is formed by melting refractory oxides in an electric furnace followed by casting in appropriate molds.

electroceramics. Ceramic products which are formulated and designed for use as insulators for power lines and other electrical applications.

electrofusion. Melting by electrical means, particularly in an electric furnace; in some instances, melting may be augmented by the use of electrodes submerged in the molten bath. Also see fusion casting.

electroluminescence. Luminescence resulting from a high-frequency discharge through a gas or by applying an alternating current to a phosphor.

electrolyte. A chemical, usually an inorganic salt, added to porcelain-enamel and ceramic slips to control the suspension and flow properties of the slips.

electrolytic pickling. A process in which the pickling of metal for porcelain enameling is enhanced by passing an electric current through the pickling bath in which the metal serves as an electrode.

electromagnet. A core of magnetic material which is magnetized when surrounded by a coil of insulated wire through which an electric current is flowing.

electromagnetic. A term describing a property of a material in which electricity and magnetism are related.

electromagnetic test. A nondestructive measure of some property, such as the thickness of a porcelain-enamel coating, by the appropriate use of electromagnetic energy.

electronic ceramics. (1) Ceramic products which display dielectric, semiconductor, magnetic or similar other properties which are useful in the production of electronic devices, such as transistors, solid-state devices, electron tubes, magnetic amplifiers, etc.

(2) Ceramic materials which amplify or control voltages or currents without mechanical or other non-electrical direction.

electron microscope. An electron-optical instrument in which a beam of electrons is focused on a replica of a specimen by means of an electron lens to produce an enlarged image of a minute area of the specimen on a photographic plate or a fluorescent screen.

electron microscope, scanning. An electron microscope in which a beam of electrons sweeps over a specimen, measuring the intensity of the secondary electrons generated at the point of impact of the beam on the specimen, relaying the signal to a cathode-ray display which is scanned in synchronism with the scanning of the specimen.

electrophoresis. The movement of colloidal particles or macromolecules through a solution under the action of an electromotive force applied through electrodes in contact with the solution.

electrostatic spraying. A coating process in which the coating particles being sprayed are given an electrostatic charge opposite to that of the item being coated, causing the particles to be attracted to the surface of the item with a minimum of overspray as in the electrostatic spraying of porcelain enamels on metal.

elephant ear. A type of flat, fine-grained sponge used in finishing pottery surfaces before firing.

elephant trunk. A series of conical sections of steel pipe forming a flexible downspout to control and confine the vertical flow of concrete during pouring.

elevator. A mechanical device employed to convey materials from one level to another.

elevator kiln. A type of kiln in which ware is placed on a refractory platform which then is raised by elevator into the firing chamber of the kiln stationed immediately above.

elutriate. To separate suspended solid particles according to size or density by washing, decanting, and settling.

embed. To enclose in a surrounding matrix.

emboss. To ornament or decorate with a raised pattern on the surface of ware.

embossing, double. The process in which a design worked on a glass surface previously treated with white acid is given an additional acid treatment to produce a three-shade effect.

embossing, single. The process in which a design worked on a glass surface previously treated with white acid is given an additional acid treatment to produce a two-shade effect.

embossing, white acid. The treatment of a glass surface with acid, such as hydrofluoric acid, to produce full obscuration.

embossment. A raised-pattern decoration on the surface of an item of ware.

emery. An impure, natural corundum, Al_2O_3, pulverized for use in grinding and polishing.

emery obscured. A glass surface obscured by grinding with a fine grade of emery.

emery paper. An abrasive paper or cloth with a tightly bonded coating of emery for use in cleaning and polishing operations.

emery stone. A sharpening device or grinding stone made of bonded emery.

emery wheel. A grinding wheel composed of tightly bonded emery which is used on mechanical grinding and polishing machines.

emissivity. (1) The ratio of the radiation given off by the surface of a body to the radiation given off by a perfect black body at the same temperature.

(2) The capacity of a body to radiate heat.

empirical. Based on experience, observation, or measurement rather than theory.

emulsification. The process of dispersing an immiscible liquid in another liquid, such as oil in water.

emulsifier. A surface-active agent which will promote the dispersion of an immiscible liquid in another liquid.

emulsion. A stable mixture of two immiscible liquids in which one is dispersed as droplets throughout the other.

emulsion, wax. A colloidal suspension of wax in a solvent used as a binder, lubricant, or suspension agent in ceramic bodies and glazes.

enamel. A glassy or vitreous coating applied and fired on metal, glass, or other ceramic ware; the coating on metal frequently is identified as porcelain enamel.

enamel, acid-resisting. A porcelain enamel resistant to household, fruit, and cooking acids.

enamel, alkali-resisting. A porcelain enamel resistant to soaps and household alkalies.

enamel, aluminum. A porcelain enamel formulated specifically for application to aluminum.

enamel-backed glass tubing. Glass tubing with a black or colored coating on the back segment of the perimeter to facilitate the observation and reading of liquid levels in the tubing, such as is used in thermometers, barometers, manometers, etc.

enamel beading. (1) Porcelain enamel, usually of a contrasting color, applied as a decoration or reinforcement to the edge or rim of porcelain-enameled articles.

(2) The process of removing excess slip from the edge of freshly dipped or coated porcelain-enameled ware.

(3) A heavy bead of porcelain enamel along the edge of porcelain enameled cast-iron ware.

enamel, blackboard. A porcelain enamel having a matte, slightly roughened surface when fired on a metal sheet for use in the construction of chalk boards; available in various colors.

enamel, blown. An imperfection sometimes occurring during the spraying of porcelain enamel in which the wet coating is blown to a ridged or wavy surface.

enamel, blue. A wet or dry process porcelain-enamel coating of insufficient thickness to hide the substrate, resulting in a bluish color.

enamel brick clay. A clay similar to that used in the manufacture of buff-colored face brick sometime applied as a coating and fired to vitrification.

enamel, cast-iron. A porcelain enamel formulated specifically for application to cast iron. See cast-iron enameling.

enamel, chalkboard. See enamel, blackboard.

enamel clay. A ball clay used to promote the suspension of porcelain enamels and glazes in aqueous slips. See ball clay.

enamel colors. Inorganic compositions employed to impart an unlimited range of colors to porcelain enamels as well as to glazes, glass, and ceramic bodies.

enamel, copper. A porcelain enamel of high-thermal expansion designed for use on copper.

enameled brick. A hard, smooth-surfaced brick with a fired wash-type coating, the coating frequently being in color.

enamel fineness. The degree to which the particle size of porcelain-enamel frit has been reduced by milling, usually expressed as grams residue retained on a sieve of specified mesh size from a 50- or 100-gram sample.

enamel firing. The process of fusing porcelain-enamel coatings to a metal base.

enamel, glass. A finely powdered mixture of a low-melting flux, frit, or calcined ceramic pigment contained in a suitable vehicle which may be applied and fired to a smooth, hard coating on glass at a temperature below the softening point of the glass.

enameling, cast-iron. (1) A porcelain-enameling process in which cast iron is heated to a temperature above the maturing temperature of the coating; a powdered coating is applied to the hot article by means of a vibrating screen or by immersing the hot article in the powdered material and fired to smoothness.

(2) A relatively low melting, long firing porcelain enamel designed specifically for cast-iron which is sprayed over a previously applied ground coat and fired to maturity over a relatively long-firing time.

enameling, dry-process. See enameling, cast iron (1).

enameling iron. A low-carbon, low-metalloid, open-hearth, cold-rolled steel designed specifically for use as a base for porcelain enameling.

enameling kiln. Any furnace in which porcelain enamels are fired.

enameling, wet-process. A porcelain-enameling process in which the enamel is applied to the base metal in slip form by dipping, spraying, or other technique, and dried and fired to a smooth, impervious, glassy finish.

enamel, jeweler's. A porcelain enamel, usually lower melting than conventional porcelain enamels, formulated specifically for use on a particular metal such as silver, gold, iron, etc., in the production of jewelry, artware, insignia, and similar items.

enamel, matte. A fired porcelain enamel of little or no gloss.

enamel, molybdenum. A porcelain enamel, usually white or in pastel colors, in which molybdenum oxide or other molybdenum compound is employed as the adherence-promoting agent.

enamel, nitty. An imperfect porcelain-enamel coating containing minute surface pits which are perceptible only on close examination.

enamel oxides. A wide range of inorganic oxides used as colorants in porcelain enamels.

enamel, porcelain. A substantially vitreous, inorganic coating bonded to a metal base by fusion at temperatures above 425°C (800°F).

enamel, reclaim. Porcelain-enamel overspray and scrapings from spray booths, dip tanks, and other sources which are suitable for reconditioning and use.

enamel, refractory. A porcelain enamel of special composition used to protect metals from attack by hot and corrosive gases.

enamel, retouch. A fine overspray or brushed-on coating of porcelain enamel applied to cover or protect areas of potential imperfections.

enamel scrapings. Porcelain enamel recovered from spray booths, dip tanks, settling tanks, and other sources, which are suitable for reconditioning for future use.

enamel, self-cleaning. Porcelain enamels containing additions of selected materials which, when applied to culinary ovens, will promote oxidation of grease and oven spills continuously during oven use.

enamel, sheet-steel. Any porcelain enamel designed and formulated specifically for use on sheet iron and steel.

enamel, spongy. A porcelain-enamel imperfection consisting of a myriad of small bubbles in a localized area producing a sponge-like appearance.

enamel, tin. A white porcelain enamel or glaze opacified by additions of tin oxide to the slip at the mill.

enamel, vitreous. A synonym for porcelain enamel.

encapsulation. A process in which a substance is enclosed, encased, or encapsulated in a protective medium or film.

encaustic. A type of decoration in which the pigment to be applied to the surface of an item is carried in hot wax, or colored clays are inlaid in the clay forming the body.

encaustic tile. Ceramic tile in which a design is inlaid and fired with clays of a color different from that of the body.

encircling coil. An electromagnetic coil surrounding a specimen or part.

encrustation. The formation and accumulation of a slag or other substance on the inside of a kiln or furnace.

end-arch brick. A brick with the faces sloped toward each other so as to provide wedge-shaped ends of the same dimensions for use in the construction of the crown of an arch.

end construction. A type of construction in which structural tile and block are laid with the hollow cells of tile or block placed in a vertical position.

end-construction tile. A load-bearing, hollow concrete or fired ceramic tile designed to receive its principal stress parallel to the axes of the cell in construction applications.

end-cut brick. Brick extruded in a position so that the wire-cut sections form the ends of the brick.

end-feather. The sharp edge of the end of a brick cut lengthwise from one corner to the other corner diagonally opposite to produce a brick of a triangular cross section.

end-feed centerless grinding. A grinding process in which the item being ground is fed through the grinding and regulating wheels of a centerless grinder to an end stop.

end-fired furnace. A furnace in which the fuel or air-fuel mixture is introduced through an end wall.

endothermic reaction. A reaction characterized by the absorption of heat.

end-port furnace. A furnace in which the ports for the introduction of fuel or fuel and air mixtures are located in an end wall.

end runner. A refractory shape designed to channel and transport molten metal from a feeder head to an ingot mold.

end skew. A brick with one end inclined at an angle other than 90 degrees to the two largest and parallel faces.

endurance, thermal. The ability of a glass or other ceramic product to withstand thermal shock or to withstand deterioration during exposure to high temperatures.

energy, internal. The sum of the intrinsic energies of individual molecules, kinetic energies of internal motions, and the contributions of interactions between molecules minus the potential or kinetic energy of the system as a whole.

energy, kinetic. The energy possessed by a body in motion equals one-half its mass times the square of its speed.

engineered brick. A brick having nominal dimensions of $3.2 \times 4 \times 8$ inches.

engine-turned ware. Bisque ware lined or fluted in a special lathe.

English bond. Alternate courses of headers and stretchers, the joints in alternate courses being lined up vertically.

engobe. A slip coating applied to ceramic bodies to mask the color and texture of the body, and to impart color or opacity, and which may or may not be covered subsequently with a glaze.

engraving. The process of carving or grinding designs on glass or other ceramic products.

enriched uranium. Uranium containing a concentration of U^{235} greater than its normal concentration, 0.711 wt %.

enstatite. $MgO.SiO_2$; mol. wt. 100.4; sp. gr. 3.2–3.5; hardness (Mohs) 5–6.5; used in electronic ceramics as a replacement for talc to minimize shrinkage.

enthalpy. The sum of the internal energy of a body and the product of its volume multiplied by the pressure exerted on the body by its surroundings. Also known as sensible heat, total heat, heat content.

entrain. To introduce air or a gas into a slurry by vigorous stirring or agitation.

entrained air. Minute bubbles of air formed in concrete by the introduction of an air-entraining agent to the batch.

entrainment. The process of introducing air or other gaseous bubbles into a body by physical means, such as by mechanical agitation, or by chemical means which will generate bubbles that will be retained by the body.

entrapped air. Air voids in concrete generally larger than one millimeter in diameter which are mechanically formed and entrapped in the batch.

entropy. A thermodynamic study of heat and work and the conversion of energy from one form to the other.

envelope kiln. (1) A kiln in which the firing zone, positioned immediately above a loaded refractory platform, is lowered to surround the work being fired.

(2) A box-type kiln in which ware placed on kiln cars is pushed into the entrance end of the kiln, thereby displacing cars of fired ware at the other end.

epitaxy. The growth of a crystal on the surface of another crystal; the growth of the deposited crystal being oriented by the lattice structure of the recipient crystal.

epoxy resin. A high-strength, low-shrinkage, polyether resin employed as a bonding agent in glass-fiber strands and cloth because of its excellent resistance to acids and alkalies and for its excellent electrical properties.

epsom salts. $MgSO_4.7H_2O$; employed as a set-up agent to adjust the viscosity and improve the application and flow properties of slips. See set, (1).

equation, van der Waals. See van der Waals equation.

equilibrium. A state in which no change occurs in a system if no change occurs in the surrounding environment.

equilibrium adsorptive capacity. The quantity of a component adsorbed per unit of an adsorbing substance at equilibrium temperature, concentration, and pressure.

equilibrium diagram. A phase diagram of the equilibrium relationship which exists between composition, temperature, and pressure of a system.

equilibrium eutectic. The composition within which any system of two or more crystalline phases will melt completely at a minimum temperature or at which the composition, per se, will melt.

equivalent. See equivalent weight.

equivalent, boron. The absorptive capacity for thermal neutrons of weights of various elements expressed in terms of the weight of natural boron.

equivalent boron content. The concentration of natural boron which will provide a thermal neutron cross-section equivalent to that of a specific impurity element.

equivalent boron-content factor. A factor employed to convert the concentration of an impurity element to a neutron cross section equivalent to that of natural boron.

equivalent uranium content. A concentration of U^{236} which will provide a fast neutron cross section equivalent to the specific impurity element.

equivalent weight. The weight of an element which will replace or combine with 1.008 parts of hydrogen by weight, 8.00 parts of oxygen, or the equivalent weight of any other element or compound.

erbium boride. (1) ErB_4; mol. wt. 210.98; sp. gr. 6.96. (2) ErB_6; mol. wt. 232.62; sp. gr. 5.58. See borides.

erbium carbide. (1) Er_3C; mol. wt. 515.10; sp. gr. 4.71. (2) Er_2C_3; mol. wt. 371.40. (3) ErC_2; mol. wt. 191.70; sp. gr. 7.95.

erbium nitride. ErN; mol. wt. 181.80; sp. gr. 10.35.

erbium oxide. Er_2O_3; 382.4; sp. gr. 8.64; used as an actuator for phosphors, as a nuclear poison, and as an ingredient in infrared absorbing glasses.

erbium selenide. Er_2Se_3; mol. wt. 573.0; m.p. ~ 1521°C (2770°F); sp. gr. 6.96.

erbium silicate. (1) $Er_2O_3 \cdot SiO_2$; mol. wt. 443.5; m.p. 1979°C (3595°F); sp. gr. 6.80; hardness (Mohs) 5–7. (2) $2Er_2O_3 \cdot 3SiO_2$; mol. wt. 947.10; m.p. 1899°C (3450°F); sp. gr. 6.22; hardness (Mohs) 5–7. (3) $Er_2O_3 \cdot 2SiO_2$; mol. wt. 503.6; m.p. 1799°C (3270°F); hardness (Mohs) 5–7.

erbium silicide. (1) $ErSi_2$; mol. wt. 223.82. (2) Er_3Si_5; mol. wt. 643.40. See silicides.

erbium sulfide. (1) ErS; mol. wt. 199.76. (2) Er_5S_7; mol wt. 1062.95; m.p. 1620° C (2950°F); sp. gr. 6.71. (3) Er_2S_3; mol. wt. 413.59; m.p. 1730°C (3151°F); sp. gr. 6.21.

erg. A centimeter-gram-second unit of energy or work equal to the work done by a force of one dyne acting over a distance of ore centimeter.

Erlanger blue. A general term for a variety of iron-blue pigments.

erosion. The wearing away of the surface of a material, usually by physical action rather than by chemical forces.

erosion of refractories. The wearing away of refractory surfaces by the washing action of the molten batch at high temperatures.

erosion resistance, electrical. The resistance of electrical insulating materials to erosion by the action of electrical discharges.

etch. (1) To produce a marking, decoration, or degree of obscuration on glass or other ceramic surface by chemical action, such as by hydrofluoric acid or other agents.

(2) To become weathered so that the surface texture of a body, glaze, porcelain enamel or other coating is changed or roughened.

ethyl cellulose. An ethyl ester of cellulose used as a binder for technical ceramics and pigments, and as a parting agent for thin-sheet ceramics made by the doctor-blade process. See doctor blade.

ethyl silicate. $(C_2H_5)_4SiO_4$; a liquid organic silicate used as a binder and as a preservative for brick, concrete, mortar, plaster, refractories, etc.

Etruscan ware. A type of basaltic ceramics decorated by the encaustic process. See encaustic.

eucryptite. $LiAlSiO_4$; sp. gr. 2.67; used as a source of lithium in bodies of low-thermal expansion.

European porcelain. A high-grade porcelain of good

physical strength, white color, low absorption, and high translucency; usually coated with a hard glaze.

europium boride. EuB_6; mol. wt. 216.92; sp. gr. 4.94.

europium carbide. EuC_2; mol. wt. 176.0; unstable at 21°C (70°F).

europium nitride. EuN; mol. wt. 166.01; sp. gr. 8.77.

europium oxide. (1) Eu_2O_3; mol. wt. 351.9; sp. gr. 7.28–7.99; hardness (Knoop) 435; used in nuclear control rods, fluorescent glasses, and as a red phosphor in color television tubes. (2) EuO; mol. wt. 168.0; sp. gr. 8.16. (3) Eu_3O_4; mol. wt. 520.0; sp. gr. 8.07. (4) $Eu_{16}O_{21}$; mol. wt. 2768.0; sp. gr. 6.74; used in phosphors sensitive to red and infrared radiation.

europium silicide. $EuSi_2$; mol. wt. 208.12; m.p. ~ 1499°C (2730°F); sp. gr. 5.5.

europium sulfide. (1) EuS; mol. wt. 184.06; sp. gr. 5.75. (2) Eu_3S_4; mol. wt. 584.24; sp. gr. 6.27.

eutectic. A mixture of two or more components having a melting point lower than the melting points of the individual components.

eutectic, deformation. The minimum temperature at which a mixture of two or more components will develop sufficient liquid to permit deformation when heated under specific controlled conditions.

eutectic, equilibrium. The minimum temperature at which a mixture of two or more components will melt completely.

eutectic point. The point in a phase diagram indicating both the composition and minimum melting temperature of a mixture of two or more components.

eutectic temperature. The minimum temperature at which a mixture of two or more components will melt completely.

evacuator. A mechanical device which produces a vacuum for the removal of moisture from a body or system.

evaporator. A shallow pan, container, or other device in which the liquid in a slurry or solution is converted to the vapor state by the application of heat, sometimes facilitated by action of a vacuum.

Ewing theory of magnetization. Each atom is considered to be a permanent magnet which can rotate about its center when subjected to applied magnetic fields.

excess air. The amount of air introduced into a combustion process that is greater than that theoretically required to obtain complete combustion.

exchange capacity, anion. The ability of a clay or other substance to exchange or adsorb ions, expressed as milligram equivalents of anions per 100 grams of substance.

exchange capacity, cation. The ability of a clay or other substance to exchange or adsorb cations, expressed as milligram equivalents of cations per 100 grams of substance.

exchanger, heat. A device employed to transfer heat from a fluid flowing on one side of a barrier to a fluid flowing on the other side, such as a muffle in a furnace or kiln.

exfoliate. (1) To expand and separate into parallel layers or sheets under the influence of heat.
(2) To flake or peel from a surface.

exhaust. A duct, flue, chimney, or opening designed for the escape or removal of gases, fumes, vapors, or odors from a room or enclosure, sometimes facilitated by means of a fan.

exhaust system. Any system by which gases are removed from a dryer, kiln, furnace, or other confined area.

exothermic reaction. A reaction characterized by, or formed with, the evolution of heat.

expanded aggregate. A lightweight cellular material formed by heating the material to a temperature at a rate which will cause the material to bloat; used in the production of lightweight cement and in other products and applications as a thermal insulator.

expanded bed. A bed of activated carbon or other granular material through which a fluid flows upward at a rate sufficient to raise and separate the particles in the bed without change in their relative positions.

expanded blast-furnace slag. A lightweight cellular material produced by the treatment of the molten slag with water, high-pressure steam, air, or a combination of those treatments; used in the production of lightweight concrete blocks and similar products.

expanded clay. A lightweight cellular clay heated suddenly to a temperature sufficient to cause bubbles to be formed and retained in the clay particles; a bloated clay.

expansion. The process of increasing the volume of a constant mass of a material, such as by heating, water absorption, etc.

expansion, coefficient of. The incremental change in either the length or volume of a body per degree of temperature change at a constant pressure.

expansion coefficient, thermal. The fractional change in the length or volume of a material per degree of temperature change.

expansion joint. A joint in a concrete or masonry unit which permits the units to expand and contract without damage or without the introduction of excessive stresses.

expansion, moisture. An increase in the volume of a substance or product caused by the absorption of, or reaction with, water or water vapor.

expansion, secondary. The permanent expansion exhibited by fireclay brick in a furnace or kiln during service.

expansion, thermal. The change in the dimensions of a substance or product under the influence of heat, the change being either reversible or permanent.

expansion, water. The increase in the dimensions of a body resulting from the absorption of or reaction with water.

expansive cement. A type of high-sulfate and alumina-bearing cement which expands after hardening to compensate for shrinkage during drying.

explosive forming. The shaping of ware in dies in which the forming pressure is generated by an explosive charge.

exposed aggregate. A type of concrete construction in which the upper surfaces of the aggregate particles are exposed for special architectural effects.

exposed finish tile. A combed, roughened or smooth-faced building block, the exposed surfaces of which may be painted or left exposed.

expression. A process in which plastic clay bodies are extruded through an aperture to form symmetrical shapes, such as brick, hollow tile, pipe, and the like which are cut in desired lengths as they emerge from the die.

extender. An inactive or inert material added to another material or body composition to serve as a filler, diluent, modifier, or adulterant.

external grinding. The process of grinding or polishing the exterior of a rotating item.

external load-crushing strength. The ability of a concrete pipe to resist crushing forces which are applied externally in specified locations and directions on a specified length of pipe.

external seal. A metal collar or flange sweated around a ceramic shape, the metal having a slightly greater coefficient of expansion to produce a hermetic or near-hermetic seal. See sweat.

extra-duty glazed tile. A ceramic floor or wall tile suitable for use in light-duty floors and similar applications where impact, wear, and abrasive forces are not excessive; may be white or colored.

extrinsic semiconductor. A semiconductor whose electrical characteristics are due to added impurities.

extrudate. The product produced by an extrusion process.

extrude. To shape a plastic body by forcing the body through a die.

extruder. A device, such as a pug mill, which forces plastic bodies through a die of appropriate shape and size in a continuous column.

extrusion. The process of shaping a plastic body or molten glass by forcing the body or glass through a die.

eye. The opening through which the flame enters the bottom of a glass-melting pot.

eykometer. An apparatus to measure the yield point of clay suspensions. See yield point.

F

°F. Symbol for degrees Fahrenheit.

fabrication. The production or assembly of components into a unit or structure.

facade. The front or face of a structure.

face. (1) The work face of a grinding wheel.
(2) The exterior surface of a structure or wall.

face brick. Brick designed for use on the exterior or facing of a structure or wall; the exposed area of the brick sometimes may be textured.

faced wall. A wall to which an aesthetic facing has been bonded or attached; the facing may or may not be load bearing.

face milling. Machining of surfaces to a desired finish by means of a cutting or milling tool.

facet. The plane surface of a crystal.

facing. A fine sand applied as a facing to a casting mold.

facing brick. See face brick.

facing tile. Tile designed for use on interior or exterior walls for aesthetic or functional purposes.

facing wall. A concrete wall serving as a barrier to prevent movement of earth in embankments and excavations.

factory. A building or group of buildings in which materials or products are manufactured.

fade. The attack on glass surfaces by substances which produce an oily or whitish appearance.

fading. The loss of color or brilliance due to deleterious conditions of surface exposure during processing or service.

Fahrenheit. A temperature scale in which water freezes at 32° and boils at 212° under 1 atmosphere of pressure.

faïence. An earthenware, frequently soft and porous, which may be coated with a transparent or opaque glaze; typical products are figurines, pottery, tile, beads, and mosaics.

failure. A condition in which a product can no longer fulfill its intended purpose.

falling slag. A high-calcium blast-furnace slag sometimes used as an aggregate i[illegible]crete.

falling-sphere viscometer. An instrument used to determine the viscosity of a liquid by measuring the rate of fall of a standardized sphere through the liquid under standardized conditions.

false header. A half brick used to complete a row of brick in a facing wall.

false indication. An erroneous test result; usually due to improper sample preparation.

false set. The premature and erratic hardening of freshly mixed concrete, mortar, or cement paste usually due to the presence of unstable gypsum in the cement; the plasticity may be restored by mixing without the addition of more water.

Falter apparatus. An instrument to determine the softening point of glass in which the elongation of glass fibers is measured under specified conditions of temperature and tensile stress.

famille rose. A series of red colors for porcelain and chinaware produced from mixtures of gold and tin salts.

famille verte. A series of green colors for porcelain and chinaware produced by blends of chromic oxide.

family. (1) A group of materials of similar chemical or physical properties.

(2) A complete series of materials necessary to perform a specific process or to produce a specific product.

fan. A mechanical device designed to produce a current of air, gas, or vapor in a furnace, kiln, dryer, or other area as a means of delivery, circulation, or exhaust within the area.

fantail. The flue joining the slag pocket to a regenerator in an open-hearth furnace.

Faraday effect. Rotation of polarization of a beam of linearly polarized light passing through a substance, such as glass, in the direction of an applied magnetic field.

farren wall. A hollow wall, four inches in thickness; common in houses.

fat. A rich, plastic, cohesive concrete mix.

fat clay. A highly plastic clay. Also known as plastic clay, ball clay, or long clay. See ball clay.

fatigue. The tendency of a material to fail under cyclic stress, usually by cracking.

fatigue limit. The maximum stress a specimen of a material can withstand over an infinite number of specified test conditons or test cycles without failure.

fatigue strength. The maximum stress a specimen of a material can withstand over a specific number of specified test conditions without failure.

fatigue test. A test to determine the ability of a material to withstand conditions of alternating stress without failure.

fat mortar. Mortar containing a high proportion of cementitious material and which adheres to a trowel.

fault. An imperfection or defect.

fayalite. Fe_2SiO_4; mol. wt. 203.7; m.p. 1205°C (2202°F); sp. gr. 4.1; formed in aluminosilicate refractories when fired with iron-bearing slags under reducing conditions.

feather. (1) An imperfection of feather-like appearance in glass caused by seeds produced by dirt and foreign matter introduced during the casting or shaping process.

(2) A defect in wire glass resulting when transverse wires are deformed.

(3) A projecting strip, flange, rib, or fin.

feather brick. A brick cut diagonally from one end or side to the opposite end or side to form a shape of triangular cross section.

feather combing. A decorative technique in which a tool containing many sharp points is drawn across superimposed layers of damp slips of various colors for artistic effect.

featheredge. (1) A thin sharp edge.

(2) A sharp edge such as is produced when a brick is cut to form a brick of triangular cross section.

(3) A level-edged tool used to straighten angles in finish-coat plaster.

featheredge brick. See feather brick.

feathering. A devitrified imperfection usually occurring in lime-rich glazes.

feed. The process or the material supplied to a processing unit for treatment.

feed, drip. A technique of introducing oil, such as paraffin oil, into a furnace to serve as a lubricant for the parts of moving equipment.

feeder. A device designed to deliver materials to a processing unit, such as raw batch to a melting unit, or to deliver gobs of molten glass to a forming machine.

feeder, apron. A belt-like conveyor used to convey materials to a process or packaging unit.

feeder, batch. A mechanical device which introduces raw batch to a melting unit or to other processing unit.

feeder, channel. The channel which conveys molten glass from the working end of a glass-melting tank to the feeder.

feeder connection. The opening in a furnace wall through which the feeder channel is placed to convey molten glass from the melting tank to the feeder.

feeder, constant-weight. A feeder which delivers a specified and constant weight of a material to a processing unit per unit of time.

feeder, cylindrical-screen. A device consisting of a cylindrical screen in which clay is shredded and mixed, and through which the clay is forced by a bladed shaft, and then delivered to a subsequent processing unit.

feeder, disk. A feeder in which a rotating disk beneath the opening of a bin delivers material at a specified rate by controlling the rate of rotation and the size of the gate opening of the bin.

feeder gate. (1) A device, such as a sliding plate or valve, which controls the passage of a material from one location to another such as from a bin to a truck.

(2) The refractory shape which controls the rate of flow of molten glass in or through a feeder channel.

feeder nose. The end of the forehearth of a glass-melting tank containing the orifice ring of the feeder.

feeder opening. The feeder connection of a glass tank.

feeder plate. A type of conveyor consisting of overlapping plates between the roller chains which deliver materials to a process or packaging unit.

feeder plug. A shaped refractory which controls the rate of glass flow in the feeder channel of a glass tank.

feeder process. A process in which a gob of glass is delivered to the forming unit.

feeder, reciprocating. A reciprocating gate at the bottom of a hopper or bin to control the transfer of material from the hopper or bin to a transfer car or to a processing unit.

feeder, screw. A device in which a rotating screw moves powdered or granulated material forward through a channel to a processing unit.

feeder sleeve, feeder tube. A cylindrical tube containing the plunger in a glass-forming machine.

feeder, vane. See feeder, screw.

feeder, vibrating. A bin or hopper equipped with a vibrating mechanism to control and expedite the outward flow of powdered or granulated material from the bin or hopper.

feed, gravity. The movement of material from one container to another container or location by force of gravity.

feed grinding. The manner or rate at which work is fed to a grinding wheel.

feed, index. A technique by which the rate a work piece being fed to a grinding wheel is indicated by a dial mounted on the grinding equipment.

feed lines. The pattern formed on a work piece during machining or grinding.

feed rate. (1) The amount of material delivered to a process per unit of time.

(2) The cutting or grinding speed of a grinding or machining operation.

feed, ribbon. The batching procedure in the production of concrete in which the batch ingredients are fed into the mixer essentially simultaneously.

feed shaft. Vertical shafts under the fire holes in top-fired kilns for the combustion of fuel and dispersion of heat through the setting.

feed-through coil. A conducting coil, usually copper, surrounding a specimen in electromagnetic testing.

feed wheel. A wheel on a centerless grinder which regulates the speed and pressure on work during grinding.

Feine filter. A type of filter in which parallel strings are employed, instead of a filter cloth, to remove a filter cake from the drum.

feldspar. A group fo aluminous silicate minerals of potassium, sodium, and calcium, the principal types being orthoclase, microcline, albite, and anorthite of the general formula $K_2O{\cdot}Al_2O_3{\cdot}6SiO_2$ or its equivalent; m.p. 1100°–1532°C (2012°–2790°F); sp. gr. 2.56–2.63; hardness (Mohs) 6–6.5; used widely in all types of porcelain, tile, dinnerware, and other whiteware bodies, glass, glazes, porcelain enamels, and similar ceramic products, generally as a flux.

feldspar, white. A milky white or colorless variety of sodium feldspar.

feldspathic. Containing feldspar, such as a body or glaze.

feldspathoid. A mineral containing aluminous silicate, but insufficient silica to be a true feldspar.

felt, asbestos. An underlayment or overlayment of sheet asbestos saturated with asphalt for use with asbestos-cement products, sometimes of the breather type to permit transmission of water vapor.

female end of pipe. The end of a pipe which overlaps a portion of the end of an adjoining pipe.

Feret's law. The strength of cement or concrete is related to the mixing ratio of the volume of the cement, water, and air contained in the mix.

fernico. An alloy composed of iron, cobalt, and nickel employed in the production of glass-to-metal seals.

ferric chloride. $FeCl_3$; mol. wt. 162.3; m.p. 300°C (572°F); used in the development of gold lusters in glass, glazes, and porcelain enamels.

ferric fluoride. FeF_3; mol. wt. 112.8; sp. gr. 3.18; employed in porcelain and pottery, primarily as a flux and slight opacifier.

ferric oxide. Fe_2O_3; mol. wt. 159.7; m.p. 1565°C (2849°F); sp. gr. 5.12–5.24; used in the manufacture of ferrites and as a pigment to produce various colors in glazes and glass; also employed as a polishing material for glass and other substances; also known as rouge, red iron oxide.

ferrite. Any ferromagnetic material having high electrical resistivity of the general formula MFe_2O_4 in which M is a divalent metal, such as cobalt, copper, magnesium, manganese, nickel, and zinc.

ferroconcrete. A concrete in which some form of iron or steel is employed as a strengthening agent.

ferroelectric. A crystalline substance, such as the titanates and zirconates of barium, calcium, magnesium, and the like, which is used in ceramic capacitors, transducers, amplifiers, and other similar applications, and which exhibits spontaneous electric polarization, electric hysteresis, and piezoelectricity.

ferroelectric crystal. See ferroelectric.

ferrorelectricity. Ferromagnetism; spontaneous electric polarization in a crystal.

ferromagnesite. An iron-bearing magnesite employed in refractories for its strong bonding properties at elevated temperatures.

ferromagnetic. A permanent magnet composed of mixtures of ceramic and magnetic powders which have been pressed together and sintered.

ferromagnetic material. Any material displaying ferromagnetism; that is, having an abnormally high magnetic permeability, a definite saturation point, and appreciable residual magnetism and hysteresis.

ferrosilite. Iron silicate, $FeSiO_3$.

ferrospinel. Any spinel of the general formula MFe_2O_4 in which M may be barium, calcium, cobalt, copper, magnesium, manganese, nickel, strontium, or zinc; used as a refractory because of its high resistance to attack by molten glass and slags.

ferrous. Containing iron in any form.

ferrous oxide. FeO; mol. wt. 71.8; m.p. 1420°C (2588°F); sp. gr. 5.7.

ferrous sulfate. $FeSO_4 \cdot 7H_2O$; mol. wt. 278.0; m.p. 64°C (147°F); sp. gr. 1.89; used as a red ceramic colorant. Also known as iron sulfate, copperas, green copperas, green vitriol, iron vitriol.

ferrous titanate. $FeO \cdot TiO_2$; generally identified as a titanium ore containing iron oxide. Also known as ilmenite.

ferruginous. Containing iron, particularly iron oxide, Fe_2O_3.

fettle. To remove rough edges, mold marks, fins, and other irregularities from dry or nearly dry ceramic ware, usually by cutting, scraping, or abrasion.

fettling knife. A sharp knife or instrument used to fettle or trim ceramic greenware. See fettle.

fiber. A long, pliable filament made by attenuating some slags, or other molten or highly plastic material.

fiber, asbestos. Milled asbestos having a minimum length to maximum transverse dimensional ratio of at least 10 to 1.

fiber, basic. Unprocessed glass fiber as obtained from the forming equipment.

fiber bundle. A bundle or package of parallel, long, thin, flexible glass fibers employed to transmit images from one end of the bundle to the other in fiber-optics applications.

fiber, ceramic. Filament-like strands of ceramic materials, such as alumina and silica, used in lightweight units for electrical, thermal, and sound insulation, high-temperature filtration, packing, and reinforcement of metal and other ceramic products.

fiber, glass. Filaments of glass used loosely or in woven form as an acoustic, electrical, or thermal insulating material, in air and other filters, in nonceramic products such as a reinforcement in rubber, plastic, and similar products.

fiber, graphite. Graphite in filament form; frequently used in the manufacture of graphite cloth.

fiber, optical. A fiber of highly transparent material, such as glass or plastic, which transmits light by a series of internal reflections.

fiber optics. See fiber bundle.

fiber, organic. A filament of a natural or manufactured organic material.

fiber, staple. A glass or other fiber of relatively short length, generally less than 17 inches.

fictile clay. A moldable clay suitable for the making of pottery and earthenware.

field, bipolar. A longitudinal magnetic field with a part having two poles.

field, circular magnetic. A magnetic field surrounding any electrical conductor or part resulting from a current being passed through the conductor or part from one end to the other.

field-cured specimen. A test sample of concrete cured at the pouring site under conditions supposedly the same as that of the concrete employed in the structure being built.

field drain pipe. Pipe used to drain surface and subsurface water from fields.

field effect. The change in the properties and characteristics of a material or a part when subjected to the influence of an electric or a magnetic field.

field, longitudinal magnetic. A magnetic field in which the flux lines traverse the component in a direction parallel with its longitudinal axis.

field, magnetic. The space within and surrounding a magnetized part, or a conductor carrying current, in which magnetic force is exerted.

field, magnetic leakage. The magnetic field that leaves or enters the surface of a part at a discontinuity or at a change in section configuration of a magnetic part.

field meter, magnetic. An instrument which measures the strength of the magnetic field.

field, residual magnetic. The magnetic field remaining in a piece after the magnetizing force is removed.

field strength, magnetic. The measured intensity of a magnetic field.

figured glass. Flat glass having a pattern etched or ground on one or both surfaces.

figured rolled glass. A translucent rolled glass, one surface of which has a pattern in consequence of which vision is not clear and, in some instances, is almost completely obscured.

filament. A long, flexible thread of small cross section, usually extruded or drawn, such as glass or metal.

fill. The unit charge of batch introduced into a melting tank, pot, or other processing unit.

filler. (1) A chemically inert material used to fill holes in a surface prior to the application of a subsequent coating.
(2) An inert extender to a composition which does not add or detract from the intended properties of the composition.

filler, joint. A material for filling and sealing joints in concrete, but which allows movement of the joint.

fillet. A concave transition surface between two surfaces, which otherwise would meet at an angle, as a means of lessening the danger of cracking.

filling. The clogging of an abrasive product, such as emery cloth or a grinding wheel, by chips, shavings, and fine particles which have been removed from a piece during grinding.

filling point. The point of normal capacity of a glass bottle.

film. A thin coating or layer of a substance over the surface of another material.

films. Trimmed mica split to specific ranges of thickness under 0.006 in. (0.15 mm).

film strength. The resistance of films and coatings, such as glazes and porcelain enamels in the unfired state, to disruption and mechanical damage.

filter. A porous material through which a fluid is passed to remove matter in suspension.

filterability. The adaptability of a material in suspension in a slurry to separation from the slurry by means of a semipermeable medium or filter.

filter, black light. A filter than transmits ultraviolet radiation while suppressing the transmission of visible light.

filter block. A hollow, rectangular, vitrified clay masonry unit, sometimes salt glazed, used in trickle-type floors in sewage disposal plants; the block is designed with apertures connecting with drainage channels through the upper surface, and are arranged to form aeration and drainage grilles to pass air into, and liquids from, overlying filter media; the drainage channels convey liquid away from the filter bed.

filter cake. The solid or semisolid residue remaining on a filter after filtration, particularly the product from a filter press.

filter candle. A porous ceramic tube employed as a filter medium.

filter cloth. A cloth employed as a filtering medium in a filter press for the removal of water from clay slips and slurries.

filter, infrared. A material or product which is transparent to infrared radiation, but is opaque to other wavelengths.

filter medium. Closely woven textile or metal cloth; used as filter cloth, which see.

filter press. A device consisting of iron frames or plates suspended on a metal rack with a filter cloth stretched between each frame, the entire assembly then being pressed together and tightened by means of a screw mechanism; the slurry to be filtered is pumped through the assembly to remove excess water; the resultant filter cake collected on each filter cloth then is removed for further processing.

filter, quartz-crystal. An electronic filter in which a quartz crystal is the essential component.

filtration. (1) The act or process of passing a gas or liquid through a porous article or mass to separate out matter in suspension.

(2) The act or process of suppressing or minimizing waves or oscillations of certain frequencies of light, electricity, or sound by passing them through a suitable material or device.

fin. A thin, feather-edge protrusion or projection from a surface such as a casting, on flat glass after cutting, or on pressed or blown ware at the seam formed between two parts of a mold.

final set. The time required for cement or concrete to harden to the point beyond which plastic deformation will not occur.

fine aggregate. Sand; the portion of the aggregate in concrete or mortar passing a 3⁄16 inch (4.75 millimeter) sieve.

fine annealing. The heat treatment of glassware to an extremely low internal stress to improve its resistance to breakage, and to obtain a uniform index of refraction to improve its brilliance.

fine grinding. The milling of materials to particle sizes less than 100 mesh.

fineness. A measurement number designating the particle size of a material, usually reported as passing a screen of a particular standard size.

fineness modulus. An empirical factor designating the fineness of an aggregate as a percentage of the total sample retained on each of a series of screens of decreasing sizes.

fineness, porcelain enamel. The particle size of milled porcelain enamel frit; reported as grams of dry residue retained on a designated screen size from a measured sample.

fines. The portions of a powder composed of particles smaller than a specified size.

fine sand. Sand grains having a diameter between 0.0098 and 0.0049 inch (0.25 and 0.125 millimeter).

finial. An ornamental projection or end of fired clayware such as is used on spires or the ends of roof ridges.

fining. The process or period in glass making during which glass becomes essentially free from bubbles and undissolved gases.

finish. (1) The quality, appearance, or condition of a surface.

(2) A material applied to a surface for decorative, protective, or other functional purposes.

(3) The stage in the processing of molten glass when the glass appears to be free of seeds.

(4) The portion of a bottle designed to receive a cap or other closure.

finish, combed. A finish formed on the surface of ware consisting essentially of parallel scratches or scarfs. See combed finish.

finisher. (1) A workman who completes or perfects the final operation of a manufacturing operation.

(2) A workman supervising the melting and fining of glass.

(3) A person or machine that prepares the bed or finishes the surface of freshly poured concrete.

finish, fire. A glass surface polished by heating, as in a flame.

finish grinding. The completion of a grinding operation to obtain a desired surface appearance or accurate dimensions.

finishing. (1) Completion of an operation or process.

(2) Completion of a grinding operation or surface treatment.

finishing lime. Any white, plastic, hydrated lime suitable for use in finish-coat plaster.

finish mold. The neck mold of a bottle.

finish, natural. Ware fired to the natural color of the body or glaze ingredients.

finish, offset. An unsymmetrical finish relative to the axis of a glass bottle or other container.

finish, sand. A ceramic surface having faces covered with sand, applied either to the clay column in the stiff-mud process or as the mold lubricant in the soft-mud process.

finish, scored. Grooved surfaces on ware as removed from a die.

finish screen. A screen for the removal of dirt and undersized particles from coarse aggregate before it enters the bins at a concrete batching plant.

finish, short. An imperfection on the surface of plate glass due to incomplete or imperfect grinding and polishing.

finish, stippled. A mottled or pebbly textured porcelain enameled or glazed surface, frequently multicolored. See stippled finish.

finish tile. Tile employed in construction with the glazed face exposed to finish a wall.

finish, velvet. A finish produced by acid treatment to provide a velvety, embossed appearance and full obscuration on glass surfaces.

fire. The process whereby ceramic and nuclear-fuel bodies and shapes are densified by the application of heat.

fire, annealing. (1) The heat treatment of glass and metals to remove internal stresses.

(2) The heat treatment of metal shapes, prior to cleaning and pickling for porcelain enameling to burn off scale, dirt, and other contaminants and sometimes to temper the metal.

fire, bisque. Kiln firing of ceramic ware prior to glazing.

firebox. The section of a furnace or kiln in which combustion of fuel takes place.

firebrick. Any refractory brick capable of withstanding high temperatures without fusion. Used to line furnaces, fireplaces, chimneys, etc.; usually made of fireclay and contains less than 50% of alumina.

firebrick, insulating. Refractory brick of low-thermal conductivity and low-heat capacity.

fire bridge. A low wall separating the hearth and grate of a reverberatory furnace.

fire check. A crack resulting from thermal stress developed in ware during firing.

fireclay. Clay containing only small amounts of fluxing ingredients, but high in alumina and silica, capable of withstanding high temperatures without becoming glassy; used in the production of refractory brick, kiln and furnace linings, glass-melting pots and tanks, crucibles, etc.

fireclay brick, high-duty. Fireclay brick having a pyrometric cone equivalent greater than cone 31½ and less than cone 33.

fireclay brick, low-duty. Fireclay brick having a pyrometric cone equivalent greater than cone 15 and less than cone 29.

fireclay brick, medium duty. Fireclay brick having a pyrometric cone equivalent greater than cone 29 and less than cone 31½.

fireclay brick, semisilica. Fireclay brick having a silica content of not less than 72%.

fireclay brick, siliceous. Any fireclay brick having appreciable quantities of uncombined silica and which is low in fluxing agents.

fireclay brick, super duty. Fireclay brick having a pyrometric cone equivalent greater than cone 33, linear shrinkage less than 1% when reheated to 1598°C (2910°F), and less than 4% loss in weight during a panel spalling test (preheated at 1649°C (3000°F).

fireclay cement. A cement composed of dry fireclay and sodium silicate; used in the repair of saggers, refractories, kiln cracks, etc.

fireclay, flint. A hard, smooth, flint-like, nonplastic fireclay. See fireclay.

fireclay, ground. A fireclay or mixture of fireclays subjected only to weathering and milling, but to no other treatment.

fireclay mortar. A mortar composed of finely ground fireclay and water.

fireclay, nodular. Rock containing aluminous or ferruginous nodules, or both, bonded by fireclay.

fireclay, plastic. Naturally plastic fireclay capable of bonding nonplastic materials for use in refractories.

fireclay, plastic refractory. Water-tempered fireclay of suitable physical and thermal properties which may be rammed into place to form a monolithic furnace lining.

fireclay, silica. A refractory mortar or cement composed of finely divided mixtures of fireclay, quartzite, and silica.

fire crack. A crack resulting from thermal stresses developed in ware during firing.

fire, decorating. The firing of ceramic or metallic decorations on glazed ceramic ware.

fire finish. A surface finish or polish on glassware produced by heat treatment, such as in a flame.

fire, glost. The process of firing a glaze on bisque ceramic ware. See bisque.

firemarks. (1) A surface imperfection resulting from contact with a flame.

(2) Shallow pinhole-like indentations on the surface of porcelain enamel.

fire over. The idling of a glass-melting tank at operating temperature.

fire polish. To produce a smooth, glossy, or rounded glass surface by heating in a fire.

fireproofing. (1) To render incombustible.

(2) Any material used to protect against fire.

fireproofing tile. Tile, usually hollow, employed to protect members of a structure against fire.

fire resistant. Resistant to combustion for a specified time under standard conditions of heat intensity without burning or structural failure.

fire sand. A highly refractory sand consisting of coarse quartz grains in combination with alumina and a clay bearing sand; used primarily in foundries.

fire, sharp. Combustion of fuel with excess air and a short flame.

fire, short. An air-deficient flame.

fire, single. Maturing a body and its glaze in one firing operation.

fire, soft. An air-deficient flame.

firing. (1) The process of igniting a mixture of fuel and air in a kiln or furnace.

(2) The heat treatment of ceramic ware or products in a kiln or furnace to develop desired physical and chemical properties.

(3) The fusion of a porce;ain enamel or ceramic coating by heat.

firing behavior. The changes in the appearance and properties of ceramic products during firing or thermal treatment.

firing chamber. Any chamber or enclosure in which fuel is burned to provide heat.

firing curve. A chart recording the time and temperature conditions during a firing operation.

firing cycle. The time required for one complete firing operation.

firing, direct. The firing of ware in direct contact with the products of combustion in a furnace or kiln.

firing, draw. The removal of ware from a furnace for a short period of time prior to completion of the firing operation to permit all areas of the ware to attain an essentially equalized temperature; the ware may or may not be returned to the furnace, depending on the maturity of the coatings.

firing expansion. The increase in the dimensions of a substance or product during thermal treatment.

firing fork. A long-handled, two-pronged tool used to charge and remove ware in furnaces.

firing, glost. The firing of bisque ware to which a glaze has been applied.

firing, open. Firing in which the flames play through the exposed wares stacked in a kiln.

firing range. The time-temperature intervals in which bodies and coatings attain the respective desired maturities or properties.

firing shrinkage. The contraction or decrease in the dimensions of a substance or product during thermal treatment, calculated by the formula:

$$\text{Linear firing shrinkage, \%} = \frac{L_d - L_f}{L_d} \times 100$$

in which L_d is the length of the dry, but unfired, specimen, and L_f is the length of the fired specimen,

$$\text{Volume firing shrinkage, \%} = \frac{V_d - V_f}{V_d} \times 100$$

in which V_d is the volume of the dry, but unfired, specimen, and V_f is the volume of the specimen after firing. Also known as linear burning shrinkage and volume burning shrinkage.

firing, sky. Completing the firing of ware in an updraft bisque kiln by inserting and burning slivers of wood in the top of the kiln to increase the draft.

firing temperature. The peak temperature reached during the firing of a porcelain enamel.

firing time. The time porcelain-enameled ware remains in the firing zone of a furnace to attain coating maturity.

firing, vacuum. The firing of ware in a vacuum to reduce porosity by the removal of gases and to prevent reaction with gases or vapors in a normal furnace atmosphere.

firing zone. The section of a furnace or kiln in which ware is subjected directly to the major influences of heat, as in a continuous furnace or kiln.

first-quality ware. Products which meet specified standards and are free of imperfections or defects.

first side. The surface of plate glass which is first ground and polished.

fisheye. A glass bubble on the fired surface of a glaze or porcelain enamel.

fishscale. A half-moon fracture, resembling the scale of a fish, on the surface of porcelain enamel caused by the presence of small pockets of hydrogen or other substance at the interface between the coating and the steel.

fishscale, delayed. Fishscale occurring on porcelain enameled surfaces after completion of the firing operation, sometimes several days after firing.

fishscale, process. Fishscale occurring on porcelain enameled surfaces during the drying or firing operation.

fission. The division of an atomic nucleus into parts of comparable mass, usually with the release of energy and one or more neutrons.

fissionable material. A material whose nuclei are capable of undergoing fission, such as the heavier isotopes of uranium, plutonium, or thorium, with the emission of a large amount of energy.

fission, controlled. Fission under conditions of continuous adjustment of control rods and other control devices in a reactor which compensates for changes in reactivity from high-power operation and from reactor-temperature fluctuations.

fission products. Nuclides produced by fission or by radioactive decay of the fission products.

fissures. Surface defects consisting of narrow openings or cracks.

fit. The stress or dimensional relationship between a coating and its substrate.

fitting. An accessory part used in the assembly of a system, such as tees, wyes, elbows, and adaptors.

fixed bed. A bed of powdered or granular material through which a fluid may flow without substantial movement of the bed.

fixed-feed grinding. The process of feeding a material to be ground to a grinding wheel at a given rate or in specific increments.

flake enamel. Porcelain-enamel frit available in egg-shell-thin fragments or flakes.

flaking. The breaking of small chips or thin fragments from the surface of a refractory, glaze, or porcelain enamel.

flambé, rouge. A red flow glaze containing reduced copper which produces a varigated appearance on pottery products.

flame. The hot gaseous part of a fire.

flame annealing. The heating of a glass or metal part in a flame to remove stresses and to make the glass or metal less brittle.

flame cleaning. Removing scale, rust, and dirt from metal surfaces with a broad flame.

flame-proof ware. Ware capable of withstanding extreme thermal shock.

flame, reducing. An oxygen deficient flame, often producing a reducing atmosphere in a furnace or kiln.

flame spraying. Deposition of a coating on the surface of a product or material by feeding the coating material through a spray gun into a gas flame to impinge and fuse molten particles on the work.

flange. (1) A rim designed to strengthen a metal part or to facilitate assembly to another part.

(2) The circular metal plates that drive a grinding wheel.

flanged bottom. An imperfection consisting of an offset on the bottom of a bottle.

flange, safety. A flange with tapered sides designed to keep a grinding wheel intact in the event of accidental breakage.

flash. (1) A thin film of another glass, frequently colored or opaque, applied and fused to the surface of sheet or other clear glass.

(2) A film of different color or texture on clayware.

flashed brick. Brick subjected to reducing conditions near the end of the firing cycle to produce a desired color.

flashing. (1) The process of applying a flash of glass. See flash (1).

(2) A thin sheet of material placed at the junction of exterior building surfaces to make the joint watertight. See flash.

flash magnetization. Magnetization by a current flow of brief duration.

flash marks. (1) The discoloration of a brick surface due to the presence of fly ash during firing.

(2) Cross-set marks in sections of brick due to flashing reduction; that is, the sections were subjected to reducing conditions during firing, causing color differences in the brick.

flashover. An electric discharge over or around the surface of an insulator.

flash point. The lowest temperature at which vapors from a volatile liquid will ignite, at least momentarily, on contact with a small flame.

flash set. Rapid and permanent hardening of fresh mortar, concrete, or cement paste with the evolution of heat.

flash wall. A refractory wall in a kiln placed so as to prevent impingement of flames on the setting of ware being fired.

flat arch. An arch in a furnace or kiln in which both outer and inner surfaces are horizontal; the inner arch may be arched with a large radius. Also known as jack arch.

flat-drawn process. A process in which sheet glass drawn vertically from the molten bath is passed between rollers to solidify the sheet to a prescribed thickness.

flat glass. Sheet glass, which see.

flat sheets, type F (flexible). Smooth, dense, flexible, flat asbestos-cement sheets of low moisture-absorption properties for use in exterior and interior applications, particularly where fire resistance is a factor.

flat sheets, type U (utility). Flat asbestos-cement sheets having sufficient strength for general utility and applications in exterior and interior construction.

flat slab. A reinforced concrete plate or slab designed to span in two directions, such as in flooring.

flatware. A generic term for flat items of dinnerware such as plates, meat platters, saucers, bread and butter plates, and the like.

flatwork. Concrete items such as sidewalks, floor, and flat slabs.

flaw. An imperfection or defect.

flaw indication, magnetic particle. The accumulation of ferromagnetic particles in the area of discontinuities or flaws due to the distortion of the magnetic lines of force.

Flemish bond. Courses of brick consisting of alternate headers and stretchers, the headers being centered on the stretchers above and below.

flexural strength. The ability of a material to withstand bending or a transverse load without fracture.

flint. A finely crystalline form of natural silica or quartz used as an abrasive, balls and liners for ball mills, and as a component in glass manufacture; sp. gr. 2.6–2.65; hardness (Mohs) 6.5–7.

flint clay. A hard, smooth, nonplastic fireclay. See fireclay.

flint-enameled ware. A semivitreous or earthenware type of pottery flecked in yellow, brown, and blue colors.

flint fireclay. A hard, smooth, nonplastic fireclay. See fireclay.

flint glass. (1) A heavy, colorless, brilliant lead-bearing glass; often used as an optical glass.

(2) A clear, colorless bottle glass.

(3) Any glass of high quality.

flint mill. A ball mill in which flint pebbles are used as the grinding media; sometimes the mill lining also may be constructed of flint blocks.

flint optical glass. A glass of high index of refraction and high dispersion used for optical applications. See optical flint glass.

flint shot. A hard, coarse, sharp-edged sand used in sandblasting.

flint, white. A colorless glass having high light-dispersing qualities; used in optical instruments.

float. (1) A flat wood or metal finishing tool for cement which is used after screeding and before troweling.

(2) A rectangular piece of wood with a handle attached to the under side used to apply and smooth coats of plaster. See screed; trowel.

floater. A floating clay or refractory shape, usually a ring, to skim foreign materials from the surface of molten glass in a glass tank, and to control their passage from one section of the tank to the next.

floater hole. An opening in a glass-melting tank through which floaters are placed into a tank. See floater.

float finish. A rough concrete surface produced and smoothed by the use of a wooden float during the finishing operation. See float.

float glass. Flat glass formed on a bath of molten metal, such as tin.

flocculant. A reagent or electrolyte added to a colloidal suspension to cause the particles to aggregate or coalesce and settle.

flocculation. The addition of an electrolyte to thicken a porcelain-enamel slip.

floc test. A test of the durability of hydraulic cement in which one gram of cement is shaken in a test tube containing 100 milliliters of water and allowed to stand for seven days; if the amount of floc, or suspended particles, is small the cement is considered to be durable.

flooding. The flowing of water over an unfired porcelain enamel surface to produce a coating having a water-streaked appearance when fired.

floor brick. Smooth, dense, abrasion-resistant brick used in floors.

floor brick, industrial. A brick having extremely high resistance to wear, mechanical damage, chemicals, and conditions encountered in industrial and commercial installations.

floor, hot. A concrete floor in which steam pipes are embedded.

floor-stand grinder. A grinding machine mounted on a stand or base attached to the floor.

floor tile. An abrasion-resistant ceramic tile used in floor construction; some large hollow tile also are used in roof construction.

floor topping. A thin layer of high-quality, high-strength concrete applied to a concrete slab as a finished floor.

Florida kaolin. A very clean, ball-type kaolin of high purity, fine particle size, and white burning; employed to promote refractoriness, plasticity, bonding strength, and suspension power in many types of ceramics.

floss. Molten or solid slag floating on the surface of molten metals and glasses.

flotation. A process employed to separate particles in a liquid bath in which one group of particles is caused to float and other groups to settle.

flow button. A pellet of frit or dried slip of porcelain enamel or glaze employed to evaluate the flow characteristics of the materials at fusion temperatures by comparison with standardized pellets.

flow coating. The process of applying a coating to an object by pouring a slip or liquid over the surface of the object and allowing it to drain.

flow, fusion. The relative flow of various glasses and frits in the molten state.

flow hole. The submerged passage between the melter and refiner of a glass-melting tank.

flow line. A line formed on a molded item at the point where two input-flow fronts meet during the molding process.

flow, plastic. The flow behavior or deformation of a material under a sustained load or stress.

flow process. A process in which a gob of glass is delivered to the forming unit.

flow test. A test to determine the flow characteristics of concrete or other plastic mass in which a measured volume is vibrated or jolted on a flat surface and its tendency to flow is observed.

flow, uniform. A flow of gas, liquid, or a powdered or granulated solid at a constant velocity or volume.

flue. A passage to exhaust combustion gases and dust from a kiln, furnace, or other combustion chamber.

flue dust. Particles of dust exhausted from a furnace, kiln, or other combustion chamber.

flue gas. The gaseous products of combustion from a furnace, kiln, or other combustion chamber.

flue-gas analyzer. An instrument, such as an Orsat analyzer, which analyzes and sometimes monitors the composition of flue gases and the air-fuel ratio in the combustion chamber of a furnace or kiln.

flue lining. The refractory shapes used to line the flues and exhaust passages of furnaces and kilns.

fluid. A substance such as a liquid or gas having low resistance to flow and a tendency to assume the shape of its container.

fluid bed. A bedding of fine particles or granules which behave in a fluid-like manner when moved by a rising stream of gas or air.

fluid carrier. A fluid in which particles are suspended to facilitate their movement or application; for example, water in which glaze and porcelain-enamel compositions are suspended.

fluid-energy mill. A size-reduction apparatus in which grinding is achieved by the collision of the particles being ground in a high-velocity stream of air, steam, or other fluid.

fluidity. (1) The property of a substance to flow like a liquid or gas.

(2) The workability or consistency of a material or mixture to flow, such as wet concrete, a glaze, or a porcelain enamel.

fluidized bed. An apparatus in which powdered or granular material is contained and suspended in a rising stream of hot air or gases as a means of drying, heating, calcining, coating, or quenching; the powdered or granular material in the suspended state behaves much like a liquid.

fluidized-bed coating. A coating applied to an article while it is immersed in a fluidized bed of the coating material or while the article is suspended in a flowing gas stream of the coating material.

fluidized-bed combustion. The combustion of particulate matter in a fluidized bed which has an excess of air passing through the bed.

fluobarite. A mixture of fluorspar, CaF_2, and barite, $BaSO_4$, used as a flux in glass manufacture.

fluor. Synonym for fluorspar, CaF_2, which see.

fluor crown glass. An optical crown glass of low index of refraction and dispersion containing substantial amounts of fluorine to serve as a flux.

fluorescence. The emission of visible radiation (glow) from an object during its period of absorption of radiation from some other source; fluorescent materials often are known as phosphors.

fluorescent magnetic inspection. The use of a finely divided ferromagnetic fluorescent medium as an inspection technique.

fluorescent penetrant. An inspection penetrant which fluoresces or glows in black light.

fluorescent pigment. A pigment which will give off light (glow) during exposure to radiant energy such as ultraviolet light.

fluorination. A chemical reaction occurring when fluorine is introduced or comes in contact with a receptive product.

fluorine. F; at. wt. 19; introduced in ceramic compositions in the form of various fluorides as a flux or opacifying ingredient.

fluorite. Synonym for fluorspar, CaF_2, which see.

fluorspar. CaF_2; mol. wt. 78.1; m.p. 1350°C (2462°F); sp. gr. 3.2; hardness (Mohs) 4; used as a flux and opacifier in ceramic glazes, porcelain enamels, and glass; as a flux in emery-wheel binders; as a component in certain cements; and as a major component in crucibles used for the melting of uranium for nuclear applications.

fluosilicate. Any salt of fluosilicic acid, H_2SiF_6, used as a source of silica and fluorine; fluosilicates of barium and zinc also are used as cement hardeners.

fluosilicic acid. H_2SiF_6; a colorless liquid used in the etching of glass.

flush tank. A ceramic container designed to supply water to a sanitary water closet.

fluting. A machine operation to form grooves parallel to the axis of taps and drills.

flux. (1) Any substance which promotes the fusion and flow of a ceramic or glass mixture when subjected to heat.
(2) A clear porcelain enamel containing no coloring oxide; used in artware. Also known as fondant.

flux block. Refractory shapes used in glass-melting tanks in areas of contact with molten glass.

flux density, magnetic. The strength of a magnetic field.

flux factor. A factor evaluating the quality of silica refractories used in steel manufacture; calculated as the percentage of alumina in the brick plus twice the percentage of alkalies. (The flux factor of first quality (Type A) brick must not exceed 0.50)

fluxing. The fusion or melting of a substance resulting from the combined influence of chemical reaction and heat.

fluxing agent. Any substance, such as lead oxide, borax, and lime, which will promote fusion of ceramic materials.

flux line. (1) The line at the surface of molten glass in a glass tank or pot where attack on the refractories is most severe.
(2) Imaginary magnetic lines indicating the behavior of a magnetic field.

flux, magnetic leakage. The excursion of magnetic lines of force from the surface of a specimen.

flux, neutron. The number of neutrons passing through an area of one square centimeter per second.

flux penetration, magnetic. The depth to which a magnetic flux is generated in a component.

fly ash. Fine particles of matter in flue gases, usually resulting from the combustion of fossil fuels; sometimes used as a pozzolan or as a filler in some cements.

foam. A froth or layer of bubbles on the surface of molten glass.

foamed clay. Lightweight cellular clay formed by the rapid heating of selected clays to form a bubbled internal structure; used as thermal and acoustic insulation.

foamed concrete. Concrete containing purposefully entrained air or gas bubbles which have been introduced either mechanically or chemically.

foamed glass. Cellular glass of high insulating value, noncombustible, moisture proof, buoyant, and odorless; produced by adding powdered carbon or other gas-forming material to crushed glass and fired in a manner to entrap the evolving gas bubbles; used as insulation for walls, floors, roofing, industrial and domestic equipment and appliances, piping, low-temperature apparatus, etc.

foam line. The line dividing the foam-covered area of a glass-melting tank from the clear area of the tank.

fold. (1) An imperfection on the surface of glassware caused by incorrect glass flow during forming.
(2) An abrasive or a tool used for lapping and polishing.

fondant. A clear porcelain enamel containing no colorants; used in artware.

foot. The base of an article.

force. An influence which tends to cause motion or a change in motion.

force, coercive. An opposite magnetic force required to return ferromagnetic materials to their original nonmagnetic state.

forced draft. Air under positive pressure produced by fans at the entrance to a furnace or combustion chamber.

Ford cup. A viscometer in which the time required for a measured quantity of liquid or slurry to flow through an orifice of specified size is taken as an indication of the flow characteristics of the material.

forehearth. The section of a furnace from which molten glass is taken for forming.

fork. An apparatus consisting of two or more prongs which may be raised or lowered, and which is employed to charge and to remove ware from a box furnace.

fork-lift truck. A machine equipped with two or more parallel prongs which can be raised or lowered, and which may be inserted under stacked materials for transport.

form griding. The shaping of a product by use of an abrasive wheel contoured to the reverse shape of the desired form.

forming. The shaping or molding of molten glass or plastic ceramics by the application of pressure, by casting, by hand shaping, or by other means.

forming, explosive. Shaping of ware by the use of an explosive force as a means of generating the forming pressure.

forming hood. The chamber of the forming equipment in which glass fibers are formed and collected.

forming rolls. Rolls employed in the forming of flat glass.

form oil. A material applied to the surface of molds and forms to prevent concrete from sticking.

formula. A recipe of ingredients used in the preparation of a desired composition expressed in fixed proportions.

forsterite. $2MgO \cdot SiO_2$; mol. wt. 140.7; m.p. 1900°C (3450°F); sp. gr. 3.21; used in electronic ceramics, ceramic-metal seals, refractories, and cements because of its high-thermal expansion and low-loss dielectric properties.

forsterite porcelain. A vitreous ceramic in which forsterite, $2MgO \cdot SiO_2$,is the essential crystalline phase. See forsterite.

forsterite whiteware. Any ceramic whiteware in which forsterite, $2MgO \cdot SiO_2$, is the essential crystalline phase.

fossil fuel. Any natural hydrocarbon, such as coal, petroleum, or gas that may be used for fuel.

foundation seal. A sand slab, or a slab of concrete, placed at the bottom of a wet excavation to serve as a seal to facilitate subsequent work.

foundry. A building or structure in which glass and metal castings are produced.

foundry clay. A refractory fireclay. See fireclay.

foundry engineering. The science and practice of melting and casting glass and metal.

foundry sand. Sand used to make molds for metal castings; characterized by refractoriness, cohesiveness, and durability.

Fourcault process. A procedure for making flat glass in which the molten glass is drawn upward from a melting tank in ribbon form through a slotted refractory block, rolled flat, annealed, and then cut to the desired size and shape.

fractionation, elastic. A process in which soft aggregate is separated from harder aggregate by hurling the composite aggregate against a steel plate, the hard particles rebounding farther from the plate than the soft, more friable particles.

fractography. The study of fractures by microscope.

fracture. A crack caused by mechanical failure due to stress.

fracture, spontaneous. Cracking or chipping which occurs without immediately apparent external causes.

fracture, thermal shock. Cracking or chipping by sudden cooling.

fracture wear. The wear of the grains of a grinding wheel due to fracture.

fragility. The property of being easily broken.

free blown. Blown glassware formed by hand without the use of a mold.

free carbon. Elemental carbon present in a composition or body in an uncombined state.

free crushing. The process of crushing friable materials in a manner that the fines separate from the coarse particles and thereby avoid further grinding.

freehand grinding. The process of grinding an item or shape by hand without the use of guides.

free moisture. The quantity of uncombined water in a body or composition which can be removed by conventional drying.

free silica. Silica in clay or glazes which remains chemically uncombined with other elements of the composition.

free water. See mechanical water.

freeze. The premature setting of a concrete in a pump, drill rod, etc., before it can be placed in its intended site.

freeze-thaw test. An accelerated test to indicate the resistance of brick, concrete, and similar products used in construction to cycles of freezing and thawing such as may be encountered in service.

freezing point. The temperature at which a liquid becomes a solid.

French (hexagonal) roofing. Asbestos-cement roofing chipped at three corners and laid with the diagonals perpendicular to the eaves to form a hexagonal pattern.

fresco. The process of decoration in which slurries of pigment and a suitable binder are applied on a previously dried but wetted plaster wall.

fresh concrete. Concrete that has not reached its initial set.

Freundlich isotherm. See isotherm, Freundlich.

friability. The ease with which a material may be broken or pulverized.

friable alumina. Alumina which is more friable than normal alumina and less friable than white alumina.

frisket. A mask or stencil used to protect an area of ware from a subsequent application of glaze or slip.

frit. A glass which has been melted and quenched in water or air to form small friable particles which then are processed by milling for use as the major constituent of porcelain enamels, fritted glazes, and frit chinaware.

frit china. A thin, highly translucent whiteware of zero water absorption composed of a body containing substantial amounts of frit, and coated with a soft glaze.

frit, clear. A frit which produces a clear, transparent porcelain enamel or glaze.

frit, colored. A frit containing a colorant which produces a colored porcelain enamel or glaze.

frit seal. A hermetic seal for ceramic packages of integrated circuits produced by fusing metal and glass powders.

fritted glass. Glass of controlled porosity formed by sintering powdered glass.

fritted glaze. A glaze in which part or all of the fluxing ingredients have been fused or quenched to form small friable particles before incorporation into the glaze slip.

fritting. The process of melting and quenching glassy or molten materials to form small, friable particles.

frog. A depression on one or both larger faces of a brick or block; so designed to reduce weight and to facilitate the keying in of mortar.

frosted. The surface treatment of glass to produce a frosty appearance or a degree of obscuration, usually by chemical action or light sandblasting.

frost glass, frosted glass. Very thin crushed glass used as a decorative material of tinsel-like appearance when distributed and fused over the surface of a glass article.

froth flotation. A materials beneficiation process for finely divided materials in which a slurry is caused to foam by the addition of a foaming agent; select particles adhere to the resultant bubbles and are removed with the froth and thus are separated from the materials remaining in the slurry.

fuchsine dye. An analine dye dissolved in alcohol; used to test the porosity of electrical porcelains and other ceramic bodies.

fuel. A material that is burned to produce useful heat.

fuel-air ratio. The proportions of air and fuel employed in the combustion process.

fuel assembly. Any device containing source and special nuclear materials which occupy individually controlled positions in the core of a nuclear reactor, plus structural materials which facilitate assembly of the reactor.

fuel bed. The layer of burning fuel on the floor of a cupola or other furnace.

fuel element. Cylinders, rods, plates, tubes, or other shapes into which nuclear materials are formed for use in a reactor.

fuel gas. Any gaseous material employed to provide heat or power by combustion.

fuel oil. Any oil employed to provide heat or power by combustion.

fuel, spent. Nuclear-reactor or other fuel which is no longer effective.

fuller's earth. A nonplastic clay-like material, composed largely of attapulgite with some montmorillonite, having high natural absorptive power.

full-trimmed mica. Mica with all cracks and cross grains or reeves removed from all sides by trimming.

funk. A capricious form of pottery.

furnace. An enclosed structure in which elevated temperatures are produced for the firing of ware to obtain desired physical and chemical changes in the ware.

furnace, acid-refractory. A furnace or cupola lined with an acid-type refractory such as silica brick.

furnace, arc-image. A furnace in which high temperatures are produced by focusing radiation from high-temperature arcs into the furnace chamber.

furnace, basic open-hearth. An open-hearth furnace constructed of basic refractories covered with magnesite or burned dolomite, and which is employed in the production of basic pig iron. See open-hearth furnace.

furnace, box. A periodic furnace in which ware is charged, fired, and removed before the introduction of another charge into the furnace.

furnace, continuous. A furnace into which ware is introduced, fired, and removed continuously by means of some type of conveyor system.

furnace, cross-fired. A furnace in which fuel is introduced into the firing chamber from side ports.

furnace, direct-fired. A furnace which has neither a recuperator nor a regenerator; that is, neither the air nor the fuel is preheated prior to charging into the firing chamber of the furnace. See furnace, regenerative; furnace, recuperative.

furnace, end-fired. A furnace into which fuel is supplied from openings in the end wall.

furnace, end-port. See furnace, end-fired.

furnace, hairpin. A continuous porcelain-enameling furnace constructed in the shape of a hairpin, the firing zone being located in the turn. Also known as a U-type furnace.

furnace, high-frequency induction. A furnace in which heat is generated within the furnace charge or the charge container, or both, by means of currents induced by a high-frequency magnetic flux produced by a surrounding coil.

furnace, image. A furnace in which high temperatures

are generated by focusing radiation from a high-temperature source such as the sun or an electric arc.

furnace, indirect arc. A furnace in which ware is heated indirectly by radiant heat from a high-temperature electric arc.

furnace, induction. A furnace in which ware is heated by electromagnetic induction.

furnace, Kryptol. A furnace in which heat is generated by passing an electric current through a rammed mixture of carbon, silicon carbide, and clay which is characterized by high electrical resistivity.

furnace lining. The exposed interior of a furnace, kiln, or smelter which is constructed of high temperature-resistant, chemical resistant, and abrasion-resistant refractory materials.

furnace, low-frequency induction. An induction furnace in which the current flow at commercial power-line frequency is induced into the material or the article to be heated.

furnace, low-shaft. A short-shafted, refractory-lined blast furnace used to produce low-grade products by using low-grade fuels.

furnace, luminous wall. A furnace in which the heat required for firing is generated by projecting hot flames on the furnace walls which, in turn, become incandescent.

furnace, open arc. A furnace heated by an electric arc positioned above the furnace charge.

furnace, open hearth. A steel-making furnace of the reverberatory type in which the charge is laid on a shallow hearth over which the flames of burning gas and hot air are played.

furnace, periodic. A furnace in which ware is placed, fired, and sometimes cooled prior to the introduction of a subsequent charge.

furnace, pot. A furnace in which the material to be melted is contained in pots or crucibles which may be open or covered.

furnace, radiant-tube. A porcelain-enameling furnace heated by radiant tubes in which the combustion of fuels takes place without the combustion gases entering the firing chamber.

furnace, recuperative. A furnace equipped with a heat exchanger in which heat is conducted from the combustion products through a system of ducts or through flue walls in a manner so as to preheat the air as it enters the burner to unite with the fuel.

furnace, regenerative. A furnace having a cyclic heat exchanger which alternately receives heat from gaseous combustion products and transfers heat to the air or gas of the fuel mixture before combustion takes place.

furnace, resistor. An electric furnace in which heat is generated by the passage of an electric current through a resistor element which is not in contact with the charge.

furnace, reverberatory. A furnace or kiln in which the fuel is burned at one end with the flame passing between the charge in the furnace and the furnace roof, the heat being deflected downward from the roof through the charge.

furnace, rocking. A horizontal melting furnace designed to rock back and forth as a means of mixing the batch ingredients during melting, thereby producing more uniform melts.

furnace sand. A relatively pure, coarse type of sand used as a refractory material for hearths and for foundry molds.

furnace, semi-muffle. A furnace equipped with a partial muffle to protect ware being fired from direct impingement of the flame, but which permits the products of combustion to come into contact with the ware.

furnace, side-fired. A furnace in which the fuel is introduced into the firing chamber through ports on the sides of the furnace.

furnace, side-port. See furnace, side-fired.

furnace, solar. An image-type furnace in which solar radiation is focused into a relatively small area as a source of heat producing extremely high temperatures. See furnace, image.

furnace, tank. A refractory-lined tank or receptacle in which glass is melted, the glass batch being charged continuously at a rate essentially equal to the amount being continuously withdrawn.

furnace, thermal gradient. A tubular furnace in which a controlled temperature gradient is maintained along its length.

furnace, U-type. See furnace, hairpin.

furnace, vacuum. A furnace or heating device constructed so as to permit the firing chamber to be evacuated and fired under a vacuum.

furniture, kiln. Small refractory shapes of many different designs, such as stilts, pins, posts, spurs, cranks, saddles, etc., upon which ware is placed and supported during firing.

furring. (1) Wood or metal strips applied to the wall or ceiling of a building to level the surface, to provide a means of attaching plasterboard to the wall, and to permit an air space between the plasterboard and the wall structure.

(2) The bristling of magnetic particles due to excessive magnetization, resulting in a fuzzy appearance.

furring brick. A type of hollow brick which has been grooved to receive and retain a coating of plaster in the construction of walls.

furring tile. A nonload-bearing tile used as an unexposed lining in interior walls, sometimes made with a furrowed or grooved face to receive and retain a coating of plaster.

fuse. To melt or join by the use of heat.

fused alumina. A form of alumina produced by heating a mixture of calcined bauxite or Al_2O_3 and iron borings to a temperature in excess of 3600°C (6512°F) in an electric arc furnace; used in applications where high resistence to abrasion is required, such as in bearings, spindles, etc.

fused-grain refractories. Refractories made predominantly from refractory substances which have solidified from a fused or molten condition.

fused-grain refractories, rebonded. Fired refractory bricks or other shapes made predominantly or entirely from fused refractory grain.

fused quartz. A pure silica glass made by melting crushed crystals of natural quartz or silica sand; used in apparatus and equipment requiring materials having low-thermal expansion, a high-melting point, high-chemical resistence, and high transparency. Also known as silica glass.

fused refractories. Cast or molded refractory shapes which have been formed from molten refractory compositions. Also known as fusion-formed or fusion-cast refractories.

fused silica. A transparent or translucent glass consisting almost entirely of silica formed by the flame hydrolysis of silicon tetrachloride.

fused-silica refractory. A refractory product composed essentially or entirely of fused, noncrystalline silica.

fusible, fusibility. Capable of being softened or melted by heat.

fusible clay. A clay which will vitrify and lose its shape at temperatures of 1200°C (2192°F) or lower.

fusion. (1) The process of melting, frequently with interaction of two or more materials, to form a more or less homogeneous mass.
(2) Joining by the use of heat.

fusion casting. The process of forming items by casting molten material in a mold.

fusion flow. The property of a material, such as a glass, frit, or metal, to flow while in the molten state.

fusion-flow test. Any test which will measure or compare the flow characteristics of a material or materials under the influence of heat, such as by heating uniform pellets of a glass or frit on a panel for a period sufficient to cause softening, and then raising the panel to a vertical position to permit the molten pellets to flow down the vertical surface.

fusion, heat of. The measure of the heat required to convert a unit weight of a solid substance to its liquid state.

fusion joint. The line at which the surfaces between two solids are joined together more or less permanently by the use of heat.

fusion point. The temperature or range of temperatures at which melting or softening of a composition will occur.

fusion test. Any test to determine the temperature or range of temperatures at which fusion takes place, or to determine the flow or other properties of a material at fusion temperatures.

fuzzy texture. An indistinct or fuzzy-appearing imperfection occurring on porcelain-enameled ware due to the presence of minute closed and broken bubbles, dimples, and the like, at the surface.

G

gable. The triangular wall section at the ends of a pitched roof, bounded by two roof slopes and the ridge pole.

gable roof. A pitched roof that ends in a gable.

gable tile. A roofing tile having the same length, but 1½ times the width of the tile used elsewhere on the roof, used to complete the alternate courses of the gable of a tiled roof.

gable wall. (1) The wall of the charging end of a glass-melting furnace.
(2) A wall crowned by a gable.

gadget. An instrument to hold the foot of a glass during hand-finishing of the bowl.

gadolinium aluminate. $Gd_2O_3 \cdot Al_2O_3$; mol. wt. 464.42; m.p. 1982°C (3600°F).

gadolinium boride. (1) GdB_4; mol. wt. 200.54; sp. gr. 6.48. (2) GdB_6; mol. wt. 222.18; m.p. 2099°C (3810°F); sp. gr. 5.28; hardness 2340 Vickers.

gadolinium carbide. (1) Gd_3C; mol. wt. 483.78; sp. gr. 8.70. (2) Gd_2C_3; mol. wt. 350.52; sp. gr. 8.02. (3)

GdC_2; mol. wt. 181.26; m.p. > 2204°C (4000°F); sp. gr. 6.90. See carbides.

gadolinium ferrite. $Gd_2O_3 \cdot Fe_2O_3$; mol. wt. 522.22; m.p. ~ 1649°C (3000°F).

gadolinium nitride. GdN; mol. wt. 171.27; sp. gr. 9.10. See nitrides.

gadolinium oxide. Gd_2O_3; mol. wt. 362.52; m.p. 2330°C (4226°F); sp. gr. 7.41; hardness 486 Knoop; used in special glasses, ceramic dielectrics, neutron shields, and phosphor activators.

gadolinium selenide. GdSe; mol. wt. 236.46; m.p. 1863°C (3385°F).

gadolinium silicate. (1) $Gd_2O_3 \cdot SiO_2$; mol. wt. 422.62; m.p. 1899°C (3450°F); sp. gr. 6.55; hardness (Mohs) 5–7. (2) $2Gd_2O_3 \cdot 2SiO_2$; mol. wt. 905.34; sp. gr. 6.29; hardness (Mohs) 5–7. (3) $Gd_2O_3 \cdot 2SiO_2$; mol. wt. 482.72; m.p. 1720°C (3128°F); sp. gr. 5.34; hardness (Mohs) 5–7.

gadolinium silicide. (1) $GdSi_2$; mol. wt. 213.38; m.p. ~ 1538°C (2800°F); sp. gr. 6.4. (2) Gd_3Si_5; mol. wt. 612.08.

gadolinium sulfide. (1) GdS; mol. wt. 189.32; sp. gr. 7.26; (2) Gd_2S_3; mol. wt. 410.70; m.p. 1885°C (3425°F); sp. gr. 6.15. (3) GdS_2; mol. wt. 221.38; sp. gr. 5.98.

gadolinium telluride. GdTe; mol. wt. 284.76; m.p. 1871°C (3400°F).

gaffer. The head workman, foreman, or blower in a hand-glass factory.

gahnite. $ZnO.Al_2O_3$; mol. wt. 183.3; m.p. 1950°C (3542°F); a spinel sometimes used in refractories.

gaize cement. A cement consisting of a finely ground mixture of a pozzolanic material and hydrated lime or a mixture of finely ground pozzolanic material and portland cement.

galena. PbS; mol. wt. 239.3; sp. gr. 7.4–7.6; hardness (Mohs) 2.5; used in glazing of pottery as a flux substitute for lead oxide.

gall. (1) Molten sulfate floating on the surface of molten glass in a pot or tank.
(2) To fret or wear away by friction.

galleting. To fill in fresh mortar joints and to level roofing tile with chips of stone, chips of roofing tile, etc.

galleyware. A synonym for delftware, a type of tin-glazed ware.

gallium. Ga; at. wt. 69.7; m.p. 30°C (86°F); b. p. 1983°C (3601°F); sp. gr. 5.91; used in high-temperature thermometers, as a metallic coating for ceramics and backing for optical mirrors, and as a heat-exchange medium for nuclear reactors.

gallium antimonide. GaSb; mol. wt. 191.5; m.p. 706°C (1303°F); a semiconductor material.

gallium arsenide. GaAs; mol. wt. 144.6; m.p. 1240°C (2264°F); used as a microwave diode, high-temperature resistor and rectifier.

gallium ferric oxide. Crystals are magnetic below −13°C (19°F); and piezoelectric from room temperature to −195°C (−319° F).

gallium nitride. GaN; mol. wt. 83.7; m.p. 800°C (1472°F); used as a semiconductor.

gallium oxide. Ga_2O_3; mol. wt. 187.4; m.p. 1900°C (3452°F); sp. gr. 6.44.

gallium phosphide. GaP; mol. wt. 100.7; m.p. 1350°C (2462°F); used as a semiconductor.

gallium sulfide. Ga_2S_3; mol. wt. 235.47; m.p. ~ 1248°C (2280°F); sp. gr. 3.5.

galvanic action. The generation of direct current electricity by chemical action.

gamma activity. The spontaneous emission from a nucleus of high-energy, short wavelength electromagnetic radiation.

gang, draw. A group of workmen employed to cut and handle glass coming from the lehr.

gangue. Accessory and valueless minerals associated with relatively valuable minerals.

ganister. Highly refractory siliceous rock used in the manufacture of refractory brick, particularly for use in metallurgical furnaces.

ganister, bastard. A mineral having the appearance of ganister, but differing in properties.

gap-sized grading. The removal of particles of intermediate sizes from a brick clay to produce brick of high-bulk density.

garden tile. Molded tile used as stepping stones in gardens or patios.

Gardner mobilometer. An instrument to measure the flow characteristics of porcelain-enamel slips in which the time required for a solid or perforated disk mounted on the bottom end of a weighted plunger to move a specified distance through a cylinder of slip is taken as a measure of the mobility of the slip.

garnet. A generic term for a group of minerals consisting of silicates of calcium, magnesium, iron, manganese, boron, chrome, or titanium; sp. gr. 3.5–4.3; hardness (Mohs) 6.5–7.5; used as an abrasive.

garspar. A feldspar substitute consisting of finely ground quartz and glass produced as a by-product in the grinding and polishing of plate glass.

gaseous inclusion. A round or elongated bubble, blister, or seed in glass, porcelain enamels, and glazes.

gaseous inclusion, open. A broken bubble forming a pit or cavity at the surface of glass, porcelain enamels, or glazes.

gas pickling. Pickling of metal shapes for porcelain enameling in a gaseous atmosphere of hydrochloric acid.

gas producer. A furnace employed for the gasification of coal by burning in an atmosphere of steam and air.

gasification. Conversion of a substance to a gas by burning or by reaction with oxygen and superheated steam.

gassing. The formation of gas bubbles in porcelain-enamel slips.

gassy surface. An imperfection characterized by poor gloss and fuzzy texture on a porcelain-enamel surface.

gas turbine. An air-breathing internal combustion engine consisting of an air compressor, combustion chamber, and turbine wheel in which the gaseous products of combustion are used as a means of generating power through a rotating shaft; used more for propulsion than for generating power.

gas-turbine nozzle. The component of a gas turbine in which hot high-pressure gas expands and accelerates to high velocity.

gate. (1) A movable refractory barrier for shutting off the flow of molten glass in the forehearth channel of a glass tank.
(2) The opening in a casting mold through which molten metal is poured.
(3) An electronic circuit having an output and two or more inputs arranged so that the output is energized only when the two input wires receive pulses.

gate level. The value established for a test signal above or below which electromagnetic test specimens may be rejected or distinguished from other specimens.

gather. (1) The mass of glass picked up on a punty or blowing iron by a hand-blowing operator.
(2) To collect molten glass on a punty or blowing iron from a pot or tank.

gathering hole. The opening in a glass pot or tank through which molten glass is gathered on a punty or blowing iron.

gathering iron. A hollow iron tube on which molten glass is collected at one end for blowing.

gathering ring. A refractory clay ring placed on a bath of molten glass to collect scum and surface impurities; glass of high purity is drawn from the center of the ring.

gauge. (1) A measure of thickness of sheet metal, rod, or wire.
(2) The minimum screen size through which an aggregate will pass.
(3) The exposed length of roofing tile as laid.

gauged brick. (1) A tapered arch brick.
(2) A brick produced to accurate dimensions by grinding or other procedure.

gauge, dry. Cullet produced by cooling molten glass from a melting unit into water. See cullet.

gauge glass. A glass tube attached to the outside of a container to measure the liquid level in the container.

gault clay. A marl containing up to 30% of calcium carbonate; used to produce yellow and buff brick and pottery.

gehlenite. $2CaO \cdot Al_2O_3 \cdot SiO_2$; mol. wt. 274.2; m.p. 1593°C (2909°F); a resin-like material found with spinel.

gel. (1) A colloidal mixture of solid and liquid of jelly-like consistency.
(2) An amorphous material formed during the hardening of cement or an exudation resulting from alkali-aggregate reaction.

gelatin. A glutinous substance obtained by boiling animal tissues; sometimes used as a sizing agent for glass fibers.

gel cement. A cement containing small additions of bentonite to increase homogeneity, increase water-cement ratio, and reduce water loss to the gel-like cement mix.

gel point. The point at which a solution or slurry begins to increase in viscosity and exhibit elastic properties.

generated heat. Heat produced by friction or grinding.

generator. (1) A machine that converts mechanical energy to electrical energy.
(2) The chamber in which solid fuel is converted to producer gas by burning with steam and air.

geometry, detector. The fraction of emissions from a source (particles or photons) which impinge on a detector in the measure of radioactivity.

Georgian-wired glass. Cast or polished glass in which wire mesh of a square pattern is incorporated as a reinforcement.

germania. See germanium dioxide.

germanium. Ge; at. wt. 72.6; m.p. 959°C (1758°F); sp. gr. 5.32; used as a high-resistance element in vacuum tubes, and in electronic devices such as transistors and diode rectifiers.

germanium detector, lithium drifted. A semiconductor detector used in high-resolution gamma-ray spectrometry.

germanium dioxide. GeO_2; mol. wt. 104.6; m.p.

1115°C (2039°F); sp. gr. 4.25; used as a replacement for silica as the glass former in glazes and bodies and in glasses of high refractive index; a semiconductor.

germanium nitride. Ge_3N_4; mol. wt. 273.8; decomposes at 1000°C (1832°F); a nonconductor. See nitrides.

germanium oxide. See germanium dioxide.

gibbsite. $Al_2O_3.3H_2O$; the major component of bauxite; used as a refractory binder for china clays and also as a bat wash.

glarimeter. An instrument designed to measure the loss of gloss of an abraded porcelain-enamel or glaze surface as an indication of the resistance of the surface to abrasive wear.

glass. (1) Any of a large class of amorphous, rigid, inorganic, nonmetallic materials of widely variable compositions, mechanical properties, and optical characteristics that solidify from the molten state, usually without crystallization; typically, the compositions include silica, boric oxide, alumina, basic oxides of sodium, potassium, or other such ingredients; the products may be transparent, translucent, or opaque, colorless, or in a wide variety of colors, and often are regarded as supercooled liquids rather than true solids.

(2) A term used for porcelain enamel frit or fired coatings.

glass, alabaster. A milky-white, translucent glass. See alabaster.

glass, annealing. The process of heating and, particularly, cooling of glassware in accordance with a prescribed schedule to reduce residual thermal stresses to a specified level and in some instances to modify the structure of the glass.

glass, antique. (1) An item of old glass, particularly one of particular value because of its age or historic background.

(2) A type of glass similar in character and appearance to the medieval glasses used in stained glass windows; these usually are produced in the form of hand-blown cylinders which are cut while in the soft or plastic state and allowed to sag to flatness on a suitable smooth or textured surface.

glass armor. Protective barriers composed of or containing glass of high strength, low-thermal or radiation permeability, or other desired property.

glass, barium crown. An optical crown glass in which barium oxide is a major component as a partial replacement for calcium oxide.

glass, barium-flint. An optical flint glass containing barium oxide as a major component. See optical flint glass.

glass blower. A craftsman engaged in the blowing of glassware.

glass blowing. The shaping of viscid glass by air pressure.

glass, blown. Glass formed by air pressure such as by blowing by mouth or by the use of compressed air.

glass, Bohemian. A hard, brilliant glass employed in table and chemical ware; usually a lime-potash glass with a high-silica content.

glass-bonded mica. An insulating material consisting of mixtures of powdered glass and powdered mica formed under pressure at elevated temperatures.

glass, borate. A glass in which boric oxide in combination with silica is employed as the major glass-forming ingredient.

glass, borax. (1) Vitreous anhydrous borax used as a flux and glass former in glass, glazes, and porcelain enamels.

(2) Glass in which borax is used as a major glass-forming ingredient in combination with silica.

glass, borosilicate. A silicate glass containing not less than 5% of boric oxide.

glass, borosilicate crown. An optical crown glass containing substantial quantities of silica and boric oxide. See crown glass, optical.

glass brick. A hollow glass block with plain or patterned surfaces used in the construction of walls, partitions, and windows.

glass brush. A bunch of glass fibers bound together with cord; used to polish exposed metal on porcelain-enameled artware.

glass,bullet-proof. A laminated, many layered glass with a thickness of ¾ inch, to several inches. Used in banks, military equipment, some pressure cookers, airplanes, automobiles, etc.

glass capacitor. A capacitor in which glass is employed as the dielectric material.

glass, cased. (1) Glassware having a surface composition different from the glass body.

(2) Glass composed of two or more layers of different colors.

glass, cast. Glass used for large castings, such as telescopes, architectural pieces, and art ware.

glass ceramic. A predominantly crystalline product produced by the controlled crystallization of glass; characterized by low-thermal expansion and high-thermal shock resistance; used in the production of high capacitance and magnetic glasses.

glass, chemical. A chemically durable glass suitable for use in laboratory and production equipment subjected to hostile materials or environments.

glass, chemically strengthened. Glass treated by an ion-exchange process to produce a surface layer of high-compressive stress.

glass, chipped. An intentionally chipped surface on a glass article.

glass, chunk. Optical glass obtained by breaking open the pot in which it was melted and cooled.

glass, cladding. Special glasses used for curtain walls, either colored or with a colored coating fused on the surface during manufacture. See curtain wall.

glass-coated steel. Steel containers, tanks, and other equipment coated with a special type of porcelain enamel having high resistance to chemicals at high temperatures and pressures; for example, chemical reactors and hot-water tanks.

glass container. A generic term for glass bottles, jars, etc.

glass, corrugated. Sheets of glass rolled into a wavy, furrowed, or corrugated form.

glass, crackled. Glassware which has been cracked by immersion in water while hot, and then reheated and shaped. See crackle.

glass, Crooke's. A glass of low ultraviolet transmission containing cerium and other rare earths.

glass, crown. (1) A hard, easily polished, highly transparent optical glass of high refraction and low dispersion, typically containing 72% of silica, 15% of Na_2O, and 13% of CaO.

(2) A type of window glass shaped by whirling a glass bubble to form a flat circular disk with a lump in the center formed by the glassblower's rod.

glass, crown flint. An optical crown glass containing a substantial amount of lead oxide; the dispersion properties are higher than those of optical crown glass.

glass, crystal. A brilliant, clear, highly transparent, lead-bearing glass used for art and tableware; frequently deeply cut to emphasize its brilliance.

glass, cut. A deeply cut item of glassware, the figures and patterns being ground on the surface by abrasive means, followed by polishing.

glass cutter. (1) A glass-cutting instrument in which the cutting member is a hard steel wheel, a diamond point, or similar substance; used to cut glass to desired sizes and shapes or to inscribe designs on glass surfaces.

(2) A workman who cuts glass to specified sizes and shapes or who inscribes designs on glass surfaces.

glass, devitrified. A glass which has been converted from a conventional vitreous state to a crystalline state by means of a nucleating agent incorporated in the original batch; nucleated crystals are precipitated and grow when the formed and finished item is subjected to subsequent controlled heat treatment. Such glasses are strong and highly resistant to thermal shock.

glass, document. A glass of high ultraviolet-absorbing properties used to cover and protect documents and other papers of value from deterioration.

glass dosimeter. A dosimeter in which a fluorescent glass is the radiation-sensing element. See dosimeter.

glass, double strength. Sheet glass having a thickness between 0.115 and 0.145 inch.

glass, drawn. Glass drawn continuously from the melting tank for processing into a product, usually by rolling.

glassed steel. A synonym for glass-coated steel, which see.

glass enamel. A finely powdered mixture of a low-melting flux, calcined ceramic pigment, and a suitable vehicle which may be applied and fired to a smooth, hard coating on glass at a temperature below the softening point of the glass.

glass eye. A large unbroken bubble or blister occurring beneath the surface of a fired porcelain-enamel coating.

glass fiber. A thread of glass used in bulk or woven form; used as acoustic, thermal, or electrical insulation, as a reinforcement in plastic and other laminations, fireproof curtains and drapes, filter cloth, surgical sutures, and numerous other domestic and commercial applications.

glass, figured. Flat glass having a pattern etched or ground on one or both surfaces.

glass flashing. The application of a thin layer of colored or opaque glass or glass enamel, by vitrification or surface fusion, to the surface of clear glass.

glass, flat. A generic term including sheet, plate, rolled, float, or other forms of glass that are of a flat nature.

glass, flint. A glass in which lead and potassium oxides replace substantial portions of calcium and sodium in ordinary glass to give a more fusible and lustrous glass of high refraction and low dispersion; used in optical applications.

glass, float. Flat glass formed as a ribbon on a bath of molten metal, such as tin. The resulting brilliant surface usually requires no polishing.

glass, fluor crown. An optical crown glass of low-index of refraction and low dispersion containing substantial amounts of fluorine as a flux.

glass, foamed. Cellular glass of high insulating value, moisture proof, noncombustible, buoyant, and odorless produced by adding powdered carbon or other gas-forming material to crushed glass and firing in a manner to entrap the evolving gas bubbles; used as insulation for walls, floors, roofing, industrial and domestic equipment, low-temperature apparatus, etc.

glass former. Any oxide which retains its amorphous state after solidification from the molten state.

glass frost. The chemical or mechanical treatment of a glass surface, or the application of crushed glass particles to the glass surface, to obscure the glass by scattering light or to simulate the appearance of frost.

glass furnace. Any enclosed or covered furnace, tank, or pot, usually of the reverberatory type or principle, in which glass is melted; sometimes electric boosters immersed in the molten glass batch are used to expedite melting and fining. See reverberatory furnace, fining.

glass, Georgian-wired. Cast or polished glass in which wire mesh of a square pattern is incorporated as a reinforcement.

glass, green. Glass tinted green by additions of copper oxide, CuO, to a clear glass batch, the copper oxide sometimes replacing chromium oxide in the batch.

glass, heat-absorbing. Any glass capable of absorbing radiant energy in the near infrared range of the spectrum.

glass heat exchanger. A device which transfers heat from one fluid to another in which the heat-transfer medium is a glass.

glass, heat-resisting. A glass of low-thermal expansion and high resistance to thermal shock such as occurs when glass is cooled suddenly from elevated temperatures.

glass, heat-strengthened. Glass subjected to a programmed heat treatment to improve its physical strength.

glass, high-transmission. A glass of a composition which transmits an exceptionally high percentage of visible light.

glass insulator. A tempered or annealed glass shape used as an insulator for electric-power transmission lines.

glassivation. The passivation of a transistor by encapsulating the semiconductor device, complete with metal contacts, in glass.

glass, laminated. (1) A transparent safety glass in which two or more glass sheets are bonded together by intervening layers of plastic materials so that, when broken, the glass will tend to adhere to the plastic rather than fly.

(2) A diffusing glass formed by sandwiching a plastic-bonded glass fiber between sheets of ordinary glass. See safety glass.

glass, lamp working. A method of reshaping cooled glass into new forms over a blowtorch. Used mostly for creating small miniatures such as vases, sailing ships, glass eyes, radio and television parts, chemical equipment, etc.

glass laser. A solid laser in which a fluorescent glass serves to amplify electromagnetic radiation by stimulated emission of radiation.

glass, lead. Glass containing a substantial quantity of lead oxide as a flux and to give a high index of refraction, optical dispersion, and surface brilliance; for use as an optical glass.

glass, lead crown. An optical flint glass containing a substantial proportion of lead oxide as a flux to improve light dispersion and brilliance.

glass, leaded. Windows made from pieces of colored or clear glass held in place by strips of lead having an ''H'' or ''U'' cross section.

glass, light-reducing. A general term describing a flat glass having reduced light transmittance properties.

glass, lime. A glass containing a high percentage of lime as a flux, usually in association with sodium oxide and silica which increases strength, stability, permanence and hardness; used in glass products such as bottles, containers, and other products.

glass, lime crown. An optical glass containing substantial quantities of calcium oxide as a fluxing ingredient.

glass-lined steel. Chemical-processing vessels and pipes coated with a chemical-resistant coating of glass or porcelain enamel.

glass, liquid. Sodium silicate, which see.

glass, low-melting. A glass containing selenium, arsenic, thallium, or sulfur, melting at 127° to 349°C (260° to 600°F).

glassmaker's soap. A material such as a selenium compound or manganese dioxide used to remove the green color created by iron salts.

glass, milk. A translucent white or colored opal glass made by adding alumina and fluorspar to a soda-lime glass.

glass, molded. Glass formed in a mold.

glass, moonstone. An opal glass resembling the moonstone in appearance.

glass, neophane. A yellow glass tinted with neodymium oxide to reduce glare; used in automobile windshields, sunglasses, etc.

glass, neutron-absorbing. A cadmium borate glass containing additions of titania and zirconia having a high neutron-capture cross section.

glass, off-hand. Glass made by an artisan without benefit of molds.

glass, opal. A fluoride-bearing glass have a white, milky appearance, usually with a fiery translucence, due to small particles or bubbles in the body of the glass which disperse the light passing through.

glass, ophthalmic. Glass of great uniformity having specified optical and physical properties, as well as specified quality, employed in eyeglasses.

glass, optical. Glass of great uniformity and free of imperfections having specified optical properties in terms of the transmission, refraction, and dispersion of light; used in the manufacture of optical systems.

glass, optical crown. (1) Any optical glass having a low dispersion and a low index of refraction, usually forming the converging element of an optical system.
(2) Any optical glass having a nu-value of at least 55, or a nu-value between 50 and 55 and a refractive index greater than 1.60.

glass, optical flint. (1) An optical glass having a high dispersion and a high index of refraction, usually forming the diverging element of an optical system.
(2) Any optical glass having a nu-value less than 50, or a nu-value between 50 and 55 and a refractive index less than 1.60.

glass, oven. A glass of low-thermal expansion exhibiting high resistance to thermal shock, and which is employed in the manufacture of articles used in the preparation of food.

glass paper. A heat and environment resistant paper made of glass fibers; used for permanent documents.

glass, phosphate. A glass in which an essential glass-forming ingredient is phosphorous pentoxide as a partial replacement for silica, SiO_2, and which is resistant to hydrofluoric acid.

glass, phosphate crown. An optical crown glass containing a substantial quantity of phosphorous pentoxide, P_2O_5, as a glass-forming agent. See optical crown glass.

glass, photochemical. A photo-sensitive glass which can be cut by acid. Pictures, designs, etc., can be reproduced on it from photographic films, then subjected to an acid bath, which leaves a three-dimensional design.

glass, photochromic. A glass that darkens on exposure to light, but which returns to its original color and clarity when the light is removed; used in some sunglasses, automobile windshields, etc.

glass, photosensitive. A light-sensitive glass containing submicroscopic particles of gold, silver, or copper which precipitate during the photographic process to produce three-dimensional color pictures which are developed by heating to 538°C (1000°F); the precipitation of the metals permits the ultraviolet light to penetrate deeper into the shadowed areas while passing through the negative to produce three-dimensional effects.

glass, plate. High-quality flat glass with plane, parallel surfaces formed by a rolling process; both sides are ground and polished to provide undistorted vision.

glass-plate capacitor. A capacitor in which glass sheets separate the metallic plates and serve as the diodes.

glass, polished plate. See glass, plate.

glass, polished wire. Wire-reinforced glass which has been ground and polished on both sides.

glass pot. A one-piece, crucible-shaped refractory container, open or closed, in which glass is melted.

glass, pressed. Glassware formed under pressure between a plunger and mold while in the molten and plastic state.

glass, prismatic. A translucent glass consisting of parallel prisms which give three nominal angles of light passing through the glass, producing an irridescent, sometimes multicolored appearance.

glass, quartz. (1) A glass prepared by fusing pure silica or sand; characterized by a high-melting point, chemical inertness, transparency in visual and ultraviolet light, and high resistance to elevated temperatures and thermal shock; used in high-temperature devices and equipment. Also known as fused silica or fused quartz.
(2) A glass made by the flame hydrolysis of silicon tetrachloride. Also known as silica glass.

glass resistor. Tubular glass with a helical electric resistor element of carbon painted on the surface.

glass, rolled. (1) Flat glass made by passing a roller over the surface of the glass in the molten or plastic state; sometimes a design may be worked into the glass surface by a patterned roller face.
(2) Optical glass rolled into plates instead of being cooled in a melting pot and then processed.

glass, rough. Rolled glass sheets cut into workable sizes.

glass, rough-cast. Rolled glass having one textured surface made by using a roller with a patterned face.

glass, safety. (1) A glass constructed with sheets laminated with plastic films to prevent shattering in the event of breakage.
(2) A glass containing a network of wire to improve its resistance to breakage and shattering.
(3) A glass that has been tempered by heat treatment so that it will break into small fragments or grains which do not scatter into sharp pieces when broken and are less liable to cause injury than ordinary glass.

glass sand. A nearly pure quartz sand with minor amounts of the oxides of aluminum, calcium, iron, and magnesium; used in glass making.

glass seal. An airtight seal in which molten glass is the sealant; for example, the glass-to-metal seals used in electric and electronic components.

glass, sealing. A glass with special thermal expansion and flow characteristics to enable it to bond with another glass or solid.

glass, shatterproof.. See glass, safety.

glass, sheet. A generic form including sheet, plate, rolled, float, and other forms of glass that are flat in nature.

glass, shielding. A transparent glass containing quantities of the oxides of the heavy elements, such as lead, which absorb high-energy electromagnetic radiation, and which are employed to shield one region of space from ionizing radiation emanating from another such as in nuclear applications.

glass, silica. A transparent or translucent glass composed of fused high-purity quartz or sand, or which is made by the hydrolysis of silicon tetrachloride. Also known as fused silica, vitreous silica.

glass, single strength. Sheet glass between 0.085 and 0.101 inch thick.

glass, sintered. A porous article in which particles of glass of selected or random sizes are compacted and sintered to produce a bonded, but unsealed, item of a desired shape and of sufficient porosity for an intended use, such as aeration, filtration, etc.

glass, skylight. Plate glass of very poor quality.

glass, slab. Optical glass obtained by forming or cutting chunk glass into plates or slabs.

glass, smoked. Commercial glassware produced in gray or smoky brown colors, sometimes by chemical additions to the glass and sometimes by exposure to a reducing atmosphere during melting and cooling.

glass, soft. (1) A glass having a relatively low-softening point or which is easily melted.

(2) A glass which is easily scratched or abraded.

glass, solder-sealing. A sealing glass having a relatively low-softening temperature used as an intermediate bonding material.

glass, spandrel. Architectural glass used in curtain wall applications in a nonvision area and in the cladding of a building. See structural glass, and curtain wall.

glass, spun. An individual filament or a mass of fine threads of attenuated glass, often having a delicate spiral threading or filagree, with diameters less than one-thousandth of an inch.

glass, square-cut. Optical glass cut into small squares which are separated and designated by weight; used in the production of ground and polished optical units.

glass, stained. Glass colored by various means such as by incorporating colorants in the glass batch or by applying and firing a clear colored enamel to the surface of the glass; used in the production of mosaics, church windows, etc.

glass, structural. (1) Opaque or colored glass, frequently ground and polished, used for structural purposes.

(2) Glass block, usually hollow and often with patterned faces, used for structural purposes such as in walls, partitions, and windows.

glass switch. A glassy, amorphous solid-state device formulated and designed to control the flow of an electric current in electronic components.

glass tank. A large covered or enclosed refractory container or reservoir in which glass is melted and from which molten glass is drawn for fabrication into commercial and domestic products.

glass, tempered. Glass that has been cooled from near its softening point to room temperature under rigorous control to increase its mechanical strength and thermal endurance by the formation of a compressive layer at its surface.

glass, tempered safety. A glass that has been tempered so that it will break into granular instead of jagged fragments as a result of a particular stress pattern created in the glass by a rigidly controlled heat treatment. See glass, tempered.

glass, textiles. Glass fibers woven into textile fabrics for use in plastic laminates, heat-resistant insulation, filter cloth, fireproof draperies and curtains, and similar products.

glass, thermal. A low-expansion glass in which boron oxide is substituted for calcium oxide, and which is highly resistant to thermal shock.

glass thermal-expansion factors. Any of various factors employed to calculate the thermal expansion properties of glass, glazes, procelain enamels, and other ceramic compositions, particularly the properties of the oxides, their eutectics, etc.

glass-to-metal seals. Air-tight seals formed by fusing glass with metals for purposes of insulation in electrical and electronic components, the glass serving as an insulation.

glass, toughened. A glass highly resistant to mechanical and thermal shock, produced by rapid and rigid control of its cooling rate from near its softening point to room temperature to produce residual internal stresses which remain after the glass has cooled; used in windows, doors, and other installations where breakage may be dangerous.

glass, transfer. Optical glass cooled to room temperature in the pot in which it was melted.

glass, translucent. A glass transmitting light with varying degrees of diffusion and which impedes or obscures vision to the degree that objects seen through it cannot be seen distinctly.

glass, ultraviolet absorbing. Glass containing appropriate quantities of elements such as cerium, chromium, cobalt, copper, iron, lead, manganese, neodymium, nickel, sulfur, titanium, uranium, or vanadium, which absorb ultraviolet rays without appreciable effect on the transmission of visible light rays.

glass, ultraviolet transmitting. Glass containing no appreciable quantities of ultraviolet absorbing elements and which will transmit ultraviolet rays without impedance. See ultraviolet-absorbing glass.

glassware. Any glass product made of glass for use in domestic or laboratory applications, the term usually referring to tableware.

glassware, graduated. Glassware marked with divisions or units for volumetric measurements.

glassware, pressed. Glassware formed under pressure between plunger and mold while in the molten or plastic state.

glassware, volumetric. Glassware that is marked with graduations for volumetric measurements.

glass welding. Joining two or more glass components by fusion at their points of juncture or contact.

glass, window. A continuously drawn soda-lime glass in sheet form; used primarily in the construction of windows.

glass, wired. See glass, wired safety.

glass, wired safety. A glass with an embedded network of wire which resists shattering when broken.

glass, Wood's. Glass having a high transmission factor for ultraviolet radiation, and which is almost opaque to visible light.

glass wool. A randomly oriented, fleecy mass of glass fibers used for acoustic and thermal insulation, air filters, packing, and similar applications.

glassy state. A vitreous state in which the atoms or molecules are not oriented in a regular order or pattern.

glass, zinc crown. An optical glass containing substantial amounts of zinc oxide as an auxiliary flux. See optical crown glass.

glaze. A glassy coating fired on a ceramic article, or the mixture of ingredients from which the coating is made.

glaze, aventurine. A glaze containing colored opaque spangles of nonglassy materials, such as copper, gold, chrome, or hematite, which give the glaze a shimmering appearance.

glaze, bright. A white, colored, or clear glaze having a high gloss.

glaze, Bristol. An unfritted zinc-bearing glaze for stoneware, terra cotta, and similar ceramic bodies.

glaze, celadon. A grayish green, semiopaque glaze fired in a reducing atmosphere, reduced iron serving as the colorant.

glaze clays. Fine-grained clays, containing considerable amounts of colloidal organic matter, which are introduced into glaze batches as suspension and binding agents, and which become an integral part of the glaze during firing.

glaze, crystalline. A glaze containing microscopic crystals which have grown during the cooling period following a firing operation.

glaze, cut. A glazed area in which the coating is of insufficient thickness for good coverage.

glazed ceramic mosiac tile. Ceramic mosaic tile that has been glazed on the face or exposed surface.

glazed interior tile. A nonvitreous tile body that has been glazed and which is suitable for normal, relatively mild conditions of interior use.

glazed pot. A glass-melting pot coated with a vitreous coating or glaze as a protection against reactive batch ingredients encountered during the glass-melting operation.

glazed, short. See glaze, cut.

glazed tile. A vitreous, semivitreous, or nonvitreous tile coated with an impervious, colored or uncolored, ceramic glaze.

glazed tile, eggshell. A tile coated with a glaze having a semimatte, eggshell-like texture.

glazed tile, extra duty. A ceramic floor or wall tile suitable for use in high-duty floors and similar applications where impact, wear, and abrasive forces are not excessive; may be white or colored.

glaze fit. The stress relationship between the glaze and body of a fired ceramic; that is, the degree to which the coefficients of expansion of the glaze and body are matched.

glaze flow. (1) The property of a glaze slip to flow over the surface of a ceramic body to form a smooth, uniform coating.

(2) The property of glaze ingredients to flow together to form a smooth, impervious coating during firing.

glaze, fritted. A glaze in which part or all of the fluxing ingredients have been fused or quenched to form small friable particles before incorporation into the glaze slip.

glaze, jardinier. A type of unfritted, hard or soft glaze containing the oxides of lead, aluminum, calcium, potassium, silicon and zinc; used as a decorative glaze on products, such as flower pots, etc.

glaze, lead. A glaze containing a substantial amount of lead oxide as a flux to lower the fusion temperature and viscosity, to improve the flow properties during firing, and to improve the brilliance, luster, resistance to water solubility, and resistance to chipping.

glaze, leadless. A glaze containing only an imperceptible amount of lead in any form.

glaze, low solubility. A lead-bearing glaze in which no more than 5% of the lead oxide is soluble.

glaze, majolica. A glossy, tin oxide-opacified, white or colored overglaze decoration fired at a relatively low temperature. See majolica.

glaze, matte. A fired glaze having little or no gloss.

glaze, opalescent. A ceramic glaze containing fluoride compounds firing to a milky or irridescent appearance.

glaze, opaque. A nontransparent, white or colored ceramic coating of bright satin or glossy finish on the surface of a ceramic product.

glaze, raw. A glaze compounded entirely of raw materials and containing no prefused ingredients.

glaze, salt. A lustrous glaze produced on ceramic surfaces toward the end of the firing cycle by throwing salt into the firebox, the salt volatizing and the resultant fumes then entering into a thermochemical reaction with the silicates and other components of the ceramic.

glaze, semiconducting. A ceramic glaze containing metal oxides in sufficient quantities to promote a degree of electrical conductivity to prevent surface discharge or flashover.

glaze, semimatte. A ceramic glaze exhibiting only a moderate degree of gloss that is considered to be between high gloss and matte in appearance.

glaze, short. An area on the surface of glazed ware to which insufficient glaze was applied for good coverage.

glaze, slip. A glaze consisting primarily of readily fusible clay or silt and other ingredients blended into a creamy consistency with water.

glaze, slop. A homogeneous slurry of glaze ingredients and water applied to ware by dipping, spraying, or brushing.

glaze, snakeskin. A decorative effect on pottery obtained by using glazes of high surface tension or very low expansion, causing the glaze to crawl during firing to produce an appearance resembling snakeskin.

glaze stains. Calcined ceramic pigments, usually metal oxides, incorporated in a glaze slip to produce a coating of uniform color; some serve essentially as pigments, some as precipitates, and some go into solution in the fired glaze.

glaze, starred. A partially devitrified glaze in which star-shaped crystals develop at the surface during firing.

glaze, starved. A glaze applied to ware to an insufficient thickness to obtain good coverage.

glaze, tea-dust. An opaque, iron oxide bearing stoneware glaze of greenish color.

glaze, transmutation. A glaze which may be changed in color by the intentional or accidental introduction of another colorant or impurity into the batch, or by melting a fritted glaze in a crucible in which a composition of a different color previously has been melted.

glaze, vapor. A glaze composed of lead, sodium, and boric oxides which will volatilize from a melt during firing, and then will condense and reliquefy on a ceramic surface on cooling.

glaze, vellum. A semimatte glaze having a satin-like appearance due to the presence of minute crystals of zinc silicate, zinc titanate, or lead titanate in the fired glaze surface.

glazing. (1) The application of a glaze to ceramic ware.
(2) The cutting and fitting of glass panes into frames, and the application of a caulking compound to seal the panes in place.

glazing size. The dimensions of a glass pane cut for glazing. See glazing (2).

glory hole. (1) A furnace for the reheating and fire polishing of hand-made glassware.
(2) The opening exposing the hot interior of a furnace in which glass is reheated for hand-working.

gloss. The polish, luster, or brilliance of a fired porcelain-enamel or glaze coating; the ratio of specularly reflected light to the total light reflected by a surface.

gloss, low. The dullness or lack of normal gloss on a fired porcelain-enamel or glaze coating.

gloss, specular. The ratio of specularly reflected light to incident light.

glost. A synonym for ''glazed.'' See glost firing.

glost firing. A firing of bisque ware to which glaze has been applied.

glucose. A monosaccharide, $C_6H_{12}O_6$, used as a binder for nonplastic materials.

glue, chrome. A glass cement or waterproofing agent made by mixing glue with ammonium or potassium dichromate or with chrome alum.

gneiss. A metamorphic rock usually composed of quartz, feldspar, and other mineral silicates.

gob. (1) A mass of molten glass gathered on a punty or blow-pipe for the hand-making of glassware.
(2) A mass of molten glass delivered by a feeder to a forming process.

goblet. A ceramic or glass drinking cup or vessel having a stem and base, but usually without handles.

gob process. The process by which a gob or mass of molten glass is delivered to a forming operation.

gold. Au; at. wt. 197.2; m.p. 1063°C (1945°F); sp. gr. 19.3; a brilliant glass and ceramic decoration applied as a powder suspended in essential oils and burnished after firing.

gold, acid. A form of gold decoration in which the surface to be glazed first is etched with hydrogen fluoride (HF).

gold, bright. An inexpensive luster of gold resinate in combination with other metal resinates and a flux; used as a ceramic decoration when fired on glass, porcelain enamel and glaze surfaces.

gold, burnished. A durable type of gold applied to glazed ware as a suspension in oil, fired, and then rubbed with agate or other polishing material to a bright finish.

gold chloride. $AuCl_3$ mol. wt. 303.6; m.p. 354°C (669°F); sp. gr. 3.9; used with a mixture of stannous and stannic chlorides to produce purple of cassius for the coloring or decoration of glass and ceramic ware, also used to produce ruby reds in glasses, glazes, and porcelain enamels.

gold hydroxide. $Au(OH)_3$; mol. wt. 248.2; employed in the decoration of ceramics.

gold, liquid. See liquid gold.

gold oxide. Au_2O_3; mol. wt. 442.4; employed in the decoration of ceramics.

gold-potassium chloride. $AuCl_3 \cdot KCl \cdot 2H_2O$; used in the decoration of glass and ceramics.

gold silvering. A process in which gold is deposited on a glass surface and coated with a protective medium.

gold sodium chloride. $AuCl_3 \cdot NaCl \cdot 2H_2O$; used in the decoration of glass and ceramics.

Goldstone glaze. An aventurine glaze composed of basic lead carbonate, $2Pb(OH)_2$, feldspar, $K_2O \cdot Al_2O_3 \cdot 6SiO_2$, silica, SiO_2, ferric oxide, Fe_2O_3, and whiting, $CaCO_3$.

gold-tin purple. A mixture of gold chloride and brown tin oxide used in coloring porcelain enamels, manufacturing ruby glass, and painting porcelain.

gold-titanium. (1) AuTi; mol. wt. 245.30; m.p. 1488°C (2710°F); (2) Au_2Ti; mol. wt. 442.50; m.p. 1465°C (2670°F).

gold-uranium. Au_3U; mol. wt. 829.77; m.p. > 1449°C (2640°F).

gold zirconium. Au_3Zr; mol. wt. 682.82; m.p. 1560°C (2840°F).

goniophotometer. An instrument measuring light reflected from a surface from different angles.

gouge test. A test to evaluate the wear resistance of porcelain enamel in which a small steel ball is rolled across the enamel surface under increasing loads, the degree of wear being determined by the loss of gloss.

grab sample. A sample taken at random from a large mass, or a large number of items being examined.

gradation. (1) A code or designation of the quality, composition, properties, or type attached to a product by a manufacturer such that the product may be reproduced by the manufacturer.

(2) To sort or classify in steps or degrees by established criteria such as by particle size, color, or other property.

(3) The strength of bond or hardness of a grinding wheel, particularly in terms of the resistance of the grains to being torn or split from the wheel during use.

grading. The process of sorting to some specified category of classification.

grading, gap-sized. The removal of particles of intermediate sizes from a brick clay to produce brick of high-bulk density.

graduated glassware. Glassware marked with divisions or units for volumetric measurements.

grain. Any small hard particle such as an abrasive grain or a grain of sand.

grain fineness. (1) The average particle size of a granular material.

(2) The maximum particle size of a granular material passing through a sieve of specified mesh size.

graining. The process of producing a decorative finish imitating the grained appearance of wood or other substance on porcelain enamels and glazes, usually by means of a rubber-roll transfer process.

graining oxides. Mixtures of ceramic pigments containing small amounts of fluxing ingredients used in graining pastes.

graining pastes. Oil suspensions of ceramic color oxides and fluxes used in the rubber-roll process of decorating porcelain-enamel and glaze surfaces.

graining roll. A special type of rubber roll used to transfer graining paste from a pattern surface to the surface of a porcelain enamel or glaze.

grain magnesite. Dead-burned magnesia in granular form suitable for refractory purposes. See dead-burned magnesite.

grain size. The average size of particulate materials used in the production of ceramic products, usually determined by screen analysis.

grain spacing. The relative density of abrasive particles in a grinding wheel.

gram. A metric unit of weight equal to 1/28 ounce or

nearly equal to the weight of one cubic centimeter of water at its maximum density.

granite. An igneous rock composed of orthoclase or albite feldspar, and mica; the pulverized product is similar to china clay in properties and use.

graniteware. A one-coat porcelain-enameled article, such as an item of kitchenware having a mottled appearance produced by controlled corrosion of the metal base prior to firing.

graniteware, white. A term designating an exceptionally strong earthenware body.

granular. Consisting of or appearing to consist of granules.

granular activated carbon. Activated carbon in particle sizes predominately greater than 80 mesh.

granulate. To form powdered materials into larger, free-flowing particles.

granulated blast-furnace slag. Glassy, granular blast-furnace slag produced by quenching molten slag in water. See slag.

granulator. A machine, such as a rotating cylinder, employed to produce granules from materials of fine particle sizes.

granules. (1) Small grains or pellets.
(2) Small ceramic grains or pellets applied to asbestos cement to add color to the surface.

granules, roofing. Approximately 8-mesh particles of crushed slag, slate, rock, tile, porcelain, or other substance used in the production of asphalt roofing and shingles.

graphite. A soft form of carbon; at. wt. 12.00; m.p. above 3500°C (6332°F); sp. gr. 2.09–2.3; hardness (Mohs) 1–2; used in crucibles and other refractories for arc furnaces, missiles, and other high temperature products; as a lubricant in both low- and high-temperature applications, and as a moderator in nuclear applications.

graphite-base carbon refractory. A manufactured refractory composed essentially of graphite.

graphite brick. A refractory ceramic brick of coke and pitch, heat-treated to form a graphitic crystal structure.

graphite fabric. Woven cloth of graphite fibers.

graphite fibers. Graphite in filament form, frequently used in graphite cloth.

graphite, manufactured. A bonded granular form of carbon whose matrix has been subjected to temperatures between 900° and 2400°C (1652°–4352°F).

graphite, pyrolytic. A form of graphite of high purity having high thermal and electrical conductivity; used in high-temperature functions.

graphite refractory. Any refractory product composed essentially of graphite.

graphite, synthetic. Crystalline graphite made by processing carbon at high temperatures and pressures.

graphitic carbon. Tiny flakes or pure carbon which form in pig iron during cooling which tend to weaken the metal; will cause blistering in porcelain enamels.

grappier cement. A cement made by using underburned or overburned slaked lime which has been finely ground.

gravel. Loose, rounded rock used as aggregate in concrete.

gravel, crushed. The product resulting from the artificial crushing of gravel with all fragments having at least one face resulting from fracture; used as an aggregate in concrete.

Gravé sandblast. A sandblasted decorative design of varying depths on glass surfaces.

gravimetric analysis. A quantitative chemical analysis based on reactions that produce a material to be weighed.

gravimetric factor. The ratio of the atomic or molecular weight of an element or compound to the molecular weight of the compound in which it is a component.

gravity. (1) A term used in the porcelain-enameling industry for the specific gravity of a milled porcelain-enamel slip in which water is equal to 1.00.
(2) The force of gravity.

gravity bed. A moving-bed technique or process in which the solid particles of a material being processed move downward through the liquid, or conversely the molten phase of the material moves upward.

gravity feed. The movement of materials from one location to another by force of gravity.

gravity separation. The separation of mixtures into layers, in accordance with their respective densities, usually in a stream of air, in a liquid suspension, by means of a vibrated sloping shaker-table, or other technique.

gravity, specific. The ratio of the weight of a unit volume of a substance to that of some other standard material under standard conditions of pressure and temperature; the specific gravity of solids and liquids is based on water as the standard.

green brick. Formed, but unfired brick.

green carbon. A formed but unfired carbon body.

green concrete. A concrete which has set, but which has not developed appreciable strength.

green copperas. See ferrous sulfate, $FeSO_4 \cdot 7H_2O$.

green cracks. Shrinkage cracks developed in concrete while in the green state.

green glass. Glass tinted green by additions of copper oxide, CuO, to a clear glass batch, the copper oxide sometimes replacing chromium oxide in the glass batch.

green, malachite. $CuO \cdot CO_3 \cdot Cu(OH)_2$; used as a green colorant in stoneware and as a green dye to indicate the absorption characteristics of ceramic bodies.

green pellet. A pellet which has been pressed, but has not been fired.

green silicon carbide. A friable form of silicon carbide. See silicon carbide.

green spot. An imperfection consisting of a prominent green spot in ceramic bodies caused by copper or copper-bearing impurities in the raw materials.

green strength. The ability of an unfired ceramic to resist mechanical damage, particularly impact and transverse loads.

green vitriol. Ferrous sulfate, which see.

greenware. A formed but unfired ceramic body.

greenware storage. An area or room in which greenware is stored, and air dried, prior to firing.

grindability. The ease with which a material can be ground, or milled to a smaller particle size.

grindability index. A numerical indication of the ease with which a material can be ground.

grinder. (1) A machine which pulverizes and reduces the particle size of materials by impact and friction.
(2) A machine equipped with an attachment, such as a grinding wheel, abrasive belt, or disk, used in mechanical shaping, grinding, sharpening, cutting, polishing, honing, buffing, or lapping operations.

grinder, bench. A grinding machine consisting of one or two abrasive wheels revolving on a single horizontal axle or spindle mounted on a stand, bench, or table, and which is used essentially for off-hand grinding. See grinding, off-hand.

grinder, dish. A grinding machine equipped with a dish-shaped abrasive wheel as the work mechanism.

grinder, disk. A grinding machine equipped with a disk-shaped abrasive wheel as the work mechanism.

grinder, impact. A device for reducing the size of solid materials by shattering blows, such as by falling balls in a ball mill.

grinder, swing-frame. A grinding machine suspended above the work piece by a chain at its center of gravity so that it may be turned and swung in any direction for the in-place grinding of work too heavy for manual handling.

grinding. (1) Reducing the particle size of a material by mechanical means.
(2) Removing excess material from a workpiece by means of an abrasive wheel.

grinding aids. Materials added to a ball mill or to the mill charge, such as a liquid, to accelerate the pulverizing process.

grinding ball. A hard, dense, abrasion-resistant sphere used as a crushing body in a ball mill; usually of flint, dense porcelain, alumina, steel, or heavy alloy composition.

grinding burn. The localized overheating of work during abrasive grinding.

grinding, centerless. Grinding the inside or outside diameter of a cylindrical piece which is supported on a work blade instead of between centers, and which is rotated by a regulating wheel.

grinding, closed-circuit. A continuous grinding or milling process in which particles of suitable fineness are removed from the grinding system by a screen or cyclone separator, while oversized particles are returned to the pulverizer for further processing.

grinding cracks. Cracks appearing on the surface of a workpiece during grinding due to overheating or overgrinding.

grinding, cross-feed. The controlled movement of a grinding wheel over a horizontal work piece resting on a work table, the grinding being done at a prescribed rate of depth.

grinding, cylindrical. Grinding the outer surface of a part which either is rotated on centers or is centered in a chuck.

grinding, dry. Reducing the particle size of a material by milling without the use of a liquid medium.

grinding feed. (1) The rate at which a material is fed automatically to a cylindrical grinder.
(2) The rate at which solid material is introduced into a continuous pulverizing mill.

grinding fluid. A cutting and cooling liquid, such as water or high heat-conducting oil, used in the abrasive grinding of solid surfaces to prevent grinding burns on ware being processed.

grinding, form. The shaping of a product by the use of an abrasive wheel contoured to the reverse shape of the desired form.

grinding, freehand. The grinding of an item or shape by hand without the use of guides.

grinding machine. Any machine equipped with an abrasive grinding wheel.

grinding machine, universal. A machine on which cylindrical, internal, or face grinding may be done as required.

grinding marks. A pattern of fine striation or score marks, usually directional, resulting from grinding and machining operations.

grinding media. The porcelain, flint, or steel balls, rods, rolls, and other materials used in grinding mills to reduce the particle size of solid substances.

grinding mill. (1) Any machine, such as ball, tube, and rod mills, employed to reduce the particle size of minerals, ceramic materials, cement clinkers, and other solid substances for commercial and domestic use.
(2) A lapidary lathe or wheel.

grinding, off-hand. Freehand grinding of work held in the hand of the operator, usually without the use of guides or patterns.

grinding pebbles. Flint and small porcelain balls employed as grinding media in ball mills, particularly in the milling of materials in which iron contamination should be avoided.

grinding, plunge. Grinding and polishing operations in which the grinding wheel rotates radially toward the work.

grinding, portable. A grinding machine which is supported and manipulated manually by the operator.

grinding, precision. Machine grinding of an item to specified and precise dimensions.

grinding ratio. The ratio of the volume of material removed from a workpiece during grinding to the volume removed from the grinding wheel.

grinding relief. The groove at the edge of a workpiece which overhangs the corner of the grinding wheel.

grinding, rough. The grinding of a glass, metal, ceramic, or other surface without regard to the quality of the finish.

grinding sensitivity. The susceptibility of a material to damage during grinding.

grinding, side. The practice of grinding on the side of an abrasive wheel.

grinding stress. The residual stress, tensile or compressive, or a combination of both, generated in a workpiece during the grinding operation.

grinding, surface. The abrading or grinding of a plane surface.

grinding, thread. The cutting of threads on a part by the use of a bonded abrasive wheel designed for the purpose.

grinding, wet. (1) The milling of ceramic bodies, glazes, and porcelain enamels in a liquid medium, usually water.
(2) The application of a liquid coolant to a workpiece and the grinding wheel during abrasive grinding.

grinding wheel. A bonded abrasive wheel or disk mounted on a mechanically actuated axis for use in grinding and polishing operations.

grinding wheel, reinforced. A grinding wheel containing mechanical reinforcement for additional strength and safety during use.

grinding wheel, resinoid bonded. A grinding wheel bonded with a thermosetting resin.

grindstone. A grinding wheel cut from natural sandstone; used for grinding, sharpening, smoothing, and shaping.

grisaille. A type of porcelain-enameled artware made by firing various thicknesses of white enamel over a black background to produce a monochromatic decoration in shades of gray.

grit. Coarse-grained, sharp, angular granules of sand, garnet, alumina, or other substance of synthetic origin used primarily as an abrasive.

grit blasting. A surface treatment in which grit is impinged on the surface of an item to clean and roughen or polish the surface.

grit number. A number designating the particle size of grit and abrasive grains based on sieve analyses.

grit size. The particle size of grit and abrasive grains based on sieve analyses.

grizzly. A screening device, consisting of parallel iron or steel bars, for the separation of coarse lumps of raw materials from smaller sizes.

grizzly chute. A chute equipped with grizzlies of decreasing size, each grizzly separating coarse lumps of raw materials from smaller lumps in decreasing size classification. See grizzly.

grizzly crusher. A type of crusher consisting of rods or bars which crush and separate lumps of raw materials according to size.

grog. A ground mixture of refractory materials such as firebrick, clinkers, pottery, sand, saggers, crucibles, and the like added as raw materials to refractories, saggers, acid-proof ware, terra cotta, high-temperature porcelain, stoneware, vitreous china sanitaryware, sewer pipe, and similar products to improve working and service properties.

grog-fireclay mortar. A refractory mortar consisting of a mixture of raw fireclay, calcined fireclay or broken fireclay brick, or both, milled to a workable fineness.

grooved pipe. The grooved portion of the end of a pipe, regardless of its shape or dimensions, which overlaps a portion of the end of an adjoining pipe.

gross weight. The total weight of a material and its container.

ground coat. The first coat of porcelain enamel applied to metal when subsequent coats are to be applied.

ground-coat boiling. The undesirable evolution of gas during the firing of porcelain-enamel ground coat resulting in a variety of imperfections such as blisters, pinholes, black specks, dimples, or spongy surfaces.

ground fireclay. Milled fireclay or mixtures of fireclays subjected to no treatment other than weathering.

ground fireclay mortar. A mortar of workable consistency composed of finely ground fireclay and water.

groundhog kiln. A type of art-potter's kiln constructed partly in a hillside.

ground laying. The application of a uniform color, usually by dusting a powdered ceramic color over ware or an area of ware previously painted with an adherent oil.

grout. A mixture of portland cement, lime, aggregate, and water of a troweling or pouring consistency which is flowed into vertical open joints or troweled into open spaces on horizontal courses of masonry.

grouting, intrusion. The technique of placing the grout components in position in an area and then adding water to convert the mixture to concrete.

guard. (1) A shield around a grinding wheel to protect a workman from injury.
(2) Any attachment or cover placed on a machine to protect an operator or other person in the vicinity.

guard ring. A ring-shaped device surrounding a test specimen to ensure an even distribution of heat in heat-flow experiments.

guillotine cutter. A mechanically or manually operated heavy steel knife used to cut through and trim material.

gum arabic. A water-soluble gum from acacia trees used as a binder in bodies, and in glaze and porcelain-enamel slips.

gum set. The abnormal, erratic, quick setting of cement in concrete.

gum tragacanth. A mucilagenous exudate, part soluble and part insoluble, of Asian shrubs used as a binder in glaze and porcelain-enamel slips, and as an adhesive to bond dry-process enamels to metals.

gunite. A mixture of cement, sand or crushed slag, and water sprayed or applied in place pneumatically.

gutta percha. A rubber-like substance obtained from tropical trees; used as a waterproofing and insulating material.

gypsum. CaSO; mol. wt. 172.1; decomposes on heating; sp. gr. 2.3; hardness (Mohs) 2; used in the production of plaster of paris, and as a fining agent in glass.

gypsum board. A flat paper-covered board such as is used in the construction of walls.

gypsum cement. A group of cements and plasters made principally from gypsum (calcium sulfate) and which is produced by mixing it with selected additions, such as sand, alum, borax, and potassium carbonate with sufficient water to produce a trowelable consistency.

gypsum lath. A flat, paper-covered plaster board which has been treated to receive a plaster coating for use in the construction of walls.

gypsum plank. A precast, wire-mesh reinforced gypsum product made with tongue and groove steel edgings for use as roofing, ceiling, and flooring in buildings.

gypsum plaster. A plaster composed essentially of gypsum mixed with water to a troweling consistency.

gypsum wallboard. A plaster board covered with paper or other fibrous material suitable for painting or papering.

gyratory crusher. A large primary crusher consisting of a rounded crushing head mounted on a vertical shaft in a conical shell, the unit rotating on an eccentric axis.

gyratory screen. A vertical nest of horizontal screens of decreasing mesh sizes rotating on an eccentric axis employed to determine the particle size distribution of powdered or granular materials, or to separate and collect quantities of a material of specified maximum and minimum sizes.

H

habit. The characteristic crystalline form or other structure exhibited by a mineral.

hack. A more or less orderly stack of newly formed brick set on boards to dry.

hacking. (1) The replacement of a single course of masonry with two or more lower courses.
(2) The laying of brick with the bottom edge set in from the plane surface of a wall.
(3) The process of stacking brick in a kiln or on a kiln car for firing.

hackle marks. Fine ridges on a glass surface parallel to the direction a fracture is propagated.

hafnia. See hafnium oxide.

hafnium beryllide. (1) Hf_2Be_{17}; mol. wt. 513.6; sp. gr. 4.78. (2) Hf_2Be_{21}; mol. wt. 550.4; m.p. > 1927°C (3500°F); sp. gr. 4.26. (3) $HfBe_{13}$; mol. wt. 298.2; m.p. 1427°C (2600°F); sp. gr. 3.93. See beryllides.

hafnium boride. (1) HfB_2; mol. wt. 200.2; m.p. 3000°C (5540°F); sp. gr. 11.2; used for high-temperature resistant products for nuclear applications.(2) HfB; mol. wt. 189.42; m.p. 2899°C (5250°F); sp. gr. 12.80.

hafnium carbide. HfC; mol. wt. 190.6; m.p. above 3890°C (7034°F); sp. gr. 12.2; hardness ~ 2800 Knoop; used in control rods for nuclear reactors.

hafnium-iron. HfFe; mol. wt. 290.28; m.p. 1649°C (3000°F). See intermetallic compound.

hafnium-manganese. $HfMn_2$; mol. wt. 288.46; m.p. 1642°C (2990°F); see intermetallic compound.

hafnium-molybdenum. $HfMo_2$; mol. wt. 370.60; m.p. 2299°C (4170°F). See intermetallic compound.

hafnium-nickel. $HfNi_2$; mol. wt. 295.98; m.p. 1788°C (3250°F). See intermetallic compound.

hafnium nitride. HfN; mol. wt. 192.6; m.p. 3300°C (5972°F); sp. gr. 14.0; hardness (Mohs) 8–9. See nitrides.

hafnium oxide. HfO_2; mol. wt. 210.6; m.p. 2790°C (5054°F); sp. gr. 9.7; used in refractories to lower thermal expansion, increase inversion temperature, and reduce volume change during inversion.

hafnium phosphide. HfP; mol. wt. 209.63; sp. gr. 9.78.

hafnium silicate. $HfSiO_4$; mol. wt. 270.7

hafnium silicide. (1) Hf_5Si_3; mol. wt. 977.18; m.p. 2299°C (4170°F). (2) HfSi; mol. wt. 206.66; m.p. 2099°C (3810°F). (3) $HfSi_2$; mol. wt. 234.72; m.p. 1699°C (3090°F); sp. gr. 8.03; hardness (Vickers) 865. See silicides.

hafnium sulfide. (1) HfS; mol. wt. 210.66; m.p. ~ 2149°C (3900°F). (2) Hf_2S_3; mol. wt. 453.38; sp. gr. 7.50. (3) HfS_2; mol. wt. 242.72; sp. gr. 6.03. (4) HfS_3; mol. wt. 274.78; unstable above 871°C (1600°F); sp. gr. 5.70.

hafnium titanate. $HfO_2 \cdot TiO_2$; mol. wt. 290.5; m.p. approx. 2200°C (3992°F); sp. gr. 7.21.

hafnium-vanadium. HfV_2; mol. wt. 280.52; m.p. 1499°C (2730°F). See intermetallic compound.

hair cracks. A pattern of hair-like cracks in concrete which occur when the surface layer of concrete dries more rapidly than the interior.

hairlines. A porcelain-enamel imperfection consisting of a series of small hair-like cracks which appear to follow the strain pattern in the metal and which are visible after the coating has been fired.

hairpin furnace. A continuous porcelain-enameling furnace constructed in the shape of a hairpin, the firing zone being located in the turn.

half-bat. A building brick one-half the length of a conventional brick, approximately 4 inches.

half-finish. The first cover coat of a two-coat porcelain-enamel system.

half-life. The period of time in which one-half of the radioactive atoms of a given radionuclide will decay.

half-timbered. A type of building construction in which stucco, brick, plaster, or other masonry is applied between exposed load-bearing timbers.

half-trimmed mica. Mica trimmed on two sides, two thirds of which are trimmed adjacent sides and the balance on parallel sides, all of which are crack free.

Hall effect. The electromotive force generated when a current carrying material is placed in a magnetic field which usually is perpendicular to the direction of current flow and the electric field which usually is perpendicular to both.

halloysite. $Al_2O_3 \cdot 3SiO_2 \cdot 2H_2O$; mol. wt. 318.1; m.p. above 1500°C (2732°F); a kaolin-like mineral used in the production of dinnerware and refractories.

Hamburg blue. A general term for a variety of iron-bearing pigments.

hammer, klebe. A falling weight device used under standardized conditions to compact or densify specimens. See klebe hammer.

hammer mill. An impact mill or crusher consisting of rotating hammers in a rigid metal casing; used to crush ores and other large solid masses, usually preparatory for further milling.

hammer test. Any of a series of tests in which weights are dropped on specimens until fracture or deformation occurs in the specimens.

hand blown. Glassware formed at the end of a blowpipe with air supplied by mouth, the ware being shaped by hand manipulation.

hand feed. To introduce or advance a material into a process by hand, such as, for example, a grinding or machining operation.

hand jig. A moving-screen jig operated by hand which is used to treat small batches of ore; the jig is attached to a rocking-type beam moving in a tank of water.

hand-made brick. Brick shaped in a mold by hand manipulation; the shape may or may not be subjected to subsequent mechanical pressing.

hanging rack. A heat-resistant metal frame suspended in a conveyor system on which porcelain-enameled ware is hung and transported during processing and firing operations.

hard. (1) Resistant to abrasion, scratching, cutting, etc. (2) Having a higher than conventional softening or fusion temperature.

hard-burned brick. Any brick, usually a refractory brick, fired at a high temperature, sometimes higher than normal.

hardening on. The process of volatilizing oils from decorating liquids and pastes applied to bisque ceramic ware and then fusing or hardening the decoration just enough to permit the application of a glaze or other treatment without damage to the decoration.

hard-finished plaster. Overburned gypsum treated with a solution of alum and then recalcined; used in special cements. See parian and Keene's cements.

hard-fired ware. Ceramic ware fired to a high temperature, usually to produce a product of high physical strength and low water-absorption properties.

hard glass. A glass having a high-temperature softening point, high viscosity at elevated temperatures, or high resistance to abrasion, scratching or other mechanical damage, or any combination of these properties.

Hardinge mill. A continuous-type ball mill of tri-cone construction in which each successive cone has a steeper wall from the feed to the discharge end; the mill sometimes is equipped with a cyclone separator to return oversized particles for additional grinding.

hard mica. Mica which does not laminate when bent.

hardness. (1) The relative resistance of a body surface to wear, abrasion, or similar physical damage. (2) The relative refractoriness of a glaze, glass, or porcelain enamel.

hardness, Brinell. A hardness measurement obtained by pressing a steel ball one centimeter in diameter into a material under a standard force, usually 3000 kilograms, and dividing the surface area of the indentation into the load.

hardness, Knoop. A relative hardness measurement of a material determined by the depth to which a diamond indenter having a rhombic base penetrates the material under specified procedures. See diamond indenter.

hardness, Mohs. An empirical scale of hardness in which the scratch resistance of a material is rated on a scale of minerals arranged in order of increasing hardness: 1. talc; 2. gypsum; 3. calcite; 4. fluorite; 5. apatite; 6. feldspar; 7. quartz; 8. topaz; 9. corundum; 10. diamond.

hardness, Rockwell. A hardness rating based on the resistance of a material to indentation by a steel ball or a rounded diamond point of various dimensions, under prescribed static or dynamic loads.

hard paste. A high-fired china body containing substantial amounts of feldspar.

hard porcelain. A porcelain body highly resistant to thermal shock.

hard solder. A solder that melts at temperatures above 370°C (700°F); used in brazing metallized ceramics in the production of glass-to-metal seals. See brazing.

Harkort crazing test. A crazing test for glazes in which a specimen is heated to 120°C (248°F) and plunged into cold water; the test then is repeated by increasing the specimen temperature in increments of 10°C until visible crazing occurs.

harsh. An unworkable, nonplastic, noncohesive mix which tends to segregate during working, particularly a concrete mix.

Hartman dispersion formula. The relationship of the index of refraction and the wavelengths of incident light of a glass expressed as $n = n_o a(\lambda - \lambda_o)$ in which n is the index of refraction, λ is the wavelength, and n_o, a, and λ_o are empirical constants.

Hastelloy. A proprietary nickel-base alloy of high-chemical resistance, heat resistance, and mechanical strength used in agitators, autoclaves, heat exchangers, dryers, burners, blowers, pickling equipment, furnace parts, and similar applications where resistance to corrosion and physical strength at elevated temperatures are required.

haunch. The section of the arch of a furnace or kiln located between the crown and skewback. See crown; skewback.

haydite. Expanded clay, shale, slate, or similar material employed as an aggregate in the production of lightweight concrete and concrete products. See expanded clay.

Haynes stellite. An alloy of cobalt, tungsten, and chromium with minor additions of iron which is characterized by extreme hardness; used in the manufacture of cutting and machining tools.

header. A brick laid with an end exposed and its length perpendicular to the face of a wall.

header course. A type of construction in which an entire row, or course, of brick is laid as headers. See header.

header, false. A half brick used to complete a row of brick in a facing wall.

header, snap. A building brick one-half the normal length, roughly 2⅜ × 4 × 4 in.

header tile. A tile designed to provide recesses for header units in masonry walls.

head lap. The distance between the lower edge of an overlapping asbestos-cement shingle or sheet and the upper edge of the lapped shingle in the second course below.

head space. The unfilled space in closed bottles or other containers.

head, wheel. The outer or upper surface of a grinding wheel.

healing. The process or the ability of a porcelain enamel, glaze, or other ceramic coating to flow and cover surface imperfections during the firing operation.

hearth. The refractory floor of a furnace, kiln, or cupola upon which a charge is placed for melting, sintering, or other heat treatment.

hearth furnace. A type of furnace in which a charge is heat-treated while resting on the furnace floor, or hearth.

hearth roasting. A process for the heat treatment or roasting of ores and other materials on the hearth of a furnace with an excess of air, without fusion, to bring about useful changes in the physical properties of the materials.

heat. Any form of energy causing a rise in temperature or which may be translated into some form of work involving mechanical energy, fusion, evaporation, expansion, etc.

heat-absorbing glass. Any glass capable of absorbing radiant energy in the near infrared range of the spectrum.

heat, available. The amount of heat per unit mass of a substance that can be converted to some form of useful work.

heat balance. The equilibrium existing in a body when the heat gain and the heat loss from all sources are equal.

heat barrier. Any material of low-thermal conductivity used to prevent the transfer or movement of heat from a source to another part or substance.

heat capacity. The amount of heat required to raise the temperature of a substance one degree, usually under some constant condition such as volume or pressure.

heat conduction. The transfer or movement of heat between two parts of a system which does not require movement of the system or any of its parts.

heat content. The sum of the internal energy contained in a body or system and the product of its volume multiplied by the pressure.

heat convection. The movement or transfer of heat by means of a circulating liquid or gas.

heater. Any device designed to produce and transfer heat.

heat exchanger. A device used to transfer heat from a fluid flowing on one side of a barrier to a fluid flowing on the other side of the barrier; for example, steam dryers, muffle furnaces, water-cooled furnaces, and nuclear reactors, etc.

heat flow. The movement of heat through a substance or the transfer of heat from one substance to another, usually reported as the quantity of heat moved per unit of time.

heating chamber. The section of a furnace or kiln in which ware is subjected to heat during a firing operation.

heating, induction. The development of heat in a body by means of an induced electric current when the body is moved through a nonuniform magnetic field or is subjected to a change in magnetic flux.

heating, radiant. Any heating system in which heat is transmitted by radiation as opposed to convection or conduction.

heating, radio-frequency. Electronic heating of a body by means of a radio-frequency current produced by an electron-tube oscillator or similar device.

heat insulator. A material of low-thermal conductivity; for example, foamed clays and concrete, glass wool, mineral wool, foamed glass, etc.

heat, latent. The amount of heat absorbed or evolved per unit mass of a substance during a change of state at constant temperature; for example, the change of a solid to a liquid or a liquid to a gas or vapor.

heat of adsorption. The measure of heat evolved during the adsorption of one mol of a material by another substance at constant pressure.

heat of adsorption, differential. The measure of heat evolved during the adsorption of an incremental quantity of an adsorbate at a given level of adsorption.

heat of adsorption, integral. The sum of the differential heats of adsorption from zero to a given level of adsorption.

heat of combustion. The amount of heat evolved by the combustion of one gram of a substance.

heat of fusion. The measure of heat required to convert a unit weight of a solid to the liquid state.

heat of fusion, latent. The increase in the enthalpy accompanying the conversion of a unit mass of a solid to a liquid at its melting point at a constant temperature and pressure.

heat of hydration. The amount of heat evolved during the hydration of a substance or mixture such as occurs during the hardening or curing of high energy-strength and other concretes.

heat of vaporization, latent. The quantity of energy required to evaporate a unit mass of a liquid at a constant temperature and pressure.

heat-resistant glass. A glass of low-thermal expansion and high resistance to thermal shock such as occurs when the glass is cooled suddenly from an elevated temperature.

heat, sensible. The heat that raises the temperature of a body in which it comes in contact; the sum of the internal energy of a body or system and the product of its volume multiplied by the pressure.

heat-setting mortar. A finely ground refractory mortar which develops its strength at elevated temperatures.

heat-setting refractory. Finely ground refractory material which develops a ceramic-type bond at elevated temperatures.

heat shield. A layer of a substance which provides protection from heat.

heat, specific. The quantity of heat required to raise the temperature of a unit mass of a substance one degree without chemical or phase change.

heat-strengthened glass. Glass subjected to a programmed heat treatment to improve its physical strength.

heat transfer. The movement of heat within a body or from one body to another body.

heat treatment. The process of subjecting a material or body to controlled conditions of heating and cooling to develop specific properties in the material or body such as strength, thermal-shock resistance, etc.

heavy aggregate. Aggregate having a high specific gravity, such as steel punchings, magnetite, barium compounds, etc., used in the production of heavy concrete for special applications such as counter weights, nuclear shielding, and other specialized applications.

heavy concrete. A concrete in which part or all of the conventional aggregate is replaced by metal punchings, magnetite, barium compounds, and similar materials to produce a concrete of high density for use in the production of counterweights, nuclear shieldings, and other specialized applications.

heavy media. Any fluid of high density used in flotation processes for the removal of low-density aggregate particles from mineral raw materials.

heavy spar. Barium sulfate, which see; sometimes used as a flux in stoneware bodies and glazes.

heavy water. Water containing substantial amounts of deuterium, an isotope of hydrogen having an atomic weight of 2.014; used as a moderator in some nuclear reactors.

hectorite. A clay mineral composed of a hydrous silicate of lithium and magnesium, and which is of the montmorillonite family.

heel tap. An imperfection in glass bottles characterized by a bottom of uneven thickness.

height. The vertical dimension or the distance from the bottom to the top of an item.

helical. A cylindrical spiral, such as a thread on a bolt.

helicoid. Having the shape of a flattened coil.

hematite. Red iron ore composed essentially of Fe_2O_3; mol. wt. approx. 159.7; sp. gr. 4.9–5.3; hardness (Mohs) about 6; the most important source of iron.

hematite (black). $BaMn_9O_{16}(OH_4)$; a mineral source of manganese; sp. gr. 3.7–4.7; hardness (Mohs) 5–6.

hematite, (brown). $FeO(OH) \cdot n(H_2O)$; sp. gr. 3.6–4.0; hardness (Mohs) 5–5.5; a minor ore of iron sometimes used as a yellow ceramic pigment.

hemming machine. (1) A device employed to form an edge on a metal sheet by bending the edge of the metal back onto itself for increased edge strength.

(2) A machine designed for the grinding of flat surfaces such as knife blades, skate runners, etc.

hercynite. $FeO \cdot Al_2O_3$; a spinel in which magnesium is replaced by ferrous iron; m.p. 1780°C (3236°F); sp. gr. 4.39; also known as iron spinel. See spinel.

Hermansen furnace. A glass-melting pot furnace of a recuperative design.

Herreshoff furnace. A mechanical, multiple-deck muffle furnace cylindrical in shape.

heterogeneous. Consisting of a mixture of dissimilar ingredients.

high-alumina brick. A refractory brick containing substantial amounts of alumina that react with silica to form mullite when fired to high temperatures; used in applications where unusually severe temperature or load conditions exist.

high-alumina cement. (1) A refractory hydraulic cement made by sintering mixtures of bauxite and limestone; will set to high strength in 24 hours.

(2) A hydraulic cement of high alumina content.

high-alumina refractories. Aluminum silicate refractory compositions in which the alumina content is 45% or more.

high-carbon steel. Steel containing more than 0.5% of carbon.

high-duty fireclay brick. Fireclay brick compositions having a pyrometric cone equivalent not less than cone 31½ nor more than cone 33.

high-early-strength concrete. Concrete which will develop a crushing strength greater than 1700 pounds per square inch when aged in moist air for 24 hours, and greater than 3000 pounds per square inch when aged for 24 hours in moist air followed by immersion in water for 48 hours.

high-energy fuel. Any fuel which produces greater energy than conventional carbonaceous fuels during combustion.

high-frequency furnace. An induction furnace in which heat is generated in a substance or its container, or both, by currents induced by a high-frequency magnetic flux produced by a surrounding electric coil.

high-frequency heating. The development of heat in a body by means of an induced electric current when the body is moved through a nonuniform magnetic field or is subjected to a change in magnetic flux.

high-heat cement. A cement which liberates a high amount of heat during curing.

highlight test. A method of evaluating the resistence of a glaze, porcelain enamel, glass, or other surface to acids, alkalies, and other corrosive and erosive conditions as indicated by a decrease in the sharpness or integrity of an image observed in a direct beam of light.

high quartz. Quartz that was or can be formed at high temperature.

high-speed cement. A fast-setting cement.

high-temperature cement. A refractory cement which will not soften, fuse, or spall at elevated temperatures.

high-temperature glaze. A glaze which matures at temperatures above 1200°C (2192°F).

high-temperature material. Any material which can be used in high-temperature environments, such as furnaces, kilns, roasters, smelters, etc.

high-transmission glass. A glass which transmits an exceptionally high percentage of visible light.

high-velocity burner. A burner which introduces combustible mixtures into the firing chamber of a furnace or kiln at a very high rate of speed.

high-velocity thermocouple. A thermocouple device which will measure the temperatures of flowing gases in an area where the surroundings are of a different temperature.

hindered settlement. A classifying process in which fine aggregate is separated from coarse particles in a water suspension in which a rising current of water hinders the fall of the fine particles while the coarse particles settle to the bottom of the apparatus.

hinge joint. A joint in a pavement or other concrete structure which will permit adjacent sections or slabs to expand, shrink, and move independently of each other and thereby reduce the possibility of uncontrolled breakage of the structure in use.

hip and rib shingles. Rectangular roofing shingles cut and installed with a side lap so as to conceal the joint of the shingles meeting at the hip and ridge of a roof.

hip roof. A roof having four sloping sections, the shorter slopes being triangular in shape.

hip tile. Specially shaped roofing tile used to form the junction of two faces of a roof.

Hispano-moresque ware. A type of luster or tin-enameled pottery.

Hoffman kiln. A multi-chambered, periodic kiln in which the chambers are connected so as to permit the use of combustion gases to dry and preheat ware in the adjacent sections before firing.

Hoffmeister series. An arrangement of anions and cations in the order of decreasing ability to produce flocculation when introduced into clay slips.

hog-back tile. A particular type of roofing tile which is not quite half-round; used along the edges of a pitched roof.

hoist, skip. A mechanical, elevator-type of apparatus consisting of a container mounted vertically or on an incline on wheels, shafts, or rails, and which is raised by chains or cable; employed to raise materials to an elevated level for storage or use.

Holdcroft bars. Bars of selected mineral compositions designed to soften at different temperatures for use as pyroscopes. See pyroscope.

holding room. An area in which ware is stored prior to subsequent processing or shipment.

hole. (1) A depression or void in a body, the bottom of which is not visible under normal vision under 200 foot-candles illumination.

(2) An atom in a semiconductor in which an electron is missing.

holes, sand. Small fractures in the surface of glass produced during grinding and which are not removed during subsequent fine grinding and polishing operations.

hollow block. A relatively large, hollow, structural clay or concrete building block which is used in the construction of walls, floors, and roofs, sometimes with metal reinforcements.

hollow casting. The forming of ceramic ware by pouring a body slip into a porous mold and then draining the mold of the excess slip when the cast article has attained the desired thickness.

hollow-clay blocks. Fired, hollow, structural-clay

building blocks used in the construction of walls, partitions, floors, and roofs of buildings.

hollowing. The process of forming a cavity in a ball of plastic clay on a potter's wheel.

hollow tile. A hollow building unit formed of concrete or fired structural clay; used in building and other construction.

hollow wall. A masonry wall in building construction with a substantial air space between the wall faces; the dead air space provides improved thermal and sound insulation.

holloware. Ceramic and porcelain-enameled ware of significant depth and volume such as bowls, cups, pots, pans, and kettles.

holmium boride. (1) HoB_4; mol. wt. 206.68; sp. gr. 6.84. (2) HoB_6; mol. wt. 228.32; sp. gr. 5.52. See borides.

holmium carbide. (1) Ho_3C; mol. wt. 502.20; sp. gr. 9.43. (2) Ho_2C_3; mol. wt. 362.80; sp. gr. 8.89.(3) HoC_2; mol. wt. 187.40; sp. gr. 7.70. See carbides.

holmium nitride. HoN; mol. wt. 177.41; sp. gr. 10.26. See nitrides.

holmium oxide. Ho_2O_3; mol. wt. 374.8; m.p. 2360°C (4280°F); sp. gr. 8.35; used in the production of special high-temperature refractories.

holmium silicide. $HoSi_2$; mol. wt. 219.52. See silicides.

hologram. A three-dimensional picture or image produced by reflected laser light on a photographic plate or film illuminated from behind.

homogeneous. Consisting of a uniform composition or structure.

hone. (1) A fine-grit stone or block of abrasive used for sharpening and fine grinding.

(2) A rotating tool with an abrasive tip used for enlarging and polishing holes and internal cylindrical surfaces to precise dimensions.

honeycomb. (1) A body with a cellular internal structure resembling a honeycomb, and which is used as a structural material of light weight and high strength.

(2) A poorly filled, insufficiently compacted, or porous concrete mass.

honing. To smooth and polish a surface with a fine-grained stone or abrasive.

hood. (1) A guard around a grinding wheel serving as protection against breakage, sparks, released and flying particles, and dust.

(2) A metal covering or cowl covering a hearth or other work area in an exhaust system for the removal of dust and fumes.

(3) A refractory form partially immersed in a molten glass batch to protect the gathering area from furnace gases and floating scum.

hooded pot. A glass-melting pot in which the interior and its contents are protected from combustion gases by a refractory cover or by careful design of the pot, and also provided with an opening for the charging and gathering operations.

hook. A curved, heat-resistant alloy upon which porcelain-enameled ware is suspended for transport through a furnace.

Hooke's law. The ratio of the stress to the strain in a solid body is constant.

hooped column. A reinforced concrete column in which the steel-rod reinforcements placed vertically in the shaft are enclosed in steel hoops to tie the rods together.

hopper. A large container in which bulk materials are stored prior to use.

horizontal-cell tile. A hollow building unit of fired structural clay in which the axis of the interior cell is in a horizontal position when placed in a wall.

horizontal crusher. A type of crushing or milling device in which the crushing stone is mounted on a horizontal shaft to minimize headroom requirements.

horizontal retort. A vessel of highly siliceous composition employed in the production of zinc metal and in the gasification of coal.

hornblende. $(Ca, Na)_2(Mg,Fe,Al)_5(Al,Si)_8O_{22}(OH,F)_2$; a common mineral present in clays and feldspathic materials; sp. gr. 3.0–3.47; hardness (Mohs) 5–6.

horse. A slightly convex rack on which drying roofing tiles are placed and permitted to sag to a slightly curved shape.

hospital. An area in a factory where defective ware is repaired.

hot area. The section of a laboratory in which highly radioactive materials are handled.

hot-blast circulating duct. A large-diameter, refractory-lined pipe which surrounds and delivers hot air to the tuyeres of a blast furnace. See tuyeres.

hot-blast main. A refractory-lined pipe which delivers hot air from a hot-blast stove to the hot-blast circulating duct of a blast furnace.

hot-blast stove. A refractory-lined apparatus in which hot air is produced for delivery to the tuyeres of a blast furnace. See tuyeres.

hot draw. The removal of a material from a furnace or kiln while hot.

hotel china. A hard-glazed, vitreous dinnerware of high strength, usually thicker than household china; used by commercial institutions.

hot end. The finishing end of a glass manufacturing operation, including the forming of the molten glass and the annealing of the formed ware.

hot floor. A floor, particularly the floor of a dryer, heated by steam pipes or other source of heat.

hot-floor dryer. An enclosed chamber or room for the drying of ware in which heat is supplied by steam pipes or other heat source embedded or contained in the floor.

hot-metal ladle. A large, refractory-lined ladle employed to convey molten metal from a blast furnace to a subsequent processing operation.

hot-metal mixer. A refractory-lined holding furnace for molten pig iron.

hot mold. A hot coated or uncoated mold in which glass or other ceramic ware is formed.

hot patch. A refractory slurry which is applied by spraying to repair a damaged, hot refractory lining of a furnace.

hot-pressed abrasives. Bonded abrasive products formed in a mold by pressing at appropriate high temperatures.

hot pressing. (1) A jiggering process employing a heated profile tool or plunger.

(2) The forming of ware by pressing in a mold at an elevated temperature.

hot spot. The area of highest temperature in a glass-melting furnace.

hot top. A special refractory shape placed on the top of an ingot or casting mold so that the riser and sinkhead will form above the casting. See riser; sinkhead.

hot zone. The area in a continuous furnace or kiln where the most intense heat is supplied to the ware being fired.

household china. Vitreous ceramic dinnerware, usually thin and of high translucency, and generally considered for domestic use.

HTI. Acronym for high-temperature insulating refractory.

humidifier. An apparatus designed to introduce water vapor into the atmosphere of an area, such as a controlled-humidity dryer.

humidity. The degree of dampness or the amount of water vapor contained in the atmosphere.

humidity, absolute. The weight of the water vapor per unit of dry air in an air-water vapor mix.

humidity dryer. A dryer in which the humidity of the atmosphere is controlled.

humidity, relative. The ratio of the amount of water vapor in the air at a specific temperature to the maximum water-vapor capacity of the air at that temperature.

hump. A large ball of clay centered on a potter's wheel from which several small pots are thrown.

Huntington dresser. A star-shaped rotating cutter tool employed to dress and true abrasive grinding wheels.

hydrate. A compound containing water in a definite ratio, the water supposedly being retained in its molecular state as H_2O.

hydrated alumina. $Al_2O_3 \cdot 3H_2O$; mol. wt. 155.9; sp. gr. 2.42; used as a component in glass and sintered ceramic bodies, and as a coating for refractory setters to prevent ware being fired from sticking to the setters during the firing operation.

hydrated lime. Quicklime to which sufficient water has been added to convert the oxides to hydrates.

hydration. (1) The reaction between a hydraulic cement and water during which new compounds are being formed, most of which have strength-producing properties.

(2) The chemical process by which cement paste is hardened.

(3) The incorporation of water molecules into a compound to form a hydrate.

hydration, heat of. The heat generated during the hydration or setting of cement.

hydration resistance. The degree to which a material, particularly a refractory material, resists chemical combination with water.

hydration, water of. Combined water in a substance which can be removed by heating without changing the composition of the substance.

hydraulic adsorption. The adsorption of a weakly ionized acid or base formed by the hydrolysis of some types of salts in aqueous solutions.

hydraulic cement. A cement that sets and hardens by chemical interaction under water; some types will set under water.

hydraulic cement, air-entraining. A hydraulic cement containing a sufficient amount of air-entraining agent to cause air to be entrained in the mortar.

hydraulic lime. Calcined limestone which absorbs water without swelling or heating and which produces a cement which hardens under water.

hydraulic press. A press actuated by a liquid under pressure.

hydraulic pusher. A hydraulically actuated mechanism designed to push loaded cars through a tunnel kiln.

hydraulic refractory cement. A composition of ground refractory materials, some of which react chemically to form a strong hydraulic bond at room temperature.

hydraulic structure. Any structure, including concrete, used to convey water from one location to another, or which may be exposed to water for substantial periods of time as in canals, sea walls, etc.

hydroabietyl alcohol. $C_{19}H_{31}CH_2OH$; m.p. 32°C (89°F); sp. gr. 1.007; used to control the drying, flowout, and viscosity of screen-process inks.

hydrochloric acid. HCl; mol. wt. 36.5; sp. gr. 1.19; widely used in the pickling of metal for porcelain enameling.

hydrodynamics. The study dealing with the motion of fluids and the forces acting on bodies immersed in fluids.

hydrofluoric acid. HF in aqueous solution; used in the polishing, frosting, and etching of glass surfaces; sometimes used to clean brick.

hydrogen defects. Imperfections in porcelain enamels particularly fishscaling, due to the presence of hydrogen when the atomic hydrogen is converted to molecular hydrogen in voids causing pressure to develop at the interface between the metal and the solidified coating after firing.

hydrolysis. The chemical decomposition or alteration of a substance by reaction with water.

hydrometer. A direct-reading floating instrument employed to measure the specific gravity or similar properties of liquids and slurries.

hydrophilic. Having an affinity for water.

hydrophobic. Having an aversion to water.

hydrostatic press. A press actuated by water, oil, or other liquid under pressure.

hydrostatic pressing. The process of forming and compacting ceramic bodies contained in a thin rubber or plastic envelope which is placed in a die and surrounded by a fluid and then subjected to high pressures, the pressures being equal in all directions on the specimen.

hydrostatics. The study of the effects of pressure in a liquid or exerted by a liquid on an immersed body.

hydrostatic strength. The property of a pipe or other shape to withstand the internal pressures of liquids of specific pressure magnitudes.

hydrothermal. Relating to hot water; for example, the formation or metamorphism of minerals by the action of hot solutions rising through the earth's crust.

hygrometer. Any of several instruments that measure the humidity of an atmosphere.

hygroscopic. Pertaining to the property of a substance to take up and retain water, particularly moisture from the atmosphere.

hygroscopic water. (1) Water taken from the atmosphere by a body, and which can be removed by simple drying.

(2) Chemically combined water in a substance or body which can be removed only by heating to elevated temperatures.

hysteresis. The lag or failure of a property that has been changed by an external agent, such as mechanical, magnetic, or electrical stress or to some influence occurring during the history of the material, to return to its original value when the cause of the change is removed.

hysteresis, loop. The divergence between the paths of the adsorption and desorption isotherms.

hysteresis loss, incremental. The loss in hysteresis when a body of magnetic material is subjected to a pulsating magnetic force.

hysteresis, magnetic. The lag in values of resulting magnetization in a material due to changes in magnetic force.

I

IACS. The international standard of electrical conductivity for annealed copper.

ice. Coarse-grained clear, white, or colored compositions of high fluxing characteristics which are applied and fired on glassware to produce a variety of frosted or pebbled effects on the ware.

IC silicon carbide. Silicon carbide impregnated with carbon.

ID coil. An electromagnetic coil inserted inside a hollow test specimen.

ID grinding. Internal grinding of a hollow body such as a pipe, cylinder, or similar structure.

idle. To run without load.

idle time. The elapsed time equipment is unused.

igneous. Rocks solidified in nature from a molten state.

ignition. The process of starting a fuel or fuel mixture to burning.

ignition arch. The section of a kiln in which fuel mixtures are preheated to expedite ignition.

ignition coil. A type of coil employed in ignition systems to ignite a fuel mixture. In practice, the coil stores energy in a magnetic field which is released suddenly by signal to ignite the fuel.

ignition, loss on. The loss in weight observed in a sample heated to a high temperature, the sample first having been dried above 100°C (212°F).

ignition temperature. The lowest temperature at which combustion of a material will occur and continue burning when heated in air or under specified conditions.

illite. A group of micaceous clay minerals ranging between montmorillonite and muscovite in composition and structure; sometimes employed as a clay addition in ceramic bodies.

illumination. The process by which light is dispersed on the surface of a beam.

ilmenite. $FeO \cdot TiO_2$; mol. wt. 151.74; m.p. 470°C (878°F); sp. gr. 4.3–5.3; hardness (Mohs) 5.5–6; employed as a source of titania in special glasses, as an opacifier in glazes and enamels, as a black coloring agent in brick coatings, as a speckling agent on ceramic tile, and as a component in some ceramic dielectrics.

image furnace. A device for producing extremely high temperatures in concentrated areas by focusing rays from the sun or electric arcs by means of lenses or mirrors.

imbition. The absorption of a liquid by a solid or semisolid material, frequently accompanied by an increase in the volume of the solid absorbent; for example, porous clays, graphite, and silica gel.

immersion cleaning. The removal of surface contamination on an object by dipping in a cleaning liquid.

immersion coating. The application of a coating to an object by dipping the object into a coating solution or suspension; for example, pottery, glazes, porcelain enamels, etc.

immiscibility. The property exhibited by liquids which will not mix with each other, as for instance, water and oil.

impact. The collision of bodies with sufficient force to cause appreciable change in the momentum or condition of the colliding bodies, such as a change in direction, a change in speed of movement, or breakage of one or both bodies.

impact crusher. A crushing device which breaks down solid materials by shattering blows imposed by rotating hammers, bars, or steel plates.

impact grinder. A device for reducing the size of solid materials by the fall of crushing bodies, such as the balls in a ball mill.

impact mill. A crushing device employed to reduce the size of minerals and rocks in which the material is thrown against steel plates by rapidly rotating blades.

impactor. See impact mill, hammer mill.

impact pressing. The process for the forming of refractory shapes in which the ground particles of refractory material are packed closely together by rapid and strenuous vibration in a mold.

impact resistance. The resistance of a body or coating to breakage, deformation, or other damage when subjected to sharp blows or shock loading.

impact strength. The property of a material to resist physical breakdown or damage under conditions of shock loading.

impact stress. The stress imposed on a solid body by a suddenly applied force; reported as force per unit of area.

impact test. A procedure to evaluate the resistance of a material to physical damage under specified conditions of impact or shock loading.

imperiale. A ceramic or glass bottle of 6 liter capacity for table wines, or 4.5 liter capacity for sparkling wines.

imperial red. A family of red colors produced by ferric oxide pigments.

impermeability. The property of a body, glaze, porcelain enamel, or other material to resist the entry or passage of fluids and gases.

impervious. A term denoting the degree of vitrification of a ceramic or ceramic coating as determined by its resistance to dye penetration; usually a visual observation.

impervious carbon. A dense, impervious, bitumen-bonded carbon body formed by pressing followed by sintering to an essentially pore-free brick; used to line chemical process and storage vessels.

impregnation. The process of forcing a liquid substance into the pores of a solid.

impressed decoration. A decoration stamped under pressure into a plastic clay body.

impurity. (1) An undesired foreign material in or on a substance.

(2) A material introduced in small amounts into a semiconductor to develop or improve desired properties.

in-and-out bond. A type of masonry construction consisting of alternate courses of headers and stretchers, brick, stone, concrete block, etc.

incandescent. Emitting visible light as a result of heat; for example, the filaments in an electric-light bulb, the walls of a kiln or furnace in use, or other object heated to visible radiation temperatures.

incise. The process of decorating ware by cutting, carving, or indenting the surface of the ware with a sharp tool.

inclusion. A particle of foreign material embedded or entrapped in a body or coating other than materials comprising the normal composition.

inclusions, air. (1) Gaseous inclusions in mica which appear as grayish areas in transmitted light and as silvery areas in reflected light.

(2) Small bubbles of air or other gas enclosed in glass, glaze, porcelain enamels or bodies which become evident after firing; usually a defect.

inclusions, clay. (1) Earthy inclusions in mica which appear in various colors when observed in any type of light.

(2) Unreacted clay or other solid material remaining in a glaze or porcelain enamel after firing.

inclusions, gaseous. Round or elongated bubbles, blisters, or seeds in glassware, porcelain enamels, glazes, etc.

inclusions, gaseous open. Broken gas bubbles and blisters forming pits and cavities on the surfaces of glass, porcelain enamels, and glazes.

inclusions, mineral. Inclusions of concentrated metallic oxides in mica which appear as areas of deep, distinct, and highly saturated colors in transmitted light.

inclusions, nonmetallic. Nonmetallic particles embedded in steel which result in defects in porcelain enamels.

inclusions, smoky. A misnomer describing dispersed metal oxides included in mica which appear as pastel colors of low to medium saturation in transmitted light.

inclusions, vegetable. See smoky inclusions; the term "vegetable" is a misnomer.

incombustible. Any material which will not burn or support combustion when exposed to fire in air at 648°C 1200°F).

incompatibility, thermal. A condition in which part of the aggregate in portland cement has such different thermal properties, particularly coefficient of thermal expansion, when compared with other portions of the aggregate or the cement paste, that distress or damage to hardened concrete may result, particularly crumbling.

incomplete combustion. A burning process in which oxidation of the fuel is incomplete, sometimes resulting in reducing atmospheres in direct-fired furnaces and kilns.

incongruent melting. Dissociation on heating of a compound to form a liquid and another compound of different composition from the original compound.

incremental hysteresis loss. Loss in hysteresis when a magnetic material is subjected to a pulsating magnetizing force.

incremental permeability. The ratio of a change in magnetic induction to the corresponding change in magnetizing force when the mean induction differs from zero.

indenting. The omission of brick from a masonry construction in such a spacing that the brick may be inserted later.

index grinding feed. A mechanical procedure for feeding internal and other grinding devices in which the rate or amount of feed is indicated by means of a dial or similar gauge.

index of refraction. The ratio of the velocity of light, or the sine of the angle of incidence, in a material to the velocity of light in a specified medium such as air; or the sine of the angle of refraction, as determined by immersing particles of the material in liquids of known refractive index.

index of workability. A measure of the consistency and forming characteristics, particularly moldability, of plastic materials.

Indiana measure of air entrainment. A procedure for estimating the quantity of air entrained in concrete in which differences in unit of weights of samples with and without air are reported.

Indian red. A red ferric oxide pigment prepared by calcining ferrous sulfate, $FeSO_4 \cdot 7H_2O$.

indication. (1) In magnetic testing, a discontinuity identified as a magnetic-particle buildup resulting from interruption of the magnetic field.

(2) In ultrasonic testing, determination of the presence of a flaw by detection of a reflected ultrasonic beam.

(3) In general, that which indicates the presence of a flaw or discontinuity in a substance.

indication, diffuse. An indication which is not clearly defined.

indication, false. An indication resulting from improper processing, as opposed to a nonrelevant indication.

indication, nonrelevant. An indication which cannot be attributed to or associated with a flaw or discontinuity.

indirect arc furnace. A refractory-lined furnace in

which ware is heated indirectly by an electric arc struck between electrodes.

indium. In; at. wt. 114.8; m.p. 156°C (313°F); a ductile metal used in glass-sealing alloys.

indium antimonide. InSb; mol. wt. 236.5; m.p. 535°C (990°F); an intermetallic compound having a small energy gap and very high electron mobility; useful in photodetector applications and in magneto-resistance and Hall-effect devices.

indium arsenide. InAs; mol. wt. 189.7; m.p. 943°C (1725°F); an intermetallic semiconductor useful as an infrared photoconductor and also in magneto-restrictive or Hall-effect devices.

indium brazing. Various alloys of indium and other elements used in ceramic-metal seals to produce vacuum-tight bonds; variable solidus temperatures to 315°C (600°F).

indium nitride. InN; mol. wt. 128.8; a semiconductor compound having a resistance of 4.0×10^3 ohm-cm.

indium oxide. In_2O_3; mol. wt. 277.64; m.p. 1910°C (3470°F); sp. gr. 7.179; N-type semiconductor useful as a resistance element in integrated circuitry.

indium phosphide. InP; mol. wt. 145.7; m.p. 1070°C (1958°F); a semiconductor useful in rectifiers and transistors, particularly at intermediate temperatures.

induced draft. A draft or current of air produced by suction fans or stream jets in the flues of industrial kilns.

induction furnace. An electric furnace in which heat is generated by electromagnetic induction or eddy currents.

induction heating. Raising the temperature of an electrically conducting material by induced electric currents of high or low frequencies.

induction method of magnetization. Magnetization of a material by a circulating current induced in a ring component by the influence of a fluctuating magnetic field.

industrial floor brick. A brick having extremely high resistance to wear, mechanical damage, chemicals, and temperature conditions such as may be encountered in industrial and commercial installations.

infrared dryer. A dryer in which heat is supplied by infrared radiation, such as infrared lamps and incandescent burners.

infrared filter. A material or product which is transparent to infrared radiation but opaque to other wavelengths.

infrared laser. A laser which emits infrared radiation.

infrasizer. An instrument for the fractionation of powders by air classification according to their density and size; the powder sample is carried in an air stream and collected in a series of tapered cylinders of the same length but of decreasing diameters, the fines being collected in a bag at the end of the system.

infusorial earth. An incorrect term for the siliceous remains of diatoms.

inglaze decoration. A ceramic decoration applied on the surface of an unfired glaze and fired simultaneously with the glaze.

ingot. A solid metal casting of size and shape suitable for remelting or working.

ingot mold. A mold in which ingots are cast.

initial permeability. The slope of an induction curve at zero force of magnetization as a specimen is removed from a demagnetized condition.

initial rate of absorption. The weight of water absorbed by a brick when partially immersed in water for one minute; expressed as grams or ounces per minute.

initial set. The period of elapsed time between the mixing of water in a mortar or cement and the moment the mixture starts to set or lose plasticity.

initial softening. The time and temperature at which a ceramic or ceramic coating begins to show evidence of flow.

injection. A pressure process of forcing a filler material into cracks, cavities, and pores.

injection molding. The forming of ceramics by the injection of a measured quantity of a body into a mold where it is densified by pressing.

ink, ceramic. An ink composed of a powdered ceramic pigment dispersed in a carrier medium such as an oil or varnish; the dispersion may be applied as a decoration on ceramic bodies and coatings by stamping, screening, brushing, or other means, and then fired to develop its final color and permanence.

ink, stamping. See ink, ceramic.

inlay, rolled. A process for the decoration of pottery by which colored clays are rolled into the surface of a clay item as a form of decoration while the item is still plastic; after drying, the item is fired.

inrush current. A transient current which exists at the instant an electrical contact is closed, and which continues briefly.

inserted coil. A coil or coil assembly which is inserted into a test specimen for test purposes, such as in magnetic testing.

inside coil. See inserted coil.

inspection. The examination of a product or specimen by visual, mechanical, electrical, or other means to determine its quality in terms of prescribed standards or specifications.

inspection, magnetic particle. A nondestructive technique for the detection of surface and subsurface discontinuities in ferromagnetic materials in which finely divided magnetic particles are applied over the surface of the specimen where they collect in visual quantities in the areas where discontinuities and defects exist.

inspection, spark-gap. A technique to detect pinholes and cracks in glass-coated iron or steel products by fanning a high-frequency discharge from a spark generator across the surface of the coating; the fanned discharge will converge to form a bright spot at the point of a pinhole or fracture.

insulated conductor. A conductor which is coated or surrounded by a nonconducting material to prevent or retard the transfer of electric current, heat, sound, or other phenomenon of concern.

insulating cement. A cement or concrete product in which a substantial quantity of an insulating material such as asbestos has been incorporated, or a lightweight concrete of relatively low density; used as thermal insulation and fire protection in structures.

insulating concrete. See insulating cement.

insulating firebrick. A refractory firebrick of low-thermal conductivity and low-heat capacity.

insulating material. Any material which will prevent or retard the transfer of electric current, sound, heat, or other form of energy, depending on the specific property desired.

insulation capacity. The property of masonry to store heat as a result of its mass, specific heat, density, etc.

insweep. The lower portion of a glass container which tapers inward toward the base.

intaglio. A depressed surface decoration in which the design is engraved on the ware.

integral heat of adsorption. The sum of the differential heats of adsorption of an adsorbate from zero to a given level of adsorption.

integral waterproofer. A material or mixture of materials added to concrete to reduce the capillarity or flow of water through the concrete.

intercalary decoration. Bits of gold foil, colored glass, or enamel applied to the surface of a glass object that is then encased in another layer of hot glass, trapping the design between the layers.

interface. The surface forming a common boundary between two substances in contact with each other, such as solid to solid, liquid to solid, gas to solid, gas to liquid, etc.

interferometer. An optical instrument which will split a beam of light into two or more beams and then reunite the beams traveling over different routes; a means of determining the expansion, contraction, fusion, and other properties of materials.

interior tile, glazed. A glazed tile, usually consisting of a nonvitreous body, which is suitable for interior use, but not resistant to impact or severe conditions of weathering when used in exterior applications.

interlayer. The plastic reinforcing material used in the production of laminated glasses and plastic-bonded glass fibers.

interlayer water. Water which enters and is held between the layers of weakly bonded solid materials, sometimes resulting in swelling; such water may be removed by heating.

interlocking tile. Roofing tile designed to interlock with adjacent tile.

interlock, mechanical. (1) The situation in which, because of roughness or other surface conditions, electrical contacts fail to separate.

(2) The process of joining two components together by means of hooks, keys, dovetails, dowels, etc.

intermediate crusher. A crushing or milling device which will reduce materials to intermediate sizes, approximately 1 to 5 mm. in cross-sections, usually before the materials are subjected to additional grinding or processing.

intermediate-duty fireclay brick. A fireclay brick having a pyrometric cone equivalent not lower than cone 29, or more than 3% deformation at 1350°C (2460°F).

intermediate piece. The refractory channel between the spouts of a glass tank and a pot.

intermetallic compound. An alloy of two metals, which includes the borides, hydrides, nitrides, silicides, beryllides, and aluminides of the transition elements in the fifth and sixth group in the periodic sequence and semimetallic elements of small diameter, in which a progressive change in composition is accompanied by a progression of phases differing in crystal structure and properties. Some are extremely hard, approaching the hardness of tungsten carbide. They are used as refractories for high-temperature applications, and as filters to increase the strength of metallic, ceramic, and metal-ceramic compounds.

intermittent kiln. Any kiln in which a batch of ware is placed, fired, cooled, and removed before a subsequent batch is placed in the kiln.

intermittent moving bed. An adsorbent bed of activated carbon in which spent carbon periodically is replaced by virgin carbon, the spent carbon being removed from the bottom of the bed and virgin or reprocessed carbon being introduced at the top.

internal energy. The sum of the intrinsic energies of individual molecules, kinetic energies of internal motions, and the contributions from interactions between molecules minus potential or kinetic energy of the system as a whole.

internal grinding. The grinding and polishing of the surfaces on the inside of holes, cylinders, and tubular products.

internal seal. A ceramic-to-metal seal in which a ceramic surrounds the metal portion or pin in a cylindrical or similar unit.

internal stress. The stress existing in a solid body which is independent of external forces; for example, the stresses remaining in a glass induced by a particular heat treatment.

internal vibrator. A vibrating apparatus placed in freshly placed concrete to render the mixture into a quasi-liquid state to attain maximum consolidation of the concrete in the forms.

interpretation. A clarification and explanation of the meaning and significance of data and related observations, particularly from the standpoint of their relevance to a situation.

interstice. A space between atoms, groups of atoms, or grains in a solid structure; the space in a lattice structure.

interstitial. Related to or occurring in interstices, which see.

intrusion grouting. The technique of placing the grout components in position in an area and subsequently converting the mixture to concrete by the addition of water.

intumescence. The property of a material which causes it to bloat or swell to a permanent vesicular structure on heating; a technique to induce sound and thermal insulation, as well as fire resistance.

inversion. The change in the crystal structure of a material, as between two or more forms of polymorphic crystals (a crystal having two or more reversible forms), without change in chemical composition, such as the inversion of quartz by thermal treatment. See quartz inversion.

inversion point. The temperature at which an inversion will occur as, for example, the change of α-quartz to β-quartz at 575°C (1070°F), and vice versa.

inversion, quartz. A change in the crystal form and properties of quartz by heat treatment, but without change in composition; for example, α quartz is converted to β quartz at 573°C (880°F) and the change is reversible. At higher temperatures, quartz is converted to cristobalite or tridymite, or a mixture of both.

invert. (1) The bottom, or floor, or the lowest point of the internal surface of the transverse cross section of a channel or pipe in which water or other fluid is conducted.

(2) To reverse an order, position, or condition, such as to turn upside down or inside out, or to revert from one form to another.

investment casting. A technique for the production of small or relatively small items of high-dimensional precision by casting in a refractory mold which itself was formed by slip casting a refractory body around a wax replica of the item of manufacture. After the mold has set, the wax is melted out, and the body is fired to produce a mold without joints.

investment compound. A mixture of refractory powder, binder, and liquid employed in the production of molds for investment castings. See investment casting.

inwall. The refractory lining of the stack of a blast furnace.

inwall brick. Fireclay brick used in the lining of the inwall section of a blast furnace. See inwall.

ion. An electron, positron, atom, group of atoms, or molecule which carries a positive or negative charge as a result of having lost or gained one or more electrons.

ion exchange. A reversible chemical reaction between a solid and a liquid, as exhibited in a mixture of colloidal inorganic substances such as clays and water, in which adsorbed ions may be replaced or exchanged with other ions in the mixture or slurry.

ion, network-forming. An ion which, in combination with other ions, forms the network of a glass structure.

ion, network-modifying. An ion which serves a modifying but not a network forming function in the structure of a glass.

iridescence. The diffraction of light reflected from a surface to produce a rainbow color effect.

iridium oxide. Ir_2O_3; mol. wt. 434.2; highly infusible; used as an underglaze black pigment. Also known as iridium sesquioxide.

iridium potassium chloride. $IrCl_4 \cdot 2KCl$; mol. wt. 484.1; used as black pigment in the decoration of porcelains.

iridium sesquioxide. See iridium oxide.

iridium silicide. (1) Ir_3Si; mol. wt. 607.36. (2) $IrSi$; mol. wt. 221.18.

Irish moss. A gelatinous or mucilaginous material sometimes used as a suspending agent for solids in aqueous slurries. Also known as chondrus, carrageen,kileen, rock-salt moss, pearl moss, and pig-wrack; a sea weed.

iron. (1) The various iron, steel, and cast iron structures or products on which porcelain enamels are employed as decorative and protective coatings, such as household appliances, architectural panels, signs, kitchenware, sanitaryware, some glass-to-metal seals, glass-forming molds, and the like. The sheet metals usually are of the low-carbon cold-rolled steels, low-metalloid enameling irons, and the decarburized steels. The cast products usually are gray iron castings.

(2) Compounds of iron are the most useful and versatile of all metals as coloring agents in clays, pigments, and glazes.

iron aluminate. $FeO{\cdot}Al_2O_3$; mol. wt. 173.7; m.p. 1438°C (2620°F); sp. gr. 4.35.

ironarc process. An extremely high-temperature smelting process used in the production of zirconia and other refractory materials, usually by plasma chemistry.

iron beryllide. FeB_2; mol. wt. 74.24; m.p. 1483°C (2700°F).

iron, blowing. A long iron pipe used by a glassmaker for the gathering and blowing of glassware by mouth.

iron boride. (1) Fe_2B; mol. wt. 122.50; m.p. 1371°C (2500°F); (2) FeB; mol. wt. 66.66; m.p. 1538°C (2800°F); sp. gr. 7.15.

iron chromate. $FeO{\cdot}Cr_2O_3$; mol. wt. 222.9; m.p. > 1771°C (3220°F); sp. gr. 5.08; employed as a brown and black pigment in engobes, glazes, glasses, and porcelain enamels.

iron-constantan. A bi-metal device used in the thermocouple functions for temperature measurements up to 1000°C (1832°F).

iron, enameling. A very low-carbon, low metalloid, open-hearth cold-rolled sheet steel used as a base metal for porcelain enameling.

ironing. A discoloration due to the crystallization of cobalt silicate in glazes.

iron modulus. The ratio of $Al_2O_3:Fe_2O_3$ in a hydraulic cement.

iron molybdenum. FeMo; mol. wt. 151.84; m.p. 1538°C (2800°F); sp. gr. 8.53; hardness 76.5 Rockwell A.

iron niobium. (1) Fe_3Nb_2; mol. wt. 353.72; m.p. 1655°C (3010°F); sp. gr. 7.89; hardness 76 Rockwell A. (2) Fe_2Nb; mol. wt. 204.78; m.p. 1626°C(2960°F).

iron-ore cement. A cement in which iron ore (Fe_2O_3) is employed as a replacement for clay, shale, or alumina; sp. gr. about 3.31; more resistant to some corrosive environments, particularly seawater, than portland cement.

iron oxide (ferric). Fe_2O_3; mol. wt. 159.68; m.p. 1565°C (2850°F);sp. gr. 5.12–5.24; used in the production of ferrites and magnetic ceramics, ferrospinels, ceramic glazes and body stains, and polishing compounds.

iron oxide (ferrous). FeO; mol. wt. 71.84; m.p. 1420°C (2588°F); sp.gr. 5.7.

iron oxide (ferrous-ferric). $FeO{\cdot}Fe_2O_3$; mol. wt. 231.52; m.p.1587°C (2890°F); sp. gr. 4.8–5.1; hardness (Mohs) 5.5–6.5; used as a black ceramic pigment and polishing compound. Also known as black iron oxide.

iron phosphide. (1) Fe_2P; mol. wt. 142.7; m.p. 1290°C (2354°F); sp.gr. 6.56. (2) FeP_2; mol. wt. 117.9; sp. gr. 5.12. (3) FeP; mol. wt. 86.87; sp. gr. 6.90. (4) Fe_3P; mol. wt. 198.55; decomposes at 1204°C (2200°F); sp. gr. 7.21.

iron red. Any of the family of red pigments made from the red varieties of iron oxide.

iron rhenium. Fe_3Re_2; mol. wt. 242.06; m.p. 1521°C (2770°F).

iron saffron. See Indian red.

iron scurf. A mixture of ground stone and iron particles used as a blue pigment in the coloring of brick; obtained by grinding and polishing gun barrels with silicious abrasives and grindstones.

iron silicate. $2FeO{\cdot}SiO_2$; mol. wt. 203.7; m.p. 1198°C (2190°F); sp. gr. 4.24; hardness (Mohs) 5–7.

iron spangles. Magnetic iron oxide (Fe_2O_3) used in the production of aventurine-type glazes. See aventurine.

iron, spathic. $FeCO_3$; mol. wt. 115.84; sp. gr. 3.83–3.88; hardness (Mohs) 3.5–4.5; used to produce a wide range of colors in ceramic bodies and glazes.

iron spinel. $(FeMg)Al_2O_4$; sp. gr. 4.39; employed as a refractory.

iron spot. A discoloration in refractory brick resulting from a concentration of iron-bearing impurities.

ironstone china. A generic term for a fine, hard earthenware of high strength and durability.

ironstone clay (brown). A natural iron oxide or brown ironstone clay of variable composition but usually reported as $FeO(OH){\cdot}nH_2O$.

ironstone clay (red). A mineral composed of ferric oxide and clay or sand.

iron sulfate. $FeSO_4{\cdot}7H_2O$; see ferrous sulfate.

iron sulfide. FeS; mol. wt. 87.90; m.p. 1749°C (3180°F); sp. gr. 4.84.

iron tantalum. Fe_2Ta; mol. wt. 237.34; m.p. 1777°C (3230°F).

iron titanium. Fe_2Ti; mol. wt. 159.78; m.p. 1516°C (2760°F).

iron tramp. A piece of unwanted iron, such as a nail,

bolt, or iron trimming, which finds its way into a bulk material or batch.

iron tungsten. Fe_7W_6; mol. wt. 1494.88; m.p. 1637°C (2980°F).

iron vitriol. $FeSO_4 \cdot 7H_2O$; see ferrous sulfate.

iron zirconium. Fe_2Zr; mol. wt. 202.90; m.p. 1642°C (2990°F); sp. gr. 7.70; hardness (Vickers) 436.

irreversible adsorption. Adsorption in which the desorption isotherm is displaced toward higher equilibrium adsorption capacities from the adsorption isotherm.

irridizing compound. A strongly adherent film or coating of a metal oxide or other compound on glass or vitreous surfaces as a decoration or to impart a desired surface property such as controllable electrical conductivity or resistance.

irrigation pipe. A conduit of concrete, tile, or other material employed in the transport of water for agricultural irrigation.

isobar. (1) A graphic indication of the quantity of a substance adsorbed by a material, such as activated carbon, plotted against equilibrium temperature at a constant pressure or concentration.

(2) Any of two nuclides having the same mass number, but different atomic numbers.

isomer. A compound of the same composition and molecular weight as another, but exhibiting different chemical or physical properties.

isometric. Relating to minerals which crystallize in the cubic system having three equal axes at right angles and having identical properties with respect to these axes.

isomorphic. Having identical crystalline structures but different chemical compositions.

isomorphous mixture. A type of solid solution in which minerals of analogous chemical composition and closely related crystal habit crystallize together in various proportions.

isostatic pressing. A technique for compacting powders into shapes of high-uniform density in which a flexible mold containing the powder is sealed in an impermeable envelope and subjected to high and equal pressures from every side.

isostere. A graphic presentation of equilibrium concentration or pressure against temperature when the quantity adsorbed per unit of a material, such as activated carbon, is held constant.

isosteric. Of equal or constant specific volume with respect to either time or space.

isotherm. (1) A line on a chart representing the relationship or changes in volume or pressure at constant temperature.

(2) A plot of the quantity of a material adsorbed per unit of another material, such as activated carbon, against equilibrium concentration or pressure at constant temperature.

isotherm, Freundlich. An equation which states that the volume of gas adsorbed on a surface at a given temperature is proportional to the pressure of the gas raised to a constant power as determined by the formula x/m equals KC^n in which x is the quantity adsorbed, M is the quantity of the adsorbent, C is the concentration, and K and n are constants.

isotherm, Langmuir. A plot of isothermal adsorption data by the formula $f = ap/(1 + ap)$ in which f is the fraction of surface covered, p is the pressure, and a is a constant.

isotone. Nuclides having the same number of neutrons but a different number of protons in their nuclei.

isotope. Nuclides having the same atomic number (number of protons) but a different mass number (number of neutrons).

isotopic assay. The determination of the percentage, by weight or by atoms, of isotopic components in source or special nuclear materials.

isotopic composition. The relative amounts of the various isotopes of an element in a sample or material, expressed as atom or weight percent.

isotopic ratio. The relative amounts of two isotopes of an element in a sample or material, expressed as a ratio.

isotropic dielectric. A dielectric polarized in the direction parallel to an applied electric field, and a magnitude which does not depend on the direction of the electric field.

Italian red. One of the several shades of ferric oxide red pigments.

Izod impact test. A measure of the impact strength of a material in which the height of a pendulum swing after striking a specimen, usually notched, is reported as the energy required to fracture the specimen.

J

jack. The model from which working molds are made. Also known as case molds.

jack arch. A sprung arch in which the outer and inner surfaces are constructed along horizontal planes, or in which the inner surface is constructed with a relatively large radius.

jack brick. A type of refractory brick employed as the base on which glass-melting pots are placed, and which is designed with openings or holes to accommodate the fork of a fork-lift truck or similar device for easy transport of the pots from one location to another.

jacket. A reinforced covering providing environmental and mechanical protection for the insulation, core, shield, or armor of a cable.

jacket, primary. A layer of insulating material applied over the primary insulation of electrical wire to provide protection from mechanical damage.

jamb. The vertical structural member forming the side wall of the opening or port of a furnace superstructure carrying the port crown load.

jamb brick. A brick modified so that the corner of one end and side is rounded to provide a radius approximately equal to the width of the brick; used to construct curved walls and other curved structures.

jamb wall. (1) The side wall of a furnace or kiln between the flux block and crown, but not including the ends.
(2) The refractory wall between the pillars of a glass-melting pot furnace and in front of or surrounding the front of a pot.

jam-socket machine. A machine designed to shape the sockets of clay sewer pipe.

Japanese porcelain. A porcelain similar to Chinese porcelain, but fired at a lower temperature to provide a softer but better-appearing finish.

jar crusher. See jar mill.

jardiniere glaze. A type of unfritted glaze, either hard or soft, containing the oxides of lead, aluminum, calcium, potassium, silicon, and zinc; used as a decorative glaze on products such as flower pots.

jar mill. A small rotating closed cylinder of porcelain or porcelain-lined steel containing pebbles or porcelain balls, and in which materials are ground or blended; a laboratory mill.

jasper ware. A vitreous, opaque, colored, unglazed stoneware developed by Josiah Wedgwood, and which is characterized by relief decorations of white or contrasting colors and containing a substantial amount, approximately 50%, of barite, $BaSO_4$.

jaw breaker. See jaw crusher.

jaw crusher. A crushing or fragmenting machine consisting of a moving jaw, hinged at one end, which swings toward and away from a stationary jaw in a regular oscillatory cycle; in some designs, both jaws may be actuated.

Jena glass. An early variety of chemically and sometimes thermally resistant optical glasses having good resistance to thermal and mechanical shock.

jeroboam. A wine bottle having a capacity of about ⅘ of a gallon.

jet. A strong well-defined stream of gas or fluid emanating from an orifice or moving in a contracted duct.

jet dryer. A dryer in which ware is dried by jets of warm air, steam, or both, injected into the drying chamber.

jet mill. An efficient and effective milling device, producing solids of extremely small and frequently controlled sizes, in which the particles are actuated by high-pressure air, steam, or other medium and are fragmented by mutual collisions at high speeds.

jet nozzle. A specially shaped refractory nozzle employed in the production and exhaust of extremely high-temperature jet streams.

jet ware. A pottery-type ware fabricated from a red-clay body and coated with a black, manganese-bearing glaze.

jewel. In the ceramic context, a synthetic alumina-bearing gem, such as a ruby or sapphire, which is used as a bearing material in watches and delicate instruments.

jewelry enamel. A specially formulated porcelain enamel, frequently melting at temperatures lower than those of conventional porcelain enamels, employed on gold, silver, iron, etc., in the manufacture of jewelry, art objects, insignia, and similar products.

jig. A device employed to hold and position work during manufacture or assembly.

jigger, jiggering. A mechanically operated device similar to a potter's wheel on which ceramic ware is formed from a plastic body by the differential rotation of a profile tool and mold, the mold having the contour of one surface, and the profile tool having the contour of the other surface of the ware.

jiggerman. (1) The operator of a jigger.
(2) The workman who returns the glass residue from a ladle to the charging end of a glass-melting tank.

jig, hand. A moving-screen jig operated by hand and used to treat small batches of ore; the jig box is attached to a rocking-type beam and is moved in a tank of water.

jigging. A process for the separation and beneficiation of concrete aggregates and other particulate materials on the principle of hindered settlement, and in which the aggregate is passed over a perforated plate in a tank of water and subjected to vertical pulsations by air jets or by vibrating diaphragms; lightweight materials are floated off and discarded.

job-cured specimen. A specimen of concrete cured at the site of use and under the presumed same conditions to which the commercial concrete installation is, or will be, exposed; such specimens may be tested or retained for future reference.

job shop. A factory which produces parts or ware for use or for sale by another organization, frequently under the purchasing organization's trade name.

jockey pot. A glass-melting pot of such size and shape that it may be supported in a furnace by two other pots.

joggle. A plaster or brass insert serving as a key to insure the correct alignment and adjustment of two halves of a plaster mold. Also known as a natch.

joint. (1) The point, position, or surface at which two or more things, such as mechanical or structural components, are joined.

(2) The interstice between masonry units.

(3) A connection between two pipe sections, made either with or without the use of additional parts.

joint assembly. An assembly of dowels and supporting framework for holding the dowels in place during the placing of concrete, particularly in the construction of pavements.

joint, cold. The surface between two successive runs of concrete in which the first concrete placed has passed into its final set and no longer can be blended into the second run.

joint, construction. A plane surface between two increments of concrete, the second increment being placed on or abutted to the first after the first has hardened.

joint, contraction. A groove formed in fresh concrete, or sawed in hardened concrete, to create a plane of weakness in a slab or panel so as to cause the concrete to crack at the weakened plane instead of at random during drying or in service.

joint, control. An expansion or contraction joint, which see.

joint, dipped. A thin joint made by dipping brick, before laying, into a mortar of thin consistency or by pouring a thin mortar over a course of brick before laying the next course.

joint, dummy. A type of preformed contraction joint employed in a concrete pavement where movement will cause a crack to form in a determined location and in a prescribed pattern instead of an undesired or random pattern.

joint, expansion. A joint in an installation of concrete, sometimes filled with an elastic material, which will permit expansion, contraction, or other movement without the development of excessive stresses or damage to the installation.

joint, filler. Premolded strips of a bituminous material or asphalt cement containing a filler, self-expanding cork, fibrous material, sawdust, felt, or similar material saturated with a bituminous substance, which are manufactured in suitable dimensions and inserted in concrete or other joints to permit movement of the joint without damage to the structure.

joint, hinge. A joint in concrete pavement which permits movement of adjacent slabs, thereby minimizing breakage.

jointing. (1) The filling or caulking of masonry joints.

(2) The process of striking, slicking, or raking the joints between masonry units to provide a desired surface appearance and to improve the tightness and strength of the joint.

jointing yard. An area situated between the grinding and polishing operations in the continuous manufacture of plate glass in which plaster joints holding the glass are remade. Also known as a laying yard.

joint, lap. A simple joint between two items or sheets at the point where one sheet overlaps the edge of the other, as in roofing tile.

joint line. The seam, mold mark, or line reproduced on glass and cast ceramic ware by the joint between two mold parts.

joint, warping. A hinge joint in concrete pavement designed to allow movement of adjacent slabs, and minimize uncontrolled breakage.

joint, weakened-plane. A groove formed in fresh concrete, or sawed in hardened concrete, to provide a line of weakness in a slab or panel which will cause the concrete to crack in a prescribed pattern instead of at random.

jolly, jollying. The process of forming or shaping ceramic holloware by means of a machine in which a rotating plaster mold has the contour of the bottom surface and a profile tool, lowered on the body from an otherwise stationary position, forms the other surface of the body.

jolt molding. A process for the shaping of refractory forms in which the plastic body is subjected to mechanical or manual jolting or jerky movements; a mold plate actuated under pressure sometimes may be employed to shape the top of the body.

jug. An earthenware, glass, or metal container for liquids, usually having a short neck and small mouth, stopper, and handle.

jumbo brick, jumbo block. A generic term indicating a brick or building block larger than standard in size; sometimes produced to specifications. See brick, standard.

jumper, jumping. A defect occurring in porcelain enamel ground coats characterized by the spontaneous popping of relatively small circular-shaped flakes of ground coat from the ware. The defect usually appears in random areas, and may appear in the ground coat or the first cover-coat of porcelain enamel on sheet steel, particularly on a radius or edge.

K

kaki. A reddish-brown, opaque stoneware glaze resembling the color of persimmon; produced when a layer of iron oxide crystals spreads over the surface of the glaze.

kaolin. A group of refractory white or nearly white-burning clays having the approximate compositon, $Al_2O_3 \cdot 2SiO_2 \cdot 2H_2O$ plus small amounts of alkalies and iron; sp. gr. 1.28–2.6; pyrometric cone equivalent 34–35; employed in ceramic bodies to impart high strength, plasticity, and workability during forming; used extensively as a body and glaze ingredient in a wide variety of products, including cements.

kaolin, calcined. Heat-treated kaolin in which mullite, $3Al_2O_3 \cdot 2SiO_2$, and an amorphous siliceous material are the major components; sp. gr. 2.67; softening point above cone 35, 1770°C (3218°F), with initial deformation being observed between cones 34 and 35, 1750°–1770°C (3182°–3218°F). Employed in refractories to impart resistance to corrosion by slags, glasses, glazes, and porcelain enamels, high-mechanical strength, and resistance to conditions of thermal shock.

kaolin, Florida. A ball-type, very clean, uniform, sedimentary kaolin of high purity, fine particle size, and white-burning properties; employed to promote refractoriness, plasticity, workability, high-bonding strength, and suspending power in a variety of ceramic products.

kaolinite. Hydrated aluminum disilicate, $Al_2O_3 \cdot 2SiO_2 \cdot 2H_2O$; the most common constituent component of kaolin and a constituent in most clays.

kaolinization. Conversion of aluminum silicate and other clay minerals into kaolin by weathering.

Kavalier glass. A high-potash-bearing, chemical-resistant glass.

Keene's cement. Anhydrous, calcined gypsum; a hard, white, finish plaster containing additions of materials, such as potassium sulfate or potash alum to accelerate the set.

Kelley consistency test (concrete). A test in which a metal ball of prescribed dimensions and weight is placed on the surface of freshly prepared concrete and the depth of penetration is taken as a measure of the consistency of the mix.

Kelvin temperature scale. An absolute temperature scale in which the degree intervals are the same as those of the Centigrade (Celsius) scale, and 0°K is the equivalent of −273.16°C, or absolute zero.

keratin. A protein extract from hair and horny substances employed as an addition to retard the set of plaster.

kerf. A slit or notch made in a body by a saw or cutting torch.

Kerr constant. The difference between usual and unusual indices of refraction divided by the product of the wavelength and the square of the electric field.

Kerr effect. The double refraction of light in glass and other substances produced by an electric field.

Kessler abrasion tester. A testing device designed to evaluate the resistance of surfaces to abrasion in which corundum of specified size is fed at a specified rate between the surface of an inclined specimen and a notched revolving steel wheel suspended so as to provide a constant, specified weight on the specimen.

Ketteler-Helmholz equation. A formula for calculating the optical dispersion of glass stated

$$n^2 = n_a{}^2 + \Sigma\, Mm\, (\lambda^2 - \lambda^2 m)^{-1}$$

in which n is the refractive index for a wavelength λ, n_a is the index for an infinitely long wavelength, and $\Sigma\lambda m$ are the wavelengths of the absorption bands for each of which there is an empirical constant, Mm.

kettle. (1) A container for molten glass.

(2) A metallic container in which gypsum is converted to plaster of paris.

key. (1) An elevation or depression formed in a concrete joint surface to provide shear strength across the joint.

(2) A device designed to lock mechanical or structural parts together.

key brick. A wedge-shaped brick placed at the crown of an arch to close and tighten the arch.

K feldspar. A potassium-bearing feldspar, $KAlSi_3O_8$; see feldspar.

kibble. To grind or divide materials into relatively large particles or pellets.

kibbler rolls. Toothed rolls used in roll-crushing devices to reduce clays and other minerals to sizes and shapes more amenable to further grinding and use.

Kick's law. The energy required to crush a solid substance to a specified fraction of its original size is the same regardless of the original size of the feed material.

kick wheel. A potter's wheel operated by a foot pedal.

kidney. A kidney-shaped instrument of rubber, plastic, polished wood, or leather used to smooth the surface of pressed, unfired ceramic bodies.

kieselguhr. Diatomaceous earth; a finely divided sedimentary siliceous material composed essentially of the skeletal walls of diatoms; employed as a filtration medium, abrasive, and aggregate for lightweight concrete, and as a component in brick.

kiln. A structure in which a material or product is fired, calcined, or otherwise subjected to elevated temperatures.

kiln, ACL. A traveling-grate preheater employed to reduce kiln time in the production of portland cement by preheating limestone and clay mixtures before introduction into rotary cement kilns.

kiln, annular. A kiln in which ware is placed in stationary compartments, and the firing zone travels around the kiln by advancing the zone in which the fuel is introduced.

kiln, archless. An updraft kiln, having no permanent parts, which is constructed with walls of either burned or unburned brick; after loading, the kiln is covered with brick, earth, or ashes and fired with solid, liquid, or gaseous fuels.

kiln, bank. A kiln constructed on a slope or bank of earth, the incline serving in the place of a flue for the exhaust of combustion and other gases.

kiln, Belgian. A longitudinal arch kiln which is side fired to grates spaced at regular intervals along the bottom.

kiln block, rotary. A modified circle brick, usually with a 9-inch chord and a smaller inside chord, 6 or 9 inches in radial length and 4 inches thick, used to line rotary and circular kilns.

kiln, bottle. An updraft kiln shaped in the form of a tapered bottle, the neck of the bottle serving as a flue.

kiln, Bull's. A clamp kiln in which brick are placed and fired in trenches below ground level. See kiln, clamp.

kiln car. A movable carriage or truck with one or more platforms on which ware is placed for transport through a kiln.

kiln, cement. A kiln, usually of the rotary type, in which limestone and clay are fused or clinkered to produce portland cement.

kiln, chamber. A kiln consisting of one or more separate compartments sometimes arranged so that each compartment may be heated separately for the firing of ware.

kiln, circular. A tunnel kiln shaped in the form of a circle in which the loading and unloading stations are side by side; that is, the entrance and exit of the kiln are in the same location.

kiln, clamp. A periodic, updraft, open-top kiln of semipermanent construction which is similar to a scove kiln except that it has walls containing fire arches laid up with scove brick.

kiln, continuous. A kiln into which ware is fed and passed through continuously, usually without interruption.

kiln, continuous-chamber. A chamber kiln in which the arched roof is constructed in a position transverse to the length of the kiln.

kiln cycle. The time and temperature conditons employed in a firing operation.

kiln, direct-fire. A kiln having neither recuperator nor regenerator, and in which the products of combustion come into direct contact with the ware. See recuperator, regenerator.

kiln, down-draft. An enclosed, periodic kiln, round or rectangular, in which hot gases from the fireboxes pass to the crown and then are pulled downward through the ware by the draft and discharged into a stack.

kiln, dry. A kiln designed to dry greenware at the lowest possible heat before it enters the firing zone.

kiln, elevator. A kiln in which the ware is placed on a refractory pallet and then raised into a refractory fire zone positioned immediately above the pallet.

kiln, envelope. (1) A kiln in which the firing zone, positioned immediately above a loaded refractory platform, is lowered to surround the ware being fired; also known as a top-hat kiln.

(2) A box-type kiln in which the ware to be fired is placed on kiln cars and pushed into the entrance end of the kiln, displacing cars of fired ware at the exit end.

kiln furniture. Small refractory shapes, such as stilts, pins, spurs, cranks, saddles, etc., and slabs, posts, and setters of various sizes and shapes upon which ware is placed for firing.

kiln, groundhog. A kiln constructed on a hillside or bank of earth.

kiln, Hoffman. An efficient, periodic, multichamber kiln in which the chambers are so constructed that combustion gases and cooling air may be used to dry and preheat the ware and subsequently fire the ware with a minimum loss of heat.

kiln, intermittent. Any kiln of the batch type in which the ware is placed, fired, cooled, and removed before placement of a subsequent charge.

kiln, lime. A furnace, frequently a long, tilted, rotating cylinder, in which calcium carbonate is heated to temperatures above 900°C (1652°F) to produce lime.

kiln, longitudinal-arch. A kiln in which the arch extends parallel to the length of the kiln.

kiln marks. Deformation of a brick resulting from slumping of the brick under load during firing.

kiln, muffle. A kiln in which combustion of the fuel takes place in refractory muffles which, in turn, conduct heat into the chamber in which ware is being fired, thus protecting the ware from the influence of the combustion gases.

kiln, multipassage. A kiln consisting of more than one tunnel or passage for the concurrent firing of ware.

kiln, periodic. Any kiln which must be loaded, fired, cooled, and unloaded to complete a firing cycle before it is loaded and fired again.

kiln, pusher. A kiln, usually small, in which ware placed on a suitable platform is pushed, manually or mechanically, through the firing zone singly or one after the other.

kiln, roller-hearth. An efficient tunnel-type kiln through which ware is transported by means of rollers.

kiln, rotary. An inclined, tube-like, refractory-lined kiln or furnace which is rotated about its horizontal axis and fired by a burner placed at one end; used for melting, clinkering, or calcining materials.

kiln, round. A periodic kiln with a circular firing chamber with a series of fireboxes stationed around the periphery of the structure. See circular kiln.

kiln run. Brick, tile, or other product from a kiln which has not been sorted or graded for size, uniformity, color variation, or other property.

kiln, sandwich. A kiln in which heat is supplied to ware from both top and bottom simultaneously.

kiln, scove. An updraft kiln, usually having no permanent parts, constructed of unfired bricks, and fired with gaseous, liquid, or solid fuels.

kiln scum. Discoloration of the surface of a body, such as brick or roofing tile, caused by the migration of soluble salts from the interior to the surface of the body or by the reaction of kiln gases with surface constituents during the drying and firing operations.

kiln, shaft. A furnace for heating lump, granular, or powdered materials consisting essentially of a vertical, refractory-lined shaft or a refractory cylinder; material is fed into the top of the kiln, passing through hot gases flowing upward from a flame positioned near the bottom, and emerging as a calcined product at the bottom of the kiln.

kiln, shuttle. An intermittent box or batch-type kiln sealed at each end by refractory-lined doors. Loaded cars are pushed into the kiln from one end and fired. At the conclusion of the firing operation the cars are displaced by loaded cars introduced from the opposite end of the kiln. The process is continued in a shuttling fashion.

kiln, sliding bat. A kiln in which ware placed on a refractory platform, or bat, is caused to slide through the firing zone by manual or mechanical pushing.

kiln, slip. A modification of a furnace or kiln consisting of suitable containers to permit the use of waste heat to dry or reduce the water content of slurries.

kiln, smother. A kiln into which smoke can be introduced as a reducing agent for the blackening of pottery.

kiln, top-fired. A kiln in which the fuel is fed into the firing zone through apertures in the kiln roof.

kiln, top-hat. A kiln in which the firing zone, placed immediately above ware positioned on a refractory base, is lowered to surround the ware to be fired.

kiln, transverse arch. A chamber kiln in which the arch is constructed in a fashion transverse to the chambers or length of the kiln.

kiln, truck chamber. A chamber-type kiln through which ware is pushed on refractory platforms, or bats.

kiln, tunnel. Any tunnel-shaped furnace through which ware is transported on cars, passing consecutively through preheating, firing, and cooling zones in a continuous operation.

kiln, tunnel updraft. A tunnel kiln in which air and combustion gases are caused to move upward through the kiln setting to the exhaust flues.

kiln wash. A coating, usually consisting of refractory clay and silica, applied to the surfaces of kilns and kiln furniture to protect them from volatile glazes or glaze drops from ware being fired.

kiln white. A white scum which forms on the surfaces of brick and roofing tile during firing as a result of dryer scum or kiln atmosphere.

kiln, zig-zag. A kiln in which the path of the combustion gases is caused to flow in a zig-zag pattern by the staggered placement of baffles or dividing walls in the structure.

kindling point. The lowest temperature at which a material will ignite and continue combustion.

kinematic. Abstract motion without reference to mass or force.

kinetic energy. The energy possessed by a body in motion equal to one-half its mass times the square of its speed.

king closer. A brick cut diagonally, having one 2-inch end and one full-width end.

king's blue. A blue ceramic colorant composed essentially of cobalt and aluminum oxides.

kink. A type of waviness occurring from the interior surface to the edges of the surface of a coating.

kiss. Accidental contact between two glazed ceramic items in a kiln during firing, resulting in damaged glaze surfaces at the points of contact.

klebe hammer. An instrument for preparing standardized compacts of mortars and cements for use in mechanical strength determinations in which a standard weight is dropped on the material to be tested from a specified height to insure compaction to a specified density.

Klein turbidimeter. An apparatus and procedure for determining the specific surface of portland cement in which the turbidity of a sample suspended in castor oil is measured photoelectrically and the results are compared with a calibrated curve.

kneading. The manual or treading process of mixing and working plastic clay and similar materials and masses to a homogeneous texture.

knives. (1) Sharp metal blades or specially shaped knives used to advance and blend clay and water in a pug mill.

(2) Sliding blades or bars, situated a specified distance above and parallel to surfaces to be coated, employed to spread coatings of uniform thickness on the items.

knocking. The accidental chipping of glaze from a body before firing.

knockout. A piece of pressed glass or other material designed so that it may be knocked out of an item to form a hole, usually of a specified dimension.

knockup. The oversized residue remaining on a screen after ceramic slip has been screened.

Knoop hardness. A scale of the relative hardness of a material based on the depth of indentation made in a material by the diamond point of a Knoop indenter. See Knoop indenter; diamond pyramid test.

Knoop indenter. A hardness-testing instrument containing a diamond having a rhombic base with diagonals in a 1:7 ratio and included apical angles of 130° and 172°30′. See Knoop hardness.

knot. An imperfection in glass resulting from an inhomogeneity in the form of a vitreous lump of a composition different from that of the surrounding glass.

knuckling. To throw on a potter's wheel, using the knuckles of the hand on the outer surface of the body being shaped.

konimeter. An air-sampling device for measuring dust in the atmosphere of cement plants and other industrial areas in which a measured volume of air drawn through a jet is collected on a glycerine-jelly-coated glass, and the particles are counted by microscope.

kordofan gum. See gum arabic.

kovar. An iron-nickel-cobalt alloy used in glass-to-metal seals.

Kozeny-Carmen equation. An equation for streamlining the flow of fluids through a powdered bed. See Kozeny equation.

Kozeny equation. A mathematical relationship of flow network permeability to capillary pore dimensions; expressed as

$$f = K \cdot \frac{AV^3}{\mu S^2} \cdot \frac{\Delta P}{L}$$

in which f is the rate of fluid flow, μ is the fluid viscosity, L is the depth of the packed powder bed, A is the area of the bed, ΔP is the pressure difference, S is the specific surface area of the particles, and V is the unit of voids per unit of weight of the bedded particles.

Kramers-Kronig relation. A relation between the real and imaginary parts of the index of refraction of a substance, based on the causality principle and Cauchy's theorem.

Kramer's theorem. The theorem that a system consisting of an odd number of electrons in an external electrostatic field is at least two-fold degenerate.

Kreüger's ratio. A measure of the frost resistance of building brick based on the ratio of the 4-day cold water absorption of the brick to its calculated total water absorption.

Kronig-Penny model. An idealized one-dimensional model of a crystal in which the potential energy of an electron is an infinite sequence of periodically spaced square wells.

Kruetzer roof. A furnace roof characterized by an arrangement of transverse and longitudinal ribs which gives the appearance of box-like compartments.

Krupp ball mill. A grinding device consisting of chilled iron or steel balls grinding against each other in a die ring of perforated spiral plates, each overlapping the next; the ground material is discharged through a cylindrical screen.

kryptol furnace. A furnace in which heat is generated by passing an electric current through a rammed refractory (consisting of a mixture of graphite, silicon carbide, and clay) of high electrical resistance.

Kühl cement. A portland cement type of cement in which 7% each of alumina and ferric oxide replace a portion of the silica.

kyanite. $Al_2O_3 \cdot SiO_2$; a natural silicate of aluminum; approx. mol. wt. 162.0; sp. gr. 3.56–3.66; hardness (Mohs) 5 along the length of the crystal and 7 at right angles to this direction; has the same composition as sillimanite and andalusite, but differs in crystal structure and physical properties; decomposes to mullite and cristobalite at about 1300°C (2372°F) with a decrease in specific gravity to about 3.0 and a volume expansion of about 10%; because of its mechanical strength and resistance to elevated temperatures, it is employed in refractories, sanitary porcelains, precision-casting molds, brake disks, wall tile, electrical porcelains, filters, and similar products.

L

labradorite. A lime-soda feldspar consisting of mixtures of materials of the general formulas $CaO \cdot Al_2O_3 \cdot 2SiO_2$ and $Na_2O \cdot Al_2O_3 \cdot 6SiO_2$.

lacing. A course of upright brick forming a bond between two or more arch rings.

lacquer. A glossy and quick-drying surface coating composed of natural or synthetic cellulose esters or ethers which dry by solvent evaporation.

ladder, rock. A series of inclined steps, arranged in a vertical column, designed to prevent breakage of coarse aggregate during its discharge from a conveyor or chute by controlling the distance of the aggregate fall.

ladle. A deep bowled, long-handled, spoon-like tool used to dip up, transport, and pour molten liquids, such as glass or metal; also used to fill open pots with materials to be melted.

ladle brick. A refractory brick of appropriate shape, uniform size, low porosity, and relatively permanent expansion for use in ladles for the containment of molten metal.

ladle, teapot. A type of ladle containing a refractory dam under which molten metal flows to prevent slag from reaching the ladle spout.

Lafarge cement. A nonstaining white or near white cement containing lime, plaster of paris, and marble powder; used as a mortar and grout in the setting of marble, granite, and limestone. Also called grappier cement.

lagging. Materials, such as asbestos and kieselguhr, used to insulate kilns.

laitance. A weak, light gray material, consisting essentially of cement, water, and clay or silt, formed on the surface of concrete during and immediately after consolidation, particularly when an excess of water is mixed with the cement; a form of bleeding.

Lambert's law. The ratio of the intensity of emergent light to incident light is an exponential function of thickness of the ware and a constant, depending on the nature of the ware.

lamellar clay. Clay exhibiting microscopic disk-like formations; a characteristic of plastic clays.

laminar. Arranged in or consisting of thin plates or scales. See laminate, laminated glass.

laminate. The product or process in which thin plates or sheets, such as glass or other material, are bonded together to form a panel of greater thickness for a particular use, for example, safety glass, laminated electrical contacts, laminated transformer cores, etc.

laminated glass. (1) A transparent safety glass in which two or more glass sheets are bonded together with intervening layers of plastic materials so that, when broken, the glass will tend to adhere to the plastic rather than fly.

(2) A diffusing glass formed by sandwiching a plastic-bonded glass fiber between sheets of ordinary glass. See safety glass.

laminations. Planes or contours of weakness which may develop in a structural shape during forming.

Lamotte comparator. An instrument employed to determine the relative acidity or alkalinity of pickling solutions used in preparing sheet metals for porcelain enameling by comparing the pH of the solutions with appropriate standard solutions.

lampblack. A black pigment of almost pure carbon made by burning carbonaceous materials with insufficient air; used in cements, ceramic ware, mortar, and thermal insulating compositions.

lampworking. A forming of glass articles from glass tubing or cane by manipulation in a gas flame.

Lancaster mixer. A counter-current, pan-type mixer which may be designed with various combinations of mullers, plows, doctor blades, and scrapers.

Langmuir adsorption theory. The surface of an adsorbent has only uniform energy sites, and adsorption is limited to a monomolecular layer.

Langmuir isotherm. A plot of isothermal adsorption expressed as $f = ap/(1 + ap)$ in which f is the fraction of surface covered, p is the pressure, and a is a constant.

lanthanum aluminide. (1) LaSb; mol. wt. 260.67; m.p. 1582°C (2800°F). (2) La_3Sb_2; mol. wt. 660.24; m.p. 1688°C (3070°F).

lanthanum boride. (1) LaB_3; mol. wt. 171.36; sp. gr. 4.92. (2) LaB_4; mol. wt. 182.18; sp. gr. 5.44. (3) LaB_6; mol. wt. 203.82; m.p. 2149°C (3900°F); sp. gr. 4.72; hardness (Vickers) 2770. See borides.

lanthanum carbide. (1) LaC_2; mol. wt. 162.90; m.p. 2438°C (4420°F); sp. gr. 5.00–5.35. (2) La_2C_3; mol. wt. 313.82; m.p. 2021°C (3670°F); sp. gr. 6.08. See carbides.

lanthanum ferrite. $La_2O_3 \cdot Fe_2O_3$; mol. wt. 485.80; m.p. ~ 1871°C (3400°F). See intermetallic compound.

lanthanum hafnate. $La_2O_3 \cdot 2HfO_2$; mol. wt. 747.0; m.p. ~ 1871°C (3400°F). See intermetallic compound.

lanthanum molybdate. $La_2(MoO_4)_3$; mol. wt. 757.8; m.p. 1181°C (2159°F); sp. gr. 4.77.

lanthanum nitrate. $La(NO_3)_2 \cdot 6H_2O$; mol. wt. 371.06; m.p. 40°C (104°F); b.p. 126°C (259°F); used in the production of gas mantles.

lanthanum nitride. LaN; mol. wt. 152.91; sp. gr. 6.85. See nitrides.

lanthanum oxide. La_2O_3; mol. wt. 325.84; m.p. 2000°C (3632°F); sp. gr. 6.51; used in optical glass and incandescent gas mantles.

lanthanum phosphide. LaP; mol. wt. 169.93; sp. gr. 5.22.

lanthanum-ruthenium. $LaRu_2$; mol. wt. 342.30; m.p. 1427°C (2600°F). See intermetallic compound.

lanthanum silicate. (1) $La_2O_3 \cdot SiO_2$; mol. wt. 385.86; m.p. 1929°C (3505°F); sp. gr. 5.72; hardness (Mohs) 5–7. (2) $2La_2O_3 \cdot 3SiO_2$; mol. wt. 831.78; m.p. 1749°C (3180°F); sp. gr. 4.85.

lanthanum silicide. La_2Si; mol. wt. 305.86; m.p. ~ 1521°C (2770°F); sp. gr. 5.14. See silicides.

lanthanum sulfide. (1) La_2S_3; mol. wt. 374.02; m.p. 2100°C (3812°F); sp. gr. 4.91. (2) LaS; mol. wt. 170.96; m.p. 1971°C (3580°F); sp. gr. 5.86. (3) La_3S_4; mol. wt. 544.94; m.p. 2099°C (3810°F); (4) LaS_2; mol. wt. 203.02; m.p. 1649°C (3000°F); sp. gr. 4.90.

lanthanum titanate. (1) $LaTiO_3$; mol. wt. 234.82. (2) $La_2Ti_3O_9$; mol. wt. 565.54.

lanthanum trifluoride. LaF_3; mol. wt. 195.92; used with neodymium oxide in laser systems.

lap. (1) An imperfection in glass consisting of a fold in the surface of an article caused by improper flow during forming.

(2) A rotating abrasive wheel or disk used for polishing glass, metal, stone, and other surfaces.

(3) An overlay of an item over the edge of another.

lap cement. A cementitious material used to seal the side and end laps of corrugated roofing.

lap, head. The shortest distance between the lower edge of an overlapping shingle or sheet and the upper edge of the lapped unit in the second course below; that is, the unexposed or protected portion of the shingle.

lapidary. A person engaged in the art of polishing the surfaces of solid substances, such as gems or ceramics and metals for observation and visual examinations.

lap joint. A simple joint between two items or sheets at the point where one sheet overlaps the edge of the other, as in roofing tile.

lapping. The finish-grinding or polishing operation on the surface of a solid by the use of abrasive grains usually contained in a liquid carrier or medium.

lap, side. The shortest horizonal distance between the exposed side edge of a course of asbestos-cement roofing or siding and the most proximate underlying area of roof deck or side wall not covered by the preceding adjacent course.

lap, top. The shortest distance between the lower edge of an overlapping shingle or sheet and the upper edge of the lapped unit in the first course below.

large 9-inch brick. A rectangular brick having a width 50% greater than a standard 9-inch brick; that is, a width of approximately 6¾ inches.

large calorie. A unit of heat equivalent to 1000 calories.

larnite. βCa_2SiO_4; mol. wt. 172.2; stable from 520° to 670°C (968° to 1228°F); a meta stable, monoclinic phase of calcium orthosilicate.

laser. A device that amplifies light to produce an intense monochromatic beam by the stimulation of atoms in the beam.

laser glass. A fluorescent glass which can amplify electromagnetic radiation by the stimulated emission of radiation.

laser, infrared. A laser which emits infrared radiation.

laser materials. Doped single crystals, ruby, etc., used in lasers for drilling, machining, surgery, and many similar applications. See laser.

laser semiconductor. A diode laser in which stimulated

emission of coherent light occurs at the pn junction when electrons are driven into the junction by carrier injection, electron-beam excitation, impact ionization, optical excitation, and other such means.

laser, sun-pumped. A continuous-wave laser in which the energy of the sun is focused on the laser material.

latent heat. The amount of heat absorbed or evolved per unit of mass of a substance during a change of state at constant temperature; for example, the change of a solid to a liquid or a liquid to a gas or vapor, and vice versa.

latent heat of fusion. The increase in enthalpy accompanying the conversion of a unit mass of a solid to a liquid at its melting point at constant pressure and temperature.

latent heat of vaporization. The quantity of energy required to evaporate a unit mass of a liquid at constant pressure and temperature.

laterite. A weathered material composed of the oxides of alumina, iron, titanium, and manganese such as or similar to bauxite; sometimes used as a glaze colorant producing reds, yellows, browns, blacks, and grays, depending on the glass composition and firing atmosphere.

lath. A material in sheet form employed as a base for plaster or tile on walls and ceilings in buildings.

lath brick. A long, slender brick.

lathe. A machine in which a workpiece is held in a holding device and turned while being shaped by an appropriate tool.

lattice brick. A hollow, perforated type of building brick used as thermal insulation.

lattice structure. The regular periodic three-dimensional arrangement of atoms forming the structure of a glass.

lattice water. Water which is an integral part of a clay structure, as opposed to interlayer water, and which may be removed by heating in the range of about 450° to 600°C (842° to 1112°F).

launder. An inclined, refractory-lined channel or trough for the conveyance of molten metal, as from the taphole of a furnace to a ladle.

lawn. A fine sheer mesh of metal, natural, or synthetic fibers for use in silk-screen printing and sieves.

layer. The workman who lays plate or sheet glass in plaster on grinding and polishing tables for finishing. See laying yard.

layer, reinforcement. Circumferential reinforcement for concrete pipe that is one bar or wire in thickness.

laying yard. The site where rough plate or sheet glass is laid in plaster to hold it firmly on grinding and polishing tables for finishing.

L-D process. A process for making steel by blowing oxygen upon or through molten iron to remove most of the carbon and other impurities by oxidation.

Lea and Nurse permeability apparatus. A device for the measurement of the specific surface of a powder in which the air permeability of a prepared bed of the powder is determined by the equation

$$S - 14\ [p^3/KV(1 - p^2)]$$

in which S is the specific surface, p is the porosity of the powder bed, V is the kinematic viscosity of the flowing air, and K is a constant.

leach. To dissolve and wash soluble components from a material by passing a liquid, usually water, through the material.

lead. To begin the batching of concrete by introducing a material into the mixer ahead of another material.

lead antimonate. $Pb_3(SbO_4)_2$; mol. wt. 993.18; an orange-yellow powder used as a yellow colorant in glass and overglazes. Also known as Naples yellow, antimony yellow.

lead-barium crown glass. An optical flint glass containing a substantial quantity of barium oxide.

lead bisilicate. $PbO \cdot 2SiO_2$; mol. wt. 343.3; a low-melting stable frit used in lead-bearing glazes to minimize lead solubility.

lead borate. $PbO \cdot B_2O_3$; mol. wt. 292.82; sp. gr. 5.6; used as a flux in low-temperature frits, vitrified colors, conducting coatings, and bonded mica.

lead borosilicate. (1) A mixture of lead borate and lead silicate used in the manufacture of optical glass.

(2) Used also in glazes to produce coatings of low-lead solubility.

lead carbonate. $PbCO_3$; mol. wt. 267.22; decomposes at 315°C (599°F); sp. gr. 6.43; used in glass, porcelain enamels, and glazes as a flux and fining agent.

lead chromate. $PbCrO_4$; mol. wt. 323.23; m.p. 844°C (1552°F); sp. gr. 6.1; used as a flux and colorant.

lead crown glass. An optical flint glass containing a substantial proportion of lead oxide (PbO) to improve light dispersion and brillance.

leaded glass. Windows made from pieces of colored or clear glass held in position by strips of lead having an "H" or "U" cross section.

lead glass. Glass containing a substantial quantity of lead oxide as a flux and to give a high index of refraction, optical dispersion, and surface brilliance for use as optical glass.

lead glaze. A glaze containing lead oxide in a substantial amount as a flux to lower the fusion temperature and viscosity, to improve the flow properties during firing, and to increase the brilliance, luster, smoothness, resistance to water solubility, and resistance to chipping.

lead lanthanum zirconate titanate. A ferroelectric electro-optical ceramic having optical properites.

leadless glaze. A glaze containing only an imperceptible amount of lead in any form.

lead-lined tank. A container lined with sheet lead for the containment of corrosive liquids, such as pickling acids.

lead metaniobate. $Pb(NbO_3)_2$; mol. wt. 489.42; used in defense electronics, thickness gauges, flaw detectors, accelerometers, air-blast gauges, and other instruments requiring dielectric, piezoelectric, or ferroelectric properties.

lead metatantalate. $Pb(TaO_3)_2$; mol. wt. 666.02; a ferroelectric.

lead molybdate. $PbMoO_4$; mol. wt. 367.22; m.p. 1062°C (1949°F); employed with antimony compounds as an adherence-promoting agent in porcelain enamels.

lead oxide, red. Pb_3O_4; mol. wt. 685.66; decomposes between 500° and 530°C (932° and 986°F); sp. gr. 8.32–9.16; used extensively in glass, glazes, and porcelain enamels as a fluxing ingredient.

lead oxide, yellow. PbO; mol. wt. 223.22; m.p. 888°C (1630°F); sp. gr. 9.53; used extensively in table, optical, and electrical glassware to increase density, index of refraction, brilliance, durability, and working properties; also used as a flux ingredient in glazes and porcelain enamels. Known commercially as litharge.

lead poisoning. The prolonged ingestion or absorption of lead into the human body, primarily the lungs, resulting in anemia, colic, inflammation of the peripheral nerves, and brain impairment.

lead selenide. PbSe; mol. wt. 286.2; m.p. 1088°C (1991°F); used in semiconductor applications.

lead sesquioxide. Pb_2O_3; mol. wt. 462.4; used in glass, glazes, porcelain enamels, and ceramic cements.

lead silicate. PbO; mol. wt. 283.28; m.p. 725° to 775°C (1337° to 1427°F); employed in lead-fluxed bodies of high dielectric strength and in lead-fluxed steatite bodies of a wide firing range.

lead silicate, hydrous. $2PbO \cdot SiO_2$; mol. wt. 807.8; sp. gr. 5.8–6.5; used as a substitute for lead carbonate to minimize evolution of carbon dioxide; used as a mill addition because of good dispersion and suspension qualities and freedom from gassing.

lead stannate. $PbO \cdot SnO_2$; mol. wt. 373.91; reduces the Curie peak in barium titanate capacitors and the tendency to depolarize when used in piezoelectrics.

lead sulfate, basic. $PbSO_4$; mol. wt. 526.5; m.p. 977°C (1791°F); sp. gr. 6.92.

lead sulfide. PbS; mol. wt. 239.28; m.p. 1170°C (2138°F); sp. gr. 7.13–7.70; used in semiconductors and in the glazing of clay wares.

lead tantalate. $PbTa_2O_6$; mol. wt. 666.02; Curie temperature 260°C (500°F); a possible electroceramic and ferroelectric.

lead telluride. PbTe; mol. wt. 334.9; m.p. 902°C (1656°F); used as a semiconductor and photoconductor in the form of single crystals.

lead titanate. $PbO \cdot TiO_2$; mol. wt. 303.12; employed as an additive to barium titanate to improve piezoelectric properties.

lead tungstate. $PbWO_4$; mol. wt. 455.22; m.p. 1130°C (2066°F); sp. gr. 8.235.

lead, white. $2PbCO_3 \cdot Pb(OH)_2$; mol. wt. 776; decomposes at 400°C (752°F); sp. gr. 6.14; used in glazes primarily as a fluxing ingredient.

lead zirconate. $PbO \cdot ZrO_2$; mol. wt. 346.44; used as a component in piezoelectric bodies.

lead zirconate titanate. A ferroelectric, electro-optic material having use in piezoelectric bodies.

leakage, magnetic field. The magnetic field that enters or leaves the surface of a part at a discontinuity or a change in section configuration of a magnetic circuit.

leakage, magnetic flux. The excursion of magnetic lines of force from the surface of a test specimen.

leakage, neutron. The escape of neutrons from a reactor.

leak detector. A device for locating holes or cracks in a coating or walls of a vessel. See spark gap inspection.

leak testing. A technique to determine the presence of a void, fracture, or other discontinuity in a coating or body structure in which a penetrant is applied to one surface, and the opposite surface is observed for indications of penetration by the testing solution.

lean cement. A concrete with insufficient cement.

lean clay. A clay of low plasticity and poor green strength.

lean fuel. Fuel low in combustibles; for example, a fuel-air mixture with a low percentage of fuel and a high percentage of air.

lean gas. Fuel gas low in butane and heavier fuel gases with an excess of air at the burner.

lean lime. A lime containing an inordinate amount of impurities and which will not slake readily with water.

lean mix. (1) A concrete of low cement content.
(2) A fuel-air mixture containing a low percentage of fuel and a high percentage of air.

lean mortar. A mortar deficient in cementitious components and which is usually harsh and difficult to spread.

leather hard. Clay that is sufficiently dry and stiff enough to be handled without deformation, but sufficiently damp to be joined to other pieces with slip.

leathery texture. A rough texture on the surface of porcelain enamel which is similar to, but of larger and more coarse pattern, than eggshell.

lecithin. Waxy phosphorous-containing substances which have emulsifying and wetting properties, and which are used to lower surface tension in silk-screen media.

LeFarge cement. See grappier cement.

lehr. A heated oven in which glassware is annealed to reduce residual thermal stresses.

lehr loader. A machine which places and spaces glassware on a continuous lehr belt.

length. (1) The horizontal dimension of a unit in the face of a wall.
(2) The longest dimension of an item.
(3) The extent of a period of time.

lens. (1) A highly polished, highly transparent, defect-free, and appropriately shaped flat piece of glass (or a substance like a plastic) either or both sides of which may be flat or curved as required so as to cause transmitted light rays to converge or diverge to form an image; used in optical instruments such as eyeglasses, microscopes, telescopes, and other such devices.
(2) An architectural term for a translucent or transparent pressed glass unit, which may be square, round, or specially shaped, for use in construction.

lens-fronted tubing. Graduated glass tubing designed for the containment of liquids for use in temperature, pressure, and similar instruments of measurement, but modified so as to magnify the liquid column for easy reading.

lepidolite. $LiF{\cdot}KF{\cdot}Al_2O_3{\cdot}3SiO_2$; m.p. 1170°C (2138°F); sp. gr. 2.9; hardness (Mohs) 2.5–4.0; employed essentially as a flux in opal and flint glasses, porcelain enamels, glazes, and ceramic bodies to reduce the coefficient of expansion, brittleness, and devitrification tendency, and to increase the firing or working range, index of refraction, brightness, and surface hardness.

leucite. $K_2O{\cdot}Al_2O_3{\cdot}4SiO_2$; sp. gr. 2.5; hardness (Mohs) 5.5–6; closely related to nepheline, an essential mineral in nepheline syenite, which see.

level, acceptance. A test level of specified minimum and maximum values or other criteria between which a product is judged to be acceptable.

level, confidence. The stated proportion of times the confidence interval is expected to include the population parameter.

level, cut-off. The value established above and below which a product is rejected or distinguished from other items of the same origin.

level, gate. The rejection level, which see.

level, rejection. A test level of specified minimum and maximum values or other criteria below and above which, respectively, a product is judged to be rejectable or to be distinguished or sorted from acceptable products.

level, test quality. The rejection level, which see.

level, threshold. The rejection level, which see.

levigated abrasive. A fine, chemically neutral, abrasive powder used as a burnishing medium.

levigation. The method of refining clay and other powdered materials by carrying them in a stream of water which deposits the particles at different stages in terms of relative particle size.

Libbey-Owens-Ford sheet process. A method of making sheet glass by bending a vertically drawn sheet over a roll which establishes the definition of the draw.

license, nuclear material. A permit issued by the Nuclear Regulatory Commission or a state to receive, process, store, or ship source or special nuclear material.

lid. A movable cover or top of a container such as a dish, crucible, or other receptacle.

life test. A test, frequently accelerated, to estimate the normal service life of a product.

lift, lifting. (1) A defect characterized by the spontaneous separation of sheet-like pieces of porcelain enamel or glaze from the surface to which it has been applied.
(2) A layer or depth of concrete placed at one time.

lift-off effect. The effect observed in a magnetic test system output due to a change in magnetic coupling between a test system and a probe coil whenever the distance between them is varied.

lift-slab construction. A method in which reinforced concrete floor or roof slabs are cast on the ground floor of a building under construction and raised to final position by means of jacks on top of the building columns.

lift truck. A small truck for lifting and transporting loads.

light. (1) Electromagnetic radiation capable of inducing visual sensation through the eye; that is, the product of the visibility and the radiant power.
(2) The subclass of a lower index of refraction in optical glass.

light, black. Light in the near ultraviolet range of wavelengths (3200 to 4000 Å); shorter wavelengths than those of visible light.

light density. The quantitative measure of film blackening as calculated by the formula *D* equals log I_o/I, in which *D* equals density, I_o equals light intensity incident on the film, and *I* equals light intensity transmitted.

light-extinction method of particle suspension measurement. A technique for measuring the concentration of particles in a suspension by determining the amount of light absorbed from a transmitted beam.

light, monochromatic. Light of a single wavelength.

light oil. Oil having a boiling range of 110° to 210°C (230° to 410°F); used as a lubricant to reduce friction between moving solids in contact with each other.

light-reducing glass. A general term describing flat glass having reduced light transmittance.

lightweight aggregate. A lightweight inert material such as bloated clay, foamed slag, vermiculite, perlite, and clinker used in reinforced concrete and similar products as an aggregate to reduce weight and improve the thermal and sound insulating values of the product.

lightweight concrete. Any concrete made with low density aggregate.

lignin extract. A substance extracted from the wood wall cells of plants; used as a binder in ceramic bodies and glazes.

lignin sulfonate. Salts made from lignin; no melting point, but decomposes above 200°C (392°F); used as a binder in ceramic bodies and glazes.

lime. CaO; mol. wt. 56.0; m.p. 2750°C (4650°F); sp. gr. 3.40; a fluxing agent used in glass, pottery, glazes, and porcelain enamels, and as a component in portland cement, mortar, and plaster.

lime blowing. Chipping or popping of small fragments from the face of building brick due to freezing and thawing of absorbed water or other stress.

lime, burnt. Calcined limestone (CaO·MgO, dolomite), or CaO (calcite), or a mixture of the two.

lime-cement mortar. A masonry mortar composed normally of one part of masonry cement, one or two parts of lime putty or hydrated lime, and five or six parts of sand by volume.

lime crown glass. An optical crown glass containing a substantial quantity of calcium oxide as a fluxing ingredient.

lime, finishing. Any white, plastic hydrated lime suitable for use in finish coat plaster.

lime glass. A glass containing a high percentage of lime, usually in association with soda and silica; widely used in glass products, such as bottles and other such containers and products.

lime, hydrated. Quicklime to which sufficient water has been added to convert the oxides to hydrates.

lime, hydraulic. Calcined limestone which absorbs water without swelling or heating and which produces a cement which hardens under water.

lime kiln. A furnace, frequently a long, tilted, rotating cylinder, in which calcium carbonate is heated to temperature above 900°C (1652°F); to produce lime.

lime matte. A matte glaze caused by the crystallization of calcium silicate during firing and cooling.

lime mortar. A mixture of hydrated lime, sand, and water used in the construction of nonload-bearing interior walls in building construction.

lime pops. The spalling of brick due to the hydration and carbonization of lime particles at or near the surface of the brick.

lime putty. Hydrated lime in plastic form used as an addition to mortar.

lime-slag cement. A cement produced from a mixture of lime and granulated blast-furnace slag.

limestone. A sedimentary rock composed of more than 80% of calcium or magnesium carbonate; used as a source of calcium in glazes; also used as a building stone, in the production of cement, in the smelting of iron ore, etc.

limits, acceptance. The test levels used in sorting that establish the group or classification into which a material under test belongs.

limonite. $FeO(OH) \cdot nH_2O$; sp. gr. 3.6–4.0; hardness (Mohs) 5.0–5.5; used as a yellow to brown ceramic colorant.

Lindemann glass. A lithium-borate-beryllia glass containing no element having an atomic number greater than 8; used in applications requiring high transmission of x-rays.

linear change, permanent. The percent dimensional change in length, based on the original length, of a refractory specimen free of externally applied stresses, after being subjected to a prescribed heat treatment.

linear shrinkage. The reduction in the length of a specimen during drying and firing. See drying shrinkage, linear and firing shrinkage.

linear thermal expansion. The expansion of a body in one direction when subjected to heat.

line, flux. (1) A horizontal line occurring in a refractory pot or refractory-lined melting tank caused by corrosion and abrasion by the molten batch at the interface between the batch, air, and refractory.

(2) The imaginary magnetic lines used as a means of explaining the behavior of magnetic fields, as demonstrated by the pattern produced when iron filings are sprinkled over a sheet of paper laid atop a permanent magnet.

line, metal. (1) The upper surface line of metal or glass in a tank furnace or pot.
(2) The line of contact between the upper surface of the glass and the refractory of a melting tank or pot.
(3) A line of maximum corrosion of the refractory by the glass at the refractory-glass-air-interface.

liner. (1) The shell forming the storage chamber of a refrigerator or other appliance.
(2) The abrasive resistant material forming the lining of a grinding mill.

lining. (1) A coating or layer adhering to or in contact with the interior surface and ends of asbestos-cement pipe and related couplings, the coating or layer being more chemically resistant than the pipe and related couplings.
(2) The material or coating on the interior of an item for decorative or protective purposes, such as the lining of a container, tank, or kiln.

lining, monolithic. A furnace lining without joints, being formed of material which is rammed, cast, gunned, or sintered into place.

line of reinforcement. The circumferential reinforcement of concrete pipe, it being comprised of one or more layers.

lines. Fine cords or strings of molten glass, molten refractory, or partially molten sand; usually occurring on the surface of flat glass as an imperfection.

lines of force, magnetic. See magnetic lines of force.

Linseis plastometer. An instrument employing the tensile strength as a measure of cohesion and the capacity for the relative movement of clay particles without rupture as a measurement of the plasticity of clay.

lintel. A horizontal piece across an opening, such as a window or door, that carries the weight of the structure above it.

lip. The edge or rim of a pot or other hollow ware article, or the part which encircles an orifice.

liquefaction. The conversion of a gas or gaseous mixture to the liquid state by cooling or compression, or both.

liquefied petroleum gas. Any gas derived from petroleum, such as propane and butane, which has been liquefied and stored under pressure in suitable containers for easy transport and future use as fuel.

liquid blast cleaning. The process of cleaning surfaces by means of a high-velocity jet of abrasive suspended in water or other liquid.

liquid gold. An inexpensive gold resinate used in the decoration of ceramic ware.

liquidus. A curve on a graph showing temperature versus concentration of a material or mixture of materials which connects with temperatures at which fusion is completed as the temperature is raised.

liquidus temperature. The maximum temperature at which equilibrium exists between a molten substance, such as glass, and its primary crystalline phase.

litharge. PbO; mol. wt. 223.2; m.p. 888°C (1631°F); sp. gr. 9.3–9.7; a powerful flux, it is used extensively in optical glass, electrical glass, tableware, glazes, glass enamels, and porcelain enamels; it increases density, acid resistance, and brilliance, and lowers fusion, viscosity, and tendency to chip or craze; also used in some ceramic cements.

litharge glass. A soda-lime glass in which part of the calcium is replaced by litharge, which see.

lithia. Li_2O; mol. wt. 29.9; m.p. $>$ 1700°C (3092°F); sp. gr. 2.012; a powerful flux; used in glasses having high electrical resistivity to improve fluidity, working properties, and ultraviolet-ray transmission; used in dinnerware, electrical porcelain, and sanitaryware to improve strength and gloss; used in ceramic bodies and refractory specialties to reduce thermal expansion and improve thermal-shock resistance; used in porcelain enamels to improve workability and reduce firing temperatures.

lithia mica. Lepidolite, $LiF \cdot KF \cdot Al_2O_3 \cdot 3SiO_2$; m.p. 1170°C (2138°F); sp. gr. 2.9; hardness (Mohs) 2.5–4.0; used as a flux and source of alumina in glass, glazes, porcelain enamels, and ceramic bodies.

lithium aluminate. $LiAlO_2$; mol. wt. 65.9; m.p. $>$ 1625°C (2957°F); sp. gr. 2.55; used as a flux in refractory porcelain enamels.

lithium aluminum silicate. (1) $Li_2O \cdot Al_2O_3 \cdot 2SiO_2$; mol. wt. 251.94; m.p. 1398°C (2550°F); sp. gr. 2.36; hardness (Mohs) 5–7. (2) $Li_2O \cdot Al_2O_3 \cdot 4SiO_2$; mol. wt. 372.06; m.p. 1427°C (2600°F); hardness (Mohs) 5–7. (3) $Li_2O \cdot Al_2O_3 \cdot 6SiO_2$; mol. wt. 492.18; m.p. 1183°C (2160°F); sp. gr. 2.41; hardness (Mohs) 5–7.

lithium borosilicate. Used extensively in high-temperature, corrosion-resistant coatings.

lithium carbonate. Li_2CO_3; mol. wt. 73.9; m.p. 735°C (1350°F); sp. gr. 2.111; used as a source of lithium oxide which serves as a flux in ceramic bodies, glazes, and porcelain enamels.

lithium cobaltite. $LiCoO_2$; mol. wt. 97.9; m.p. $>$ 1000°C (1832°F); used in porcelain enamel ground coats to combine the fluxing power of lithium and the adherence promoting properties of cobalt oxide.

lithium-drifted germanium detector. See germanium detector, lithium-drifted.

lithium feldspathoids. A group of minerals including lepidolite, spodumene, and petalite, which see; used in bodies, glazes, and porcelain enamels to reduce thermal expansion and improve thermal-shock resistance.

lithium fluophosphate. $LiF \cdot Li_3PO_4 \cdot H_2O$; mol. wt. 159.8.

lithium fluoride. LiF; mol. wt. 25.9; m.p. 870°C (1598°F); sp. gr. 2.295; used as a flux and minor opacifier in porcelain enamels and glazes, and as crystals in infrared instruments.

lithium magnetite. $Li_2O \cdot MnO_2$; used as a flux in porcelain enamels and in the production of ceramic-bonded grinding wheels.

lithium metaborate dihydrate. $LiBO_2 \cdot 2H_2O$; m.p. 840°C (1544°F); used as a flux in porcelain enamels; also increases tensile strength.

lithium molybdate. Li_2MoO_4; m.p. 705°C (1300°F); used as an adherence agent for white enamels applied directly to steel.

lithium nitrate. $LiNO_3$; mol. wt. 68.95; m.p. 261°C (502°F); used as an oxidizing flux in porcelain enamels, glazes, and glasses.

lithium oxide. See lithia.

lithium silicate. (1) $Li_2O \cdot SiO_2$; mol. wt. 89.94; m.p. 1215°C (2200°F); sp. gr. 2.48; hardness (Mohs) 5–7. (2) $2Li_2O \cdot SiO_2$; mol. wt. 119.82; m.p. 1253°C (2290°F); hardness (Mohs) 5–7. Both used in porcelain enamels as a flux to improve surface texture and as a minor opacifier.

lithium tetraborate. $Li_2O \cdot 2B_2O_3$; mol. wt. 99.5; loses water at 200°C (392°F); used as a flux in glazes and porcelain enamels.

lithium titanate. $Li_2O \cdot TiO_2$; mol. wt. 109.78; used as a flux in porcelain enamels and as a mill addition in glazes.

lithium zirconate. Li_2ZrO_3; mol. wt. 153.1; used as a flux and opacifier in porcelain enamels.

lithium zirconium silicate. (1) $Li_2O \cdot ZrO_2 \cdot SiO_2$; mol. wt. 213.2; a strong flux used in porcelain bodies, porcelain enamels, and glazes. (2) $4Li_2O \cdot 3ZrO_2 \cdot 5SiO_2$; mol. wt. 789.48; m.p. 1154°C (2110°F); sp. gr. 4.02; hardness 5–7.

lithography. A technique for making ceramic decalcomanias in which a design is printed on special paper from a plane surface, such as a smooth stone or metal plate, on which the image to be printed is ink-receptive and the blank area is ink-repellant.

liver, livering. A defect in glazes and dry-process enamels characterized by a wave-like form of abnormally thick coating.

liver spotting. Stains of irregular shape occurring in silica brick as the result of the precipitation of ferric oxide from solution.

lizard skin, snake skin. A decorative mottled glaze having matte or shiny and sometimes colored spots on the surface giving the appearance of lizard or snake skin.

load. (1) The quantity of glass delivered by a furnace during a given period of time, usually 24 hours.
(2) The charge in a furnace or kiln.
(3) The mechanical force applied to a body.

load-bearing tile. A tile used in masonry upon which loads are superimposed during construction and during the life of the completed structure.

load-crushing strength, external. The ability of a concrete pipe or other product to withstand external crushing forces.

load-crushing strength, test. A test of concrete pipe in which external crushing forces are applied in specified directions and locations on a specified length of pipe; reported in pounds per square inch.

loaded concrete. Concrete containing elements of high atomic number and of high capture cross section; used as a radiation shield in nuclear reactors.

loaded wheel. A grinding wheel which has been dulled by becoming filled with particles of materials being ground.

loading. (1) Placing a charge in a furnace or kiln.
(2) The filling or clogging of the pores of a grinding wheel face with the material being ground.

loading station. A site where materials or products are loaded on a truck or other device for movement to some other location.

load-transfer device. Any device, such as a dowel or key, for improving the transference of live load across a joint of concrete; that is, to improve the shear strength of the joint, and minimize wear.

loaf. A raised decoration in the center of a plate or bowl.

loess. A fine clay-like material, such as brick clay, which is largely siliceous in composition but contains calcareous matter, is characterized by the absence of stratification, and contains sharply angular grains of quartz.

long. A comparative term denoting a slow-setting glass.

long clay. A plastic clay of high green strength.

long glass. A slow, solidifying glass.

longitudinal-arch kiln. A kiln in which the arch extends parallel to the length of the kiln.

longitudinal magnetism. A field of magnetism in which the flux lines travel in a direction essentially parallel with the longitudinal axis of the component.

loop, hysteresis. The divergence between the paths of the adsorption and desorption isotherms.

loops, looping. A defect occurring in porcelain-enamel ground and cover coats characterized by a sagged or draped appearance.

loose splittings. Heterogeneous shapes of mica packed loosely in bulk form, but arranged in no particular order.

loose splittings with powder. Loose splittings of mica dusted with mica powder; used as electrical insulation.

Los Angeles abrasion test. A test of the hardness and abrasion resistance of concrete aggregates in which a standard sample is tumbled in a standard ball mill for a certain number of revolutions.

loss-on-ignition (LOI). The loss in weight which results from heating a sample of material to a high temperature after preliminary drying at a temperature just above the boiling point of water; the loss in weight upon drying is identified as free moisture, and the loss in weight occurring above the boiling point of water as loss on ignition, and is reported as a percentage of the weight of the original dry sample.

lost-wax process. The process of preparing an investment casting mold by encasing a wax replica in a bonded refractory powder, removing the wax from the refractory by melting, and then firing the resultant mold prior to use.

lot. A quantity of material which is uniform in isotopic, chemical, and physical characteristics, and which may be composed of one or more batches, provided that the same starting material is used for all batches.

lot sample, composite. A single sample prepared from several containers or lots by combining them thoroughly in the same ratio as the net weight of the materials sampled.

low-duty fireclay brick. A fireclay brick having a pyrometric cone equivalent to not less than cone 15 nor higher than cones 28–29.

lower blades. An asbestos-cement product shaped so as to allow the flow and control of air in the ventilation of a building.

low-frequency induction furnace. A refractory-lined furnace in which the charge is heated by eddy currents.

low gloss. Dullness or lack of gloss on a porcelain enamel or glaze surface.

low-heat cement. A portland cement containing a relatively high percentage of dicalcium silicate and tetracalcium aluminoferrite, and a low percentage of tricalcium silicate and tricalcium aluminate, and having a considerably lower heat of hydration than portland cement.

low-melting glass. Glass containing selenium, arsenic, thallium, or sulfur having a melting point of 127° to 349°C (260° to 600°F).

low-shaft furnace. A short-shafted, refractory-lined blast furnace used to produce low-grade products by using low-grade fuels.

low-soda alumina. Aluminum oxide with less than 0.15% of sodium oxide, and which is used in high-grade electrical insulators and other ceramic bodies.

low-solubility glaze. A lead-bearing glaze in which no more than 5% of the lead oxide is soluble.

low-temperature glaze. A glaze which fires at a temperature of 1050°C (1922°F) or below.

LPG. Abbreviation for liquified petroleum gas.

lubricant. (1) A material such as graphite used to lubricate the die-punch interface and/or the green pellet-die interface and/or the particle interfaces within the green pellet during a pressing operation.

(2) Substances which facilitate the flow of nonplastic or poorly plastic materials in the formation of dense compacts under pressure.

(3) A solution which, when applied to glass fibers, facilitates their handling by reducing mutual abrasion.

(4) The liquid used to lubricate the work face, promote a more efficient action, and retard loading of the face of an abrasive wheel.

(5) A substance such as a lubricating oil employed to reduce friction between moving parts.

lubricant, mold. A substance such as graphite applied on or into glassforming molds to reduce friction and to prevent adhesion to facilitate release of the formed item from the mold.

lug. A protuberance or knob on an item or tool used as a handle.

lug brick. A brick formed with lugs to facilitate spacing with adjacent brick.

luminescence. The production of light without generating high temperatures or incandescence, usually resulting from electromagnetic radiations, electron bombardment, electric fields, and chemical reactions at room or ambient temperatures.

luminous-wall firing. The firing of a furnace or kiln by projecting the fuel onto an incandescent refractory surface.

lump. A raised area or projection, usually rounded, on a porcelain enamel, glaze, or other solid surface.

Lunden conducting tile. Tile, particularly floor tile, in which an electrically conducting material such as carbon or metal has been incorporated as a means of dissipating electrostatic charges.

Lurgi cement. A hydraulic cement produced by sintering the charge on a grate.

luster. An irridescent decorative surface appearance on

porcelain enamels, glass, and glazes usually produced by the application of a very thin film of metal such as gold, silver, platinum, copper, bismuth, and tin over the coating surface; the luster is applied as an oxide or resinate and fired in a reducing atmosphere.

luster, vitreous. A surface having a bright, glassy appearance.

lute. A clay or cement packed into a joint or applied as a coating over a porous surface to render the joint or surface impervious to gases and liquids.

lutetium boride. (1) LuB_4; mol. wt. 218.28; sp. gr. 7.52. (2) LuB_6; mol. wt. 239.92; sp. gr. 5.74. See borides.

lutetium carbide. (1) Lu_3C; mol. wt. 537.0; sp. gr. 10.54. (2) LuC_2; mol. wt. 199.0; sp. gr. 8.73. See carbides.

lutetium nitride. LuN; mol. wt. 189.01. sp. gr. 11.59. See nitrides.

lutetium oxide. Lu_2O_3; mol. wt. 398.0; sp. gr. 9.42.

lutetium silicide. (1) $LuSi_2$; mol. wt. 231.12. (2) Lu_3Si_5; mol. wt. 665.30. See silicides.

lutetium sulfide. Lu_2S_3; mol. wt. 446.18.

luting. The joining of two leather-hard, unfired ceramic surfaces with slip; for example, the joining of handles to cups, vases, etc., to form a monolithic structure.

M

machinability. The ease with which a material can be machined. See machining.

machine. A mechanical device designed to perform some form of useful work.

machine tool. A hard, fracture-resistant attachment to a machining apparatus used to cut, drill, shape, grind, or polish a solid product.

machining. The process of cutting, grinding, or shaping a piece of work.

Mack's cement. A quick-setting cement composed of plaster of paris with additions of calcined sodium sulfate, Na_2SO_4, and potassium sulfate, K_2SO_4.

macro. A prefix meaning large.

macropore. A pore of sufficient size that it will not retain water by capillary attraction.

macroscopic. Visible to the unaided eye.

mafic minerals. A group of magnesium, iron, and calcium silicates sometimes used as inexpensive substitutes for feldspar; m.p. approximately 1250°C (2282°F).

magnesia. See magnesium oxide, MgO.

magnesia brick. A refractory brick composed of approximately 85% of magnesium oxide, MgO, and 15% of other oxides; used in high-temperature applications where corrosion by basic slags may be severe.

magnesia cement. Magnesium oxychloride cement produced by adding a magnesium chloride solution to magnesium oxide.

magnesia, dead-burned. Dense, water-stable granules of MgO formed by heating to temperatures above 1450°C (2642°F); used as a refractory or as an ingredient in refractory products.

magnesian matte. A matte glaze containing an excess of magnesium oxide (MgO).

magnesia, refractory. A dead-burned, crystalline form of MgO having a high resistance to heat and corrosion; used as linings for furnaces and melting tanks, either in the form of brick or cement.

magnesia, seawater. Magnesia recovered from sea water by treatment with slaked lime or lightly calcined dolomite.

magnesiochromite. $MgO{\cdot}Cr_2O_3$; mol. wt. 192.3; m.p. 2250°C (4082°F); sp.gr. 4.41; a spinel component of chrome magnesite refractories.

magnesioferrite. $MgO{\cdot}Fe_2O_3$; mol. wt. 200.0; m.p. 1750°C (3128°F); sp.gr. 4.2; a spinel sometimes found in basic refractories.

magnesite. $MgCO_3$; mol. wt. 84.3; decomposes at about 350°C (662°F); sp.gr. 3–3.12; hardness (Mohs) 3.5–4.5; used as an ingredient in basic refractories and glazes as a source of magnesium oxide, MgO.

magnesite brick. A refractory brick composed of approximately 85% of magnesium oxide and 15% of other oxides; used in high-temperature kilns and furnaces where corrosion by basic slags may be severe.

magnesite-chrome brick. A refractory produced from a mixture of dead-burned magnesite and chrome ore, magnesite being the predominant ingredient.

magnesite, dead-burned. Granules of magnesium oxide produced by calcining magnesite at temperatures above 1450°C (2642°F) for use as a refractory or as a refractory ingredient.

magnesia-dolomite brick. A refractory brick produced from a mixture of dead-burned magnesite and magnesia-rich dead-burned dolomite, the magnesite being the predominant ingredient.

magnesite grain. A granular form of MgO produced by calcination of magnesite; used in refractory applications.

magnesite refractory. A refractory product in which magnesite is the essential starting raw material.

magnesite wheel. A grinding wheel in which magnesium oxychloride is employed as the bonding material.

magnesium aluminate. $MgO{\cdot}Al_2O_3$; mol. wt. 142.3; softens 1000°–1100°C (1832°–2012°F); m.p. 2135°C (3875°F); sp.gr. 3.59; hardness (Vickers) ~ 1000; a spinel used to produce refractories of high resistance to corrosion by slags, glass, etc., at elevated temperatures.

magnesium aluminum silicate. (1) $2MgO{\cdot}2Al_2O_3{\cdot}5SiO_2$; mol. wt. 584.94; m.p. 1471°C (2680°F); sp.gr. 2.51; hardness (Mohs) 5–7. (2) $4MgO{\cdot}5Al_2O_3{\cdot}2SiO_2$; mol. wt. 790.98; m.p. 1454°C (2650°F); hardness (Mohs) 5–7; employed as a ceramic binder.

magnesium boride. (1) MgB_2; mol. wt. 89.24. (2) MgB_2; mol. wt. 45.96. (3) MgB_{12}; mol. wt. 197.44. See borides.

magnesium carbonate. $MgCO_3$; mol. wt. 84.3; decomposes at 350°C (662°F); sp.gr. 3.04; used both as a low-temperature refractory and high-temperature flux in glass, porcelain, insulator bodies, vitreous and semi-vitreous ware, glazes, and porcelain enamels; also used as a setting-up agent in porcelain enamel and other slips.

magnesium chromite. Magnesiochromite, $MgO{\cdot}Cr_2O_3$; mol. wt. 192.3; m.p. 2250°C (4082°F); sp.gr. 4.39; a spinel in chrome-magnesite refractories.

magnesium ferrite. Magnesioferrite, $MgO{\cdot}Fe_2O_3$; mol. wt. 200.0; m.p. 1750°C (3182°F) sp.gr. 4.2; a spinel in some basic refractories.

magnesium fluoride. MgF_2; mol. wt. 62.3; m.p. 1396°C (2545°F); sp.gr. 3.0; used as a flux in various ceramics and glass compositions, particularly in infrared components used under severe conditions.

magnesium fluosilicate. $MgSiF_6{\cdot}6H_2O$; mol. wt. 274.4; used in ceramic coatings and as a concrete hardener; sometimes as a waterproofing agent.

magnesium flux. Magnesium fluoride, MgF_2, which see.

magnesium germanate. (1) $4MgO{\cdot}GeO_2$; mol. wt. 265.88; m.p. 1497°C (2725°F) (2) $2MgO{\cdot}GeO_2$; mol. wt. 185.24; m.p. 1854°C (3370°F). (3) $MgO{\cdot}GeO_2$; mol. wt. 144.92; m.p. 1699°C (3090°F).

magnesium lime. Lime containing more than 20% of magnesium oxide; slakes more slowly, evolves less heat, sets more rapidly with less expansion, and produces mortars of higher strength than high-calcium limes.

magnesium mica. $K_9Mg_3AlSi_3O_{10}(OH)_2$; sp.gr. 2.86; hardness (Mohs) 2.5–3; used as thermal and electrical insulation.

magnesium niobate. (1) $MgO{\cdot}Nb_2O_5$; mol. wt. 306.52; m.p. ~ 1538°C (2800°F). (2) $2MgO{\cdot}Nb_2O_5$; mol. wt. 346.84; m.p. ~ 1510°C (2750°F). (3) $3MgO{\cdot}Nb_2O_5$; mol. wt. 387.46; m.p. ~ 1538°C (2800°F). (4) $4MgO{\cdot}Nb_2O_5$; mol. wt. 427.78; m.p. ~ 1483°C (2700°F).

magnesium nitride. Mg_3N_2; mol. wt. 100.98; decomposes at 1500°C (2732°F); sp.gr. 2.71.

magnesium oxide. MgO; mol. wt. 40.3; m.p. 2800°C (5072°F); sp.gr. 3.22; hardness (Mohs) 5–7; used in refractories, crucibles, thermocouple tubing, thermal insulation, infrared windows, etc., as a viscous flux and as an opacifier; also used in the production of sorel (magnesium oxychloride) cement and some electronic components.

magnesium oxychloride cement. A mixture of magnesium oxide with an aqueous solution of magnesium chloride; used for interior flooring.

magnesium phosphate. $Mg_2P_2O_8$; mol. wt. 223.7; m.p. 1383°C (2521°F); sp.gr. 2.6; used in glazes for sanitary ware as a replacement for tin oxide to obtain improved color, opacity, brilliance, and texture.

magnesium phosphide. Mg_3P_2; mol. wt. 135.02; sp. gr. 2.02.

magnesium silicate. (1) $3MgSiO_3{\cdot}5H_2O$; sp.gr. 2.6–2.8; used as a component in glass, refractories, and other ceramic bodies. (2) $MgO{\cdot}SiO_2$; mol.wt. 100.42; m.p. 1554°C (2830°F); hardness (Mohs) 5–7. (3) $2MgO{\cdot}SiO_2$; mol. wt. 140.74; m.p. 1910°C (3470°F); sp.gr. 3.22; hardness (Mohs) 5–7.

magnesium silicide. Mg_2Si; mol.wt. 76.7; m.p. 1102°C (2033°F); sp.gr. 2.0.

magnesium silicofluoride. See magnesium fluosilicates, $MgSiF_6{\cdot}6H_2O$.

magnesium stannate. (1) $MgSnO_3$; mol. wt. 101.0; used in dielectric compositions and as a phosphor base. (2) $2MgO{\cdot}SnO_2$; mol. wt. 231.34; m.p. ~ 1949°C (3540°F); sp.gr. 4.74.

magnesium sulfate. $MgSO_4$; mol. wt. 120.4; m.p. 1185°C (2165°F); sp.gr. 2.65; used as a suspension promoting agent in slips and as a flux in glaze composition; may be a source of scum formation.

magnesium sulfide. MgS; mol. wt. 56.38; m.p. >1999°C (3630°F); sp.gr. 2.68.

magnesium titanate. (1) $MgTiO_3$; mol. wt. 120.2; m.p. 1690°C (3074°F); sp.gr. 4.0; used in dielectric compositions for its low-dielectric constant. (2) $2MgO \cdot TiO_2$; mol. wt. 160.74; m.p. 1732°C (3150°F); sp.gr. 3.52. (3) $MgO \cdot 2TiO_2$; mol. wt. 200.52; m.p. 1649°C (3000°F); sp.gr. 3.66.

magnesium tungstate. $MgWO_4$; a fluorescent pigment.

magnesium uranate. $MgO \cdot UO_3$; mol. wt. 326.52; m.p. 1749°C (3180°F).

magnesium wolframate. Magnesium tungstate, $MgWO_4$, which see.

magnesium zirconate. $MgZrO_3$; mol. wt. 163.5; m.p. 2150°C (3902°F); used in dielectric compositions; also used as a setter for firing titanates, ferrites, etc.

magnet. A body which attracts iron and produces a magnetic field external to itself.

magnetic. Exhibiting the property of a magnet. See magnet.

magnetic analysis inspection. A nondestructive test to identify variations in magnetic flux in ferromagnetic materials of constant cross section which may be caused by defects, discontinuities, irregularities, variations in hardness, etc.

magnetic ceramics. Fired mixtures of Fe_2O_3 and appropriate compounds of divalent metals such as barium, cobalt, copper, lead, magnesium, manganese, nickel, strontium, or zinc which exhibit ferromagnetic and antiferromagnetic, magneto-optical, and magnetostrictive effects; used in antennae, computer memory cores, TV yokes, telecommunication systems, etc.

magnetic field. The space in which a magnetic force is exerted by a magnet or conductor carrying a current.

magnetic field leakage. The magnetic field that leaves or enters the surface of a part at a discontinuity or at a change in section configuration of a magnetic part.

magnetic field, longitudinal. A magnetic field in which the flux lines traverse a component parallel to its longitudinal axis.

magnetic-field meter. An instrument that measures the strength of a magnetic field.

magnetic field, residual. The field of magnetism remaining in a material following removal of a magnetic force.

magnetic field, resultant. A magnetic field resulting when two magnetizing forces are impressed on the same area of a magnetizable material.

magnetic field strength. The measured intensity of a magnetic field expressed in oersteds.

magnetic filter. Any device which provides a magnetic field through which powders and slurries are passed for the removal of magnetic iron.

magnetic flaw detector. See magnetic particle inspection.

magnetic flux density. The strength of a magnetic field expressed as flux lines per unit of area.

magnetic flux leakage. The excursion of magnetic lines of force from the surface of a specimen.

magnetic flux penetration. The depth to which a magnetic flux is generated in a specimen.

magnetic hysteresis. The lagging of changes in the magnetization of a specimen following changes in the magnetizing force.

magnetic iron. Iron and the oxides of iron which are influenced by magnetic forces.

magnetic lines of force. Lines used to represent the magnetic induction in a magnetic field such as are produced when iron filings are sprinkled over a nonmagnetic sheet placed over a magnet.

magnetic-particle inspection. A nondestructive technique to detect discontinuities in ferromagnetic materials in which magnetic particles sprinkled over the surface collect at the site of a defect.

magnetic-particle inspection, fluorescent. A nondestructive technique to detect discontinuities in ferromagnetic materials in which a finely divided fluorescent ferromagnetic medium fluoresces under black light at the site of a defect.

magnetic pulley. A magnetized pulley at the discharge end of a conveyor which attracts and removes magnetic impurities from a material cascading from the conveyor.

magnetic purification. The removal of magnetic particles from a slip or slurry by means of magnets.

magnetic saturation. The point at which the field strength of a magnetized material will not be increased by application of additional magnetizing force.

magnetic separator. An apparatus in which fluid suspensions are passed over a series of magnets to remove magnetic substances.

magnetic thickness gauge. An instrument used to measure the thickness of porcelain enamel in which the magnetic force required to lift a magnet from the coating is calibrated to indicate the distance between the coating surface and the coating-metal interface.

magnetic writing. A nonrelevant indication caused when a magnetic part comes in contact with another ferromagnet.

magnetism. The property of attraction as exhibited by a magnet.

magnetism, residual. The retention of a magnetic field by ferromagnetic material after exposure to magnetic force.

magnetite. A spinel composed essentially of Fe_3O_4; mol. wt. 215.5; m.p. 1594°C (2901°F); sp.gr. 5.2; hardness (Mohs) 6; used as a colorant in the production of pale green, celadon greens and pale blues, and black glazes.

magnetization, flash. Magnetization by a current flow of short duration.

magnetization, swing-field. A magnetic field induced in two different directions to detect defects in a part which are oriented in two different directions in the part.

magnetization, yoke. A longitudinal magnetic field induced in a material, or in an area of a material, by means of an external yoke-shaped electromagnet.

magnetizing current. The flow of electric current inducting magnetism into a substance.

magnet, permanent. A strongly magnetized material which retains magnetic properties for a substantial period of time.

magnum. A two-quart or 1.5 liter ceramic or glass bottle.

main arch. The crown of a furnace or kiln.

majolica. An earthenware of relatively high absorption and low mechanical strength, usually coated with a glossy, tin oxide-opacified white glaze and colored overglaze decorations fired at relatively low temperatures.

majolica glaze. A glossy, tin oxide-opacified white or colored, overglaze decoration fired at a relatively low temperature. See majolica.

malachite green. $CuCO_3 \cdot Cu(OH_2)$; used as a green colorant in stoneware and as a green dye to indicate the absorption characteristics of ceramic bodies.

male end of pipe. The end portion of a pipe which is overlapped by the end of an adjacent pipe.

malfunction. Failure to perform in the normal or intended manner.

mallet. A hammer with a head of wood, rubber, or rawhide used to shape metal to be used in making porcelain-enamel artware.

malm. An easily crumbled limestone used in the manufacture of brick, and as an anticrazing ingredient in stoneware.

malmstone. Chert used as a foundation material in building and paving.

mandrel. (1) A steel shaft on which bonded abrasives are attached in the production of grinding wheels.

(2) A shaft inserted through a hole in a component to serve as a support during machining operations.

(3) A refractory tube used in the production of glass rod and tubing.

manganese-alumina pink. A ceramic colorant consisting of a calcined mixture of manganese carbonate, aluminum hydrate, and borax.

manganese aluminate. $MnO \cdot Al_2O_3$; mol. wt. 172.87; m.p. 1560°C (2840°F); sp.gr. 4.12.

manganese aluminum silicate. (1) $2MnO \cdot 2Al_2O_3 \cdot 5SiO_2$; mol. wt. 614.04; m.p. 1297°C (2730°F); hardness (Mohs) 5–7. (2) $3MnO \cdot Al_2O_3 \cdot 3SiO_2$; mol. wt. 494.91; m.p. 1198°C (2190°F); sp.gr. 4.18; hardness (Mohs) 5–7.

manganese arsenide. Mn_2As; mol. wt. 184.8; m.p. 760°C (1400°F).

manganese boride. (1) MnB; mol. wt. 65.75; sp.gr. 6.2. (2) Mn_2B; mol. wt. 120.68; sp.gr. 6.9. (3) Mn_3B_4; mol. wt. 208.07; sp.gr. 6.12. (4) Mn_4B; mol. wt. 230.54. See borides.

manganese carbonate. $MnCO_3$; mol. wt. 114.93; decomposes on heating; sp.gr. 3.125; used as a black, brown, and purple colorant in glazes.

manganese dioxide. MnO_2; mol. wt. 87.0; converted to Mn_2O_3 at 535°C (995°F); sp.gr. 5.0; used in glass as both a colorant and decolorizer, as a mineralizer in whiteware bodies and electrical porcelain, as a black, brown, and purple in glazes, and as an oxidizing agent, colorant, and adherence promoting agent in porcelain enamels.

manganese ferrite. $MnO \cdot Fe_2O_3$; mol. wt. 230.61; m.p. 1571°C (2860°F). sp.gr. 4.75. See intermetallics.

manganese-palladium. MnPd; mol. wt. 161.63; m.p. 1516°C (2760°F). See intermetallic compound.

manganese phosphide. (1) MnP; mol. wt. 85.96; m.p. 1098°C (2010°F); sp.gr. 5.71. (2) Mn_3P_2; mol. wt. 226.85; m.p. 1204°C (2200°F). (3) Mn_2P; mol. wt. 140.89; m.p. 1315°C (2400°F); sp.gr. 6.33. (4) Mn_3P; mol. wt. 195.82; m.p. 1229°C (2245°F); sp.gr. 6.70. See phosphides.

manganese pyrophosphate. $Mn_2P_2O_7$; mol. wt. 283.9; m.p. 1196°C (2185°F); sp.gr. 3.7.

manganese silicate. $MnSiO_3$; mol. wt. 131.0; m.p. 1323°C (2413°F); sp.gr. 3.7. (2) $2MnOSiO_2$; mol. wt. 201.92; m.p. 1340°C (2445°F); sp.gr. 4.05.

manganese silicide. (1) MnSi; mol. wt. 83.0; m.p. 1280°C (2336°F); sp.gr. 5.9. (2) Mn_3Si; mol. wt. 192.85; decomposes at 1120°C (2050°F); sp.gr. 6.60–6.71. (3) Mn_5Si_3; mol. wt. 358.83; decomposes at 1283°C (2345°F); sp.gr. 6.02. See silicides.

manganese sulfide. MnS; mol. wt. 86.99; m.p. 1615°C (2940°F); sp.gr. 3.99. See sulfides.

manganese titanate. (1) $MnO \cdot TiO_2$; mol. wt. 151.03; m.p. 1359°C (2480°F); sp.gr. 4.54. (2) $2MnO \cdot TiO_2$; mol. wt. 221.96; m.p. 1454°C (2650°F); sp.gr. 4.54.

manganese-zirconium. MnZr; mol. wt. 146.15; m.p. 1499°C (2730°F).

manganic oxide. Mn_2O_3; mol. wt. 157.86; loses oxygen at 1080°C (1976°F); sp.gr. 4.32–4.82.

manganite. $Mn_2O_3 \cdot H_2O$; mol. wt. 175.9; sp.gr. 4.2–4.4; hardness (Mohs) 4; a source of manganese dioxide, MnO_2.

manganous-manganic oxide. Mn_3O_4; mol. wt. 228.79; m.p. 1565°C (2850°F); sp.gr. 4.82; hardness (Mohs) 5–5.5. See manganese dioxide.

manganous nitrate. $Mn(NO_3)_2 \cdot 6H_2O$; mol. wt. 287.1; m.p. 26°C (79°F); b.p. 129°C (264°F); sp.gr. 1.82.

manganous oxide. MnO; mol. wt. 70.9; m.p. 1650°C (3002°F); sp.gr. 5.09–5.18; used in glass compositions, in ferromagnetic materials, and computer memory cores.

mangle. A vertical-type dryer in which ware is dried in the molds in which it was formed.

manhole. A concrete structure serving as an access to underground passages or enclosed areas.

manhole base. A concrete slab foundation of the bottom manhole riser section with or without an integrally cast concrete floor over which a manhole is constructed.

manhole reducer. A concrete pipe serving as the transition joint between manhole risers of different diameters.

manhole riser. The section of concrete pipe used in the construction of a manhole, but excluding the base, reducers, and top sections.

manhole top. The concrete slab or conical top employed to reduce the diameter of the manhole riser to that of the desired access hole.

manifold. An arrangement in which a pipe or tube with at least one inlet provides two or more outlets to other pipes, such as in the delivery of fuel from a single line to several burners in a furnace or kiln.

manometer. An instrument to measure the difference between two fluid pressures.

mantel. A horizontal structure over a space in a blast furnace to carry the weight of the refractories and the casing of the exhaust stack.

manufactured alumina. Alumina, and mixtures of alumina with other ingredients, subjected to thermal treatments sufficient to produce crystalline products for use as abrasives.

manufactured graphite. A bonded granular form of carbon whose matrix has been subjected to temperatures between 900° to 2400°C (1652° to 4352°F).

map cracking. A random distribution of cracks on the surface of concrete due to surface shrinkage or internal expansion.

marble. Any limestone or dolomite, $CaCO_3$, that may be polished; marble dust may be used as a source of calcium oxide, CaO, in glazes.

marbled ware. A surface finish on ceramic and porcelain-enameled ware produced by the irregular blending of slips of different colors, resulting in the appearance of variegated marble.

marl. A crumbling clay containing magnesium and calcium used in the production of building brick, and as an anticrazing ingredient in stoneware.

Mars pigments. A series of pigments (yellow, orange, brown, red, and violet) which are produced by calcining to different temperatures the precipitate formed by mixing solutions of calcium hydroxide and ferrous sulfate.

Martin's cement. A quick-setting gypsum cement in which potassium carbonate is used instead of alum.

marver. A flat plate of metal or stone on which hand-gathered glass is rolled, shaped, and cooled.

maser, optical. A device that utilizes the natural oscillations of atoms or molecules for generating or amplifying electromagnetic waves in the microwave region of the spectrum.

mask. A protective covering placed over portions of a surface to prevent the influence of subsequent treatments of coatings to which the ware is subjected.

masking power. The ability of a coating, such as a glaze or a porcelain enamel, to obscure the surface to which it is applied.

mason. A worker engaged in the building of structures with brick, concrete block, and stone.

masonry. A construction of brick, tile, concrete, and stone, used separately or in various combinations; usually bonded with mortar.

masonry cement. A hydraulic cement composed of a mixture of natural or portland cement, hydrated lime, and sand for use in mortars for masonry construction. See cement, masonry.

masonry, prefabricated. Masonry products fabricated in a factory or other location for rapid assembly at the construction site.

masonry, reinforced. Masonry units strengthened by the use of expanded metal, wire mesh, metal fibers, iron rods, or similar materials in the bed joints.

masonry unit. Natural or manufactured building units of fired clay, stone, glass, gypsum, or concrete.

masonry unit, modular. A masonry unit produced in modules of specified dimensions.

mass concrete. Concrete placed in large masses, such as in dams or large footings, frequently containing pozzolans and large aggregate, and set without structural reinforcement.

mass, critical. The minimum mass of fissionable material of particular shape just sufficient to sustain a nuclear chain reaction.

massicot. The mineral form of lead monoxide, PbO. See litharge.

mass number. The total number of neutrons and protons in an atomic nucleous.

mass transfer zone. The region in which the concentration of adsorbate in a fluid decreases from influent concentration to the lowest detectable concentration.

master mold. A plaster mold cast around a model, in which a case mold or replica of the model may be cast.

mastic. A paste-like material used as an adhesive in the setting of tile or similar product.

mat. (1) A felt-like product made of glass fibers.
(2) A concrete footing under a post or mesh reinforcement in a concrete slab.
(3) A surface finish of low gloss. See matte.

material balance. The comparison of input and output of material quantities for a particular process; generally, the comparison of inventory plus receipts at the beginning of a process with the inventory plus shipments at the end of the process.

material balance area. An area within a factory where material records are maintained in a manner that a balance may be taken from records to show the amount of material for which the area is responsible.

material, conducting. A material which will carry an electric current.

material, diamagnetic. A material repelled by a magnet and which will position itself at right angles to magnetic lines of force.

material, ferromagnetic. A material exhibiting high magnetic permeability, the ability to acquire a high degree of magnetization in a weak magnetic field, a characteristic saturation point, and magnetic hysteresis.

material, insulating. A material which will retard or prevent the passage of heat, sound, or electric current.

material, nonferromagnetic. A material that cannot be magnetized and which generally is insensitive to magnetic fields.

material, paramagnetic. A material that is magnetized parallel to an applied magnetic field to an extent usually proportional to the strength of the field.

material test. Any test designed to measure or evaluate the chemical, physical, or mechanical properties of a substance or product.

material transfer arc. The movement of contact material by the action of an electric arc.

mat reinforcement. Tension-zone circumferential-reinforcement secured to a cage in a concrete-pipe wall. See cage.

matrix. The solid matter in which aggregates or crystals are embedded or bonded.

matte. A dull, nonreflective surface; sometimes spelled mat, matt.

matte glaze. A fired glaze having little or no gloss.

matte, lime. A matte glaze caused by the crystallization of calcium silicate during firing and cooling.

matte porcelain enamel. A fired porcelain enamel of little or no gloss.

maturing. The final stages of processing during which ceramic bodies and coatings develop desired chemical, physical, and mechanical properties or characteristics.

maturing range. The combination of time and temperature required to develop desired chemical and physical properties or characteristics in a ceramic body, coating, or other material.

maturing temperature. The temperature at which ceramic bodies and coatings develop desired chemical and physical properties or characteristics over an appropriate time interval.

mat, vacuum. A combination screen and textile filter placed over freshly poured concrete and through which, by application of a vacuum, air and water are sucked out to produce a dense concrete.

maximum size. The smallest sieve opening through which the entire amount of a material, such as aggregate, is permitted to pass.

maximum thermometer. A thermometer that registers the maximum temperature attained during a period of time.

mazarine blue. A rich, dark blue ceramic color containing approximately 50% cobalt oxide.

measurement. The quantity, capacity, or dimensions of anything as determined by suitable technique.

mechanical analysis. Mechanical separation of particles, such as aggregate, on a nest of sieves of specified sizes to determine particle-size distribution in the parent material.

mechanical boy. A mechanism to manipulate the mold in the hand-forming of glass.

mechanical damping. Mechanical resistance which retards movement of moving parts.

mechanical interlock. (1) Failure of electrical contacts to separate due to interference or surface asperities.
(2) The joining of two components by means of hooks, dowels, keys, dovetails, etc.

mechanical press. A press in which ware is formed in a die under pressure provided by mechanical means.

mechanical properties. Properties of materials associated with elastic and inelastic reactions under the influence of an applied force or that involve a relationship between stress and strain.

mechanical separation. A process in which materials are separated into fractions by settling, filtration, or centrifugal action.

mechanical shovel. A machine consisting of a scoop which can be manipulated to pick up materials from a source and discharge the materials into a container or into a vehicle for movement.

mechanical slip. A fine surface layer produced during the smoothing of the wet surface of an article formed of clay.

mechanical spalling. Breaking or cracking of refractories in service under the influence of impact or pressure.

mechanical water. Uncombined water, usually added to a body or slip to produce plasticity or workability, and which is removed by evaporation during drying or the early stages of firing; calculated as the difference in the weight of a specimen of the plastic body and the weight of the specimen dried to constant weight at 110°C (230°F); may be reported as a percentage of either the plastic or dry weight of the specimen, the latter being preferred. Also known as free water, uncombined water.

mechanical wear. Removal of surface material due to mechanical action such as abrasion.

medina quartzite. A variety of quartz containing 97.8% silica; m.p. 1900°C (3452°F).

medium. A surrounding substance in which bodies exist and move, and through which a force acts or an effect is produced; for example, water is the medium in which glazes and procelain-enamel frits are milled.

medium-duty fireclay brick. A refractory fireclay brick with a P.C.E. value of at least 29 and no greater than 31½.

melamine formaldehyde. An amino resin made from formaldehyde and melamine used as a temperature-resistant bonding agent for glass-fiber insulation and electrical panels of good arc resistance and dielectric strength.

melt. (1) To change from a solid to a liquid state by the application of heat.
(2) A molten substance.
(3) A specific quantity of glass melted at one time.

melter. (1) The chamber of a glass tank in which a glass batch is melted.
(2) A person tending a glass tank during the filling and melting operation.

melting. The thermal process of converting glass, glazes, porcelain enamels, etc., from the solid to liquid state; fusion.

melting, boost. The process of passing an electric current through molten glass as an auxiliary source of heat in fuel-fired glass tanks.

melting, congruent. The change of a substance from the solid state to a liquid of the same composition by the application of heat.

melting furnace. A furnace in which raw batches of glass and porcelain enamels are melted.

melting, incongruent. The dissociation of a compound to form another compound and a liquid of different composition by the application of heat.

melting point. (1) The temperature at which a solid is converted to a liquid.
(2) The temperature or range of temperatures at which crystalline and liquid phases of the same composition coexist in equilibrium.

melting temperature. The temperature or range of temperatures at which heterogeneous mixtures, such as a glass batch, glazes, and porcelain enamels, become molten or softened.

melting zone. The section of a glass tank or smelter in which batches of glass or glass-forming compositions are melted.

melt spinning. The process in which molten glass is extruded through a spinneret into fibers.

membrane curing. The process of curing concrete by spraying a liquid, such as a bituminous compound, on the fresh surface to form a solid impervious layer which serves as a seal to prevent loss of moisture. See bitumen.

membrane waterproofing. A procedure for waterproofing concrete by the application of alternate layers of bitumen and felt or fabric to the concrete surface; for example, the waterproofing of basements and roofs. See bitumen.

meniscus. (1) A bulb-like mass of glass at the origin of the drawn sheet produced by the Fourcault process.
(2) The curved surface of a nonturbulent liquid at the point of contact with the wall of a container.

merch brick. Discolored, off-size, or distorted building brick.

mercurous chromate. Hg_2CrO_4; mol. wt. 516.23; a green pigment for ceramics.

mercury barometer. A barometer in which variations in atmospheric pressure are measured by the rise and fall of a column of mercury contained in a partially evacuated vertical glass tube sealed at the top, the open end resting in a reservoir of mercury exposed to the atmosphere.

merthiolate. A powerful germicide and fungicide used to prevent fermentation in porcelain-enamel, glaze, and other slips.

mesh. (1) Any of the open spaces in a screen or sieve.
(2) A woven or expanded metal resembling an open basket weave.

mesh number, mesh size. (1) A code number designating the number of openings in a screen per linear inch.
(2) The designated size of particles passed through a screen, the number being derived from the number of openings in the screen per linear inch.

metal. The body of molten glass in a melting unit.

metal base. A metal product to which a coating, such as procelain enamel, is applied.

metal blister. A blister-like bloating occurring in sheet metal; a source of defects in porcelain enameling.

metal-ceramic. A body composed of an intimate mixture of metal and ceramic produced by pressing and sintering to combine the thermal-shock resistance and tensile strength of metals with the oxidation resistance and refractoriness of ceramics; used in stator blades, turbines, brake linings, etc.

metalkase brick. Basic brick contained in a thin steel casing or box-like enclosure as protection against hostile environments, particularly corrosive atmospheres at high temperatures, as in flues.

metallic colors. A suspension of metal powders, such as gold, silver, and platinum, in an oil; employed to produce metallic-appearing decorations when fired on ceramic ware.

metallic mortar. A ceramic mortar containing substantial amounts of lead powder; used to form plasters, casting sections, and blocks for x-ray and nuclear shielding.

metal line. (1) The upper surface of metal or glass in a melting tank or pot.
(2) The line of contact between the upper surface of molten glass and the refractory of a melting tank or pot.
(3) A line of maximum corrosion of the refractory by the glass.

metallizing. The process of coating or impregnating the surface of glass or ceramic with a metal.

metallurgical coal-base refractory. A commercial refractory made of metallurgical coke.

metallurgy, powder. The production of objects by pressing, binding, and sintering powdered metal.

metal marking. A line of discoloration formed when a metallic object, such as a knife, is drawn across the surface of a ceramic body, glaze, or porcelain enamel.

metal-nitride-oxide semiconductor. A semiconductor consisting of a layer of silicon dioxide, SiO_2, over a substrate of silicon, Si, over a layer of silicon nitride, Si_3N_4.

metal oxide resistor. A resistor consisting of a film of metal oxide deposited over a ceramic substrate.

metal oxide semiconductor. A metal insulator semiconductor system in which the insulating layer is the oxide of the metal substrate.

metal oxide semiconductor field-effect transistor. A magnetic field effect transistor having an insulated gate of silicon dioxide, SiO_2.

metal tender. A workman supervising the temperature and melting operations of a glass tank.

metal transfer. The transfer of material from one electrical contact to another in a mating situation.

metameric color. A color which will appear the same under one condition or type of light, but will assume a different color or shade under light of a different type.

methallyl acetate. A colorless liquid used as a liner between sheets of glass in the production of safety glass to prevent the glass from shattering into sharp fragments when broken.

methuselah. A ceramic or glass bottle of 6 liter capacity for table wines, or 4.5 liters for sparkling wines.

methylcellulose. A cellulosic gum used as a binder, lubricant, wetting agent, plasticizer, and suspension agent in refractories, whitewares, abrasives, and structural clay products.

mica. A group of mineral silicates having similar physical characteristics and atomic structures, but of varying chemical compositions, containing hydroxyl, alkali, and aluminum silicate groups; used as electrical insulation. May be colorless, brown, red, yellow, green, or black.

mica book. Large irregular crystals of mica having cleavage plates resembling the pages of a book.

mica, crude. Mica as mined.

mica, dressed crude. Crude mica from which dirt and other contaminants have been removed.

mica, full trimmed. Mica trimmed on all sides, with cracks and cross grains removed.

mica, glass-bonded. See glass-bonded mica.

mica, half-trimmed. Mica trimmed on two sides, two-thirds of which are trimmed adjacent sides and the balance on parallel sides, all of which are crack free.

mica, hard. Mica which exhibits no tendency to delaminate when bent.

mica schist. A variety of laminated mica containing silica, feldspar, and other minerals; used in refractories and in roofing compositions.

mica, soft. Mica which exhibits a tendency to delaminate when bent.

micro. Prefix meaning small.

microcrystalline alumina. Small, rough grains of alumina, Al_2O_3, recrystallized from a molten bath for use as an abrasive.

microelectronics. The branch of electronics dealing with electronic components of miniature size.

microelement. Any electronic element (resistor, transistor, capacitor, diode, etc.) mounted on a thin ceramic wafer; microelements may be potted, stacked, interconnected, and arranged to form micromodules.

microglass. Very thin glass plates used for microscope cover slips.

micrograms per gram. A measure of the content of a substance, such as an impurity, in a material; reported as micrograms per gram of parent liquid or solid.

microline. A potassium-rich feldspar, $KAlSi_3O_8$. See feldspar.

micron. One-thousandth of one millimeter, 0.001 mm.

micropore. A pore sufficiently small that it will retain water against the pull of gravity and retard water flow.

microscope. An instrument employing a system of lenses to produce magnified images of objects too small to be seen by the unaided eye.

microscopy. The application of the microscope to study materials that cannot be seen sufficiently by the unaided eye.

microstructure. The structure of a material as revealed by a microscope.

microstructure of carbon black. The arrangement of carbon atoms in a particle of carbon black.

migration, water. The flow of water through the interstices of a body, such as the flow from the interior to the surface of a body during drying.

mil. One thousandth of an inch.

mild abrasive. An abrasive material having a hardness (Mohs) of 1–2; for example, talc.

milk glass. A translucent white or colored opal glass made by adding alumina, Al_2O_3, and fluorspar, CaF_2, to a soda-lime glass.

milkiness. A cloudy appearance in glass.

mill. A machine employed to reduce the particle size of solid substances.

mill addition. A material, other than frit, charged into a mill to complete the batch formula of a porcelain enamel or other ceramic slip.

mill, ball. A closed, horizontally rotating cylinder in which cascading pebbles or porcelain balls are used as a pulverizing medium.

mill, colloid. A high-speed mill designed to produce emulsified dispersions in liquid and paste form.

mill, edge runner. A crushing and grinding machine consisting of circular rollers rotating in a circular enclosure on a floor of stone, concrete, or metal.

millefiori. (1) Glass containing a decorative design of multicolored glass rods or shapes in a clear-glass matrix, such as a paperweight.

(2) Small cross sections of colored glass rods that can be fused into the surface of porcelain enamels in regular or random decorative patterns.

mill, flint. A ball mill in which flint pebbles are used as the grinding media; sometimes the mill lining is constructed of flint blocks.

mill, fluid-energy. A grinding apparatus capable of producing particles of small sizes in which attrition is achieved by collision of the particles in a fluid introduced at high speed.

mill, hammer. A crushing device in which materials are reduced in size by the impact of hammers revolving in vertical plane within a steel casing; used to crush ores and other large solid masses.

mill, Hardinge. A type of ball mill consisting of three cone-shaped grinding areas joined by a central cylinder, and in which the cones become progressively steeper from the feed to the discharge ends; coarse particles may be returned by a separator for additional grinding.

mill, impact. A mill in which minerals and rocks are pulverized by the action of rotating blades projecting the material against steel plates.

milling. Reduction of the particle size of substances by mechanical means.

milling, dry. Grinding of porcelain enamel frits and other substances without a liquid vehicle.

milling, wet. Grinding of porcelain enamel frits and other substances with sufficient water to form a slurry.

mill, jar. A small, closed rotating cylinder of porcelain or porcelain-lined steel containing pebbles or porcelain balls in which materials are ground or blended; a laboratory mill.

mill, jet. An efficient and effective milling device, producing solids of extremely small and frequently controlled sizes, in which particles are actuated by high-pressure air, steam, or other medium and are fragmented by mutual collisions at high speeds.

mill, pebble. An enclosed, rotating cylinder in which materials are ground by cascading pebbles or porcelain balls.

mill, pin. A disintegrating device consisting of a rotating disk equipped with pin-like protrusions which provide the disintegrating force or action.

mill, prall. An impact mill consisting of a rotating impeller, a baffle rotating in the opposite direction, and a stationary baffle.

mill, pug. A machine in which materials are mixed and tempered by the action of a bladed shaft rotating in a trough or drum.

mill, rod. A pulverizer in which materials are ground in a rotating tube by the impact of heavy metal rods in a parallel arrangement rolling over each other.

mill scale. A black, magnetic form of iron oxide, chiefly Fe_3O, formed on the surface of iron and steel before and during rolling and forging.

millstone. A hard, tough stone used to grind minerals and cements.

mill, tube. A tube-like ball mill, sometimes compartmentalized, with smaller grinding media in each successive compartment, in which materials are introduced with water at one end, and removed as slime at the other end.

mill, vacuum pug. A pug mill containing a vacuum chamber for the removal of air from a pugged clay body before it moves into the extrusion chamber.

mill, vibratory. A ball mill equipped with a vibrating device as a means of increasing the grinding rate.

mill, vibroenergy. A ball mill designed to vibrate on both horizontal and vertical planes.

mill wash. The residues obtained by washing the interior of a mill after a charge has been removed.

mineral. A naturally occurring substance of characteristic chemical composition and physical properties, usually expressed by a chemical formula; for example, sand, feldspar, limestone, ilmenite, clay, bauxite, chromite, cryolite, fireclay, etc.

mineral, economic. A mineral of commercial interest or value.

mineral green. See copper carbonate.

mineral inclusion. Foreign matter of different chemical and physical characteristics contained in a parent mineral, such as metallic oxides in mica which appear as deep, distinct, and saturated colors in transmitted light.

mineralizer. A small quantity of material, such as a flux, added to a refractory brick or other refractory composition to promote crystal growth of compound formation.

minerals, accessory. Small amounts of foreign matter or inclusions which do not affect the character or properties of a parent mineral.

mineral wool. Fibrous products of random orientation produced by blowing air or steam through a molten stream of rock, slag, or glass; used essentially for sound and thermal insulation, fireproofing, and as a filter medium.

minimum thermometer. A thermometer which registers the lowest temperature during an interval of time.

minium. Red lead, Pb_3O_4; mol. wt. 685.7; decomposes between 500° and 530°C (932° and 986°F); sp.gr. 9.0–9.2; used as a fluxing component in various glasses, glazes, and porcelain enamels.

mirror. Polished glass with an adherent coating of silver on the back side to produce a highly reflective unit.

miscibility. The ability of two or more liquids to blend or mix together uniformly.

mismatch. To match or fit inaccurately or unsuitably, such as colors, joints, mold parts, expansion characteristics, etc.

miter bevel. A bevel made for decorative or aesthetic purposes, or for making a right-angle joint such as with two pieces of glass.

miter cut. A cut made by the V-shaped edge of an abrasive wheel.

miter joint. A joint made by beveling each of two surfaces to be joined, usually at a 45° angle, to form a 90° corner.

mix. (1) To combine and blend into a homogenous mass. (2) A blended mixture or batch ready for processing.

mixer. A machine designed to mix batch ingredients.

mixer, batch. A mixer designed to mix a measured quantity of material for use at one time or produced in one operation.

mixer, dual-drum. A mixer consisting of a long drum divided into two compartments by a bulkhead with a swinging chute extending through the unit.

mixer, muller. A pan-type mixer in which wet or dry batches are blended by heavy mullers or wheels rotating within the pan. See muller.

mixer, pug. A machine for mixing and tempering plastic materials by means of a rotating auger shaft mounted in a trough or tube.

mixer, shaft. A mixer consisting of a trough or tube in which blending is accomplished by means of a rotating bladed shaft.

mixer, truck. A mixer mounted on a truck for mixing concrete in transit from a proportioning plant to a job site.

mixer, vacuum. A pug-type mixer equipped with a vacuum chamber to remove air and densify a batch as it is fed into the pugging zone.

mixer, vacuum pug. A pug mixer equipped with a vacuum chamber to deair plastic bodies before they are extruded. See mixer, pug.

mixing. The process of combining, and blending ingredients into one mass or mixture until the individual constituents are indistinguishable.

mix, lean. (1) A concrete low in cement content.
(2) A full-air mixture containing a low percentage of fuel and a high percentage of air.

mix proportions. (1) The ratio, by weight or volume, of ingredients constituting a batch.
(2) The actual amounts of ingredients in a batch.

mobility. The workability or freedom of a plastic mass to move, either in random motion or under the influence of force.

mobilometer. An instrument to evaluate the flow characteristics of slips and slurries. See Gardner mobilometer.

mockup. A scale model of an apparatus for demonstration, testing, or study.

model. A pattern or representation of an object which is to be fabricated.

moderator. A material used in nuclear applications to reduce neutron energy without appreciable capture.

modification. A change in a composition or design in equipment.

modified design. A change in design from a standard design.

modular brick. A brick of a size which will fill a 4-inch modular unit, including the mortar joint.

modular masonry unit. A masonry unit of nominal dimensions based on a 4-inch module.

module. Any in a series of units of standardized size and shape for use together, as in the design and construction of buildings.

modulus. A number or quantity expressing a measure, function, force, or effect.

modulus of elasticity. The ratio of stress to strain within an elastic range.

modulus, iron. The ratio of Al_2O_3 + Fe_2O_3 in a hydraulic cement.

modulus of rigidity. The measurement of the resistance of a material to shearing stress.

modulus of rupture. The transverse or cross-bending strength of a material, calculated by the formula

$$M = \frac{3Pl}{2bd^2}$$

in which M is the modulus of rupture expressed in kilograms per square centimeter, P is the breaking load in kilograms, l is the distance between the knife edges of the test apparatus, b is the width of the test specimen in centimeters, and d is the thickness of the test specimen in centimeters.

modulus of rupture, effective. The average modulus of rupture obtained for a number of pieces of a body or product being tested; from this average, each value less than the average is subtracted and the difference squared. (If the individual value is greater than the average, the latter must be subtracted from the individual value and the difference squared.) The effective modulus is calculated by the formula:

$$M_e = \sqrt{\frac{Sd^2}{N-1}}$$

in which M_e is the effective modulus of rupture, Sd is the sum of the differences, and N is the number of pieces tested.

modulus, silica. The ratio of SiO_2:Al_2O_3+ Fe_2O_3 in a hydraulic cement.

modulus, torsional. The ratio of the torsional rigidity of a body to its length.

modulus, Young's. The ratio of tensile strength to elongation within the elastic limit of a solid body. Also known as modulus of elasticity, which see.

Mohs hardness. An empirical scale of hardness in which the scratch resistance of a material is rated on a scale of minerals ranging from the softest to the hardest as follows: 1. talc, 2. gypsum, 3. calcite, 4. fluorite, 5. apatite, 6. orthoclase, 7. quartz, 8. topaz, 9. corundum, and 10. diamond.

moil. (1) The glass remaining on a blow-pipe or punty after a gob has been cut off, or after a piece of ware has been blown or severed.

(2) Glass originally in contact with the blowing mechanism or head, which becomes cullet after the article is severed. See cullet.

moisture barrier. A material or coating applied to retard the passage of moisture into a wall.

moisture content. The quantity of water in a substance, expressed as the percentage of water, by weight, in the mass.

moisture expansion. The increase in the dimensions or bulk volume of an article caused by the reaction with water or water vapor.

moisture, free. Uncombined moisture in a body or substance which can be removed by evaporation or drying.

mold. (1) A form in or around which an item is shaped. (2) The process of forming in or around a mold.

moldability. The capability of a material or composition to be shaped by molding.

mold, blank. The mold which forms the initial glass shape in the production of glass holloware.

mold, block. A forming mold constructed in one piece.

mold, blow. A mold in which a parison is placed and blown to final shape by air pressure. See parison.

mold brick. An insulating brick shaped to fit the top of an ingot mold.

mold, case. A plaster replica of an original model around which a working mold is cast.

mold, double cavity. A mold containing two cavities to permit the forming of two items of ware at the same time.

molded. Formed in a contoured cavity or around a model.

molded glass. Glass shaped in a mold.

mold, finish. The neck mold of a bottle.

molding, hot. Shaping of glassware in a hot unlined mold.

mold, ingot. A refractory mold in which molten metal is cast in the form of ingots.

molding, injection. The shaping of ware by injecting measured quantities of molten material into dies.

molding machine. A machine designed to compact sand around a pattern to form a mold, such as a mold for a cast-iron item.

molding pressure. The force required to press a plastic substance into all areas of a mold chamber.

molding sand. The sand applied to the surface of the wooden molds in which soft-mud brick are formed as a means of texturing the surface of the brick and to facilitate removal of the brick from the molds.

molding, shell. The precision casting of metals in a rigid, porous, self-supporting refractory mold; the mold is formed by sprinkling a mixture of molding sand and a thermoplastic material over a preheated metal pattern, followed by oven curing.

mold lubricant. A substance applied over the work surface of a mold to reduce friction, prevent adhesion, and facilitate separation of ware from the mold; for example, graphite, soap, etc.

mold mark. A seam line on ware at the juncture of mold parts.

mold, master. A plaster mold cast around a pattern or model which is to be duplicated.

mold, neck. The mold, or mold part, that shapes the neck of a bottle.

mold, parison. A mold forming the preliminary shape from which a glass article is to be formed.

mold, paste. A mold lined with an adherent coating of wet carbon into which glassware is blown, to prevent the glass from adhering to the mold and to facilitate its removal from the mold.

mold plug. A refractory clay, graphite, or metal seal for the bottom of an ingot mold.

mold, porous. A forming mold containing numerous open pores or channels through which gases and liquids can pass.

mold, ring. The metal section of a glass mold that shapes and finishes the neck of a bottle or other glass article.

mold, semi-permanent. A reusable mold.

mold, three-cavity. A mold containing three cavities for the simultaneous forming of three glass articles.

mold wash. A suspension, or emulsion used to coat the cavity of a mold to facilitate the release of ware from the mold after it has been formed.

mold, waste. A plaster mold or form for concrete which is destroyed or wasted after the concrete is set.

mold, working. A plaster-of-paris mold in which bodies are shaped by casting, jiggering, or roller-forming.

molecular weight. The sum of the atomic weights of all atoms in a molecule.

molten cast refractory. A refractory product made by casting the molten ingredients into molds.

mol. wt. Abbreviation for molecular weight.

molybdenite. MoS_2; mol. wt. 160.1; m.p. 1185°C (2165°F); sp.gr. 4.7; hardness (Mohs) 1.5; used as a lubricant and drawing compound for iron and steel ware for porcelain enameling.

molybdenum. Mo; at. wt. 95.95; m.p. 2470°C (4478°F); sp.gr. 10.2; used as winding for electric furnaces, glass-to-metal seals, electro-optical applications, and for filaments, screens, and grids in vacuum tubes. See winding.

molybdenum aluminide. (1) Mo_3Al; mol. wt. 314.8; m.p. 2150°C (3902°F); employed as a major constituent in refractory crucibles for melting titanium metal. (2) MoAl; mol. wt. 122.97; m.p. 1699°C (3090°F).

molybdenum beryllide. (1) $MoBe_2$; mol. wt. 114.40; m.p. >1871°C (3400°F). (2) $MoBe_{12}$; mol. wt. 206.40; m.p. 1649°–1705°C (3000°–3100°F); sp.gr. 3.03; hardness 950 Vickers. (3) $MoBe_{22}$; mol. wt. 298.4. See beryllides.

molybdenum borides. (1) Mo_2B; mol. wt. 202.7; m.p. 2000°C (3632°F); sp.gr. 9.3; hardness (Mohs) 8–9. (2) Alpha MoB; mol. wt. 106.8; m.p. 2350°C (4262°F); sp.gr. 8.8; hardness (Mohs) 8.0. (3) Beta MoB; mol. wt. 106.8; m.p. 2180°C (3956°F); sp.gr. 8.4. (4) MoB_2; mol. wt. 117.6; m.p. 2100°C (3812°F); sp.gr. 7.8. (5) Mo_2B_5; mol. wt. 246.0; dissociates at 2250°C (4082°F). See borides.

molybdenum carbide. (1) MoC; mol. wt. 108.0; m.p. 2690°C (4874°F); sp.gr. 8.4. (2) Mo_2C; mol. wt. 203.9; m.p. 2687°C (4870°F); sp.gr. 9.2.

molybdenum disilicide. $MoSi_2$; mol. wt. 152.1; m.p. 1870°–2030°C (3398°–3686°F); used in electrical resistors, high-temperature protective coatings, and in combination with Al_2O_3 in kiln furniture, sandblast nozzles, saggers, induction brazing fixtures, and hot-press and hot-draw dies; sometimes used to promote special porcelain-enamel adherence.

molybdenum disulfide. MoS_2; mol. wt. 160.1; m.p. 1185°C (2165°F); sp.gr. 4.7; hardness (Mohs) 1.5; used as a lubricant and drawing compound for iron and steel ware for porcelain enamel, and as a heating element for electric furnaces.

molybdenum enamel. A white or pastel-colored porcelain enamel containing molybdenum oxide as an adherence promoting agent.

molybdenum germanide. Mo_3Ge; mol. wt. 360.60; m.p. 1749°C (3180°F).

molybdenum nitride. (1) MoN; mol. wt. 110.01; m.p. ~ 748°C (1380°F); sp.gr.9.18. (2) Mo_2N; mol. wt. 206.01; m.p. ~898°C (1650°F); sp.gr. >8.04; hardness 650 Vickers. (3) Mo_3N; mol. wt. 302.01; m.p. > 593°C (1100°F). See nitrides.

molybdenum oxide. See molybdenum trioxide.

molybdenum phosphide. (1) MoP; mol. wt. 127.03; decomposes at 1483°C (2700°F); sp.gr. 7.50. (2) MoP_2; mol. wt. 158.06; sp.gr. 5.21. (3) Mo_3P; mol. wt. 319.03; sp.gr. 9.14. See phosphides.

molybdenum-rhenium. Mo_2Re_3; mol. wt. 750.66; m.p. 2499°C (4530°F).

molybdenum silicide. (1) Mo_3Si; mol. wt. 316.06; decomposes at 2171°C (3940°F); sp.gr. 8.97. (2) Mo_3Si_2; mol. wt. 344.12; m.p. ~ 2093°C (3800°F); sp.gr. 8.08.

molybdenum trioxide. MoO_3; mol. wt. 144.0; m.p. 1463°C (2665°F); sp.gr. 4.5; used as an adherence promoting agent in porcelain enamels, as an opacifier in porcelain enamels, glazes, and glass, and as a wetting agent in whiteware bodies.

molybdenum-zirconium. Mo_2Zr; mol. wt. 283.22; m.p. 1822°C (3420°F). See intermetallic compound.

monazite. Ce, La, $Th(PO_4)$; sp.gr. 4.9–5.3; hardness (Mohs) 5–5.5; used as a source of rare earths.

monel. A nickel-copper alloy of high resistance to acids used in pickle baskets.

monitoring. (1) Periodic or continuous instrumental examination to determine the amount of radiation or radioactive contamination in an area or in an individual.

(2) The continuous or periodic examination of a process.

monkey wall. The section between the front and back walls and the port side walls of the open-hearth furnace.

monochromatic light. Light of a single wavelength.

monochrome decoration. A decoration of a single color.

monoclinic. A crystal structure characterized by three axes of unequal length, two of which intersect obliquely and are perpendicular to the third.

monolithic lining. A lining of a furnace containing no joints, formed of a material which is rammed, cast, gunned or sintered into place.

monolithic refractory. A refractory which may be installed into place by ramming, casting, gunning, or sintering without the formation of a joint.

monolithic refractory construction. A joint-free refractory installation.

monomolecular layer. A layer, usually adsorbed, one molecule thick.

monorail. A conveyor system employing a single overhead rail for the transport of ware.

monticellite. $CaMgSiO_4$; sometimes formed in basic refractories containing silica and lime.

montmorillonite. Clay minerals, except vermiculite,

with an expanding structure; used as a lubricant in pottery bodies.

moonstone. A whitish to translucent form of feldspar having a pearly or opalescent appearance.

moonstone glass. An opal glass resembling moonstone in appearance. See opalizer.

m.p. Abbreviation for melting point.

morpholine. C_4H_8ONH; a hygroscopic liquid used as a rust inhibitor when grinding or lapping ceramics, and as an emulsifying agent for ceramic binders.

mortar. (1) A mixture of cement, lime, and aggregate used for the placement of brick or masonry.
(2) A hard, abrasion-resistant bowl-shaped container in which substances may be broken and powdered with a pestle.

mortar, agate. A bowl-shaped vessel formed from agate in which solids may be broken and powdered with a pestle with minimum contamination by the grinding media.

mortar admixture. A material added to mortar to control the setting rate and sometimes to serve as a water repellent or coloring agent.

mortar, fat. A mortar of high-cement content which is cohesive and plastic, and which adheres to a trowel.

mortar, grog-fireclay. A finely ground refractory mortar composed of raw fireclay mixed with calcined fireclay or broken fireclay brick.

mortar, ground fireclay. A finely ground refractory mortar made of raw fireclay.

mortar, heat setting. A finely ground refractory mortar which attains its strength at furnace or process temperatures.

mortar, lean. A mortar low in cement content which is nonplastic and difficult to spread.

mortar-mix clay. A finely ground clay used as a plasticizer in masonry mortar.

mortar, lime. A mixture of hydrated lime, sand, and water used in the construction of interior, nonload-bearing building walls.

mortar, pneumatically applied. A mortar of cement, sand, and water applied by shooting into place with compressed air.

mortar, refractory. A finely ground mortar made of refractory ingredients, for use in setting and bonding refractories in furnaces and kilns.

mosaic. A decorative design or picture made by setting small colored pieces, such as tile, in mortar.

mosaic, faïence. Glazed or unglazed tile with characteristic variations in face, edges, and glaze to give a handmade appearance, and having facial dimensions less than 6-inches square (39 cm^2), and thicknesses of $\frac{5}{16}$ to $\frac{3}{8}$ inch (8 to 9.5 mm); usually mounted on a backing to facilitate installation.

mosaic tile. Glazed or unglazed porcelain or natural clay tile shaped by dust-pressing or plastic forming to facial dimensions of less than 6-inches square and thicknesses of $\frac{1}{4}$ to $\frac{3}{8}$ inch; frequently mounted on a backing to facilitate setting.

mottled finish. A speckled finish, frequently of different colors, produced as a decorative effect on porcelain enamels and glazes.

mounted wheel. Small, variously shaped, abrasive products mounted on steel spindles or mandrels.

mouth. The opening into a machine or processing operation into which a batch is charged.

moving bed, continuous. A cracking or catalytic process in which granulated solids are introduced and withdrawn continuously from the reaction tower by mechanical means or by gravity. See cracking.

moving bed, intermittent. A cracking or catalytic process in which granulated solids are introduced and withdrawn periodically from the reaction tower. See cracking.

mud jacking. The process of raising a concrete slab by pumping a cement-soil-water slurry under the slab.

mud-up. To seal a smelter, furnace, pot, gas line, etc., by the insertion of wet clay or wet-clay ball.

muff. A blown cylinder of glass which is cut and flattened while plastic to form small segments of window glass.

muffle. A refractory enclosure or chamber in a furnace designed to protect ware from the flame and products of combustion. See muffle kiln.

muffle kiln. A kiln in which the combustion of fuel takes place within a refractory enclosure to protect ware from the flame and products of combustion, the heat being transferred by conduction to the area in which the ware is being fired.

muller. A heavy roller or wheel, usually of metal, mounted in a heavy pan for grinding, mixing, and tempering.

muller crusher. See muller.

muller mixer. See muller.

mullet. A knife-like instrument used to separate handblown glass from the blowpipe.

mulling. The wet or dry process of grinding, mixing,

and tempering substances by means of a muller. See muller.

mullite. $3Al_2O_3 \cdot 2SiO_2$; mol. wt. 425.9; m.p. 1810°C (3290°F); softening temp. 1650°C (3002°F); resistant to corrosion and heat; used as a refractory in high-temperature applications and as a strength-producing ingredient in stoneware and porcelain.

mullite porcelain. A vitreous whiteware containing mullite, $3Al_2O_3 \cdot 2SiO_2$, as the essential crystalline phase; used for spark plugs, laboratory ware, and other products where resistance to thermal shock, chemicals, and deformation under load are important.

mullite refractories. Refractory products in which mullite, $3Al_2O_3 \cdot 2SiO_2$, is the predominant crystalline phase. See mullite.

mullite whiteware. Any ceramic whiteware in which mullite, $3Al_2O_3 \cdot 2SiO_2$, is the essential crystalline phase.

multibucket feeder. A machine equipped with a series of buckets mounted on an endless chain to scoop up materials for delivery to a container or vehicle for movement from one location to another.

multimolecular layer. A film or coating more than one molecule thick.

multipassage kiln. A kiln consisting of more than one tunnel or passage for the concurrent firing of ware.

multiport burner. A burner with several nozzles for the discharge of fuel and air.

Munsell color system. A classification of colors based on a three-dimensional system of lightness, saturation, and hue.

Murgatroyd belt. The portion of the sidewall of a bottle near the bottom.

muriatic acid. Hydrochloric acid, HCL; used in the cleaning and pickling of metals for porcelain enameling.

muscovite. $3Al_2O_3 \cdot 6SiO_2 \cdot 2H_2O$; a mica found in many clays; sp.gr. 2.7–3.1; hardness (Mohs) 2–2.5. Usually very clear in color.

mushroom anvil. A steel form having a mushroom-like appearance used in shaping metal bowls for porcelain-enameled artware.

N

n, n_D, etc. An abbreviation for index of refraction, generally used with a subscript indicating the spectral line; for example, n_D = the index of refraction for the sodium line *D*.

nacrite. $Al_2O_3 \cdot 2SiO_2 \cdot 2H_2O$; mol. wt. approx. 258.1; a mineral of the kaolinite group.

nailing concrete. A lightweight concrete containing a material such as sawdust in proportions that it will receive nails.

Naples yellow. Lead antimonate, $Pb_3(SbO_4)_2$; mol. wt. 993.18; a very potent, yellow-orange powder that is used as a colorant in glass, porcelain enamels, and glazes for whiteware.

natch. See joggle.

natrium. A synonym for sodium.

natron. Sodium carbonate, which see.

natural alumina. One of two types of alumina abrasives; corundum, which is of relatively high purity, and emery, which is less pure, containing iron oxide as the major impurity.

natural cement. A hydraulic cement produced by calcining a naturally occurring argillaceous limestone at a temperature below the sintering point and then grinding it to a fine powder.

natural clay tile. A tile made by the dust pressing or plastic method of forming from clays that produce a dense body of distinctive, slightly textured appearance.

natural diamond. A mineral consisting of carbon of the isometric system; sp.gr. 3.51–3.53; hardness (Mohs) 10; the hardest mineral known. The term bort sometimes refers to all diamonds not fit for use as gems or for most other industrial applications, but are suitable for the preparation of diamond grain and powder for use in lapping or in the manufacture of diamond grinding wheels. This type of bort also is known as fragmented or crushing bort. These diamonds also are used in glass cutters, diamond drill bits, wire dies, and metal cutting tools.

natural finish. Unglazed or uncoated facing tile and other products fired to the natural color of the raw materials from which the bodies were fabricated.

natural mica. A group of minerals, all of which contain hydroxyl, aluminum silicate, and alkali. All have similar physical properties and atomic structure, all may be split into flexible elastic sheets, but may be of varying chemical compositions. Hardness (Mohs) 2.0–2.5. The general formula is (K, Na, Ca) $(Mg, Fe, Li, Al)_{2-3}$ (Al, $Si)_4O_{10}(OH,F)_2$.

natural resource. A deposit or accumulation of minerals, potable water, water power, and industrial materials occurring in nature.

natural uranium. Uranium having an isotopic composition as it occurs in nature, 0.711 weight per cent of U^{235} which has not been altered.

neat cement. A plastic mixture of portland cement, but without aggregate.

neat grout. A grout consisting only of cement and water.

neat plaster. A base-coat plaster in which sand is added at the site of use.

nebuchadnezzar. A ceramic or glass wine bottle of 15-liter capacity.

neck. (1) The constricted portion of a bottle between the shoulder and the opening or finish.

(2) The part of a tank furnace connecting the melting and working chambers.

(3) The section of a furnace structure connecting the uptake and part of a furnace where the flame is diminished before reaching the stack.

(4) The narrow section of a pot.

neck brick. A brick so modified that one large face is inclined toward one end.

neck mold. The segment of a metal mold employed to form the neck and finish of a glass bottle or other article.

needle. The vertical reciprocating refractory part of a feeder in a glass-forming machine which alternately forces glass through the orifice and then pulls it upward after shearing.

needle material transfer. The transfer of material in electrical contacts in which the buildup is needlelike in that it is of small diameter and of relatively great length.

negative. Having a negative charge as demonstrated when an electrode or point possesses a lower electric potential than another point in a system.

negative material transfer. The transfer of material in electrical contacts in which the buildup occurs on the negative contact.

neodymia. Nd_2O_3; see neodymium oxide.

neodymium. A trivalent, rare earth, metallic element employed in the production of glass filters for color television plates and in glass lasers having radiation wavelengths beyond the visible range.

neodymium boride. (1) NdB_4; mol. wt. 187.55; sp.gr. 5.83.(2) NdB_6; mol. wt. 209.19; m.p. 2538°C (4600°F) sp.gr. 4.95. See borides.

neodymium carbide. (1) Nd_2C_3; mol. wt. 324.54; sp.gr. 6.90. (2) NdC_2; mol. wt. 168.27; m.p. >1982°C (3600°F); sp.gr. 6.0. See carbides.

neodymium glass. A glass containing small amounts of neodymium oxide used in television filter plates; transmits 90% of the red, blue, and green light rays and 10% or less of the yellow.

neodymium glass laser. A glass doped with neodymium having properties similar to those of a pulsed ruby laser; the wavelength of radiation is outside the visible range.

neodymium hafnate. $Nd_2O_3 \cdot 2HfO_2$; mol. wt. 757.74; m.p. ~ 2399°C (4350°F); sp.gr. 8.38.

neodymium molybdate. $Nd_2(MoO_4)_3$; mol. wt. 768.54; m.p. 1176°C (2150°F); sp.gr. 5.14.

neodymium nitride. NdN; mol. wt. 158.28; sp.gr. 7.69.

neodymium oxide. Nd_2O_3; mol. wt. 336.5; m.p. 2271°C (4120°F); sp.gr. 7.2; employed in glass to impart a violet color (red-violet in artificial light and blue-violet in daylight) and to suppress the yellow sodium light in technical glasses; also serves as a decolorizer in heat-resisting glasses of high boric oxide content. Glasses containing neodymium are employed in the production of lasers and capacitors.

neodymium phosphide. NdP; mol. wt. 175.30; sp.gr. 5.94. See phosphides.

neodymium silicate. (1) $Nd_2O_3 \cdot SiO_2$; mol. wt. 396.64. (2) $Nd_2O_3 \cdot 2SiO_2$; mol. wt. 456.66.

neodymium silicide. $NdSi_2$; mol. wt. 200.39; m.p. ~ 1527°C (2780°F); sp.gr. 5.84. See silicides.

neodymium sulfate. $Nd_2(SO_4)_3$; used in small amounts as a decolorizer in glass, and in larger amounts as a glass colorant in tableware and in glassblowers' and welders' goggles.

neodymium sulfide. (1) NdS; mol. wt. 176.33; m.p. 2140°C (3885°F); sp.gr. 6.36. (2) Nd_3S_4; mol. wt. 561.05; m.p. 2040°C (3705°F); sp.gr. 6.02. (3) Nd_2S_3; mol. wt. 384.72; m.p. 2010°C (3650°F); sp.gr. 5.50; hardness (Vickers) 330. (4) NdS_2; mol. wt. 208.39; m.p. 1760°C (3200°F); sp.gr. 5.34.

neodymium telluride. (1) Nd_3Te_4; mol. wt. 942.81; m.p. 1862°C (3060°F). (2) NdTe; mol. wt. 271.77; m.p. 2043°C (3710°F).

neophane glass. A yellow glass tinted with neodymium oxide to reduce glare; used in automobile windshields, sunglasses, etc.

nepheline. $NaAlSiO_4$; mol. wt. 142.03; sp.gr. 2.5–2.6; hardness (Mohs) 5.5–6; a feldspathoid mineral occurring in alkali-rich volcanic rocks usually having higher alkali and alumina contents and a lower silica content than conventional feldspars; used as a substitute for feldspar because of its lower melting point.

nepheline syenite. An ingenous rock consisting of a mixture of nephelinic minerals ($K_2O \cdot 3Na_2O \cdot 4Al_2O_3 \cdot 9SiO_2$), potash feldspar, soda feldspar, and minor quantities of magnetite, hornblende, and mica; approx. mol. wt. 447; sp.gr. 2.61 (crystalline) and 2.28 (glassy); hardness (Mohs) 6; starts to sinter at cone 08 and has a PCE of about cone 7. Employed in sanitary ware, floor and wall tile, semivitreous ware, electrical porcelains, glass, porcelain enamels, and other ceramic products as a substitute for feldspar to lower firing temperature, shorten firing time, and increase firing range; to reduce warpage, expansion, and water absorption. and to increase mechanical strength; tends to increase shrinkage. Its use also results in lower fuel and refractory costs.

Nernst body. A ceramic body consisting essentially of zirconia, thoria, and yttria, plus small additions of other rare-earth oxides; employed as a resistor in laboratory-sized, high temperature furnaces.

net weight. The weight of the contents of a container, generally determined as the difference between the gross weight and tare weight of the container.

network. (1) A system of electrical components assembled to perform a specific function.

(2) A system of ions which together will form a three-dimensional noncrystalline structure as in glass.

network-forming ion. An ionic material which will form a network with other ionic materials in the structure of a glass.

network-modifying ion. An ion of low valency and of relatively large radius, such as the alkaline earths and alkali metals, which modifies but does not directly form an atomic network in the structure of glass.

network structure. See network.

Neuberg blue. A blue ceramic colorant composed of copper carbonate and a mixture of iron ferrocyanide and iron sulfate.

neutral atmosphere. An atmospheric condition that is neither oxidizing nor reducing; usually, the term is applied to the firing zone of a furnace or kiln.

neutral glass. A term employed to describe a glass that is resistant to chemical attack.

neutralizer. (1) A dilute alkaline solution employed as a treatment in the preparation of sheet-metal ware for porcelain enameling in which acids remaining on the ware following the pickling process are neutralized.

(2) An aqueous solution of a chemical or a mixture of chemicals which is alkaline in nature.

neutralizer, cyanide. A neutralizer bath employed in the treatment of metals for porcelain enameling which contains a small addition of sodium cyanide to reduce the hardness of water, to assist in further cleaning of the metal, and to aid in the complete neutralization of acids remaining on the pickled ware.

neutral refractories. Refractories that are chemically neutral and are resistant to both acidic and basic refractories, fluxes, and slags at high temperatures.

neutral solution. An aqueous solution which exhibits neither acidic nor alkaline properties.

neutral-tinted glass. A glass employed as a light filter to reduce the transmission of light with minimal selective absorption of specific wavelengths.

neutron. A fundamental atomic particle having no electrical charge and a mass slightly greater than that of a hydrogen atom or proton, or 1.00897 atomic mass units. It is a constituent of the nuclei of all atoms except those of hydrogen.

neutron-absorbing glass. A cadmium borate glass containing additions of titania and zirconia having a high neutron-capture cross section.

neutron cross section. A measure of the probability that a nucleus will capture a neutron, the cross section being a function of the neutron energy and the structure of the target nucleus.

neutron diffraction. The interference processes which will occur when neutrons are scattered by atoms in solids, liquids, and gases; the intensities of the diffracted beams are measured by means of a radiation counter or an ionization chamber.

neutron flux. The number of neutrons which pass through an area of one square centimeter per second, equal to the number of neutrons per cubic centimeter times the average neutron velocity.

neutron leakage. The escape of neutrons from a reactor.

newton. The unit of force required to accelerate a mass of one kilogram one meter per second per second; equal to 100,000 dynes.

nib. (1) A tungsten carbide die employed in the drawing of wire and similar materials.

(2) A small projecting point occurring as a defect or fault in a corner or edge of plate glass during cutting.

(3) The protrusion formed on the end of roofing tile to anchor the tile in place in roofing construction.

nibbed sagger. A series of projections fabricated on the interior walls of a sagger on which ware is placed during the firing operation.

nibber. The blade of a squeegee employed in rubbing coloring pastes and inks through a silk screen in the decoration of ware.

nickel aluminate. $NiO \cdot Al_2O_3$; mol. wt. 176.63; m.p. 2020°C (3668°F); sp.gr. 4.45.

nickel aluminide. NiAl; mol. wt. 85.66; m.p. 1640°C (2980°F); sp.gr. 5.90; x-ray density 6.05 gm./cc; hardness (Rockwell A) 68–72; coefficient of thermal expansion 15.1×10^{-6}/°C.; electrical resistivity at room temperature is 25-micron-ohm-cm; excellent oxidation and thermal shock resistance; resistant to molten glass; may be used in turbine blades, combustion chamber applications, and glass-processing equipment.

nickel beryllide. NiBe; mol. wt. 67.89; m.p. 1471°C (2680°F). See beryllides.

nickel blues. A generic term for a number of iron blue pigments made from iron ferrocyanide and iron sulfate.

nickel-bonded titanium carbide. TiC with additions of nickel to serve as a bonding agent.

nickel boride. (1) NiB; mol. wt. 69.51; m.p. above 1800°C (3272°F); sp.gr. 7.39; decomposes in water and is soluble in aqua regia and nitric acid. (2) Ni_2B; mol. wt. 128.2; m.p. 1226°C (2240°F). See borides.

nickel carbonate. $NiCO_3$; sometimes used as an ingredient in ceramic colors and glazes.

nickel-copper-gold brazing compound. An alloy having a solidus of 950°C (1742°F); used in specialized metal-joining applications.

nickel dip, nickel flash, nickel pickle. A thin film of metallic nickel deposited on the surface of steel ware to be porcelain enameled; the process involving galvanic action, reduction, or both. See nickel pickling.

nickel ferrite. $NiO{\cdot}Fe_2O_3$; mol. wt. 234.37; m.p. 1660°C (3020°F); sp.gr. 5.34.

nickel-gold brazing compound. An alloy having a solidus of 950°C (1742°F); employed in specialized metal-joining applications.

nickel nitrate. $NiNO_3{\cdot}6H_2O$; sometimes used in the manufacture of brown ceramic colors.

nickel nitride. Ni_3N; an electrical conductor having a resistivity of 2.8×10^{-3}ohm^3–cm.

nickel oxide. (1) NiO; mol. wt. 74.7; absorbs oxygen at 400°C (752°F) to form Ni_2O_3 which is reduced back to NiO at 600°C (1112°F); m.p. 1985°C (3605°F); sp.gr. 6.6–6.8; employed as an adherence-promotion agent in porcelain-enamel ground coats, as a blue, green, gray, brown, and yellow coloring agent in glazes and porcelain enamels, and as a decolorizer in glass. (2) Ni_2O_3; mol. wt. 165; sp.gr. 4.84; reduced to NiO at 600°C (1112°F); employed as a source of NiO in porcelain enamels, glass, and glazes.

nickel phosphide. (1) NiP_3; mol. wt. 151.78; sp.gr. 4.16. (2) Ni_2P; mol. wt. 148.41; m.p. 1098°C (2010°F); sp.gr. 7.33. (3) Ni_3P; mol. wt. 207.10; decomposes at 1098°C (2010°F); sp.gr. 7.66. See phosphides.

nickel pickling. A process to deposit metallic nickel on iron or steel by galvanic action, reduction, or both, to promote the adherence of porcelain enamel to the metal during firing.

nickel stannate. $NiSnO_3{\cdot}2H_2O$; mol. wt. 261.42; loses water of hydration at 125°C (257°F); employed in barium titanate bodies to reduce the Curie temperature.

nickel sulfate. (1) $NiSO_4$; mol. wt. 154.75; sp.gr. 3.4–3.7; loses SO_3 at 840°C (1544°F). (2) $NiSO_4{\cdot}6H_2O$; mol. wt. 262.44; sp.gr. 2.03–2.07; loses $6H_2O$ at 280°C (536°F). (3) $NiSO_4{\cdot}7H_2O$; mol. wt. 280.86; sp.gr. 1.98; loses $7H_2O$ at 98–100°C (208–212°F). All are employed in the nickel-dipping operation to improve the adherence of porcelain enamels to steel.

nickel-tantalum. Ni_3Ta; mol. wt. 357.57; m.p. 1543°C (2810°F). See intermetallic compound.

nickel-thorium. Ni_5Th; mol. wt. 525.6; m.p. 1532°C (2790°F). See intermetallic compound.

nickel zirconium. (1) Ni_3Zr; mol. wt. 267.69; m.p. 1600°–1749°C(2910°–3180°F); sp.gr. 8.3. (2) Ni_4Zr; mol. wt. 325.98; m.p. 1649°C (3000°F); sp. gr. 8.4. See intermetallic compound.

nicol. A prism composed of two pieces of transparent calcite bonded with Canada balsam; used to produce plane polarized light.

nine-inch brick. A rectangular brick measuring approximately 9 × 4⁷⁄₁₆ × 2½ inches and used as the standard unit of size in the refractories industry.

nine-inch equivalent. The volume of a 9-inch brick (100-in.^{3}) used to express the amount of material in a single shape, shipment, or period of production of refractory materials.

niobium aluminide. $NbAl_3$; mol. wt. 174.0; m.p. >1755°C (3190°F); sp.gr. 4.50.

niobium beryllide. (1) $NbBe_{12}$; mol. wt. 203.5; m.p. 1690°C (3070°F); sp.gr. 2.91. (2) Nb_2Be_{17}; mol. wt. 342.6; m.p. 1750°C (3100°F). (3) $NbBe_2$; mol. wt. 111.5; m.p. ~ 2080°C (3775°F). (4) $NbBe_5$; mol. wt. 139.1; m.p. ~ 1829°C (3325°F). (5) Nb_2Be_{19}; mol. wt. 361.0; m.p. ~ 1705°C (3100°F); sp.gr. 3.15. All are intermetallic compounds having good strengths at elevated temperatures. See beryllides.

niobium boride. (1) NbB; mol. wt. 103.9; m.p. above 2900°C (5252°F); sp.gr. 7.2; hardness (Mohs) 8; resistivity 32 microhm/cm. (2) NbB_2; mol. wt. 114.7; m.p. 3050°C (5522°F); sp.gr. 7.0; thermal expansion 5.9×10^{-6} parallel to a and 8.4×10^{-6} parallel to c; (3) Nb_3B_4; mol. wt. 322.6; m.p. 2700°C (4892°F) but incongruent; sp.gr. 7.3. (4) Nb_3B_2; mol. wt. 300.94; m.p. 1816°C (3300°F). See borides.

niobium carbide. NbC; mol. wt. 105.1; m.p. about 3500°C (6332°F); sp.gr. 7.82; microhardness values of 2470 and 2400 have been reported (above 9 on Mohs scale); modulus of rupture (25°C) 35,000 psi; specific electrical resistivity 147 microhm/cm. at room temperature and 254 microhm/cm. at the melting point; employed in cemented carbide-tipped tools. (2) Nb_2C; mol.

wt. 198.2; m.p. ~ 3087°C (5590°F); sp.gr. 7.85. See carbides.

niobium germanide. (1) Nb_3Ge_2; mol. wt. 424.5; m.p. 1649°C (3000°F). (2) Nb_2Ge; mol. wt. 258.80; m.p. 1910°C (3470°F). (3) Nb_3Ge; mol. wt. 351.90; m.p. 1910°C (3470°F). See intermetallic compound.

niobium nitride. (1) NbN; mol. wt. 107.31; m.p. 1800°C (3272°F); sp.gr. 7.3. Nb_2N and Nb_4N_3 compositions also have been reported. (2) Nb_2N; mol. wt. 200.21; m.p. ~ 2316°C (4200°F); sp.gr. 8.31. See nitrides.

niobium oxide; niobium pentoxide. Nb_2O_5; mol. wt. 266.6; sp.gr. 4.5–5.0; m.p. 1520°C (2768°F); coefficient of linear thermal expansion 6×10^{-7}/°C (25–400°C); Curie temperature 200–275°C (392–527°F); a ferroelectric material. (2) NbO; mol. wt. 109.10; m.p. 1945°C (3533°F); sp.gr. 6.27. (3) Nb_2O_3; mol. wt. 234.2; m.p. 1773°–1777°C (3222°–3230°F). (4) NbO_2; mol. wt. 125.1; m.p. 1914°C (3479°F).

niobium phosphide. NbP; mol. wt. 124.13; decomposes between 1660 and 1730°C (3020° and 3145°F); sp.gr. 6.40–6.54. See phosphides.

niobium silicide. (1) $NbSi_2$; mol. wt. 149.2; m.p. 2000°C (3632°F); sp.gr. 5.3. (2) $NbSi_3$; mol. wt. 177.28; sp.gr. 7.05. (3) Nb_3Si_2; mol. wt. 335.42. (4) Nb_4Si; mol. wt. 400.46; m.p. ~1949°C (3540°F); sp.gr. 8.01. (5) Nb_5Si_3; mol. wt. 549.68; m.p. ~ 1949°–1999°C (3540°–3630°F); sp.gr. 7.34–7.75.

niobium sulfide. (1) NbS_2; mol. wt. 157.22. (2) NbS_3; mol. wt. 189.28.

niobium telluride. NbTe; mol. wt. 220.60; m.p. 1649°C (3000°F).

nip. (1) A small glass bottle of approximately one-half pint capacity.

(2) The largest angle that will just grip a lump between the jaws, rolls, or mantle and ring of a crusher.

niter. Potassium nitrate, KNO_3; mol. wt. 101; sp.gr. 2.09–2.27; m.p. 334°C (635°F); decomposes at about 400°C (752°F); hardness (Mohs) 2; employed in glass, glazes, and porcelain enamels because of its powerful oxidizing and fluxing properties.

nitric acid. HNO_3; mol. wt. 63.02; m.p. −41.65°C (−41.7°F); decomposes at 86°C (186.8°F); sp.gr. 1.503; viscosity (25°C) 0.76 centipoise; used to some extent in glass-etching processes and ore flotation.

nitrides. Binary compounds of nitrogen and a metal more electropositive, characterized by the formula Me_x-N_y; the boron and silicon nitrides are stable to about 1093°C (2000°F) and nitrides of titanium, zirconium, hafnium, and tantalum are moderately stable in oxidizing atmospheres; most nitrides are stable in reducing atmospheres; boron nitride is the most important and is used in composite structures for yarns, fibers, and woven products of high strength.

nitrobarite. Native barium nitrate, which see.

nitroparaffins. Any organic compound derived from the methane series in which a hydrogen atom is replaced by a nitro group. Employed in formulations for the electrophoretic deposition of ceramic materials for a variety of technical ceramic applications.

nits, nitty enamels. Defects in dry-process porcelain enamels characterized by minute surface pits visible only on close examination.

nodular fireclay. A rock containing aluminous or ferruginous nodules, or both, bonded by fireclay.

no-fines concrete. A concrete containing no aggregate of less than ⅜ inch in maximum cross section.

noise. (1) A varying voltage across a pair of electric contacts due to conditions at their interface.

(2) Any nonrelevant signal that tends to interfere with the normal reception or processing of a desired flaw signal during electromagnetic testing. Such signals may be generated by inhomogeneities in the inspected part that are not detrimental to the end use of the part.

nominal dimension. A dimension that may be greater than the specified masonry dimension by the thickness of a mortar joint.

nominal maximum size. The smallest sieve opening through which the entire amount of aggregate is permitted to pass as designated in specifications or descriptions of the aggregate to be employed in the concrete batch.

nomogram. A graphic representation of information or data which consists of lines marked off to scale and arranged in a manner that, by using a straightedge to connect known values on two lines, an unknown value may be read at the point of intervention of another line.

nonaqueous developer. In liquid penetrant inspection, a developer consisting of fine particles suspended in a volatile solvent which helps dissolve the penetrant out of the discontinuity and bring it to the surface where it dries out, fixing the indication.

noncombustible. Any material that will neither ignite nor actively support combustion in air at 648°C (1200°F) when exposed to fire.

nondestructive measurement. A measurement that involves no loss in the utility of a material or product being measured.

nondestructive test. Any test method which does not involve or result in damage to a test sample.

nonferromagnetic material. A material that is not magnetizable or affected by magnetic fields, including paramagnetic and diamagnetic materials.

nonloadbearing tile. Tile designed for use in masonry wall or other construction carrying no superimposed loads.

nonlustrous glaze, nonlustrous finish. A glaze or finish on the surface of a product which consists of an inseparable fire-bonded ceramic glaze or enamel of low-gloss or dull appearance.

nonmetallic inclusion. A nonmetallic particle, such as sand, which is entrapped or embedded in steel during solidification from the molten state or during subsequent processing, and causes defects in porcelain enamels.

nonplastic. A solid material which exhibits no freedom of movement, either in random motion or under the influence of forces or fields.

nonplastics. A trade term referring to ceramic materials other than the plastic clays.

nonreinforced pipe. A concrete pipe designed and constructed without reinforcements.

nonrelevant indication. An indication observed in an inspection test which cannot be associated with a discontinuity or flaw.

nonself-sustaining discharge. An electrical discharge that depends, at least partially, on an independent source for the supply or generation of charge carriers.

nonshattering glass. A plastic-laminated or tempered glass which will not shatter when broken. See safety glass.

nonslip concrete. Concrete having a mechanically roughened surface or a sand-like surface made by additions of sand to the concrete surface just before it hardens; used for steps and other areas of pedestrian traffic to prevent slipping.

nonstoichiometric. A situation in which the numerical relationship or ratio of elements and compounds differ from that required by the products in chemical reactions.

nonvitreous. The degree of vitrification evidenced by relatively high-water absorption, more than 3%, except for floor and wall tile, which are considered nonvitreous when the water absorption exceeds 7%.

normal cure. A condition of curing asbestos cement at atmospheric pressure with incidental external heat.

normal-cure cure. The method of hardening or setting asbestos-cement products wherein the portland cement is allowed to hydrate at atmospheric conditions of pressure, preferably under conditions to inhibit water loss.

normal permeability. The ratio of the induction of electromagnetic materials, made cyclically to change symmetrically about zero, to the corresponding change in magnetizing force.

normal uranium. Uranium containing the same weight percentage of U^{235} as occurs in nature. It may be obtained by blending uranium of different isotopic compositions or by processing in a diffusion plant. Loosely, it means uranium as it occurs in nature.

Norman brick. A brick having nominal dimensions of 2½ by 4 by 12 inches.

Norman slabs. Square or rectangular panels of clear and colored glasses cut into special shapes used in the construction of stained glass windows.

nose. (1) The working end or refining chamber of a glass-melting tank.

(2) The refractory opening through which a steel-making converter is charged and discharged.

notch, slag. An opening in the hearth to permit the flow of slag from a blast furnace.

notch test. A test in which the transverse strength of a notched specimen is correlated with the low-temperature spalling resistance of fireclay refractories.

nozzle. (1) The opening in a ladle through which steel is teemed.

(2) The discharge opening of a spray gun in which a suspension is atomized.

nozzle brick. A tubular refractory shape with a hole through which steel is teemed at the bottom of a ladle, the upper end of the shape serving as a seat for the stopper. See teeming.

nozzle, jet. A specially shaped refractory nozzle employed in the production and exhaust of extremely high-temperature jet streams.

nozzle refractory. Any refractory shape containing an orifice for the purpose of transmitting molten metal from a refractory-lined container.

n-type semiconductor. A semiconductor containing small amounts of phosphorus, antimony, or arsenic which produce free electrons that form an electric current.

nuclear engineering. The technology dealing with the design, construction, and operation of nuclear reactors and their auxiliary facilities, the development and fabrication of materials, and the handling and processing of reactor products.

nuclear fuel. Any fissionable material, such as plutonium^{239}, uranium^{235}, and uranium^{233}, which is capable of acting as a source of energy and a source of neutrons for the propagation of a chain reaction.

nuclear grade. Material of a quality adequate for use in nuclear reactors.

nuclear-material dilution. Reduction of the weight percentage of a desired isotope or constituent by the addition of similar materials of lesser isotopic material content.

nuclear material license. A permit, issued by the Nuclear Regulatory Commission or a state, to receive, process, store, or ship source or special nuclear material.

nuclear material, special. U^{233}, Pu^{239}, uranium or plutonium containing more than the natural abundance of U^{235}; or any material artifically enriched in any of these substances.

nuclear poison. A material, such as cadmium, having a high neutron-absorption cross section which, if present in a reactor, reduces the neutron flux.

nuclear reactor. A device in which the chain reaction of neutrons can be sustained and regulated for production of heat-energy, synthetic elements, and radioisotopes.

nuclear-reactor ceramics. Ceramics which are employed in nuclear fuel elements, such as the compounds of uranium; as moderators, such as carbon and beryllium compounds; and as control materials, such as boron carbide and the rare-earth oxides.

nucleated glass. Glass containing a nucleating agent to promote the formation of a crystalline structure in the glass during cooling or subsequent heat treatment.

nucleation. The process of developing an initial fragment, seed, or nucleus of a stable phase in an encompassing material consisting essentially of a metastable phase, such as a glass, followed by the growth of larger fragments or crystals of the stable phase.

nuclei. Points at which crystals begin to grow during solidification.

nucleonics. The science of the nucleus of the atom, its components, and energies.

nuclide. An atomic species characterized by the number of protons and neutrons.

nu-value. Expressed by the Greek letter ν or by the English letter V, designating reciprocal dispersive powers by the formula:

$$\text{nu value} = \frac{n_D - 1}{n_F - n_c}$$

in which n_D is the index of refraction for the sodium line at 589.3*n*, and n_F and n_c are the indices for the hydrogen lines at 486.1*n* and 656.3*n*, respectively.

O

obscure glass. A glass which will transmit and also diffuse light so that objects beyond cannot be distinguished clearly.

obscuring process. Any process, such as acid etching, sandblasting, etc., which is designed to diffuse light and thereby obscure vision, in varying degrees, through glass.

observable quantity. A physical quantity which can be measured.

obsidian. A highly siliceous natural glass, usually of volcanic origin, which is transparent but dark in color and which resembles granite in composition.

obsolescence, obsolete. No longer in use because of improvements or revised requirements.

occlusion. The trapping of a gas or liquid within a solid, or the adherence of a gas or liquid to a solid mass.

occult material. A component of a material which cannot be observed by optical means, but which may be detected by chemical analysis.

ochre. A natural, sometimes plastic earth containing iron oxide, clay, and sand employed in engobe slips, underglaze colors, and overglaze decorations to produce yellow, brown, and red colors.

octyl alcohol. A group of isomers having the formula $C_8H_{17}OH$, employed as deairing agents during the ball milling of some ceramic slips.

OD grinding. Cylindrical grinding on the circumference of a specimen or item.

odorant. A material added to an odorless gas, such as fuel gas, to give it a readily identifiable odor for purposes of detection, identification, safety, etc.

odor test, threshold. A technique to evaluate the odor level in a fluid medium by dilution under specified conditions, and then comparing it with an odor-free fluid.

oersted. The centimeter-gram-second (cgs) electromagnetic unit of magnetic intensity equal to the intensity of a magnetic field in a vacuum in which a unit magnetic pole experiences a mechanical force of one dyne in the direction of the field.

off-beader. An operator who removes excess beading enamel from porcelain-enameled ware, or smooths porcelain enamel at the edges of coated ware prior to firing.

off-beading. The removal of excess slip from the edge of porcelain-enameled ware preparatory to the application of beading enamel.

offhand glass. Glass prepared by an artisan working without benefit of molds.

offhand grinding. Freehand grinding of work held in the hand of the operator, usually without the use of guides or patterns.

offhand process. The forming of glassware without the aid of molds.

offset. (1) An imperfection resulting when mold parts are not properly matched.
(2) A finish, or base offset from the body or neck of an item.

offset finish. A finish that is not symmetrical to the axis of a bottle.

offset lithography. A process of printing in which an inked impression from a surface is first made on a rubber cylinder and then transferred to the item or ware being decorated or printed.

offset press. A printing press in which a lithographic stone or a plate of metal, paper, or other material is used to make an inked impression on a rubber blanket which, in turn, transfers the impression to paper, such as is used in the production of decalcomanias, or other surface being printed.

offset punt. The bottom of a bottle that is asymmetric to the axis of the bottle.

off-the-shelf. A product available for immediate shipment.

oil bath. (1) A container of oil in which hot metal is immersed for tempering.
(2) A container of oil in which a mechanical device is immersed for lubricating purposes.

oil, bunker. A heavy fuel oil formed by stabilization of the residual oil remaining after the cracking of crude petroleum.

oil, bunker C. A special grade of bunker oil used as fuel by industry for large-scale heating operations.

oil burner. A liquid-fuel burning device in which mixtures of atomized or vaporized oil and air are employed for combustion.

oil emulsion. A mixture or suspension of finely divided oil minutely dispersed in a medium such as water in which the oil is insoluble.

oil filter. A device or a material employed to remove contaminants from circulating oil.

oil-fired furnace or kiln. A furnace or kiln in which oil is employed as the heat-producing fuel.

oil, form. Any liquid material applied to forms to prevent concrete from sticking; some oil forms remain liquid, while others, such as lacquers, harden before the concrete is placed.

oil, light. A colorless, oily, water-insoluble liquid of light density obtained by distillation of petroleum, and which consists of mixtures of hydrocarbons; employed primarily as a lubricant.

oil spots. Lustrous decorative metallic markings on stoneware glazes produced by excess additions of iron oxide, manganese oxide, and cobalt-oxide to the glaze.

oil, squeegee. A liquid mixture of organic material used as the vehicle in squeegee pastes for the application of colored designs on porcelain enamel, glaze, and other surfaces.

oilstone. A natural or synthetic abrasive stone, generally impregnated with oil, for putting the final edge on cutting tools by abrasion.

oil, vegetable. Hydrogenated oils of peanuts, soybeans, coconut, and the like, which are employed in the sizings for glass-textile yarns as lubricants to improve the resistance of the fibers to abrasion.

oil-well cement. A special kind of hydraulic cement which is slow setting at the temperatures encountered in oil wells; used to support pipe and to bypass unwanted areas in the wells.

oligoclase. A triclinic soda-lime feldspar.

olivine. (Mg, Fe_2)SiO_4; a group of natural minerals including chrysolite, forsterite, fayalite, peridot, monticellite, and tephroite; sp. gr. 3.2–3.6; hardness (Mohs) 6.5–7; used in refractories, cement, and foundry sand; purer grades are used in the manufacture of electronic components and ceramic-metal seals.

omission solid solution. A crystal in which certain atomic positions are unfilled.

once-fired. Ware finished in a single fire.

one-coat ware. (1) Articles finished in a single coat of porcelain enamel.
(2) A contraction of one-cover-coat ware in which a single porcelain-enamel cover coat is applied over a ground coat.

one-fire finish. A porcelain enamel or porcelain-enamel system applied to ware and subjected to a single firing operation.

one-way slab. A steel-reinforced concrete slab in which the reinforcement rods are perpendicular to the supporting beams.

on-glaze. A glaze applied and fired on a previously glazed surface of ceramic ware.

on-glaze decoration. A ceramic or metallic decoration applied and fired on a previously glazed surface of ceramic ware.

onion. A bulb-like mass of glass at the origin of a drawn sheet produced by the Fourcault process, which see.

opacifier. A material used in porcelain enamels, glazes, and glass to impart or increase the diffuse reflection, refraction, and diffraction, and to produce an opaque appearance by reducing the transparency of the product.

opacity. (1) The property of reflecting light diffusely and nonselectively.
(2) The covering power and relative ability of porcelain enamel to reflect incident light and produce whiteness.

opal. An amorphous form of hydrated silica found in nature in many varieties, colors, and iridescence.

opalescence. The quality or state of reflecting an iridescent light.

opalescent glaze. A ceramic glaze with a milky or iridescent appearance.

opal glass. Glass having a white, milky appearance, usually with a fiery translucence.

opalizer. Any fluoride compound, such as cryolite, fluorspar, sodium fluoride, etc., which will produce an opalescent appearance in glasses and glazes.

opaque glaze. A nontransparent, white or colored ceramic coating of bright satin or glossy finish on the surface of a ceramic product.

opaque medium. (1) A material which is not transparent when observed by the eye.
(2) A medium which does not transmit electromagnetic radiation.

open. To start a hollow or an opening in a ball of clay as it spins on a potter's wheel.

open-arc furnace. A furnace heated by an electric arc positioned above the furnace charge.

open cells. Bodies in which the cells are interconnected.

open clays. Porous or sandy-textured clays.

open firing. Firing in which the flame may impinge on or through the exposed ware.

open gaseous inclusion. A bubble at the surface of glass which has burst or is open, leaving a cavity in the surface of the finished article.

open-hearth furnace. A steel-making furnace of the reverberatory type in which the charge is laid on a shallow hearth over which play flames of burning gas and hot air.

open-hearth furnace, basic. An open-hearth furnace constructed of basic refractories covered with magnesite or burned dolomite, and which is employed in the production of basic pig iron.

opening force. The force available or required to open electrical contacts.

opening material. Sand, flint, grog, chamotte, pitchers, and the like added to plastic clay to increase the porosity, decrease shrinkage, and expedite drying.

open pit. An open or surface-working excavation from which minerals are removed, the opening being the full size of the excavation.

open-pit mining. The removal of minerals located near the earth's surface by first removing the overlying material, or overburden, and excavating the minerals.

open pore volume. The volume of the pores of a solid body which may be penetrated by a liquid or gaseous substance from the external surface of the body, calculated by the formula:

$$P_o = \left[\frac{1 - D_b}{D_a} \right] \times 100$$

in which P_o is the total volume of open pores reported as a percentage of the bulk volume of the specimen, D_b is the bulk density, and D_a is the apparent density of the specimen.

open pot. A glass-melting pot open to the flames and gases of combustion.

open setting. Ware placed in a kiln and fired with the flames passing over, around, and between the ware.

open storage. The storage of raw materials out-of-doors and exposed to the weather.

operation. An item of work, usually performed at one location, consisting of one or more work elements.

operation analysis. A study of all of the procedures and activities involved in the design and improvement of production, including materials, equipment, processes, inspection, and conditions of work.

operations research. The application of scientific, particularly methematical, methods to the study and analysis of complex problems.

operator. A person who operates, adjusts, and maintains a piece of equipment.

ophthalmic glass. Glass of great uniformity, having specified optical and physical properties, as well as specified quality; employed in eyeglasses.

optic. (1) A lens or prism in an optical instrument.
(2) Glassware having variations in wall thickness which produce refractive effects.
(3) Pertaining to the eye.

optical analysis. The study of the chemical composition, particle size, and other properties of a material or mixture by means of transmitted light, such as absorption, polarization, refraction, and scattering.

optical blank. Optical glass formed to the approximate

specified dimensions required, and from which final lenses are made.

optical crown glass. (1) Any optical glass having a low dispersion and low index of refraction, usually forming the converging element of an optical system.
(2) Any optical glass having a nu-value of at least 55.0, or a nu-value between 50.0 and 55.0, and a refractive index greater than 1.60.

optical fiber. A long, thin thread of a highly transparent substance, such as glass or plastic, which transmits light by a series of internal reflections.

optical figuring. The final shaping, grinding, and polishing of glass components for optical instruments.

optical flat. A polished flat glass having an overall flatness of 0.05 micrometer, used as a standard in comparative linear measurements.

optical flint glass. (1) An optical glass with a high dispersion and high index of refraction, usually forming the diverging element of an optical system.
(2) Any optical glass having a nu-value less than 50.0, or a nu-value between 50.0 and 55.0, and a refractive index less than 1.60.

optical glass. Glass of great uniformity and free of imperfections having closely specified optical properties in terms of the transmission, refraction, and dispersion of light; used in the manufacture of optical systems.

optical maser. A device that utilizes the natural oscillations of atoms or molecules for generating or amplifying electromagnetic waves in the microwave region of the spectrum.

optical pyrometer. An instrument for measuring high temperatures in which, usually, the color of an electrically heated filament in a telescope or similar device is matched (disappears) with the color of the surface of an object in a kiln; the temperature is indicated on a calibrated scale on the instrument.

optical reflectometer. An instrument which measures the surface reflection of light waves in or near the visible region.

optical spectrometer. A calibrated instrument for the measurement of the wavelengths or refractive indices of transparent materials.

optical surface. The interface between two media, such as air and glass, which is used to reflect or refract light.

optical system. A collection of mirrors, lens, prisms, and other devices placed in some specified arrangement which reflect, refract, disperse, absorb, polarize, or otherwise act on glass.

optical system numerical designation. A numerical designation based on the index of refraction for sodium line (n_d) and nu-value (ν). The unity value for the index of refraction is dropped (for example, 1.496 becoming 496) and the decimal point for the nu-value is deleted (for example, 64.4 becoming 644). Thus, a glass may be specified as 496/644 without reference to its chemical composition. The numerical designation may be preceded, if desired, by symbols to indicate composition.

optics. The science that deals with light, its origin and propagation, the effects it undergoes and produces, and other phenomena with which it is associated.

optimization. To make as near perfect, functional, or effective as possible.

optimum frequency. In electromagnetic testing, that frequency which provides the largest signal-to-noise ratio obtainable for the detection of an individual material property, each property of a particular material having its own optimum frequency.

orange peel. A pattern of roughness or waviness on porcelain-enameled, glazed, pickled, painted, or other surface which resembles the skin of an orange in texture.

ore, refractory chrome. A refractory ore consisting essentially of a chrome-bearing spinel with only minor amounts of accessory minerals, and with physical properties that make it suitable for use in refractory products.

organic bond. (1) An organic material, such as rubber, synthetic resins, or shellac, employed to bond abrasive grains in the production of grinding wheels.
(2) A gum, starch paste, or similar material incorporated in a ceramic body or glaze to increase its unfired strength.

organic fiber. A natural or synthetic fiber having a length-to-diameter ratio of 100 to 1.

organic solvent. An organic substance, usually liquid, capable of dissolving or dispersing other substances.

orifice ring. The ring or bushing in the feeder through which glass flows to a forming machine.

o-ring. A flat ring of rubber or plastic employed under pressure to produce an air-tight vacuum or high-pressure seal.

ornamental tile. A decorative tile, frequently containing designs, applied to conventional tile surfaces, or tile of diverse sizes and shapes, to be installed in decorative patterns.

Orsat analyzer. An instrument employed in the volumetric analysis of gases in which the gases are adsorbed while passing through a series of preselected solvents.

orthoclase feldspar. $K_2Al_2O_3 \cdot 6SiO_2$; a potash-bearing feldspar employed in the manufacture of glass, electrical and other porcelains, vitreous sanitaryware, and pottery.

Orton cones. Trigonal prisms of standardized shapes, sizes, and compositions which will deform by bending under predetermined conditions of time and temperature, and which are employed to indicate the thermal history of ceramic ware during the firing operation.

osmium boride. (1) OsB_2; mol. wt. 212.44; sp.gr. 12.8–14.8. (2) Os_2B_5; mol. wt. 435.70; sp.gr. 15.2.

osmium plutonium. OsPu; m.p. 1499°C (2730°F).

osmium silicide. (1) OsSi; mol. wt. 218.86. (2) Os_2Si_3; mol. wt. 465.78.

osmosis. The diffusion of fluids through semipermeable membranes or porous partitions.

outgassing. The release of adsorbed and occluded gases or water vapor from a body or substance, such as by heating or by vacuum.

out-of-round. A defect or imperfection in glass and other products in which the degree of roundness is no longer perfect.

oven. A heated compartment or chamber in which ware is dried, fired, or otherwise thermally treated.

oven, bottle. An intermittent kiln of bottle or cone shape for draft purposes.

oven, chamber. A refractory-lined oven or kiln employed to produce gas from coal.

oven, drying. An oven or heated structure employed to evaporate water from ware or test pieces.

oven-drying loss. The reduction in weight resulting when a substance is heated in an oven under specified conditions.

oven glass. A glass of low-thermal expansion exhibiting high resistance to thermal shock, and which is employed in the manufacture of articles to be used in the preparation and cooking of food.

oven, hobmouth. A top-fired oven or kiln of cone or bottle shape.

ovenware. Ceramic whiteware or glass of high resistance to thermal shock, such as casseroles, ramekins, etc., used in culinary utensils.

overburden. Topsoil, sand, gravel, or silt overlaying a bed of clay, shale, or other material of commercial significance.

overfire, overfiring. (1) Heating ware to a temperature sufficient to cause pronounced deformation, bloating, or other defect.

(2) The firing of porcelain enamel at temperatures too high or for periods too long, resulting in pinholes, pitting, or undesired dull finishes.

overflow. An excess of a material, usually liquids, exceeding the capacity of its container.

overflush. A fault in glassware caused by an excessive flow of glass at a mold joint.

overglaze. A glaze coating applied over a previously glazed surface of ceramic ware.

overglazed. Ware coated with a glaze.

overglaze decoration. A ceramic or metallic decoration applied and fired on a previously glazed surface of ceramic ware.

overgrinding. A reduction in the particle size or particle-size distribution of a material, or mixture of materials, by grinding or milling to a degree less than desired or required.

overlay. A concrete topping employed to repair worn concrete surfaces.

overpickling. Pickling of metal shapes for porcelain enameling for periods too long, at temperatures too high, or in solutions too strong, resulting in blistering or fishscaling of the finished ware.

overpress. An imperfection in glassware consisting of a projection or fin of excess glass due to an imperfect closing of mold joints.

oversanded. An excess of fine aggregate in a concrete mix.

oversize. In the mixing of concrete, the aggregate material retained in the maximum specified sieve.

overspray. (1) The portion of the slip from a spray gun which passes by or which is not deposited on ware during the spraying operation.

(2) The application of a second coat of porcelain enamel over a previously applied, but unfired, coating which, in effect, produces two coats of enamel requiring a single firing operation.

Owens process. A bottle-making process in which the blank or parison mold is filled by suction.

oxidation. (1) A chemical reaction in which the oxygen content of a compound is increased.

(2) A chemical reaction in which an atom or ion loses electrons or in which the positive valence is increased.

oxide. A binary compound of oxygen with an element or radical.

oxide ceramics. Ceramics made by dry-pressing or slip casting essentially pure oxides, such as alumina, beryllia, magnesia, thoria, and zirconia, followed by sintering at high temperatures.

oxide, color. A material which may be added as a batch ingredient or as a mill addition to produce color in porcelain enamels, glazes, or other ceramics.

oxide-fuel reactor. A nuclear reactor in which the fuel is uranium oxide (UO_2) or plutonium oxide (PuO_2).

oxide, graining. A ceramic pigment or mixture of pigments, sometimes containing fluxing ingredients, which are incorporated in a suitable medium for the transfer of a decorative finish to porcelain-enamel and glaze surfaces by means of rolls.

oxide mineral. A naturally occurring mineral in which the components are in essentially oxide form, such as SiO_2, Al_2O_3, CaO, Fe_2O_3, etc.

oxide nuclear fuel. Fissionable uranium or plutonium oxide.

oxides. (1) Compounds containing oxygen in combination with one or more elements, particularly metals.
(2) Materials added to the mill in the preparation of porcelain enamels and glazes which will produce color or opacity in the fired finish.

oxides, surface. Oxygen-containing compounds and complexes formed at the surface of activated carbon.

oxidizing agent. A compound which removes or displaces hydrogen in another compound, attracts negative electrons, or gives up oxygen easily.

oxidizing atmosphere. An atmosphere in which an oxidation reaction takes place.

oxidizing condition. The presence of air in a kiln in excess of that needed for complete combustion.

oxidizing flame. A flame in which oxygen is present in amounts greater than is required for complete combustion.

oxidizing period. The portion of a firing operation during which any combustible or carbonaceous material in a ceramic composition is burned out.

oxidizing temperature. The temperature at which the rate of oxidation of carbon, or other element, or compound in a product becomes appreciable.

P

pack. (1) The quantity of ware contained in a package.
(2) The ratio of ware contained in a package to that theoretically possible.

packaged brick. One or more brick encased in a package to facilitate handling and to minimize breakage.

packerhead. A mechanical device in which concrete pipe is formed by compacting the concrete against a stationary, outside, vertical mold with a revolving shoe or "packerhead."

packing density. The density of an aggregate, expressed as grams per milliliter or pounds per cubic foot, packed in a container under specified and controlled conditions.

pad. The refractory brickwork floor under the molten iron in a blast furnace.

paddle and anvil. A procedure for the shaping and decoration of plastic pottery bodies by means of a textile- or cord-wrapped paddle on which a design may be carved; the anvil, a smooth piece of wood, stone, or stiff leather, is held against the interior wall to resist the beating of the paddle on the outside.

paddling. The rough shaping of a piece of glass by paddles or tools prior to the pressing of optical-glass blanks.

Padmus method of expansion measurement. A technique for calculating the coefficient of thermal expansion based on the birefringence resulting from the stress generated at the juncture when the glass and another glass of known expansion and a similar transformation temperature are fused together.

paillons. Small pieces of metal foil over which a porcelain enamel is applied and fired as a form of artware; for example, jewelry, mobiles.

painting. The process of applying a pigmented coating or design to ware by means of a brush, silk-screen, roller, spray gun, dipping, or other technique for the purpose of decoration or protection, or both; a design so applied.

pale glass. A pale, usually green-colored, glass.

pale oxide of iron. Fe_2O_3; used as a pigment; normally a red color.

palette. A small board upon which an artist lays and mixes color for use.

palette knife. A blunt spatula used for blending colors.

Palissy ware. A type of fine faïence coated with a bright-colored tin-enamel glaze.

palladium. Pd; at. wt. 106.7; m.p. 1554°C (2830°F); sp.gr. 12.0; hardness (Brinell) 109; used in the manufacture of aircraft sparkplugs.

palladium aluminide. PdAl; mol. wt. 133.67; m.p. 1642°C (2990°F).

palladium beryllide. (1) PdBe; mol. wt. 115.9; m.p.

1465°C (2670°F). (2) $PdBe_{12}$; mol. wt. 217.10; sp.gr. 3.18.

palladium chloride. $PdCl_2$; mol. wt. 177.6; decomposes at 501°C (934°F); sometimes used in porcelain compositions.

palladium uranium. Pd_3U; mol. wt. 558.27; m.p. 1638°C (2980°F).

pallet. (1) A low, portable platform upon which materials or products are stacked for easy handling, movement by fork-lift truck, and storage.

(2) A tool used by potters for smoothing and rounding plastic-clay surfaces.

(3) A tool used by glassmakers for shaping the foot of stemware.

pallet dryer. A periodic dryer in which ware, stacked on pallets, is charged, dried, and removed.

palletizing. The stacking of brick, materials, and other products on a platform to facilitate handling and moving.

pall ring. A slotted ceramic cylinder used as packing in chemical distillation columns.

palygorskite. A family of tough, fibrous, lightweight clays related to attapulgite in which an extensive amount of magnesia is replaced by alumina; used as a source of both magnesia and alumina.

PAM. An abbreviation for pneumatically applied mortar.

pan crusher. A large crushing device consisting of a pan in which one or more mullers or grinding wheels roll over the material being ground.

panel. (1) A brick with a depression in the bed surface to improve its adherence with mortar.

(2) A large, but relatively thin sheet of material, such as plasterboard, used in construction.

panel brick. A long silica brick employed as the refractory in the wall lining a coke oven.

panel spalling test. A test in which the loss in weight, by fragmentation, of a refractory panel subjected to a series of heating and cooling cycles is taken as an indication of spalling behavior in service. See spalling.

panel wall. A non-loadbearing wall.

pan, sludge. A container in which sludge is collected for subsequent recycling or disposal.

pan, tempering. A mechanical, pan-type mixer in which clays and bodies are blended with water to working consistencies.

pan tile. S-shaped roofing tile which interlock with the sides of adjacent tile.

pan, wet. A cylindrical pan in which damp materials are blended by rotating mullers or wheels. See muller.

paper resist. A decoration process in which paper, cut in the desired design or configuration, is smoothed tightly on the surface of an item being decorated to prevent the deposition of colors, glazes, or slips in the covered area.

paramagnetic material. A material which has a relative magnetic permeability slightly greater than unity and is essentially independent of the magnetizing force.

parameter. A quality or constant whose value varies with the conditions of its application.

paramorphism. A structural or physical change in a material without change in the chemical composition which determines the characteristics or behavior of the material.

parget. (1) A rough ornamental plaster on walls, or a rough-cut plaster used to line chimneys.

(2) A cement mixture sometimes used to waterproof outer walls of constructions.

parian cement. A gypsum cement to which borax is added to produce a hard finish.

parian paste. A body composed of two parts of feldspar and one part of china clay; fired at approximately 1200°C (2192°F).

parian ware. A soft, usually unglazed, porcelain resembling white marble in appearance, and which is composed of two parts of feldspar and one part of china clay; used in making figurines and statuettes.

parison. (1) A preliminary shape or blank from which a glass article is formed.

(2) A hollow tube from which a glass bottle or other hollow glass object is blow molded.

parison mold. A metal mold that first shapes glass, in the manufacture of hollow ware, prior to finishing shaping of the item.

parison swell. The ratio of the cross-sectional area of a parison tube to the opening of the die in which an item is to be blow molded.

Paris white. $CaCO_3$; mol. wt. 100.1; decomposes at 825°C (1517°F); sp.gr. 2.7–2.95; used in portland cement, soda-lime glass, pottery, and coatings for printed circuits and capacitors.

particle density. The weight of a unit volume of a substance under specified conditions, including its pore volume but excluding interparticle voids. Also called block density.

particle inspection flaw indications, magnetic. A test process in which there will be an accumulation of ferromagnetic particles sprinkled along the areas of invisible flaws and discontinuities in magnetic materials due to the distortion of magnetic lines of force in those areas. See particle inspection, magnetic.

particle inspection, fluorescent-magnetic. An inspection test process for magnetic materials employing a finely divided fluorescent ferromagnetic medium that fluoresces under black light.

particle inspection, magnetic. A nondestructive method for detecting cracks and other discontinuities at or near the surface of ferromagnetic materials in which finely divided magnetic particles sprinkled over a magnetized surface are attracted to areas of nonuniformities, such as discontinuities and defects.

particle orientation. The arrangement of a particle or mass of particles in a definite position or pattern.

particle shape. The surface or spatial configuration of a particle.

particle size. (1) The general dimensions of the particles of a granular or powdered substance or mixture, usually assuming all particles to be spherical in shape.
(2) The controlling lineal dimensions of a particle or mixture of particles as determined by a sieve analysis or other means.

particle-size analysis. The determination of the proportion of particles of defined sizes contained in a powdered or granular sample.

particle-size distribution. The percentage of each size fraction into which a powdered or granular sample may be classified.

particle sizing. The separation of the particles of a powdered or granular sample into defined size fractions.

parting compound. A powdered or colloidal material applied to a mold to facilitate the separation of a molded material from the mold.

parting line. The line or seam on glass, ceramic, or other molded product caused by the joints of the mold parts.

parting wheel. A thin abrasive wheel, usually organic bonded, used to cut, slice, or slot a material.

partition tile. Tile used in the construction of nonload-bearing partitions.

parts per million. The measurement of the number of parts of a substance, such as an impurity, per million parts of the parent material, usually expressed as micrograms per gram.

passivation. To render passive or inactive.

paste. (1) The clay body used in the fabrication of ware.
(2) The cementing ingredient in concrete consisting of cement and water.

paste, graining. A well-blended oil suspension of finely milled color oxide or porcelain enamel, usually colored, used as decoration in the rubber-roll transfer process.

paste, hard. A relatively high-firing porcelain body.

paste mold. A carbon-lined mold used in the forming of blown glassware.

paste, parian. A body composed of two parts of feldspar and one part of china clay fired at approximately 1200°C (2192°F).

paste, porcelain. Porcelain bodies, both hard and soft, usually in the plastic state.

paste, screening (squeegee). A suspension of finely milled color oxides in squeegee oil used in the decoration of porcelain enamel, glazes, and other ceramic ware by the silk-screen process.

paste, soft. A low-firing porcelain body, usually containing a substantial amount of fritted glass.

patch, hot. A refractory-lining composition used in the repair of hot furnaces.

patching cement. (1) A mixture of portland cement and fine aggregate used to repair concrete.
(2) A fireclay cementitious material used to patch furnace walls, the bottoms of glass molds, and to make corrections in molds.

pâte dure. Ceramic whitewares fired at relatively high temperatures.

pâte sur pâte. A technique for the decoration of ceramic ware in which a relief pattern is built up mostly by hand with successive layers of slip.

pâte tendre. Ceramic whitewares fired at relatively low temperatures.

patina. A thin, usually decorative film with a colored or metallic sheen, formed in various ways on the surface of ware during firing, frequently in a reducing atmosphere.

pattern burnishing. Special effects obtained on the surfaces of clay vessels by polishing the leather-hard clay, or overglaze gold, with a stone, sand, or steel tool.

pattern cracking. A random distribution of cracks on the surface of concrete as a result of surface shrinkage or internal expansion.

pat test. An estimate of the soundness of concrete in which thin cylinders of concrete are submerged in either boiling or cold water for specified periods of time, and then examined for cracking, warping, and disintegration.

pavers. Unglazed porcelain or natural clay tile formed by the dust-process method, and similar to mosaics in composition and physical properties, but relatively thicker, with 6 square inches or more of facial area.

paving brick. Low-absorption vitrified brick of high strength, usually with spacing lugs, produced with smooth or wire-cut surfaces; used in the construction of roads, driveways, sidewalks, etc.

paving-brick clay. Impure refractory fireclays and shales which are used to form paving brick of high tensile strength and durability.

paving train. A battery of road-construction equipment on a road-paving job.

PCE. Abbreviation for pyrometric cone equivalent, which see.

pearl ash. Commercial potassium carbonate, K_2CO_3; mol. wt. 138.3; m.p. 909°C (1668°F); sp.gr. 2.3; used in glass, glazes, and porcelain enamels as a flux.

pearlite. (1) A siliceous glassy rock composed of small spheroids of varying sizes, usually less than one centimeter in diameter, and having a water content of 3 to 4%; when heated to a suitable temperature, it will expand to form a light glassy material with a cellular structure.

(2) A lamellar aggregate of almost pure iron and cementite produced in cast iron and carbon steels.

pebble mill. A rotating steel, ceramic, or ceramic-lined cylinder in which materials are pulverized by cascading flint pebbles or porcelain balls.

pebble mill, vibrating. A pebble mill in which conventional milling is combined with a vibratory or bouncing action of the mill to obtain more efficient and rapid grinding.

pebbles. Hard flint, porcelain, or other heavy abrasive-resistant material used as grinding media in ball mills.

peeling. A defect characterized by the separation of flakes of a porcelain enamel, glaze, or engobe from the base to which it was applied, usually as a result of poor adherence or to critical compressive stresses.

peephole. A small opening in the door or wall of a furnace or kiln to permit observations into the interior of the structure. Sometimes called peepdoor.

pegmatite. A rock consisting essentially of feldspar, quartz, and mica used as a source of lithia, zircon, tin, tungsten, tantalum, or uranium.

pellet. A small compacted shape, usually cylindrical, formed by pressing a powdered or granulated material in a die, by casting or by other technique; used for test or reference purposes.

pellet, green. A pellet which has been pressed but has not been fired.

pelletize. To form powdered or granulated materials into pellets.

pelletizing. The process of forming pellets.

pellet, sintered. A pellet in which adjacent surfaces of particles are bonded together by pressing followed by heat treatment.

penetrant. A liquid capable of permeating a body through openings or discontinuities in the body; usually employed as a test of the surface porosity of a body.

penetrant, fluorescent. A penetrant which will emit radiation, usually as visible light, during the absorption of radiation from a source such as black light. See penetrant.

penetrant, postemulsifiable. A penetrant containing an emulsifying agent to render it water-washable for easy removal from a test specimen. See penetrant.

penetrant, visible. A penetrant, usually colored, which may be detected visually in a substance of contrasting color. See penetrant.

penetration. The process by which a penetrant enters or impregnates a substance. See penetrant.

penetration, depth of. (1) The depth to which a penetrant permeates a body.

(2) The depth at which a magnetic field of induced eddy currents has decreased to 37% of its surface value. See penetrant.

penetration, effective depth of. The minimum depth beyond which a test system no longer can detect a further increase in the thickness of penetration into a specimen. See penetrant.

penetration indication. An observation which marks the presence of a discontinuity.

penetration, magnetic flux. The depth to which a magnetic flux is generated in a specimen.

penetration time. The total time, including application and draining, in which a penetrant is in contact with the surface of a specimen. See penetrant.

peppered sandblast. A finely textured, mottled appearance produced on the surface of a substance by sandblasting; usually a decorative treatment.

peptize. (1) To convert to a colloidal solution.

(2) To liquefy a colloidal gel to form a colloidal solution.

(3) To deflocculate a slurry or slip.

perforated brick. A building brick containing symmetrically arranged holes parallel with the face of the brick to reduce the weight of the brick.

performance test. A test to evaluate the ability of a product to meet prescribed conditions of service.

periclase. Natural MgO used in refractories; mol. wt. 40.3; sp.gr. 3.56; hardness (Mohs) 5.5.

periodic. Occurring repeatedly, at regular intervals.

periodic dryer. A dryer in which ware is placed, dried, and removed prior to the introduction of a subsequent batch.

periodic furnace. A furnace in which ware is placed, fired, sometimes cooled, and removed prior to the introduction of a subsequent charge.

periodic kiln. A kiln in which ware is placed, fired, sometimes cooled, and removed prior to the introduction of a subsequent charge.

peripheral speed. The rate of movement of a point on the circumference of a revolving wheel, determined as the product of the circumference and the rate of revolution, and expressed as a unit of distance per unit of time.

perish. To disintegrate or to be destroyed under conditions of exposure, such as dampness or high temperature.

perlite. A glassy rock consisting of 65 to 75% of silica, 10 to 20% of alumina, 2 to 5% of water, and small amounts of soda, potash, and lime; expands on heating to form a light, fluffy material used as a lightweight aggregate in concrete and plaster, and as heat and acoustic insulation.

permanent linear change. The percentage change in the original length of a specimen free of applied stresses, after the specimen is subjected to a prescribed heat treatment; the change is irreversible.

permanent magnet. A strongly magnetized material which retains its magnetic properties for a substantial period of time.

permanent mold. A reusable mold.

permeability. (1) The property of a body that permits liquids or gases to seep into minute openings of the body.

(2) The ease with which a material can become magnetized.

(3) The ratio of flux density to a magnetizing force.

permeability, effective. The magnetic permeability exhibited by a specimen under prescribed physical conditions.

permeability, incremental. The ratio of change in magnetic induction to the corresponding change in magnetizing force when the mean induction differs from zero.

permeability, initial. The slope of the induction curve at zero magnetizing force as the test specimen is being removed from a demagnetized condition.

permeability, normal. The ratio of the induction to the corresponding change in magnetizing force.

permeability of refractories. The capacity of a refractory to transmit a liquid or gas through the pore structure.

permeability test. A test to determine the movement of a liquid or gas through a body under a hydraulic or pressure gradient.

permeability variations. In electromagnetic inspection, changes in the ability of a material to be magnetized that occur along the body of a test specimen; the variations may or may not be indicative of the physical conditions in the part that are detrimental to its end use.

pernetti. (1) Small iron pins or tripods used to support ceramic ware in the kiln during firing.

(2) Marks on a fired ceramic caused by the ware sticking to the supporting pins during firing.

perovskite. $CaTiO_3$; mol. wt. 136.7; m.p. 1915°C (3479°F); sp.gr. 4.0; hardness (Mohs) 5.5; used in stannate, titanate, and zirconate dielectric bodies.

perpend. A brick extending through a wall from one side to the other, serving to bind two segments of a wall together.

Persian red. Red pigments derived from ferric oxide or basic lead chromate.

perthite. An intergrowth of sodium and potassium feldspars.

pestle. A relatively small club-like instrument, usually composed of porcelain, quartz, or hard metal alloy, the working end being rounded and slightly roughened, for use in pounding and grinding solid substances in a mortar; may be manipulated manually or by machine.

petalite. $Li_2O \cdot Al_2O_3 \cdot 8SiO_2$; mol. wt. 612.3; m.p. 1400°C (2552°F); sp.gr. 2.39–2.46; hardness (Mohs) 6–6.5; a lithium feldspathoid used as a source of lithia in porcelain enamels, glass, glazes, and specialty bodies as a flux to promote fusion, to reduce thermal expansion, and to improve thermal-shock resistance.

petrography. The science dealing with the description and classification of rocks or the mineral composition of a ceramic body.

petroleum-coke-carbon refractory. A refractory composed substantially of calcined petroleum coke.

petrology. The science concerned with the origin, occurrence, structure, and composition of rock minerals, particularly in terms of their use in ceramic compositions.

pH. A term used to express the hydrogen ion activity or the relative acidity and alkalinity of a solution, neutral solutions being numerically equal to 7, decreasing with increasing acidity and increasing with alkalinity.

phase. A separate, but homogeneous, fraction of a system.

phase diagram. A graphical representation of the equilibrium relationships between different compounds, mixtures, and solid solutions under varying conditions of temperature.

phase equilibria. The equilibrium relationships between gas, liquid, and solid states of a compound, mixture, or solid solution under varying conditions of temperature, composition, and pressure.

phase, primary. The first crystalline phase to appear during the cooling of a liquid.

phase rule. The number of degrees of freedom in a material system at equilibrium is equal to the number of components minus the number of phases plus the constant 2.

phase transition. The change of a substance from one phase to another, such as from a solid to a liquid.

phenol-formaldehyde. A thermosetting resin used as a bonding agent for fiberglass insulation and glass-fiber cloth laminates used in electrical applications.

phenolic laminate. A glass-fiber laminate bonded with a thermosetting phenolic resin; used in electrical, structural, and mechanical applications.

phi scale. A scale for particle-size determination in which the diameter value of a sedimentary particle is replaced by the negative logarithm to the base 2 of the particle in millimeters.

phlogopite. Magnesium mica; sp.gr. 2.86; hardness (Mohs) 2.5–3.0; used in electrical and thermal insulators.

phosphate. A generic term frequently used to indicate a phosphorous-bearing compound, such as bone ash, calcium phosphate, potassium phosphate, or similar material used in glass, ceramic bodies, and glazes.

phosphate crown glass. An optical crown glass containing a substantial amount of phosphorous pentoxide, P_2O_5, as a glass-forming agent. See optical crown glass.

phosphate glass. A glass in which an essential glass-forming ingredient is phosphorous pentoxide, P_2O_5, as a partial replacement for silica, SiO_2, and which is resistant to hydrofluoric acid.

phosphate slag. A phosphate-bearing slag used in glass making. See slag.

phosphides. Binary compounds of phosphorus and metals having a potential use as semiconducting and ferroelectric materials; not so hard as the corresponding carbides, but generally more stable thermally than the nitrides.

phosphor. A luminescent material used to produce fluorescent colors, usually red, green, or violet, in porcelain enamels and other ceramics when they are subjected to ultraviolet light.

phosphorescence. Luminescence caused by the absorption of radiation such as black light and which persists after the exciting source is removed.

phosphoric acid. H_3PO_4; a rust-proofing agent for metals; sometimes present as an auxiliary opacifier in glazes and porcelain enamels.

phosphorite. A more or less impure source of calcium phosphate, phosphate tribasic, $Ca_3(PO_4)_2$, which see.

photoceramic process. A process in which an emulsion is applied to a ceramic or porcelain-enamel surface and is developed to produce a positive photographic print which subsequently is made permanent by firing. See photographic emulsion.

photochromic glass. A glass which darkens on exposure to light, but which returns to its original color and clearness when the light is removed; used, for example, in sun glasses.

photoconductor. A material in which the conductivity is increased when it is exposed to electromagnetic radiation.

photoelasticity. A technique for measuring the stresses and strains in a glass by observing the change in the double refraction of the glass when it is subjected to stress.

photoelectric colorimeter. An instrument which classifies color by means of a photocell or phototube, a set of standardized color filters, an amplifier, and a metering device.

photoelectric pyrometer. A photoelectric instrument for measuring high temperatures based on the radiant energy given off by a heated object.

photographic emulsion. A light-sensitive silver halide suspended in a gelatinous film on glass, porcelain-enamel, or glaze surfaces as a means of producing a photographic print which may be made permanent by firing.

photometer. An instrument for measuring light and electromagnetic radiation in the visible range. See disappearing-filament pyrometer.

photon. The quantum of electromagnetic energy, generally regarded as a discrete particle having no mass, no charge, and an indefinitely long lifetime.

photosensitive glass. A light-sensitive glass containing submicroscopic particles of gold, silver, or copper which precipitate during the photographic process to produce three-dimensional color pictures which are developed by heating to 538°C (1000°F); the precipitation of the metals permits the ultraviolet light to penetrate deeper into the shadowed areas while passing through the negative to promote the three-dimensional effect.

physical adsorption. The binding of an adsorbate to the surface of a solid by forces whose energy levels approximate those of condensation.

physical property. A property of a substance which may be changed without change in its chemical composition.

physical separation. The separation of solid particles by mechanical means, such as by screening.

physical stability. The ability of a solid substance to resist change in its physical characteristics under conditions of service.

pickle acid. The acid, usually sulfuric or hydrochloric acid, used to pickle iron and steel for porcelain enameling.

pickle basket. A woven or perforated, corrosion-resistant metal container in which ware is placed for cleaning and pickling prior to porcelain enameling.

pickle, pickling. The process of cleaning and etching iron and steel in an acid bath preparatory for porcelain enameling.

pickle pills. Small gelatinous capsules containing prescribed amounts of appropriate chemicals which are used to measure the strength of pickling solutions, the strength being estimated by the color of the solution in which a capsule is dissolved.

pickle stain. The discoloration of metal following the pickling operation; usually the result of inadequate washing, improper neutralization, insufficient drying, or undue exposure to the atmosphere.

pickling, anodic. The pickling of metal by electrolysis in which the metal is the anode.

pickling, gas. The pickling of metal shapes for porcelain enameling in a gaseous atmosphere of hydrochloric acid.

pickling, nickel. A process to deposit metallic nickel on iron or steel by galvanic action, reduction, or both, to promote the adherence of porcelain enamels to the metal during firing.

pick-up. The amount of porcelain enamel retained on dipped ware per unit of area, usually expressed as ounces per square foot.

picotite. A chrome spinel frequently occurring in basic refractory slag; sp.gr. 4.08; hardness (Mohs) around 8.

piezoelectric. A ceramic material such as barium titanate, lead zirconate-titanate, etc., which generates mechanical force when electrical force is applied; used in sonar, ultrasonic devices, phonograph cartridges, etc.

pig. A rest for a blowpipe or punty used during the glass-gathering operation for glass blowing.

pig-iron. Relatively impure, high-carbon iron produced in a blast furnace during the reduction of iron ore.

pigment. A solid powder employed to give black, white, or other color to bodies and coatings by reflecting light of certain wavelengths and absorbing light of other wavelengths.

pigskin. A porcelain-enamel or glaze imperfection in which the surface resembles pigskin in appearance.

pile. (1) A column of concrete or other material placed in the ground to support a vertical load or to resist lateral pressure.

(2) Nuclear material contained in a reactor in a quantity and order so as to sustain nuclear fission.

Pilkington process. A glass-making process in which molten glass is poured continuously from the tank and passed between rolls to form a continuous sheet of prescribed thickness.

pillar. (1) A column for supporting a section of a superstructure, such as in a kiln or steel furnace.

(2) The upright post used in conjunction with cranks to provide support for dinnerware, tile, and other ware during firing.

pilot plant. A small version of a planned industrial plant employed to evaluate materials and processes prior to their use on a production scale.

pimple. A small rounded or conical defect occurring on the surface of porcelain enamels, glazes, and other coatings during firing.

pin. An item of kiln furniture consisting of a triangular refractory bar or peg employed as a support for ware during firing.

pinch effect. The crazing of tile due to the contraction of the setting medium.

pinholes. Imperfections occurring in porcelain enamels, glazes, and ceramic bodies having the appearance of pin pricks, burst bubbles, or small conical holes.

pinite. A form of mica, chiefly muskovite, used in the production of dense, abrasion-resistant refractories.

pin marks. Visible imprints or marks on the back of porcelain-enameled ware caused by the firing tools.

pin mill. A disintegrating device consisting of a rotating disk equipped with pin-like protrusions which provide the disintegrating force or action.

pinning. The arranging of pins, such as posts, preparatory for the placement of ware in a kiln for firing.

pin scratching. The forming of lines or designs in porcelain enamels and glazes by scratching the coating with a sharp instrument before firing.

pin seal. A wire positioned and sealed through the inside diameter of a ceramic bushing for use in electrical and electronic applications.

pip. A type of kiln furniture which consists of a rounded refractory with a protruding point upon which ware is rested during firing.

pipe. (1) A tubular structure of concrete, metal, or other substance used to convey gases, liquids, and finely divided solids.

(2) A cavity formed in metal by contraction of the metal during solidification.

pipe, agricultural. A conduit employed to facilitate drainage of water from agricultural lands.

pipe blister. A blister-like formation in hand-blown glassware caused by an unclean or scaled blowpipe.

pipe, blow. A long metal pipe used by the glassmaker to gather and blow molten glass.

pipe, bustle. A large refractory pipe surrounding a blast furnace through which the hot-air blast is supplied to the furnace.

pipe clay. A fine-grained plastic clay marl, or fireclay, but usually a ball clay, containing little or no iron.

pipe, culvert. A drain, usually concrete, crossing under a road, railroad, or other passageway.

pipe diameter. The inside diameter of a pipe.

pipe, drain. A pipe for collecting and conveying surface and subsurface water.

pipe, field-drain. See pipe, drain.

pipe, grooved. The grooved portion of the end of a pipe which overlaps a portion of the end of an adjoining pipe.

pipe, irrigation. Pipe used for the transport and distribution of gases, liquids, and finely divided or granulated solids.

pipe, modified-design. A concrete pipe of a design different from standard.

pipe, modified-groove. The enlarged end of a pipe into which the normal end of an adjoining pipe is inserted.

pipe, modified-tongue. The normal end of a pipe which is inserted into the enlarged end of an adjoining pipe.

pipe, nonreinforced. A concrete pipe fabricated without reinforcement.

pipe-rack dryer. A steam-heated dryer in which ware is placed directly on the steam pipes for drying.

pipe, reinforced. Concrete pipe in which the concrete is strengthened by the incorporation of a reinforcing material.

pipe section. A single pipe, usually of standard or specified length.

pipe, sewer. A pipe, usually concrete, used to convey sewage and wastewater.

pipe, tamped. A concrete pipe fabricated by tamping a somewhat dry, nonslumping concrete mixture into a rotating vertical mold.

pipette, pipet. A graduated, tubular-glass device employed to transfer small measured volumes of liquids.

pipe, vibrocast. A concrete pipe fabricated by placing the concrete mix into a stationary form or mold, and densified by vibration.

pipe, vitreous clay. A clay pipe fired in a kiln to induce vitrification and which then is glazed to assure water tightness for use in drainage applications.

pipe, wall. The structural element composed of concrete or concrete and steel between the inside and outside surfaces of a concrete pipe.

pipe, well-hole. A refractory pipe or tube directing the flame upward from the well of a melting furnace.

pisé. An adobe-type construction in which walls are formed by pounding or stamping straw-tempered clay in place.

piston extruder. A machine in which clay is forced through a die by a mechanically operated cylinder.

pit. A small shallow depression or dimple in the surface of a porcelain enamel, glaze, or ceramic body.

pitch. (1) The dark, highly adhesive, sticky residue remaining from the distillation of tar or petroleum.

(2) The distance between the centerpoints of adjacent crests of a corrugated product.

(3) The ratio of the rise of a roof to its span.

pitchblende. A dark, thick, sticky substance distilled from coal tar, wood tar, or petroleum.

pitch-bonded basic refractories. Unburned basic refractory shapes bonded with pitch; if the shapes subsequently are heat treated to minimize softening of the bond on reheating, they are identified as a tempered product.

pitch-bonded basic refractories, tempered. Pitch-bonded basic refractories which are heat treated to minimize softening of the bond on reheating.

pitchers. Fragments of broken pottery, sometimes ground to a powder, for use as an ingredient in bodies, glazes, and coloring compounds.

pitch-impregnated refractories. Burned basic refractories which subsequently are impregnated with pitch after they have been fired.

pitch polishing. A glass-polishing operation in which pitch is employed as the carrier of the polishing agent instead of felt.

pit, rouge. An imperfection in the surface of incompletely polished glass containing traces of the rouge employed as the polishing agent.

pit run. Aggregate in its natural state, as excavated.

pit, sludge. An area in which sludge is collected for subsequent recycling or disposal.

pit, soaking. A conditioning furnace in which molten glass is brought to a uniform temperature for casting.

Pittsburgh sheet-glass process. A procedure for making sheet glass in which the glass is drawn vertically from the surface of the molten bath through a drawing slot of the desired thickness, the edges of the resultant sheet being formed by rollers.

place. (1) To pour concrete.
(2) To pack ware in saggers for firing.

place brick. An underfired, relatively soft brick of generally poor quality, frequently of salmon color; used in temporary or noncritical installations.

placing. The setting of ware in kilns for firing.

placing sand. Silica sand used in the placement of ware in kilns to prevent the ware from sticking to shelves, setter plates, etc., during firing.

plagioclase. A sodium-calcium feldspar.

plain. Molten glass relatively free of seeds and bubbles.

plain concrete. Unreinforced concrete, but sometimes containing light steel to minimize temperature cracking and shrinkage.

planches. A support used in the firing of porcelain-enamel artware.

planing. The smoothing of plaster molds and other surfaces by means of a tool equipped with a cutting edge.

plant. The building, machinery, tools, fixtures, instruments, and other equipment and facilities employed in a manufacturing operation.

plant layout. The arrangement of production and other facilities of a factory.

plant test. A production trial of a material or process before adoption as a standard procedure.

plaque. A flat piece of refractory ceramic upon which pyrometric cones are placed, frequently in triangular indentations, for placement in kilns with ware to be fired. See pyrometric cone.

plasma. An ionized gas generated by heating the gas to extremely high temperatures or by passing high-energy electrons through the gas.

plasma gun. A device which converts gases into high-velocity plasma. See plasma.

plasma spraying. The application of a refractory ceramic or metallic coating to a surface by means of a plasma gun. See plasma, plasma gun.

plaster. (1) Plaster of paris.
(2) Any of a group of plastic pastes made by mixing materials such as gypsum or lime with sand and water for use in construction or other applications.
(3) A mold for the casting or jiggering of ceramic bodies.

plaster, aridized. Plaster treated chemically to improve its uniformity and strength.

plaster-base finish. A rough, scored, or combed surface on the back of ceramic tile to enhance its adherence in wall and other installations.

plaster bat. A flat working surface of plaster on which clay is worked and formed.

plasterboard. A flat wall board consisting of a hardened gypsum plaster core encased in an envelope of paper, felt, or pulpboard for use as a substitute for plaster in construction; generally available in panels 4 x 8, 4 x 10, or 4 x 12 feet in surface area, and ⅜, ½, or ⅝ inches in thickness.

plaster, casting. A high-grade white gypsum product used in making castings and carvings.

plaster coat. A layer of plaster applied as a coating on walls and ceilings.

plaster ground. A section of wood, usually placed around doors, windows, archways, and at the floor in building construction as a control for plaster thickness.

plaster, gypsum. Plaster made principally from gypsum, $CaSO_4 \cdot 1/2H_2O$.

plaster, hard-finished. Over-burned gypsum treated with a solution of alum and then recalcined; used in special gypsum cements, such as Parian cement and Keene's cement.

plaster of paris. Calcined gypsum, $CaSO_4 \cdot 1/2H_2O$ or $2CaSO_4 \cdot H_2O$, which forms a quick-setting cement when mixed with water; used in building construction, as a casting and mold material, as a bedding and leveling agent in the grinding and polishing of glass, and sometimes as a batch ingredient in glasses and glazes.

plaster retarders. Substances which slow the setting rate of plasters, such as dextrin, glue, hair, and blood.

plastic. A pliant substance capable of being molded.

plastic cement. A cementitious material used to seal holes and openings in concrete.

plastic clay. Any clay which will form a moldable mass when blended with water.

plastic cracks. Cracks developing in concrete which is still in the green state.

plastic deformation. Permanent change in the size or shape of a body under stress, without fracture.

plastic fireclay. Fireclay which, when tempered with water, can be molded, extruded, or tamped into shapes or forms. See fireclay.

plastic flow. The flow of concrete or wet ceramic material under a sustained load.

plasticity. The property of a material or mixture to change permanently in size or shape when subjected to a measurable force exceeding its yield value.

plasticity, water of. The quantity of water required to convert a clay or ceramic mixture to the plastic state; reported as the percentage by weight of the dry body.

plasticizer. A substance which will impart or increase plasticity in a material or mixture.

plastic pressing. The forming of plastic bodies in dies under pressure.

plastic refractory. A water-tempered refractory which can be molded, extruded, or tamped into a desired shape or form.

plastic shrinkage. The shrinkage of concrete while in the plastic state or before the development of appreciable strength after the concrete has become rigid.

plastometer. A device to evaluate the flow properties of plastic materials or mixtures.

plate feeder. A type of conveyor consisting of overlapping plates between roller chains which delivers pulverized materials to a process or packaging unit.

plate glass. High-quality flat glass with plane, parallel surfaces formed by a rolling process; both sides are ground and polished to permit undistorted vision.

plate glass, polished. Plate glass ground and polished on both surfaces.

plate mounted. A bonded abrasive product attached to a steel wheel or plate for use on a grinding machine.

plate, screen. A perforated metal plate used in conjunction with hammer, dry-pan, and other mills to sort materials according to particle size.

platinum. Pt; used in the production of metallic colors.

platinum beryllide. $PtBe_{12}$; mol. wt. 305.63; sp.gr. 4.53. See beryllides.

platinum, liquid bright. Liquid organic mixtures containing platinum with additions of palladium, gold, or bismuth, and which fire out to a silvery finish on pottery, glass, tile, etc.

platinum-rhenium. PtRe; mol. wt. 381.45; m.p. 2449°C (4440°F). See intermetallic compound.

platting. A layer of fired brick forming the top of a scove kiln.

plinth. (1) A flat member forming the base of a column or pier.

(2) The base of a figurine or vase.

plique-a-jour. A decorative form of porcelain enamel in which colored enamel is fused between metal partitions and subsequently polished, giving an appearance of a stained-glass window.

plucking. (1) A blemish in glazed ware where the fused coating, adhering to the points of the firing supports, is broken during removal.

(2) A surface defect in flat glass caused by the glass adhering to the rollers.

plug. (1) The reciprocating part of a glass-blowing machine which forces molten glass into the mold, or forms the initial cavity in a blank mold for subsequent blowing.

(2) A wad of plastic fireclay used to seal the tap hole of a smelter.

plug clay. A damp plastic clay used to seal the tap hole of a smelter.

plug feeder. A shaped refractory controlling the rate of glass flow in the feeder channel of a glass tank.

plugging compound. A mixture of inorganic materials, such as powdered frit, clay, and water, of putty-like consistency used to fill holes and to provide a smooth surface in cast iron prior to porcelain enameling.

plug, mold. A refractory clay, graphite, or metal seal for the bottom of an ingot mold.

plumbago. Graphite or clay-graphite refractories used as linings in metallurgical furnaces.

plumbing fixtures, sanitary. China and porcelain-enameled ware employed for personal hygiene and sanitary purposes, such as lavatories, bathtubs, sinks, and toilet bowls.

plunge grinding. Grinding and polishing operations in which the grinding wheel rotates and moves radially toward the work.

plunger. The reciprocating section of a feeder which forces molten glass into a mold or forms the cavity in a blank mold for subsequent blowing.

plutonium. Pu; a radioactive metallic element, similar chemically to uranium, that is formed as the isotope 239 by the decay of neptunium which undergoes slow disintegration with the emission of a helium nucleus to form uranium 235, and is fissionable with slow neutrons to yield atomic energy. Used as a nuclear fuel and in the production of radioactive isotopes.

plutonium beryllide. $PuBe_{13}$; m.p. 1699°C (3090°F); sp.gr. 4.36. See beryllides.

plutonium boride. (1) PuB; sp.gr. 14.10. (2) PuB_2; sp.gr. 12.81. (3) PuB_4; sp.gr. 9.36. (4) PuB_6; sp.gr. ~7.25. See borides.

plutonium carbide. (1) PuC; m.p.~1655°C (3010°F); sp.gr. 13.5–14. (2) Pu_2C_3; sp.gr. 12.7. See carbides.

plutonium nitride. PuN; m.p. 2554–2749°C (4630–4980°F); sp.gr. 14.25. See nitrides.

plutonium oxide. (1) PuO; sp.gr. 13.9. (2) Pu_2O_3; m.p. 2216°C (4020°F); sp.gr. 10.2–11.2. (3) PuO_2; m.p. 2241°C (4065°F); sp.gr. 11.46.

plutonium phosphide. PuP; sp.gr. 10.18.

plutonium-rich particle. Any area of a sample which has a plutonium content greater than the surrounding matrix.

plutonium silicide. (1) Pu_3Si_2; sp.gr. 11.98. (2) $PuSi_2$; sp.gr. 9.12–9.18. (3) $PuSi_3$.

plutonium sulfide. (1) PuS; sp.gr. 10.60. (2) Pu_2S_3; m.p.~1721°C (3130°F); sp.gr. 8.41.

plyglass. A generic term for a colored sandwich-like structure consisting of a layer of glass fibers between two layers of sheet glass; usually employed in decorative applications, such as light fixtures.

pneumatically applied mortar. A concrete mortar of cement, sand, and water driven into place by compressed air.

pneumatic clay. Natural clay which has been subjected to hot liquids and gases during formation. Also known as pneumatolytic clay.

pock. A partially closed cavity on the surface of a ceramic or ceramic coating.

pocket, air. An isolated cavity of air occurring in clay bodies during the working process.

pocket, rock. Voids occurring in concrete as a result of incomplete consolidation of materials.

pocket setting. A technique of hand-placing refractory shapes in a kiln to minimize deformation and the development of stresses in the ware during firing.

pocket, skimming. A recess in a glass tank into which impurities, floating on the surface of the molten batch, may be collected and removed.

pocket, side. A slag pocket in glass tanks to trap slag and dust.

pocket, slag. A refractory-lined receptacle or pit for the collection and removal of slag from an open-hearth furnace. See slag.

point bars. A rack fabricated of a high-temperature alloy which contains points upon which porcelain-enameled ware is placed for firing.

pointing. The insertion of mortar into unfilled masonry joints, such as brickwork.

point mark. A small fracture on the back of porcelain-enameled ware occurring at the point of contact with the burning tools during firing.

poise. A measure of viscosity equal to one dyne per second per square centimeter.

poison, burnable. A neutron adsorber, such as boron, included in a reactor to assist in the control of long-term reactivity changes during progressive burnup.

poison, nuclear. A material having a high neutron-absorption cross section which reduces neutron flux in a reactor.

Poisson ratio. The ratio of the transverse contracting strain to the elongation strain in a bar or rod when forces are applied at the end parallel to its axis.

polar crystal. A crystal of ferroelectric material.

polarization. The action or process affecting light or other transverse wave radiation so that the wave vibration is confined to a single plane.

polarized ceramic. A substance, such as barium titanate, having high electromechanical conversion efficiency; used as a transducer in an ultrasonic system.

pole. (1) The part of a magnet toward which the lines of magnetic flux converge or from which they diverge. (2) A terminal of a battery.

poling. The mechanical stirring of molten glass or porcelain enamel with a metal rod to facilitate removal of gases from the molten batch.

polish. To render a surface smooth and glossy by rubbing it with a finely milled abrasive, such as rouge, cerium oxide, or a similar material.

polished plate glass. Plate glass ground and polished to render both surfaces flat and parallel to minimize reflection and visual distortion.

polished section. A small sample of a substance, highly polished on one surface, for microscopic examination.

polished wire glass. Wire-reinforced glass which has been ground and polished on both sides.

polishing, acid. The polishing of a surface by acid treatment to minimize roughness.

polishing, edge. The polishing of the edges of plate glass after cutting.

polishing, pitch. A glass-polishing operation in which pitch is employed as the resilient carrier for the polishing agent instead of the conventional felt.

polishing, surface. The further polishing of previously ground plate glass to remove slight defects.

polishing wheel. A fine-grained abrasive wheel or disk used for mechanical polishing.

polonium. A radioactive element chemically similar to tellurium and bismuth.

polyacrylamide. A water-soluble polymer, $(CH_2CHOONH_2)_x$, employed as a suspension agent or thickening material for ceramic slips and slurries.

polybasic. A chemical compound which, in solution, will yield two or more hydrogen ions per molecule; for example, sulfuric acid (H_2SO_4).

polychrome. A multicolored decoration.

polycrystalline. Composed of variously oriented crystals.

polyester laminates. Sheets, bars, and structural shapes made by impregnating glass fibers and fabrics with polyester-resin solutions, followed by curing.

polyester resin. A class of thermosetting resins produced by esterification of polybasic organic acids with polyhydric alcohols; in the cured state, they have high strength and resistance to moisture and chemicals; used as a bonding agent for glass fibers and laminated products.

polyethylene glycol. A family of colorless water-soluble liquids and solids used as binders, lubricants, and emulsifying agents.

polyethylene resins. A family of tough, water-repellent, thermoplastic materials composed of polymers of ethylene used as protective coatings for glass bottles, glass fibers, and fabrics of glass fibers; also used as an injection molding material for ceramics.

polymers, acrylic. Thermoplastic acrylic derivatives used as binders and laminating adhesives in the production of glass-fiber and sheet products.

polymorphism. The crystallization of a compound, such as silica, into two or more distinct forms.

polyvinyl acetate. A thermoplastic polymer, insoluble in water, employed as a binder in sizing compounds for glass-fiber textiles and as an adhesive for ceramic materials.

polyvinyl alcohol. A water-soluble polymer used as an addition to glazes and bodies to improve dry strength prior to firing, as a sizing and adhesive for glass fibers, and as a thickening and suspension agent for ceramic slurries.

polyvinyl butyral resin. Used as a plasticizer for the inner layer of laminated glass.

polyvinyl chloride. Used as a coating for glass bottles, glass-fiber fabrics, and as a component of molding compounds to minimize damage by abrasion.

pontil. An iron rod to which glassware is attached and held for easy manipulation during fire polishing or finishing. Also known as a punty.

pop-off. A porcelain-enamel defect in which segments of ground coat separate spontaneously from the base metal.

popout. A blemish in concrete in which a conical piece is pushed out of the surface due to expansion of an aggregate particle at the apex of the cone.

poppers. A porcelain-enamel defect in which small detached disks of ground coat appear in sheet-steel cover coats.

popping. (1) The rapid expansion of aggregate materials in light-weight cellular products.
(2) The fracture of small segments from the face of building brick.

pops, lime. The spalling of brick due to the hydration and carbonation of lime particles at or near the surface of the brick.

porcelain. A generic term for a glazed or unglazed ceramic whiteware of high quality, high strength, low absorption, and often good translucency in thin sections.

porcelain capacitor. A capacitor in which the dielectric is a high-quality porcelain fused to alternate layers of silver electrodes to form an essentially monolithic unit requiring no hermetic seal.

porcelain cement. A cement, such as a mixture of gutta-percha and shellac, used to bond porcelain to porcelain.

porcelain, chemical. A high-quality, vitreous porcelain having high resistance to chemicals.

porcelain clay. Kaolin, which see.

porcelain, cordierite. Any porcelain in which cordierite, $2MgO{\cdot}2Al_2O_3{\cdot}5SiO_2$, is the essential crystalline phase. See cordierite.

porcelain, dental. A dense, tinted, specially shaped porcelain used in prosthetic applications.

porcelain, eggshell. A generic term for very thin translucent porcelain.

porcelain, electrical. Any of several vitreous porcelain bodies used as insulators in electrical applications.

porcelain enamel. A substantially vitreous or glassy, inorganic coating applied to a metal surface and subsequently fired to temperatures above 425°C (800°F) to develop a bond between the coating and the metal.

porcelain enamel, aluminum. A porcelain enamel designed for application to aluminum, firing at 520° to 540°C (970° to 1000°F), having excellent resistance to weathering and chemical attack, impact, thermal shock, and salt corrosion, and having good dielectric strength and color retention.

porcelain-enamel fineness. The particle size of porcelain enamel frit reported as grams of dry residue retained on a designated screen size from a measured sample.

porcelain enamel frit. Selected ingredients which are mixed, melted, and then quenched in water or air to form small friable particles which are processed by milling with clay, electrolytes, and color oxides for application as a coating for metal.

porcelain enamel sanitary ware. Porcelain-enameled ware, such as bathtubs, lavatories, sinks, and other products used for sanitary and hygenic purposes.

porcelain, forsterite. Any porcelain in which forsterite, $2MgO \cdot SiO_2$, is the essential crystalline phase.

porcelain insulator. Any electrical insulator made of porcelain, the body and glaze frequently being fired simultaneously.

porcelainite. Mullite, an aluminum silicate, $3Al_2O_3 \cdot 2SiO_2$. See mullite.

porcelain, mullite. Any porcelain in which mullite is the essential crystalline phase. See mullite.

porcelain paste. Unfired porcelain bodies, usually in the plastic state.

porcelain process. The process of producing porcelain ware in which the body and glaze are fired simultaneously.

porcelain, Réaumur. A porcelain in which a fritted or devitrified glass is the major constituent.

porcelain, semi-. A generic term for semivitreous dinnerware.

porcelain, steatite. Any porcelain in which steatite, $MgO \cdot SiO_2$, is the essential crystalline phase. See steatite.

porcelain tile. A dense, fine-grained, smooth, usually impervious tile having a sharp face; generally produced by dust pressing.

porcelain, titania. Any technical porcelain in which titania, TiO_2, is the essential crystalline phase. See titanium dioxide.

porcelain, zircon. Any technical porcelain in which zircon, $ZrO_2 \cdot SiO_2$, is the essential crystalline phase. See zircon.

porcelain, zirconia. Any porcelain in which zirconia, ZrO_2, is the essential crystalline phase. See zirconia.

porcelaneous. Resembling unglazed porcelain. See porcelain.

pore. (1) An internal cavity in a solid substance, usually one which can be exposed by cutting, grinding, or polishing.

(2) Voids between grains of a grinding wheel. See porosimeter.

pore diameter. The average diameter of pores in a solid substance.

pore diffusion. The passage of a gas or liquid into and through the porous structure of a solid.

pores, sealed. Pores or small bubbles in a ceramic body which have no outlet to the exterior of the body. See sealed pores.

pore size. The average volume of pores contained in a solid substance.

pore-size distribution. The range of the size variation of pores contained in a solid.

pore volume. The total combined volume of open and sealed pores contained per unit of volume or weight of a solid, calculated by the formula:

$$P_t = \left[1 - \frac{D_b}{D_t}\right] \times 100$$

in which P_t is the total volume of pores reported as a percentage of the total bulk volume of the specimen, D_b is the bulk density, and D_t is the true density of the specimen.

pore volume, closed. See pore volume, sealed.

pore volume distribution. The distribution or arrangement of pore volume among pores of varying dimensions.

pore volume, open. See open pore volume.

pore volume, sealed. The ratio of the pores or small bubbles entrapped in a ceramic body which have no outlet to the exterior of the body, to the bulk volume of the body; expressed as a percentage of the bulk volume. See sealed pores.

pore water. The tempering water contributing to the pore structure in clays and bodies during and after working and forming.

porosimeter. An instrument used to measure porosity in a solid substance.

porosity. (1) The state of being porous.

(2) The ratio of the volume of pores, both open and closed, to the total volume of a body.

(3) The ability of fired ware to absorb water or other liquid by capillary action.

porosity, apparent. The ratio of the open pore volume to the total bulk volume of a solid. See apparent porosity.

porosity, sealed. The ratio of the volume of pores sealed within a solid to the total volume occupied by the solid.

porous. Containing or being filled with pores.

porous area. A three-dimensional area in a body into

which a dye will penetrate through the surface if an opening should exist.

porous carbon. An item fabricated from carbon particles pressed together without use of a binder; greater in strength but less resistant to oxidation than porous graphite.

porous graphite. An item fabricated from graphite particles pressed together without use of a binder; lower in strength but more resistant to oxidation than porous carbon.

porous mold. A forming mold containing numerous open pores or channels through which gases and liquids can pass as a means of removal from a formed body.

porous wheel. A vitrified or resin-bonded grinding wheel having a porous structure.

port. An opening in a furnace wall serving as an entrance for fuels or flames, and as an exit for exhaust gases.

portable grinder. A grinding machine which is supported and manipulated manually by an operator.

portland blast-furnace slag cement. A hydraulic cement consisting of an intimately ground mixture of portland-cement clinker and granulated blast-furnace slag. See slag, portland cement.

portland cement. A hydraulic cement produced by finely pulverizing a calcium aluminum silicate clinker together with additions of gypsum or other forms of calcium sulfate.

portland cement, white. Finely milled white cement made from pure calcite limestone and a white-burning clay.

portland-pozzolan cement. A hydraulic cement consisting of a mixture of portland and pozzolan cements.

positron. The positive counterpart of an electron, having approximately the same mass and magnitude of charge.

post. An upright member of a kiln-furniture assembly which holds the pins upon which ware is placed for firing.

postcleaning. The removal of a penetrant from a specimen following a penetration-porosity study, usually by wiping or washing.

postemulsification cleaning. The removal of a penetrant from a specimen by means of an emulsifying agent following a penetration-porosity study.

post-tensioned concrete. A prestressed concrete to which a tensile stress is applied to the prestressing tendons after the concrete has attained sufficient or maximum strength.

pot. A rounded refractory container or crucible in which glass is melted and refined.

pot arch. A furnace for the preheating or firing of a glass-melting pot.

potash. Potassium carbonate, K_2CO_3, which see.

potassium acetate. $KC_2H_3O_2$; used as a flux in the production of crystal glass.

potassium aluminate. $K_2O{\cdot}Al_2O_3$; mol. wt. 196.14; m.p. >1649°C (3000°F).

potassium aluminum silicate. (1) $K_2O{\cdot}Al_2O_3{\cdot}SiO_2$; mol. wt. 256.20. (2) $K_2O{\cdot}Al_2O_3{\cdot}2SiO_2$; mol. wt. 316.26; m.p. 1749°C (3180°F); sp.gr. 2.6; hardness (Mohs) 5–7. (3) $K_2O{\cdot}Al_2O_3{\cdot}4SiO_2$; mol. wt. 436.50; m.p. 1688°C (3070°F); sp.gr. 2.47; hardness (Mohs) 5–7. (4) $K_2O{\cdot}Al_2O_3{\cdot}6SiO_2$; mol. wt. 556.50; m.p. 1149°C (2100°F); hardness (Mohs) 5–7.

potassium bifluoride. KHF_2; employed as a glass etchant.

potassium carbonate. K_2CO_3; mol. wt. 138.2; m.p. 909°C (1668°F); sp.gr. 2.3; used as a fluxing agent, glass former, and sometimes as an opacifier in glass, glazes, and porcelain enamels.

potassium chloride. KCl; mol. wt. 74.6; m.p. 776°C (1429°F); sp.gr. 1.98; used as a set-up agent in porcelain enamel slips.

potassium chromate. K_2CrO_4; mol. wt. 194.2; m.p. 971°C (1781°F); sp.gr. 2.73; used as a yellow or orange pigment in porcelain enamels and glazes.

potassium cyanide. KCN; mol. wt. 65.1; m.p. 634°C (1173°F); sp.gr. 1.52; used as a neutralizer in the pickling of metals for porcelain enameling.

potassium dichromate. $K_2Cr_2O_7$; mol. wt. 294; m.p. 396°C (745°F); sp.gr. 2.69; used in glass for aventurine effects and for the production of green colors; used in glazes for the production of chrome-tin pinks, low-fire greens, and purplish red colors.

potassium feldspar. A potassium-bearing feldspar of the general formula $KAlSi_3O_8$; see feldspar.

potassium fluoride. KF; mol. wt. 58.1; m.p. 800°C (1472°F); sp.gr. 2.5; employed as a glass etchant and as a flux in the preparation of ferroelectric crystals of barium titanate.

potassium fluosilicate. See potassium silicofluoride.

potassium metatantalate. $KTaO_3$; mol. wt. 268.5; used in special ferroelectric and ferromagnetic applications.

potassium niobate. $4K_2O{\cdot}3Nb_2O_3$; mol. wt. 863; a ferroelectric compound having a Curie temperature of 420°C (788°F).

potassium nitrate. KNO_3; mol. wt. 101; m.p. 337°C (639°F); sp.gr. 2.1; employed in glass, glazes, and porcelain enamels as a flux and oxidizing agent.

potassium nitrite. KNO_2; mol. wt. 39.1; m.p. 297°–450°C (567°–842°F); sp.gr. 1.9; employed as a color stabilizer, antitearing agent, and set-up addition in porcelain enamels.

potassium orthophosphate. K_3PO_4; mol. wt. 212.3; m.p. 1340°C (2444°F); used as a suspension and dispersing agent in porcelain enamel and glaze slips.

potassium oxide (potash). K_2O; mol. wt. 94.2; decomposes on heating; sp.gr. 2.32; used as a flux and color stabilizer in glass, glazes, and porcelain enamels, and as a deflocculating agent in engobes and in casting and glaze slips.

potassium pyrophosphate. $K_4P_2O_7 \cdot 3H_2O$; mol. wt. 374.4; m.p. 1090°C (1994°F); sp.gr. 2.33; used as a suspension and dispersing agent in porcelain enamel and glaze slips.

potassium silicate. $K_2O \cdot SiO_2$; mol. wt. 154.3; sp.gr. 1.25–1.39; used as a source of potassium and silica and as an antiblooming agent. See bloom.

potassium silicofluoride. K_2SiF_6; mol. wt. 224.2; sp.gr. 3.0; used as a fluxing ingredient in porcelain enamels.

potassium sulfate. K_2SO_4; mol. wt. 174.3; m.p. 1072°C (1968°F); sp.gr. 2.66; used as a raw material in glassmaking.

potassium tantalate. $KTaO_3$; mol. wt. 268.5; used in ferroelectric and ferromagnetic applications.

potassium titantate. K_2TiO_3; mol. wt. 174.1; m.p. 1370°C (2498°F); sp.gr. 3.2; used in the production of thermal insulating fibers.

potassium uranate. $K_2O \cdot UO_3$ mol. wt. 380.37; m.p. 1620°C (2950°F).

potassium zinc silicate. $K_2O \cdot ZnO \cdot SiO_2$; mol. wt. 235.64; m.p. 1297°C (2370°F).

pot bank. (1) A battery of glass-melting pots or crucibles.
(2) A pottery factory.

pot, cannon. A small glass-melting pot or crucible.

pot clay. Refractory clays used in the manufacture of glass-making pots and crucibles.

pot, closed. A glass-melting pot in which the batch is protected from the furnace atmosphere by means of a refractory cover or crown.

potentiometer. An instrument for the measurement of temperature by means of electromotive forces.

potentiometer, sliding. A potentiometer which employs a sliding contact along a length of resistance wire to regulate the voltage in the wire in temperature measuring and control instruments.

pot furnace. A furnace into which pots and crucibles are placed for the melting of porcelain enamels, glass, and glazes.

pot glass. Glass melted in pots or crucibles.

pot, glass. A one-piece crucible-shaped refractory container, open or closed, in which glass is melted.

pot, glazed. A glass-melting pot in which the interior is coated with a protective glass layer before use.

pot, jockey. A glass-melting pot of such size and shape that it may be supported in a furnace by two other pots.

pot life. The length of time, or the number of cycles, a pot is in actual use before it is discarded.

pot, open. A glass-melting pot in which the batch is exposed to the furnace atmosphere.

pot, revolving. A shallow, slowly rotating, refractory pot from which molten glass is drawn in the Owens bottle-making process.

pot ring. A floating refractory ring on the surface of glass melted in a pot, to prevent the accumulation of scum in the gathering area.

pot, skittle, A small pot or crucible in which glass is melted.

pot spout. A connecting refractory shape through which molten glass is transferred from a glass tank to a revolving pot.

potter. A workman, usually an artisan, engaged in the production of pottery and similar artware.

potter's clay. Any ball clay used in the production of pottery.

Potter's red cement. A cement composed of a mixture of portland cement and sintered red clay which has been crushed; used as a decorative cement.

potter's wheel. A rotating wheel or disk, powered manually or mechanically, upon which pottery is shaped by manual manipulation.

pottery. (1) A generic term denoting ware, such as vases, bowls, plates, and pots, shaped from moist clay and hardened by firing.
(2) The building or establishment in which pottery is made.
(3) The craft or occupation of a potter.

pottery-body stains. Finely ground pigments used in coloring terra cotta, tile, abrasives, and other ceramic products where the pigment becomes part of the body.

potting. (1) An embedding process in which an electronic assembly is encased in a thermosetting material to protect the assembly from the effects of vibration, shock, moisture, and corrosive agents.

(2) The process of making pottery and similar artware.

potting material. The thermo-setting insulating material used to protect potted electrical components from shock, vibration, air, moisture, etc.

pot wagon. A cart-like vehicle used to transport pots from a pot arch to a pot furnace.

pot warping. The distortion of pots during drying or firing.

pour. (1) To empty a pot, crucible, or other container of its contents, usually in a stream, by tipping the container.
(2) The batch removed from a pot, crucible, or other container by pouring.
(3) To place concrete.
(4) A batch of concrete in a single continuous placement.

pour density. The weight of a powdered or granular material poured into a graduated container divided by its volume.

pouring-pit refractory. (1) The refractory shapes used in the flow control of steel between the furnace and the mold.
(2) A refractory used in casting molten metal.

pouring, top. The direct transfer of molten steel from a ladle into ingot molds, usually by means of refractory nozzles.

pour point. (1) The lowest temperature at which a liquid will flow.
(2) The optimum temperature for the pouring of a molten substance or batch, such as glass or a glassy composition.

powder. Dry, finely divided particles of a solid substance.

powder blue. A mixture of cobalt oxide, silica, and potassium carbonate or other flux used as a colorant in glass, glazes, and porcelain enamels.

powder density. The density of a material in powder form, including all pores; calculated as the ratio of the mass of the material to its true volume.

powder, diamond. Micron-size diamonds, usually dispersed in a paste or other carrier, employed as an abrasive or polishing material to produce finely finished surfaces.

powdered activated carbon. Activated carbon predominately of 80-mesh and smaller particle size.

powdering. (1) The process of reducing the particle size of a substance to powder form.
(2) The process of applying powdered coatings or decorations to pottery or other ceramic ware.

powder metallurgy. The production of objects by pressing, binding, and sintering powdered metals.

powder, separating. A powder applied to a surface, as in a mold, to facilitate the removal of ware after forming.

powder, sinterable. A powder or compact of powders which can be formed and then bonded by heat treatment.

power factor. The ratio of electric power dissipated in a material to the effective voltage and current, expressed in watts.

pozzolan. A material such as certain fly ashes and blast-furnace slags which, in finely divided form, will exhibit cementitious properties when mixed with lime and water.

pozzolan cement. A cement produced by grinding portland cement with a pozzolanic material or a mixture of pozzolanic material with hydrated lime. See pozzolan.

Prague red. A red pigment consisting essentially of red iron oxide.

prall mill. An impact mill consisting of a rotating impeller, a baffle rotating in the opposite direction, and a stationary baffle.

prase. A form of natural silica or quartz.

praseodymium boride. (1) PrB_4; mol. wt. 184.20; sp.gr. 5.20. (2) PrB_6; mol. wt. 205.84; sp.gr. 4.86. See borides.

praseodymium carbide. (1) Pr_2C_3; mol. wt. 317.84; sp.gr. 6.62. (2) PrC_2; mol. wt. 164.92; m.p. 2535°C (4595°F); sp.gr. 5.73. See carbides.

praseodymium nitride. PrN; mol. wt. 154.93; sp.gr. 7.49. See nitrides.

praseodymium oxide. (1) Pr_6O_{11}; mol. wt. 1021.8; used in the production of yellow and green ceramic colors. (2) PrO_2; mol. wt. 172.92. (3) Pr_2O_3; mol. wt. 329.84; m.p. 2199°C (3990°F); sp.gr. 10.9.

praseodymium phosphide. PrP; mol. wt. 171.95; sp.gr. 5.72. See phosphides.

praseodymium-ruthenium. $PrRu_2$; mol. wt. 344.32; m.p. 1679°C (3055°F).

praseodymium silicate. $Pr_2O_3 \cdot SiO_2$; mol. wt. 389.90; m.p. 1398°C (2550°F).

praseodymium silicide. $PrSi_2$; mol. wt. 197.04; sp.gr. 5.64.

praseodymium sulfide. (1) PrS; mol. wt. 172.98; m.p. 2229°C (4045°F); sp.gr. 6.03. (2) PrS_2; mol. wt. 205.04; m.p. 1782°C (3240°F); sp.gr. 5.16. (3) Pr_2S_3; mol. wt. 378.02; m.p. 1793°C (3260°F); sp.gr. 5.27. (4) Pr_3S_4; mol. wt. 551.0; m.p. 2099°C (3810°F); sp.gr. 5.77.

praseodymium yellow. A glaze colorant composed of a mixture of silica, zirconia, and approximately 5% of praseodymium oxide.

precast concrete. Concrete which has been cast in molds or forms at a location other than the site of its ultimate use.

precision. The highly accurate agreement of repeated measurements, particularly measurements which vary minimally from an established standard.

precision-bore glass tubing. Glass tubing heated to softness and then shrunk over a metal mandrel or core.

precision casting. The forming of a product of precise dimensional measurements by casting in a mold.

precision grinding. The machine grinding of an item to specified and precise dimensional measurements.

precleaning. The removal of surface contamination prior to subsequent treatment or use.

precoat. The preliminary application of a refractory slurry to a casting pattern prior to application of the main slurry.

predryer. Preliminary drying of a substance prior to further treatment or processing.

prefabricated masonry. Masonry products fabricated in a factory or other location for rapid assembly at the site of construction.

preferential adsorption. The adsorption of certain materials to a greater extent or at a more rapid rate than other materials.

preform. (1) The initial fabrication of a shape.
(2) A sintered or prefired compact of powdered glass used in the production of glass-to-metal seals.

preheat. To subject to heat treatment prior to firing.

preheat zone. The section of a continuous furnace or kiln preceding the hot or firing zone.

prehnite. $Ca_2Al_2Si_3O_{10}(OH)_2$; sp.gr. 2.8–2.95; hardness (Mohs) 6.0–6.5; a natural hydrous silicate of calcium and aluminum related to the zeolites, which see.

premature stiffening. The false or erratic, abnormal, quick-setting of cement in concrete due, usually, to the presence of unstable calcium in the cement.

premix burner. A burner in which the fuel and air are mixed prior to the injection and ignition in the combustion chamber of a furnace or kiln.

preset cracks. Cracks occurring in concrete before the concrete has set.

presintering. A preliminary heat treatment prior to subsequent sintering or firing.

press-and-blow process. A process of glass manufacture in which the molten, seed-free glass is pressed into a preliminary shape and subsequently is blown to the final shape of the ware.

press cloth. The cloth, such as nylon, cotton, or jute, which is used in filter presses for dewatering slurries.

pressed brick. Brick densified under pressure before firing; usually made from clay of low-moisture content (5 to 7%).

pressed glassware. Glassware formed under pressure between a plunger and a mold while in the molten or plastic state.

press, hydraulic. A press actuated by a liquid, such as oil, under pressure.

pressing. The forming of ware under pressure, usually in a die.

pressing blank. A rough shape, particularly glass, from which a finished article is formed.

pressing die. A mold in which an item is formed under pressure.

pressing, dry. The forming of ware from powdered or granular material by pressing in a mold or die, the water content of the body ranging from 5 to 10%.

pressing, dust. The forming of ware from powdered or granular material by pressing in a mold or die, the water content of the body being 1.5% or less.

pressing, hot. The forming of plastic bodies under the influence of heat, such as by the use of heated plungers and dies or jiggering tools.

pressing, impact. The forming of ware in a die in which powdered or granular materials are compacted by rapid vibration.

pressing isostatic. The forming of ware of high density from powdered or granular material encased in a soft rubber or plastic container and subsequently subjected to high isostatic pressure in a closed cylinder or die.

pressing, plastic. The forming of plastic bodies in molds by direct application of pressure.

pressing, ram. The forming of plastic bodies by pressing in porous molds which are subjected to a vacuum to hasten the extraction of water; the formed item then is released from the mold by air blown through the mold parts.

pressing, wet. The forming of bodies in a wet plastic state by the use of pressure.

press molding. The forming of ware by pressing in absorbent plaster molds.

press, offset. A printing press which applies an inked

impression to a rubber-blanketed cylinder which transfers the impression to the item being decorated.

press, screw. A press in which the slide is actuated by a screw mechanism.

press, slug. A press employed to compact fine powders prior to granulation.

press, swing. A screw press, frequently hand operated, used to form special shapes in small quantities.

press, toggle. A mechanical press in which the slide mechanism is actuated by means of a toggle mechanism.

pressure. The compressive stress applied to a substance or item, expressed as exerted force per unit of area.

pressure casting. The forming of ware by casting, followed by the application of pressure to densify the formed item in the mold, and to minimize drying shrinkage and speed the rate of production.

pressure check. A crack in a glass article caused by the use of excess forming pressure.

pressure, contact. The force of contact between two surfaces per unit of area.

pressure density. The density of a compacted substance prior to firing or sintering.

pressure dye test. A porosity test in which a dye solution is applied to a test surface under pressure.

pressure, hydrostatic. The pressure exerted by a liquid at rest.

pressure measurement. The measurement of static or dynamic pressures in units of weight per unit of area.

pressure, molding. The force required to press a plastic substance into all areas of a mold chamber.

pressure regulator. An instrument designed to control the pressure exerted on a substance.

pressure sintering. The heat treatment of a substance under pressure to form a coherent mass, but without melting.

pressure tank. An air-tight container in which slurries and liquids are placed under pressure and forced into a spraying or other distributing system.

pressure, vapor. The pressure exerted by a vapor that is in contact equilibrium with its liquid or solid form.

press, vibratory. A press in which the top and bottom dies employ impact-type vibrations to compact powder and granulated materials.

prestress. To introduce internal stresses into a structure to counteract stresses or loads to which the structure will be subjected in service.

prestressed concrete. Concrete in which a compressive stress is applied by means of prestressed steel rods, wires, or strands incorporated in the concrete mix during the fabrication of a product.

pretensioned steel. Steel rods, wires, or strands placed in tension for use in prestressed concrete, the tension being released after the concrete has set, thus placing the concrete under compression.

primary air. The air introduced into a burner or combustion chamber together with the fuel.

primary boiling. The initial evolution of gas during the firing of porcelain enamel, sometimes resulting in blisters or other surface defects.

primary clay. A feldspathic type of weathered clay which remains geologically at the site of its formation.

primary crusher. The initial crusher of a series employed to pulverize minerals.

primary insulation. The initial layer of insulating material applied over a conductor.

primary jacket. An insulating material applied as mechanical protection over primary insulators.

primary phase. The first crystalline phase occurring during cooling of a liquid.

primary recrystallization. The initial growth of strain-free grains in a matrix which has been plastically deformed.

primary standard. A specimen in which specific properties have been measured, and these measurements have been adopted as standards for comparison.

printed circuit. A circuit for electronic apparatus consisting of a deposit of conducting material on an insulating surface in a prescribed pattern.

printer's bit. A refractory spacer used in the setting of ware in a decorating kiln.

printing ink. A mixture of ceramic pigment and liquid medium used in the decoration of ware.

printing, silk screen. A decorating process in which a design is printed on a surface through a tightly stretched screen by means of a rubber squeegee, the areas not to be coated being blocked by a suitable resist medium.

printing transfer. The marking or decoration of ware by means of decals made from engravings or lithographs.

prism. A crystal consisting of three or more faces parallel to the crystal axis.

prismatic glass. A translucent glass consisting of parallel prisms which produce an iridescent, sometimes multicolored, appearance.

probability. The occurrence or state of being probable; the relative frequency of the occurrence of an event based on the ratio between its occurrence and the total average number of cases necessary to insure its occurrence.

probe coil. A small coil or coil assembly used in electromagnetic testing which is placed near or on the inside of a test specimen.

process. A series of operations directed toward a particular end result, such as a manufacturing process or a forming process.

process control. The manipulation of manufacturing conditions to obtain an end product of desired or specified quality.

process, dry. (1) The mixing of the ingredients of a ceramic body, glaze, or other composition in the dry state, followed by the addition of a liquid as required for subsequent processing or fabrication.
(2) The application of powdered porcelain enamels to metal articles, usually cast iron, which have been preheated to temperatures above the maturing temperature of the coating.

process fishscaling. The development of half-moon or fishscale-like fractures in porcelain enamels during drying or firing of the coating.

process, wet. (1) The mixing of the ingredients of a ceramic body or coating with liquid in quantities sufficient for subsequent processing.
(2) The application of porcelain enamels or glazes in slip form.

Proctor dryer. A type of tunnel dryer in which heat for drying is obtained by circulating air over pipes containing steam or waste heat.

producer gas. The gas produced by burning a solid fuel with a restricted supply of air or by passing air and steam through an incandescent fuel under conditions to convert carbon dioxide to carbon monoxide.

producer, gas. A furnace employed for the gasification of coke. See producer gas.

production control. The planning, scheduling, routing, dispatching, and expediting of the flow of materials in an orderly and efficient manner through a manufacturing operation.

profilometer. An instrument, designed to measure the surface roughness of a flat solid, consisting of a needle drawn across the surface, irregularities being recorded by an appropriate instrument.

promethium oxide. Pm_2O_3; mol. wt. 347.0; the oxide of the most abundant promethium isotope.

proof. A sample of molten glass obtained for inspection by means of an iron rod stirred in the molten bath.

properties, mechanical. The properties of materials associated with elastic and inelastic reactions under the influence of an applied force, or that involve a relationship between stress and strain.

props. Refractory supports on which shelves are arranged for the setting of ware to be fired.

proximate analysis. A mineralogical analysis of a substance calculated on the basis of its chemical composition.

prunt. A handle or other piece fused on art, dinner, and similar glassware following the forming operation.

Prussian blue. The most common of the iron ferrocyanide blue pigments.

Prussian red. A family of red pigments made from ferric oxide or potassium ferrocyanide.

psi. Abbreviation for pounds per square inch.

psf. Abbreviation for pounds per square foot.

***p*-type semiconductor.** A semiconductor containing small quantities of boron, aluminum, or gallium which take a few electrons away from the atoms of the semiconductor to form holes which pass from atom to atom to produce an electric current. See hole.

pucalla. A tool used to widen the mouth of a goblet or other glass product during the forming operation.

pugging. The process of blending clays and water by manual or mechanical means to produce bodies of forming consistencies.

pug mill. A machine consisting of auger-like blades mounted in a trough for use in the mixing, compression, and extrusion of plastic clay bodies.

pug mill, vacuum. A pug mill containing a vacuum chamber for removal of air from a pugged clay body before it moves into the extrusion chamber.

pug mixer. See pug mill.

pug mixer, vacuum. See pug mill, vacuum.

pull. (1) The quantity of glass produced in a glass-melting tank during a designated period of time.
(2) The draft in a chimney or flue.

pulled stem. A stem of glassware pulled from the bowl while the glass is in a plastic state.

pulpstone. Sandstone cut into wheels for use in grinding and polishing operations.

pulverizer. Any machine designed to reduce solid substances to very small particle sizes.

pumice, pumicite. A lightweight, porous volcanic ash

of glassy composition and texture used as a polishing medium, as a lightweight concrete aggregate, as a sound and thermal insulator, and as a raw material in brick manufacture.

pumpcrete. (1) Concrete pumped through a pipeline.
(2) A machine which pumps concrete through a pipeline.

pumping. The loss of concrete fines through cracks and joints of a wet pavement under heavy traffic, which creates a pumping action.

pump, vacuum. A device for exhausting air or other gases from an enclosed area.

punch test. A test in which a glaze is fractured by means of a center punch to determine if the fired coating is under tensile or compressive stresses.

punchware. Thin, handblown glassware, such as tumblers.

punt. The bottom section of a glass container.

punt code. The hallmark on the bottom of a glass article.

punt, offset. The bottom of a bottle that is asymmetrical to the axis of the bottle.

punt, pushed. The concave bottom of a glass article.

punty. (1) An iron rod used to gather glass gobs for the production of pressed ware.
(2) An iron rod to which glass is attached while being shaped and fire polished. Also known as pontil.

pup. A long, refractory brick of square cross section.

purchase order. An order for the purchase or procurement of merchandise under some condition of payment.

purchaser. An individual or organization issuing an order to purchase.

pure clay. Alumino-silicic acid; a clay consisting theoretically of 39.5% of alumina, 46.6% of silica, and 13.9% of water.

purge. To sweep a furnace atmosphere of undesirable gases, usually by passing nitrogen through the chamber.

purity grade. The degree of purity of a substance, determined by chemical analysis.

purity grade, ultra-high. Materials containing no more than ultra-trace levels of impurities, designated in the region below 1 ug/g.

purple of Cassius. A ceramic pigment composed of mixtures of tin and gold chlorides.

purpling. The change of chrome-tin pink to an off-color during the firing operation due to an excess of borax and alkali and a deficiency of lime.

push-bat kiln. A kiln in which ware is pushed through the kiln on bats or refractory slabs.

pushed-down cullet. An imperfection in glassware caused by the presence of cullet in the drawing zone of a glass-melting tank.

pushed punt. The concave bottom of a glass article.

pusher. A device designed to push ware through a kiln during the firing operation.

pusher kiln. A kiln, usually small, in which ware, placed on a platform is pushed manually or mechanically through the firing zone.

push schedule. The rate at which ware is pushed through a firing kiln.

push-up. A pushed punt. See punt, pushed punt.

putty. (1) A white polishing compound.
(2) A dough-like cement composed of whiting and linseed oil used to set glass in sashes.

putty, lime. A plastic form of hydrated lime sometimes employed as an additive to mortar.

puzzolan. An alternative spelling of pozzolan, which see.

pycnometer. A container of known volume used to determine the density of a liquid, the density being calculated from the weight and volume of the liquid contained in the container.

pyrite. Fe_2S; mol. wt. 143.7; sp.gr. 4.9–5.2; hardness (Mohs) 6–6.5; used in amber glass and as a filler in resin-bonded abrasives and brake linings.

pyro. An abbreviated term for tetrasodium pyrophosphates, $Na_3P_2O_7$,which see.

pyroceram. A proprietary, hard, strong, opaque-white nucleated glass with a nonporous, crystalline structure having high-shock resistance and flexural strength.

pyrohydrolysis. The decomposition of a substance by the combined action of heat and water vapor.

pyrolucite. MnO_2; mol. wt. 86.9; sp.gr. 4.73–4.86; hardness (Mohs) 2–2.5; used as a purple or red colorant in glazes, glass, and porcelain enamels and as an adherence-promoting agent for porcelain enamels on sheet iron and steel.

pyrolytic coating. A coating formed on the surface of an article by thermal decomposition of a volatile compound, such as a coating of silica, by the decomposition of silicon tetrachloride in a vacuum.

pyrolytic graphite. A form of graphite of high purity having high-thermal and electrical conductivity; used in high-temperature applications.

pyrometer. An instrument used for the measurement of high temperatures.

pyrometer, disappearing filament. An optical pyrometer in which an electrically heated filament in a telescope disappears when focused on an incandescent surface, the temperature being read from a calibrated scale on the instrument.

pyrometer, optical. A disappearing filament pyrometer, which see.

pyrometer, photoelectric. An instrument which employs a photoelectric system to measure the radiant energy emanating from a heated object as a measurement of elevated temperature.

pyrometers, sentinal. Small cylinders of standardized compositions which melt at predetermined temperatures; used to measure and control thermal treatments of materials. See pyrometric cones.

pyrometric cone. Pyramids composed of oxide mixtures which deform at known temperatures; used to indicate the thermal history of fired ware. See cone, pyrometric.

pyrometric cone equivalent. The assigned identifying number of a pyrometric cone which deforms or bends so that its tip touches the supporting plaque or base during a firing cycle to indicate the approximate temperature and thermal history of ware during the firing operation. See pyrometric cone.

pyrometry. The science of thermal measurement.

pyrophoric. The property of igniting simultaneously.

pyrophyllite. $Al_2O_3 \cdot 4SiO_2 \cdot H_2O$; a phyllosicate mineral resembling talc; m.p. 1800°C (3272°F); mol. wt. 360.2; sp.gr. 2.8–2.9; hardness (Mohs) 1–2; employed in refractories, castables, plastic and gunning mixes, insulator bodies, and tile to reduce thermal expansion and as a source of alumina; also used as a sealer in the pressure forming of synthetic diamonds at elevated temperatures.

pyroplasticity, pyroplastic deformation. High-temperature plasticity resulting in permanent deformation of a body under stress.

pyroscope. A shaped material such as a cone, ring, bar, or pellet, which melts or softens at a definite temperature and which is placed in a kiln to serve as an indicator of temperature conditions during a firing operation.

pyroxene. Any of a group of silicate minerals containing two metallic oxides, such as calcium, iron, magnesium, or sodium.

PZT. An acronym for lead zirconium silicate which is used in piezoelectric transducers.

Q

quadrant mat. The tension-zone circumferential reinforcement secured to a cage in a concrete-pipe wall.

qualification test. A test or series of tests designed to evaluate the functional, environmental, reliability performance, and other pertinent properties or characteristics of a material, component, or system to assure the producer, supplier, and consumer that the item or product will meet specified conditions of performance.

qualitative. Concerned with or related to quality.

qualitative analysis. An analysis in which some or all of the components of a sample are identified.

quality. The degree or relative nature of a material or product in terms of excellence rather than in terms of amount.

quality assurance. Activities undertaken by a manufacturer or supplier to assure a customer or consumer that a delivered product is acceptable in all respects.

quality control. Activities undertaken by a manufacturer or supplier to attain materials or products of a specified or a satisfactory quality.

quality-control chart. A graphic or pictorial presentation of data to indicate the properties or quality of a product in the course of manufacture in order to maintain or adjust procedures of production to make certain that the product meets prescribed specifications.

quality-control tests. A test or series of tests performed to verify and to maintain a desired level of quality in a product or process.

quality verification tests. Tests performed to verify and maintain a desired level of quality in a product or process.

quantitative analysis. (1) An analysis in which the relative amounts of some or all of the components of a sample are determined.

(2) For emission spectrochemical determination of impurities, the amounts of the impurities are measured relative to standards or reference materials containing the impurity elements being determined. If the measure-

ments are made by visual comparison of the spectrographic lines, the analysis is classed as semiquantitative. If the intensities of the spectrographic lines are measured with a microdensitometer, or if the emitted light from the sample is measured with phototubes, the analysis is classified as quantitative.

quantity. Concerned with or related to the amount or number, as opposed to the nature, of a thing or substance; a specified amount in prescribed units.

quarl block. A refractory shape employed as a burner segment for the injection of gaseous or liquid fuel into a glass-melting tank.

quarry. An excavation in the earth from which building stone, limestone, slate, coal, sand, clay, gravel, or other mineral is removed.

quarry tile. An unglazed tile, usually 39 cm.^{2} (6 in.^{2}) or more in top surface area, and 13 to 19 mm. (½ to ¾ in.) in thickness, made by the extrusion process from natural clays or shales.

quarter. To divide a sample of a material, such as an aggregate or other material sample, into four equal parts to reduce the sample to a size suitable for analysis.

quartz. SiO_2; mol. wt. 60.1; m.p. 1710°C (3110°F); sp.gr. 2.65; hardness (Mohs) 7; a natural crystal appearing in many varieties such as agate, chalcedony, chert, flint, opal, etc.; it is the most abundant and wide-spread of all minerals, and is used extensively as a glass former and as a vitrification aid in ceramic compositions.

quartz, alpha. Quartz with a trigonal-trapezohedral structure which is stable at temperatures below approximately 573°C (1060°F).

quartz, beta. See cristobalite, tridymite.

quartz crystal. A natural or artificial crystal of SiO_2 having piezoelectric properties. Also known as rock crystal.

quartz-crystal filter. An electronic filter in which a quartz crystal is the essential component.

quartz-crystal resonator. A quartz plate having a natural vibration frequency such that it may be employed to control the frequency of an oscillator.

quartz, fused. Pure SiO_2 fused to yield a glass-like material when cooled; used in apparatus and equipment requiring low-thermal expansion and high thermal-shock resistance, high-melting point, high-chemical resistance and transparency.

quartz glass. A transparent or translucent vitreous silica made by the fusion of vein quartz or silica sand. Also known as fused quartz.

quartz inversion. A change in the crystal form and properties of quartz by heat treatment, but without change in composition; for example, alpha quartz is converted to beta quartz at 573°C (880°F); the change is reversible. At higher temperatures, quartz is converted to cristobalite or tridymite, or a mixture of both.

quartzite. A metamorphic or sedimentary rock consisting almost entirely of quartz grains bonded by silica, and usually formed by the metamorphism of sandstone; used as a refractory, particularly in salt-glazing kilns.

quartz, synthetic. A quartz crystal grown at high temperature and pressure around a seed of quartz which is suspended in a solution containing natural quartz crystals.

quartz, vein. Quartz occurring as gangue in a vein.

quebracho extract. A tannin-rich material extracted from the quebracho tree used as a deflocculant in dressing muds, ore flotation, and ceramic slips.

queen closer. A cut brick having a nominal 2-inch horizontal face dimension used to close courses and spaces less than normal depth in construction.

queen's ware. A variety of white or cream colored ware developed and introduced by Josiah Wedgwood in England in 1759 to 1765.

quenching. The rapid chilling of molten porcelain enamel or other glassy material in water, causing the material to shatter into small, friable particles or flakes called frit. More commonly known as fritting.

quenching of fluorescence. The extinction of fluorescence by causes other than the removal of exciting radiation such as black light.

quenching, spray. Rapid cooling of a molten material in a spray of water or other liquid.

quicklime. A calcined material, the major part of which is CaO or CaO in natural association with a lesser amount of MgO; may be slaked in water. Approx. mol. wt. 56.1; m.p. 2570°C (4658°F); sp. gr. 3.40; used in mortars and plasters, and as a refractory ingredient in other ceramic products. See slaking.

quill. A removable arbor or spindle; a hollow shaft frequently surrounded by another shaft; employed in mechanical rotating devices.

quoin. An external corner of a masonry wall; also a wedge-shaped piece employed as the keystone of an arch or vault.

Q-value. A synonym for nuclear-disintegration energy.

R

rack car. A car containing racks on which ware is placed without stacking for movement through the dryer.

rack, comb. (1) A comb-like burning tool on which porcelain-enameled ware is supported during firing.

(2) A comb-like rack on which metal shapes for porcelain enameling are supported during the cleaning and pickling operation.

rack, hanging. A rack mounted on an overhead conveyor for the transport of ware through various porcelain-enameling processes, including firing.

rack mark. An imperfection on the surface of glass due to malfunction of the rolling mechanism.

rad. A unit of energy absorbed from ionizing radiation equal to 100 ergs per gram of irradiated material.

radial brick. A brick with each end curved for use in concentric, cylinder, or circular construction.

radiant. Emitting light or heat.

radiant heating. Any heating system in which heat is transmitted by radiation as opposed to convection or conduction.

radiant-tube furnace. A porcelain-enameling furnace heated by radiant tubes in which the combustion of fuel takes place within the tubes and does not enter into the firing chamber.

radiation. The transmission of energy, such as light, heat, x-rays, electricity, etc., through space without the presence or movement of matter in or through this space.

radiation damage. Harmful changes in a substance induced by radiation.

radio-frequency heating. The electronic heating of a substance by means of induced high-frequency currents.

radome. A strong, thin, dome-like protective covering for radar equipment made of a dielectric material transparent to radio-frequency radiation.

rake. A scratch on the surface of glass caused by particles of cullet inadvertently contained in the polishing felt.

raku. A type of thick, coarse-textured pottery ware covered with a soft lead borosilicate glaze.

ramming. The process of forcing or driving bodies into place, such as by means of a pneumatic device, to form monolithic furnace linings and shapes.

ramming mix. A mixture of water-tempered refractory materials suitable for ramming into place to form monolithic furnace linings.

ramming mix, pitch or tar bearing. A refractory ramming mix, containing pitch or tar additions, having properties suitable to form rammed monolithic furnace linings.

ram pressing. A process of forming ware in plaster molds in which water removal is expedited by the application of a vacuum; ware is released from the mold by applying air pressure through the porous structure of the mold.

ram seal. A seal in which a metal sleeve is forced to form a thin circumferential line of contact over the sharp edge of a ceramic shape, and then is completed by bracing or plating a metal over the joint.

ranch-type roofing. A rectangular asbestos-cement roofing panel which is lapped at the top and one side.

random cracking. Cracks formed on the surface of concrete in a random pattern, sometimes hexagonal or square, due to surface shrinkage or internal expansion.

Rankine temperature scale. A scale of absolute temperature based on Farenheit degrees in which °R is equal to °F + 459.67.

rare earths. Lanthanum, cerium, praseodymium, neodymium, promethium, samarium, europium, gadolinium, terbium, dysprosium, holmium, erbium, thulium, ytterbium, and lutetium. Yttrium, didymium, and thorium, although not rare earths, are closely associated. The rare-earth salts are employed in numerous ceramic applications, including coloring agents, glass decolorizers, ultraviolet absorbers, polishing compounds, cores for arc carbons, incandescent gas mantles, laser glass, electronic components, magnetic compositions, phosphors, fiber-optic glass, etc.

Raschig rings. Packing rings in the shape of short pipes composed of stoneware, glass, carbon, or metal; used as columns in absorption and distillation towers.

rasorite. A natural sodium borate used as a substitute for borax.

rate of absorption. (1) The weight of water, cold or hot as required, absorbed by a partially immersed standard brick in one minute.

(2) The percentage of the weight of water absorbed by a dry specimen obtained by dividing the weight of water absorbed, in grams, by the weight of the dry test piece, in grams.

rational analysis. The mineral composition of a material calculated from its chemical composition.

rattler. A cylinder filled with steel balls in which paving brick are rotated to test the impact and abrasion resistance of the brick; calculated as the percentage loss in weight.

raw batch. (1) A furnace-charge of glass raw materials without cullet.
(2) A blend or batch of raw materials ready for processing.

raw cullet. A furnace charge of glass consisting entirely of cullet. See cullet.

raw data. Data ready for evaluation.

raw edge. The sheared edge of a porcelain-enameled panel not completely covered by the coating.

raw glaze. A glaze compounded entirely of raw materials and containing no prefused ingredients.

raw material. Crude, unprocessed or partially processed material used in a processing operation.

raw refractory dolomite. Natural limestone containing essentially equal amounts of $CaCO_3$ and $MgCO_3$; suitable for use as a refractory material.

raw shape. A metal part ready to be started through the porcelain enameling process.

Raymond concrete pile. A tapered metal shell driven into the ground and filled with concrete and used as a structural base.

reaction, surface. A chemical or physical reaction taking place only on the surface of an item.

reactivation. An oxidation process to restore the adsorptive properties of activated carbon.

reactive aggregate. An aggregate which will react chemically, such as some siliceous minerals with alkalies, after inclusion in the concrete, causing the concrete to expand and crack after it has hardened.

ready-mixed concrete. Concrete mixed by any means prior to delivery to a job site.

ream. An imperfection consisting of heterogeneous layers in flat glass.

rearing. Glazed ceramic flatware set on edge during firing.

Réaumur porcelain. A porcelain in which a fritted or devitrified glass is the major constituent.

rebar. A reinforcing bar embedded in concrete.

RBM. An acronym for reinforced brick (or clay) masonry.

reboil. (1) The appearance of bubbles in glass after it once appears to be bubble free.
(2) A fine boiling occurring in porcelain-enamel ground coat due to the evolution of gas in the metal or coating during repeated firing.

rebonded fused-grain refractory. A fired refractory brick or shape made predominantly from fused refractory grain.

recessed abrasive wheel. A grinding wheel with a contoured central recess on one or both sides.

reciprocal. The quotient of a specific quantity divided into 1.

reciprocating feeder. A tray situated at the bottom of a bin, hopper, or other container which moves back and forth in a horizontal plane as it transfers material from the container to a processing unit or transport car.

reciprocating screen. A sieve or screen which moves back and forth in a horizontal plane; used in the separation or classification of solid particles.

recirculating dip tank. A dipping tank for the application of porcelain enamels which is provided with a mechanical means or pump to keep the slip in constant circulation and the solid components in uniform suspension.

recirculating fan. A mechanical device which aids the movement of air from one location, such as a furnace, to another point of usefulness, such as a dryer.

reclaimed enamel. Porcelain enamel and glaze residues collected from spray booths, dip tanks, washed ware, etc., and reconditioned for use.

recovery time. The time required for a freshly charged periodic furnace to regain its intended firing temperature; that is, the temperature which was lost during the period the furnace was open for the discharging and charging of ware.

recrystallization. The process of producing strain-free grains in a matrix that has been plastically deformed.

recrystallization, primary. The process by which nucleation and growth of a new generation of strain-free grains occurs in a matrix which has been plastically deformed.

recrystallization, secondary. The process by which a few large grains are nucleated and grow at the expense of a fine-grained but essentially strain-free matrix.

recrystallization temperature. The minimum temperature at which ceramic particles bond together or at which phase changes occur in the solid state, which take place in sintering, precipitation, exsolution, or grain growth.

rectangular kiln. A periodic kiln of rectangular shape.

rectifier. A device which converts a-c current to d-c current, such as a diode.

recuperative furnace. A furnace having a heat exchanger in which heat is conducted from the combustion products to incoming air through flue walls or a system of ducts.

recuperator. A continuous heat exchanger in which heat is conducted from combustion products to incoming cooler air through flue walls or a system of thin-walled ducts.

recycling. The recovery of valuable materials from discarded or scrap products.

red clay. An iron-bearing clay which produces a red color when fired; used in the production of brick, roofing tile, and some types of pottery.

red earth. See ferric oxide.

red edge. Small cavities along the edges of plate glass which are filled with rouge during the polishing operation.

red heat. The temperature at which the interior of a furnace, kiln, or pot is glowing red in color, approximately 700°–750°C (1292°–1382°F).

red iron oxide. Fe_3O_4; mol. wt. 231.52; decomposes between 500°–530°C (932°–986°F); sp.gr. 9.0–9.2; used extensively in optical, electrical, and tableware glasses, and in glazes and porcelain enamels as a fluxing ingredient.

reducer section, manhole. A section of concrete pipe used as a transition joint between manhole riser sections of different diameters.

reducer, water. A surface-acting admixture which reduces the amount of water required in a batch of concrete, but which will not influence the working or performance properties of the concrete.

reducing agent. A chemical which lowers the state of oxidation of other glass, glaze, or porcelain-enamel batch ingredients when they are subjected to elevated temperatures.

reducing atmosphere. A furnace atmosphere deficient in oxygen and containing a reducing gas such as hydrogen.

reducing bushing. A device used as a liner to reduce the size of an arbor hole for an axle, shaft, or spindle.

reducing flame. A flame deficient in oxygen resulting in incomplete combustion of the fuel.

reduction. A chemical reaction in which an element gains an electron.

red ware. A type of porcelain body made of iron-bearing clay which fires to a characteristic red color.

reel cutter. A device consisting of a tightly stretched wire on a circular frame in association with a pugging machine in a manner to cut extruded clay columns into desired lengths.

reeves. Tangled laminations causing imperfect cleavage in mica.

reference coil. The section of a coil assembly that excites or detects, or both, the electromagnetic field in the reference standard in a comparative system.

reference standard. A specimen used as a basis for comparison or calibration.

refiner. The section of a glass-melting tank in which the molten glass becomes virtually free of bubbles and undissolved gases, and conditioning for subsequent processing is completed.

refining. The process in which bubbles and undissolved gases and solid particles are freed from molten glass.

refining temperature. The temperature just above the melting temperature of glass and vitreous compositions at which the molten batch is sufficiently fluid to permit the solution or escape of gaseous inclusions.

refire. A second fire given a porcelain-enameled item without complete coverage of the item with a new application of the coating; the partial coating usually is in the form of a repair or an added decoration.

reflectance. The fraction of incident light that is reflected diffusely by a surface, measured relative to magnesium oxide under standardized conditions; employed as an indication of the opacity or covering power of a porcelain enamel, glaze, or other coating, as well as the degree of obscuration of a glass.

reflectivity. The reflectance of a porcelain-enamel coating of sufficient thickness that an additional thickness will not change the reflectance value.

reflectometer. A photoelectric device measuring the reflectance of visible light from a surface. See reflectance.

refraction. A change in the direction of a propagated light wave, as when the light ray passes from one medium to another of different density, and is bent from its original path.

refractive index. The ratio of the velocity of light in a substance to the velocity of light in air.

refractories. Inorganic, nonmetallic materials which will withstand high temperatures; such materials frequently are resistant to abrasion, corrosion, pressure, and rapid changes in temperature; examples are alumina, sillimanite, silicon carbide, zirconium silicate, and the like.

refractories, acid. Refractories containing a substantial amount of silica, SiO_2, which will react with basic refractories, slags, and fluxes at high temperatures.

refractories, air-setting. Ground refractory materials

which, when tempered with water, develop a strong bond on drying; used as mortars, plastic refractories, ramming mixes, and gunning mixes.

refractories, aluminum silicate. Refractories composed essentially of alumina and silica in various combinations; prepared from such materials as bauxite, andalusite, diaspore, gibbsite, kyanite, sillimanite, and blends of alumina and silica.

refractories, basic. Refractories composed essentially of lime, magnesia, or both, which may react with acid refractories, slags, or fluxes at high temperatures.

refractories, calcined. Refractory materials which have been heat treated to remove volatile materials and materials which produce volume changes.

refractories, casting. Refractories of special shapes in which molten metals are cast.

refractories, electrocast. Refractories, such as mullite, aluminum silicate, etc., fused in an electric furnace and cast into blocks or other shapes; such products usually are vitreous, nonporous, hard, and of low coefficient of expansion.

refractories, fused grain. Refractories made predominantely from refractory materials which have solidified from a fused or molten condition.

refractories, fusion-cast. Cast or molded refractory shapes formed from molten refractory components.

refractories, high-alumina. Alumina-silica refractories containing 45% or more of alumina; raw materials include diaspore, bauxite, gibbsite, kyanite, sillimanite, andalusite, and fused alumina.

refractories, insulating. Refractory products of low-thermal capacity and low-thermal conductivity, and which frequently are porous and light weight.

refractories, mechanical spalling of. The chipping, cracking, or breaking of a refractory brick in service which is caused by stresses imposed by impact, pressure, or other forces of mechanical origin.

refractories, mullite. Refractory products in which mullite is the predominant crystalline phase. See mullite.

refractories, neutral. Refractories that are neither acidic nor basic, such as carbon, and which are resistant to attack by both acidic and basic materials, slags, fluxes, etc., at high temperatures.

refractories, permeability of. The capacity of a refractory to transmit a gas or liquid.

refractories, rebonded fused-grain. Fired refractory brick or shapes made predominantly from molten refractory components.

refractories, silicon carbide. Refractory products consisting predominantly of silicon carbide.

refractories, slagging of. A destructive chemical reaction between a refractory and another material at a high temperature, resulting in the formation of a liquid.

refractories, spalling of. The cracking and breaking of refractories in service, usually with the detachment of a portion of the unit to expose a new refractory surface.

refractories, structural spalling of. The cracking or breaking of refractories in service due to stresses created in the brick by shifts or changes in the structure of which it is a part; that is, the furnace, stack, etc.

refractories, thermal spalling of. The chipping, cracking, or breaking of a refractory brick or unit in service as a result of nonuniform stresses imposed in the unit by differences in temperature.

refractories, unburned. Refractory products of various shapes which usually are deaired to reduce voids, shaped under high pressure, and placed in service without burning.

refractoriness. The property of a material to withstand high temperatures, the environment, and conditions of use without change in its physical or chemical identity.

refractoriness, under load. The resistance of a refractory to the combined effects of heating and loading, often expressed as the temperature of shear or 10% deformation when heated under 25 or 50 pounds per square inch.

refractory. The property of being resistant to high temperatures.

refractory, aluminum silicate. A refractory composed essentially of various combinations of alumina and silica; prepared from materials such as bauxite, diaspore, gibbsite, kyanite, sillimanite, andalusite, and blends of the oxides of aluminum and silicon.

refractory, anthracite-coal-base carbon. A carbon refractory prepared essentially from calcined anthracite coal.

refractory brick. Any refractory brick which will be subjected to high temperatures during use.

refractory, carbon. A refractory shape composed essentially of carbon, including graphite, used in crucibles, stopper nozzles, and similar application.

refractory, carbon-ceramic. A refractory product composed of a mixture of carbon, including graphite, and a refractory ceramic, such as fireclay, silicon carbide, and the like.

refractory, castable. A blend of refractory grain and refractory cementitious material which, when tempered with water, will develop structural strength when cast in a mold or structural form.

refractory cement. Any of a variety of mixtures of finely ground refractory materials which, when tempered with water, become plastic and trowelable for use as a mortar for the laying, filling of cracks, and bonding of refractory brick.

refractory, chrome. A refractory product made substantially or entirely of chrome ore; used to line steel furnaces. See chrome brick.

refractory, chrome-magnesite. A burned or unburned refractory shape made essentially of mixtures of refractory chrome and dead-burned magnesite in which the chrome ore, by weight, is predominant.

refractory chrome ore. A refractory ore consisting essentially of chrome-bearing spinel with only minor amounts of accessory minerals.

refractory clay. A clay having a melting point above 1600°C (2912°F) used in the manufacture of refractory products, such as firebrick, linings for furnaces, reactors, kilns, etc.

refractory coating. A coating composed of refractory ingredients used for the protection of metals, brickwork, and other structures subjected to elevated temperatures.

refractory coating, composite. A combination of heat-resistant ceramic material applied to a metal substrate, or to a nonmetallic substrate such as graphite, and which may or may not require heat treatment prior to placing in service.

refractory concrete. A heat-resistant concrete made of a mixture of high-alumina or calcium-aluminate cement and a refractory aggregate.

refractory corrosion. The deterioration of refractory surfaces in service by chemical reaction with gases and reactive substances.

refractory dolomite, raw. Natural limestone containing essentially equal amounts of $CaCO_3$ and $MgCO_3$, which is suitable for use as a refractory material.

refractory enamel. A porcelain enamel of special composition used to protect metals from attack by hot and corrosive gases.

refractory erosion. The wearing away of refractory surfaces by the washing action of moving melts at high temperatures.

refractory, fused-grain. A refractory made of refractory grain solidified from a fused or molten state.

refractory, fused silica. A refractory product composed essentially of fused, noncrystalline silica, SiO_2.

refractory, fusion-cast. A refractory product made by casting molten refractory ingredients in a mold.

refractory, graphite. A refractory product composed essentially of graphite.

refractory, heat-setting. Mixtures of ground refractory material which require high temperatures for the development of bond.

refractory, hydraulic-cement. Mixtures of refractory materials which are tempered with water and which develop a strong bond by chemical reaction; employed as a mortar in the lining of furnaces and kilns and as a filler for holes and cracks in refractory linings.

refractory magnesia. Dead-burned, crystalline magnesium oxide, MgO, having high resistance to heat and corrosion; used in linings for furnaces and melting tanks, either in the form of brick or cement.

refractory, magnesite. A burned or chemically bonded refractory consisting essentially of dead-burned magnesite.

refractory, metal-cased. Basic brick encased in a thin metal casing; used chiefly in steel-making furnaces. Also known as metalkase brick.

refractory, metallurgical-coke-basic. A refractory produced substantially of metallurgical coke.

refractory, molten cast. A refractory product made by casting molten refractories into molds.

refractory, monolithic. An integral structure of refractory compositions installed without joints; see ramming mix; refractory, castable.

refractory mortar. A mixture of finely ground refractory materials tempered with water to produce a plastic, trowelable mortar for laying and bonding refractory brick and shapes.

refractory, mullite. A refractory product consisting essentially of mullite, $3Al_2O_3 \cdot 2SiO_2$.

refractory nozzle. A refractory shape containing an orifice through which molten metal is poured from a ladle or other container.

refractory patching cement. A finely ground mixture of refractory ingredients which become plastic and trowelable when tempered with water; used to repair damaged areas in furnaces, kilns, glass tanks, refractory molds, etc.

refractory, petroleum-coke-base. A refractory product produced substantially of calcined petroleum coke.

refractory, pitch-bonded basic. An unburned refractory shape bonded with pitch.

refractory, plastic. A water-tempered refractory material or mixture of a consistency that can be extruded and of suitable workability that it can be pounded into place to form a monolithic structure.

refractory, pouring pit. A refractory shape employed in the transfer or flow control of molten steel from the furnace to the mold.

refractory, rebonded fused grain. A fired refractory brick or shape produced essentially or entirely from fused refractory grain.

refractory, semi-silica. A refractory product containing not less than 72% of silica, the balance being other fireclay components.

refractory, sillimanite. A refractory composed essentially of sillimanite, $Al_2O_3 \cdot SiO_2$.

refractory, single-screened. A refractory material from which particles greater than a specified size have been removed, but otherwise contains the particle-size distribution as processed by crushing and grinding.

refractory, tar-bearing basic. Refractory compositions of essentially basic refractory grains to which tar has been added during processing or manufacture.

refractory, unburned. Refractory shapes installed and placed in service without prior burning.

refractory, zircon. Refractory products consisting essentially or entirely of zircon, $ZrO_2 \cdot SiO_2$.

refractory, zirconia. Refractory products consisting essentially or entirely of zirconium oxide, ZrO_2.

regeneration. The process of restoring the adsorptive or other properties of a substance.

regenerative furnace. A furnace equipped with a cyclic heat exchanger which alternately receives heat from gaseous combustion products and transfers heat to air or gas before combustion.

regenerator. A heat exchanger which utilizes heat from the combustion of fuels to preheat air or fuel entering the combustion chamber.

regular alumina. A recrystallized grade of alumina, Al_2O_3, of relatively large crystal size, the Al_2O_3 content being approximately 95 per cent.

regulating wheel. A wheel on a centerless grinder which controls the speed and pressure on an item being ground.

regulator, pressure. An open-close type of mechanism used in a system to maintain a specified gas or liquid pressure within the system.

reheat behavior. The change in the dimensions or volume of a substance when subjected to a temperature equal or greater than the temperature to which the substance previously was heated.

reheat test. A prescribed heat treatment of a fired refractory or other product to determine the changes in dimensions or volume which occur during reheating.

reinforce. To strengthen with the aid of some additional material or support.

reinforced beam. A concrete beam supported in tension, compression, or torsion by steel bars, wire, rods, or other structural material embedded in the concrete.

reinforced brickwork. Brickwork strengthened by metal bars, rods, mesh, or other material embedded in the bed joints.

reinforced center. A grinding wheel in which steel rings have been incorporated near the center to provide additional strength.

reinforced column. A concrete column in which longitudinal metal bars, sometimes with ties, circular ties, or other materials, are incorporated as reinforcing agents.

reinforced concrete. Concrete containing reinforcing steel rods, bars, wire mesh or other strengthening material.

reinforced masonry. Masonry construction in which steel bars, mesh, or similar materials are employed as strengthening components.

reinforced pipe. Concrete pipe designed with some type of reinforcement to produce a composite structure of increased strength.

reinforced products. Any products containing mechanical reinforcements to provide increased strength.

reinforced wheel. A grinding wheel containing mechanical reinforcement to provide additional strength and safety during use.

reinforcement, circumferential. A wire or bar reinforcement approximately perpendicular to the longitudinal axis of a concrete pipe.

reinforcement, layer. Circumferential reinforcement in a concrete pipe which is one bar or wire thick. See reinforcement, circumferential.

reinforcement, line. Circumferential reinforcement in concrete pipe comprised of one or more layers of reinforcement.

reinforcement, mat. Tension zone circumferential reinforcement secured to a cage in a concrete pipe wall.

reinforcing bars. Steel bars incorporated in concrete and other building materials to increase strength.

rejected material. A material or product which fails to meet specifications.

rejection level. The composition or property level above or below which a specimen or product is considered rejectable or to be distinguished or sorted from acceptable products.

relative detector efficiency. The product of the detector efficiency and the detector geometry.

relative efficiency. The ratio of the performance characteristics or property of a product or material to that of a standard reference in a defined test.

relative humidity. The ratio of the amount of water actually present in air to the maximum amount possible at the same temperature and pressure.

relative standard deviation. The standard deviation of a value expressed as a percentage of the mean value.

reliability. The probability that a material or product will satisfactorily perform its intended functions under specified conditions.

relieving arch. A sprung arch in the substance of a wall above an opening in a furnace wall; designed to support the wall, give it strength, and reduce the strain on a second arch constructed below it.

remover, detergent. A solution of detergent in water employed to remove the dye-penetrating solutions used in absorption tests.

repellent, water. Hydrophobic materials such as waxes, soaps, silicones, and the like, used to render a surface resistant to wetting by water, but not completely waterproof.

representative sample. A sample collected in a manner that every particle of a lot to be sampled is equally represented in the gross sample.

re-press. A machine employed to press previously formed blanks into shapes.

repressed brick. A brick formed by repressing blanks cut from a column of stiff or soft clay produced by an extruding machine.

reprocessing. The recovery and separation of materials for reuse.

reservoir. A place where a liquid, especially water, is collected and stored for use when wanted.

residual clay. A clay which, geologically, remains at the site of its formation.

residual magnetic field. A magnetic field that remains in a magnetized material after the magnetizing forces are removed.

residual magnetism. The retention of a magnetic field by ferromagnetic materials after exposure to a magnetic force.

residual method. The indicating material, such as iron filings, employed in magnetic particle testing which is applied after the magnetic force is discontinued.

resin. Any of a class of natural or synthetic materials of an organic composition which usually have a high molecular weight, are solid or semisolid, and have no well-defined melting point.

resinoid bond. The adhesive or bonding force produced by thermosetting resins in joining two or more solids together.

resinoid wheel. A grinding wheel in which the abrasive grains are bonded together by a thermosetting resin.

resin, polyester. A class of transparent thermosetting or thermoplastic resins of high strength which are resistant to moisture and chemicals, and which are used in bonding and laminating glass-fiber products.

resist. A patterned protective film, layer, or covering such as wax, paper, metal, foil, or plastic which is laid over an area or surface to shield the area from subsequent applications of colors and glazes, from etching compounds, from sandblasting, or other such decorative treatment.

resistance, abrasion. The property of a material to resist attrition, wear, or damage by friction.

resistance, acid. The ability of a material to resist attack or injury by acids.

resistance, alkali. The ability of a material to resist attack or injury by alkaline solutions.

resistance, apparent dc. The reciprocal of apparent dc conductance.

resistance, apparent dc surface. The reciprocal of apparent dc surface conductance.

resistance, apparent dc volume. The reciprocal of apparent dc volume conductance.

resistance, chemical. The resistance of a material, particularly glasses and porcelain enamels, to attack or injury by chemicals.

resistance, crazing. The resistance of glazes, porcelain enamels, and other ceramic coatings to cracking. See craze, crazing.

resistance, dc insulation. The reciprocal of conductance, which see.

resistance, electrical erosion. The erosion of electrical insulating materials due to the action of electrical discharges as determined under specific conditions.

resistance element. A material which resists or opposes the flow of electricity, usually with the development of heat.

resistance, furnace. An electric furnace in which heat is generated by passing an electric current through a resistor surrounding the furnace charge, through a resistor embedded in the charge, through the charge itself, or a combination of any of these procedures.

resistance heating. The generation of heat by passing an electric current through a resistor, which see.

resistance, impact. The resistance of a body or coating to breakage, deformation, or other damage when subjected to sharp blows or impact loading.

resistance material. Any material exhibiting resistance to the passage of an electric current per unit of length or volume, and which may be used as a resistor, which see.

resistance, surface. The electrical resistance of an insulating product, usually measured between the opposite sides of a square on the surface of the insulator.

resistance, thermal shock. The ability of a material to withstand sudden heating and cooling without cracking or other damage.

resistor. A device exhibiting resistance to the flow of electric current and which is used in an electric circuit for protection, operation, or current control.

resistor furnace. An electric furnace in which heat is generated by the passage of an electric current through a resistor element which is not in contact with the furnace charge.

resistor, glass. Tubular glass with a helical electric-resistor element of carbon painted on the surface.

resistor, metal oxide. A resistor consisting of a metal oxide deposited over a metal surface.

resistor oven (dryer). A dryer in which heat is generated by the passage of an electric current through a resistor.

resist, paper. A pattern, cut of paper, which is placed over an area of a surface to shield the area from subsequent applications of coloring inks and glazes.

resist, wax. A protective coating of wax placed over a patterned area of a surface to shield the area from subsequent applications of glaze or coloring inks, from etching compounds, and other such decorative treatments.

resonator, quartz crystal. An electronic filter in which a crystal of quartz is the essential component.

rest. A platform attached to a grinding wheel stand upon which work or a dressing tool is supported during a grinding operation.

resteel. Any form of steel used to reinforce concrete in a construction.

resultant magnetic field. The magnetic field resulting when two magnetizing forces are impressed on the same area of a magnetizable material.

retarder. A substance added to cement, mortar, plaster, or stucco to slow the setting rate, but which will have little or no effect on the properties of the cement after the initial set.

retentivity. (1) The residual flux density corresponding to the saturation induction of a magnetic material.

(2) The ability of a substance to resist desorbation of an adsorbate.

retentivity, water. The property by which concrete, mortar, or plaster will resist the rapid loss of water when applied to background units having high-absorption properties.

reticulated glass. An ornamental glassware containing an interlaced network of decorative lines.

retort. A closed refractory chamber in which materials are decomposed by heat.

retort carbon. (1) Carbon or graphite employed in a glaze or other ceramic to promote localized reduction during firing.

(2) A dense form of carbon or graphite formed in the upper sections of retorts used in coal-gas manufacture; useful in producing reducing atmospheres at high temperatures.

retort clay. A plastic, dense-burning, semirefractory clay used in the production of gas and zinc retorts.

retort, vertical. A vertical refractory chamber lined with silicon carbide brick; used for the smelting of zinc.

retouch enamels. A fine overspray or brushed-on coating of porcelain enamel applied to cover or protect areas of potential imperfections.

return. The actual number of cubic feet of concrete in a one-cubic-yard batch based on tests made on the fresh concrete.

reverberatory furnace. A furnace or kiln in which fuel is burned at one end with the flame passing between the charge and the furnace roof, the heat being radiated from the roof onto the charge.

reverse. To reverse the direction of gas and air flow in a regenerative furnace.

reversible adsorption. Adsorption in which the desorption isotherm approximates the adsorption isotherm.

revivification. An oxidation process to restore the adsorptive properties of activated carbon.

revolving pot. The rotating circular container from which glass is gathered in the Owens process. See Owens process.

revolving tube. A hollow cylinder, concentric with the needle of a feeder, revolving in a molten glass batch, the feeder delivering gobs of glass to a forming unit.

rhenium boride. (1) Re_3B; mol. wt. 569.48; sp. gr. 19.4. (2) Re_7B_3; mol. wt. 1336.0. See borides.

rhenium phosphide. (1) ReP; mol. wt. 217.25; m.p. 1204°C (2200°F); sp. gr. 12.0. (2) ReP_2; mol. wt. 248.28; sp. gr. 8.33. (3) Re_2P; mol. wt. 403.47; sp. gr. 16.4. See phosphides.

rhenium silicide. (1) ReSi; mol. wt. 214.28; m.p. ~1899°C (3450°F); sp. gr. 13.04. (2) $ReSi_2$; mol. wt. 242.34; m.p. ~1927°C (3500°F); sp. gr. 10.71. (3) Re_5Si_3; mol. wt. 1015.28; decomposes above 1921°C (3490°F); sp. gr. 15.44. See silicides.

rhenium-tungsten. Re_3W_2; mol. wt. 926.66; m.p. >2998°C (5430°F). See intermetallic compound.

rheostat. A resistor employed to regulate an electric current by means of variable resistances, thereby controlling the temperatures of furnaces and kilns, etc.

rhodium boride. (1) RhB; mol. wt.113.73. (2) RhB_2; mol. wt. 124.55. (3) Rh_2B; mol. wt. 216.64. See borides.

rhodium silicide. RhSi; mol. wt. 130.97. See silicides.

rib. A tool of hard wood, metal, stone, plastic, or other smooth solid used to smooth the outer surface of a pot or similar item while the item is being thrown.

ribbed rolls. A roll-type crusher in which the crushing surface of the rolls are ribbed parallel with the axis of the rolls.

ribbon. A continuous strip of glass in the plastic state during processing.

ribbon feed. A batching procedure in the production of concrete in which the batch ingredients are fed into the mixer essentially simultaneously.

ribbon process. The process of delivering molten glass to the forming operation in ribbon form.

rich clay. Plastic clay exhibiting good workability and green strength.

rich concrete. Concrete with a high cement content.

rich mixture. A mixture of air and fuel in which the concentration of the fuel component is high.

riddle. A screening apparatus to remove foreign substances from granular materials.

rider arch. An arch or series of arches supporting the checkerwork in the regenerator of a furnace.

ridge roll. A half-round section of asbestos cement applied along the hips and ridge of a roof to conceal and waterproof the apex joint of the roofing material.

rigidity. The property of being resistant to change in shape or form.

rigidity modulus. The measure of the resistance of a body or material to shear under stress.

rim. (1) Protrusion bordering a hole, pit, or pock at the surface of a body.

(2) The outer edge of a shape.

ring. (1) The part of a mold that forms the outer edge of a pressed article.

(2) A floating refractory ring on the surface of molten glass which prevents scum from collecting within the ring area from which the glass is gathered.

ring crusher. A type of hammer mill consisting of steel rings held outwardly by the centrifugal force of a horizontal shaft rotating at high speed, the feed material being crushed between the rings and the outer shell of the mill.

ringhole. An opening or hole in a glass-melting tank through which glass is gathered.

ring mold. The metal section of a glass mold that shapes and finishes the neck of a bottle or other hollow glass article.

ring, orifice. The ring or bushing in the feeder through which glass flows to a forming machine.

ring, pall. A specially shaped steel ring filled with a ceramic which is used in distillation columns.

ring, pot. A refractory ring floating atop molten glass in a pot to prevent scum from collecting within the ring, and through which glass is gathered for forming.

ring, safety. A metallic ring embedded in organic bonded abrasive wheels to contain particles fractured during grinding operations.

ring section. A narrow peripheral section cut from a glass article for inspection.

ring test. (1) A test to evaluate the expansion or contraction properties of a glaze or porcelain enamel in which ceramic or metal rings, respectively, are coated on the outside and fired; the rings then are cut open, and the distance between previously scored reference marks serve as the basis for evaluating the expansion or contraction properties of the coating.

(2) A test to determine the presence of cracks in a grinding wheel by tapping the wheel while it is freely suspended in the arbor hold or freestanding to the periphery.

ring wall. The refractory wall of the unit delivering hot air to the tuyeres of a blast furnace.

rinse. A liquid bath to remove foreign matter or solutions from the surface of an article or substance.

rinse dip. A procedure to remove foreign matter or solutions from the surface of an item by immersion in a tank of water or other cleaning solution which is sometimes agitated.

ripple. A surface imperfection characterized by uniform waviness over a substantial area of a porcelain-enameled surface.

rise. The vertical distance between a plane connecting the spring lines and the highest point on the undersurface of an arch.

riser. The projection on a casting resulting from an excess of molten metal supplied to make certain that a mold is completely filled during pouring.

riser, manhole. A section of concrete pipe forming a manhole, excluding the base, reducers, and top section.

rock crystal. (1) A transparent, colorless form of quartz used for lenses and prism components in optical instruments.

(2) Highly polished, hand-cut or engraved, blown glassware.

rocker. A glass bottle or other item with a deformed bottom which will rock when placed on a flat surface.

rocket engine. A jet engine that is propelled by the rearward reaction to discharge gases produced by burning mixtures of fuel and oxygen.

rocking furnace. A horizontal melting furnace designed to rock back and forth as a means of producing uniform melts.

Rockingham ware. An ornate earthenware or semivitreous ware coated with a brown or mottled manganese glaze.

rock ladder. A series of inclined steps in a vertical arrangement designed to break the vertical fall and to minimize breakage of concrete aggregate as it is discharged from a conveyor or chute.

rocklath. A sheet of gypsum board used as a plaster base in construction.

rock pocket. A void in concrete due to incomplete consolidation of the mass.

Rockwell hardness. A measure of the hardness of a material based on the resistance of the material to indentation by a steel ball or conical diamond, of various dimensions, with a rounded point, under prescribed static or dynamic load; reported as the depth of indentation.

rock wool. A mass of fine intertwined fibers formed by blowing air or steam through molten rock or slag; used for thermal and acoustic insulation, fireproofing, filters, and similar applications.

rod crusher. See rod mill.

rod mill. A pulverizing apparatus consisting of heavy metal rods impacting on a charge in a rotating metal cylinder.

rodproof. A sample of molten glass removed for inspection from a bath by means of a metal rod.

roll back. A form of crawling of porcelain enamels where the fired coating pulls away or rolls up at the edges of the base metal or over areas of dirt or grease.

roll, corner. A half-round asbestos-cement unit used to trim and flash corners in corrugated asbestos-cement construction.

roll crusher. A pulverizer consisting of two horizontal rolls rotating toward each other.

rolled glass. (1) Flat glass made by passing a roller over the glass in a molten or plastic state; sometimes a design may be worked into the glass surface by a patterned roller face.

(2) Optical glass rolled into plates instead of being cooled in the melting pot and then processed.

rolled inlay. A decorative process for pottery in which colored clays are pressed into the surface of the pot or the clay slab from which the pot is formed before firing.

roller. A cylinder of blown glass which is cut in the soft state and then flattened to form window glass.

roller coating. The application or transfer of designs from a pattern to the surface of ware by means of a roller.

roller conveyor. A gravity-type conveyor consisting of freely rotating, parallel, cylindrical rollers mounted in a rigid steel frame.

roller-head machine. A shaping machine for pottery flatware equipped with a revolving shaping tool having the shape of the back of the article.

roller-hearth kiln. A tunnel-type kiln through which ware is moved on parallel rollers.

roll, graining. A special type of rubber roll used to transfer graining paste from a pattern surface to the surface of a porcelain enamel or glaze.

rolling. A synonym for crawling; a defect in glazes and porcelain enamels.

rolls, compound. Rolls contained in a crushing machine consisting of two or more pairs of rolls, one pair above the other, the particle size of crushed material decreasing as it passes through one set of rolls to the next.

rolls, conical. Rolls contained in a pulverizer in which solids are crushed between rotating, tapered, or cone-shaped rolls.

rolls, kibbler. Toothed steel rolls used in crushing and pulverizing machinery to reduce clays and other minerals to sizes and shapes more amenable to further grinding and use.

rolls, ribbed. Steel rolls with a series of surface ribs running parallel to the axis of each roll; used in crushing and pulverizing machinery.

rolls, smooth. Smooth-surfaced steel rolls employed in crushing machinery.

roof, bonded. A furnace roof with staggered transverse joints.

roof, boxcar. The roof of an open-hearth furnace in which the transverse and longitudinal ribs form box-like shapes along the top.

roofing granules. Approximately 8-mesh particles of crushed slag, slate, rock, tile, porcelain, or similar substances used in the production of asphalt roofing and shingles.

roofing tile. Any of several designs of overlapping or interlocking concrete or fired-clay shapes with overlapping or interlocking edges used for roofing.

roof, Kruetzer. A furnace roof characterized by an arrangement of transverse and longitudinal ribs which gives the appearance of box-like compartments.

roof, shell. A thin, curved plate-like roof, usually constructed of concrete.

roof, sprung. A curved roof in the working zone in a furnace which is supported by abutments at the sides or at the ends of the furnace.

roof, zebra. The roof of an open-hearth furnace consisting of alternate rows of silica and chrome-magnesite brick having a zebra-like appearance.

room dryer. A room or compartment in which ware is dried and stored before firing.

roping. A slip-casting defect consisting of a rope-like formation of clay body on the side of the ware.

rotary crusher. A pulverizer in which a cone, rotating at high speed on a vertical shaft, forces solid materials against a metal encasement or shell.

rotary dryer. An inclined, rotating cylinder in which tumbling particles are dried by rising hot air.

rotary feeder. A machine in which rotating fins deliver granules or powders to an operation at a predetermined rate.

rotary kiln. An inclined rotating, refractory-lined elongated cylinder, fired from the lower end, which is charged at the upper end and discharged at the lower end; used for melting, clinkering, or calcining materials.

rotary kiln block. A curved refractory shape, usually with a 9-inch chord and a smaller inside chord, 6 or 9 inches in radius and 4 inches thick; used as segments in the lining of rotary or circular kilns.

rotary smelter. A batch-type, refractory-lined cylinder with conical ends in which porcelain enamels or glazes to be fritted are melted; the raw batch is charged at the burner end, the cylinder is rotated as melting progresses, and the mass is discharged at the flue end by tilting the furnace when melting is complete.

rotor. The rotating part of an electrical or mechanical device.

rottenstone. A soft, decomposed, siliceous limestone used as a polishing material.

rouge. Finely divided, hydrated ferric oxide used as a polishing material and as a colorant.

rouge flambé. A decorative pottery glaze containing colloidal copper which produces a typical red color when fired in a reducing atmosphere.

rouge pits. Imperfections consisting of traces of rouge entrapped in incompletely polished glass surfaces. See rouge.

rough and burred edges. Frayed or serrated edges of a metal shape after cutting, stamping, shearing, or trimming.

rough cast. A rough plaster finish such as is obtained by throwing the plaster on a wall with a trowel.

rough-cast glass. A flat glass having one textured surface made by using a roller with a patterned face.

roughened finish tile. Tile having a back surface which has been roughened by wire cutting, wire brushing, or other mechanical means before firing to obtain increased bond with mortar, plaster, or other substances.

rough glass. Rolled glass sheets cut into workable sizes.

rough grinding. The grinding of glass, metal, ceramic, or other surface without regard to the quality of the finish.

roughness. (1) The difference between the peaks and valleys of a surface.
(2) The relative degree of coarse, ragged, pointed, or bristle-like projections on a surface.

rough turning. The rapid and efficient removal of excess stock from a work piece by a grinding or milling machine without regard to the quality of the finished surface.

round kiln. A kiln constructed in the form of a circle with a series of fireboxes stationed around the periphery of the structure. See circular kiln.

round table. A table upon which plate glass is laid for grinding and polishing.

rowlock arch. An arch constructed of wedge-shaped brick arranged in concentric rings.

rowlock course. A course of brick laid on edge with the longest dimensions perpendicular to the face of a wall.

royal blue. A rich, deep blue ceramic color composed of cobalt oxide and flux.

rub. Closely arranged scratches produced simultaneously on a glass surface as a decorative treatment.

rubbed surface. A formed concrete surface rubbed with carborundum stone, or with burlap and mortar, to obtain an improved appearance.

rubber gasket. A rubber seal used in the joints of concrete pipe.

rubber set. The premature setting of cement in concrete due to the presence of unstable gypsum.

rubber wheel. A grinding wheel bonded with rubber.

rubbing brick. A block of bonded abrasive used for rubbing down castings, scouring chilled iron rolls, polishing marble, etc.

rubbing stone. A fine-grained abrasive shape used to remove imperfections from porcelain-enameled and glazed surfaces by rubbing.

rubidium carbonate. Rb_2CO_3; mol. wt. 230.9; m.p. 837°C (1538°F); used in special glass formulations.

ruby alumina. Chromic oxide-bearing corundum, Al_2O_3, red in color, used as an abrasive.

ruby glass. A glass of deep red color produced by additions of selenium or cadmium sulfide, copper oxide, or gold chloride.

ruby, selenium. A ruby-red glass containing selenium oxide, cadmium sulfide, arsenic oxide, and carbon and produced in a reducing atmosphere.

runner. Large blocks of chert used in the bottom of pan mills. See chert.

runner bar. An iron casting attached to a circular grinding head or runner for the abrasive grinding of plate glass.

runner brick. A perforated refractory brick through which molten steel is passed during the bottom pouring of ingots.

runner cut. An imperfection in plate glass resulting from the rupture of the surface by the runner bar. See runner bar.

running batch. A glass batch formulated to produce a desired composition when used with its own cullet.

running bond. A masonry bond involving the placement of each brick as a stretcher and overlapping the bricks in adjoining courses.

rupture, effective modulus. See modulus of rupture, effective.

rupture modulus. The transverse strength of a material based on the length of the specimen between supports and the cross-sectional dimensions of the specimen. See modulus of rupture.

rustication strip. A strip of wood or other material attached in a form to form a groove in concrete at a construction or panel joint.

ruthenium boride. (1) RuB_2; mol. wt. 123.34; sp.gr. 7.6–10.1 (2) Ru_2B_5; mol. wt. 257.5; sp.gr. 9.2. (3) Ru_7B_3; mol. wt. 744.36.

ruthenium silicide. (1) RuSi; mol. wt. 129.76. (2) Ru_2Si_3; mol. wt. 287.58.

rutile. TiO_2; mol. wt. 79.9; m.p. 1640°C (2984°F); sp.gr. 4.3; hardness (Mohs) 6–6.5; used as an opacifying agent and colorant in porcelain enamels and glazes, and as a component in titanate dielectrics.

R-value. A calculation of the partial light-dispersive ratio of glass based on indices of refraction determined by the equation:

$$R = \frac{n_d - n_c}{n_F - n_c}$$

in which R is the ratio, n_d is the index of refraction for the sodium line at 589.3n, and n_F and n_c are the indices for the hydrogen lines at 486.1n and 656.3n, respectively.

ryolex. A volcanic mineral of SiO_2 and Al_2O_3 used as light-weight insulation.

S

S-crack. An S-shaped laminar defect occurring in a clay body during extrusion from a pug mill.

saddle. An item of wedge-shaped kiln furniture employed as a prop between plates packed on edge in a kiln.

saddle arch. One of a series of arches which supports the checker-work in a regenerator.

saddle clay. A clay of fine particle size and high-flux content which fuses at a low temperature; used as a stoneware and electrical porcelain glaze.

saddle, serrated. A grooved or notched item of kiln furniture to support flatware on edge during firing.

safety can. A metal container of cylindrical design used for the storage and transport of hazardous materials.

safety door. A door designed to contain catastrophic conditions to a restricted area in the event of an emergency, such as fire or an explosion.

safety flange. A type of flange with tapered sides designed to hold parts of a wheel intact in the event of its breakage during use.

safety glass. (1) A glass constructed of sheets laminated with plastic films to prevent shattering in the event of breakage.

(2) A glass containing a network of wire to improve its resistance to breakage and shattering.

(3) A glass which has been tempered by heat treatment so that it will break into small fragments or grains which do not scatter when broken and are less liable to cause injury than ordinary glass.

safety glass, tempered. See safety glass (3).

safety lever. A lever which actuates safety mechanisms on hazardous machinery or mechanical devices.

safety requirements. The regulations required to protect persons against injury by failure, breakage, or other accidents of a hazardous nature.

safety ring. A metallic ring embedded in organic-bonded abrasive wheels to contain pieces if breakage should occur on the grinder.

safety valve. A pressure-activated device designed to permit the escape of steam or gases from boilers or other equipment, and from hydraulic systems, when the internal pressures exceed safe working limits.

sag. See sagging.

sagger. A fired refractory container, usually of box-like shape, in which ceramic ware may be bisque or glost fired to protect the ware from furnace gases, dirt, uneven heating, thermal shock, and physical damage.

sagger clay. A fairly uniform open-firing refractory clay which, when employed in saggers, will withstand conditions of repeated heating and cooling.

sagger maker's bottom knocker. The operator who beats out clay-grog mixes to form the bottoms of saggers.

sagger, nibbed. A sagger containing internal projections or flat ribs on which a refractory slab may be placed to permit the firing of multiple tiers of ware.

sagging. (1) A defect consisting of a wavy line or lines which flow or slide on the vertical surface of porcelain-enameled ware during firing.

(2) The process of forming glass, usually flat, by reheating until it conforms to the shape of the mold or form on which it rests.

(3) A defect characterized by the irreversible downward bending of an article insufficiently supported during the firing operation.

sagging, spray. A defect characterized by wavy lines in glazes and porcelain enamels on the vertical surfaces of ware during and after spraying, but before the coating has dried.

salmanazar. A ceramic or glass wine bottle of 9 liter capacity.

salmon, salmon brick. A relatively soft, underfired brick of salmon color.

sal soda. $Na_2CO_3 \cdot 10H_2O$; mol. wt. 286.2; m.p. 32.5–34.5°C (91–94°F); loses water at these temperatures; sp. gr. 1.44; used as an oxidizing agent and flux in glasses and porcelain enamels and as a neutralizer in the pickling of iron and steel for porcelain enameling.

salt cake. Impure sodium sulfate. Na_2SO_4; mol. wt. 142.04; m.p. 888°C (1630°F); sp.gr. 2.67; used in glazes and glass as a source of sodium and as an anti-scumming agent.

salt glaze. A lustrous glaze produced on ceramic surfaces toward the end of the firing cycle by throwing salt into the firing box, the salt volatilizing, and the resultant fumes then entering into a thermochemical reaction with the silicates and other components of the ceramic.

salt-glazed tile. Facing tile having a lustrous glaze finish resulting from the thermochemical reaction between the silicates of the clay body and the vapors of salt or other chemicals during firing.

saltpeter. KNO_3; mol. wt. 101; m.p. 334°C (630°F); decomposes at 400°C (752°F); sp.gr. 2.1; used in glass, glazes, and porcelain enamels because of its powerful fluxing and oxidizing properties.

salts, soluble. The chlorides, sulfates, and some silicates of calcium, magnesium, potassium, and sodium contained in solution in a body which before, during, or after firing may cause efflorescence or scumming on the surface of ware.

salt-spray test. An accelerated test of the resistance of a material or product to corrosion in which a specimen is subjected to a spray of a sodium chloride solution under prescribed conditions.

salt water. Molten sulfates floating on the surface of molten glass in the glass-melting unit.

salvage value. The net worth of a material recovered from a process.

samarium aluminate. $Sm_2O_3 \cdot Al_2O_3$; mol. wt. 450.80; m.p. 1982°C (3600°F).

samarium boride. (1) SmB_4; mol. wt. 193.71; sp.gr. 6.18. (2) SmB_6; mol. wt. 215.35; m.p. 2538°C (4600°F); sp.gr. 5.07; hardness 2500 Vickers. See borides.

samarium carbide. (1) Sm_3C; mol. wt. 463.29; sp.gr. 8.14. (2) Sm_2C_3; mol. wt. 336.86; sp.gr. 7.47. (3) SmC_2; mol. wt. 174.43; m.p. >2204°C (4000°F); sp.gr. 6.50. See carbides.

samarium nitride. SmN; mol. wt. 164.44; sp.gr. 8.50. See nitrides.

samarium oxide. (1) Sm_2O_3; mol. wt. 348.7; m.p. 2300°C (4172°F); sp.gr. 7.43; used in luminescent glasses, and infrared absorbing glasses, as a phosphor activator, and as a neutron absorber in nuclear applications. (2) SmO; mol. wt. 166.43.

samarium phosphide. SmP; mol. wt. 181.46; sp.gr. 6.34. See phosphides.

samarium selenide. SmSe; mol. wt. 229.63; m.p. 2093°C (3800°F).

samarium silicate. (1) $Sm_2O_3 \cdot SiO_2$; mol. wt. 408.92; m.p. 1940°C (3525°F); sp.gr. 6.36. (2) $2Sm_2O_3 \cdot 3SiO_2$; mol. wt. 877.9; m.p. 1921°C (3490°F); sp.gr. 5.77. (3) $Sm_2O_3 \cdot 2SiO_2$; mol. wt. 468.98; m.p. 1777°C (3230°F); sp.gr. 5.20.

samarium silicide. $SmSi_2$; mol. wt. 206.55; sp.gr. 6.26. See silicides.

samarium sulfate. Sm_2SO_4; mol. wt. 396.92; used in red and infrared phosphors.

samarium sulfide. A group of compounds exhibiting thermoelectric properties. (1) SmS; mol. wt. 182.49; m.p. 1940°C (3525°F); sp.gr. 6.01. (2) Sm_3S_4; mol. wt. 579.53; m.p. 1799°C (3270°F); sp.gr. 6.14. (3) Sm_2S_3; mol. wt. 397.04; m.p. 1782°C (3240°F); sp.gr. 5.83. (4) SmS_2; mol. wt. 214.55; m.p. 1730°C (3145°F); sp.gr. 5.66. See sulphides.

samarium telluride. SmTe; mol. wt. 277.93; m.p. 1915°C (3480°F). See intermetallic compound.

samel. An under-burned brick, usually fired near the outer rim of a clamp kiln where the temperature is lower than the interior of the kiln.

sample, archive. A sample retained for record or future use.

sample, composite lot. A sample prepared by combining several materials in the same ratio as the net weight of the material sampled.

sample, container. Samples obtained from individual containers or sources by the use of a sample thief or other approved or accepted means. See sample thief.

sample log. A recorded listing of samples preserved for reference purposes.

sampler. A device with which to obtain small samples of materials for analysis.

sample, representative. A mixture of samples in which every particle in the lot is equally represented in the gross sample.

sample, sampling. A number of representative specimens or fractions drawn from a lot to determine their nature, composition, properties, quality, or other attributes of concern.

sample splitter. A device employed to mix and subdivide a sample of powdered or granular material for analysis or other evaluation.

sample thief. A device of suitable design employed to remove a sample from a batch or lot of a material for subsequent evaluation or analysis.

sampling plan. A procedure that specifies the frequency and number of samples taken from a lot and the criteria for accepting or rejecting the lot.

sampling, systematic. The taking of samples from a batch or manufacturing operation at fixed time intervals or in fixed quantities, or both.

sand. Fragments of rock composed essentially of rounded grains of quartz ranging from 0.05 to 5 mm. in grain size; employed in glass, glazes, porcelain enamels, ceramic bodies, portland cement, building and construction work, and as an abrasive, as a setting medium for the firing of ceramic ware, as a core in foundry molds, and numerous other applications.

sand, bank. A sand of low-clay content used in making casting cores.

sandblast fireclay. See slag sand.

sandblast, Gravé. A sandblasted design of varying depth on glass surfaces.

sandblasting. The process of projecting sand in a stream of air or steam at high velocity to engrave, cut, obscure, or clean glass, metals, or other surfaces.

sandblast, peppered. A finely textured sandblast obscuration of a mottled appearance on the surface of a glass item.

sandblast, shaded. Surface obscuration of glass, graduating in texture from clear to full obscuration, produced by sandblasting.

sand, blending. A sand, added to normally available sand, employed in concrete for improvement of gradation.

sand creased. A type of texture produced on the surface of facing brick by sprinkling or rolling the brick in sand before molding or by texturing the face of the brick during molding.

sand finish. Structural clay products having surface faces covered with sand which is applied to the clay column in the stiff-mud process or as a lubricant to the molds in the soft-mud process.

sand, foundry. Sand used to make molds for metal castings; characterized by refractoriness, cohesiveness, and durability.

sand, furnace. A relatively pure and coarse sand used as a refractory material for furnace hearths and foundry molds.

sand, glass. A high-grade sand containing 98 to 100% of silica of medium-sized grains and less than 1% of iron oxide; mol. wt. 60.1; m.p. 1710°C (3110°F); sp.gr. 2.2–2.6.

sand holes. Small fractures in the surface of glass produced during the rough grinding operation and which were not removed during the subsequent fine-grinding or polishing operation.

sanding. (1) A surface texture produced on brick during manufacture.
(2) A bedding material for brick, saggers, earthenware, etc., in a kiln.

sand-lime brick. A brick made from a mixture of silica sand and lime, and cured under the influence of high-pressure steam.

sand, Lynn. A pure form of quartzose sand.

sand, molding. Sand used on wooden molds in the forming of brick by the soft-mud process.

sandpaper. An abrasive product in which the abrasive is bonded to paper on one side. Used for sanding and polishing.

sand, placing. Sand used for the placement of ware in kilns to prevent ware from sticking to shelves, setter plates, etc., during firing.

sand seal. A seal consisting of metal plates attached along the sides and bottom of a kiln car, and immersed in a trough of sand extending through the length of the kiln and along the car rails, to prevent hot gases from entering under the car.

sand, silica. Sand containing a high percentage of free silica.

sand slab. A slab of concrete placed in the bottom of a wet excavation to seal the bottom and facilitate subsequent work.

sand, slag. Finely crushed slag used in mortars and cements. See slag.

sandstone. A sedimentary rock consisting essentially of quartz, sometimes in combination with feldspar, mica, and other minerals, which have been united by pressure or cemented by a clay, silica, iron oxide, calcium carbonate, or other material.

sand streak. A blemish on a formed concrete surface caused by the loss of grout or mortar through cracks in the form, or by failure of the concrete to consolidate.

sand-struck brick. A wet-clay brick, containing 20 to 30% moisture, formed in a mold in which the inside of the mold is coated with sand to prevent the damp clay from sticking to the mold.

sandwich kiln. A kiln in which heat is applied to ware from both top and bottom simultaneously.

sang de boeuf. A red copper-bearing glaze fired under reducing conditions.

sanitary ware, sanitary plumbing fixtures. Glazed, vitrified whiteware or porcelain-enameled fixtures having sanitary service functions, such as sinks, lavatories, bathtubs, and similar products.

sanitary ware, vitreous. See sanitary ware.

sapphire. A form of corundum, Al_2O_3, employed as a bearing material fabricated to high precision, and as an abrasive and polishing material.

satin finish. A very smooth surface finish with low or dull reflective properties.

saturated air. Air containing the maximum amount of water vapor possible at a given temperature or barometric pressure.

saturation. The point at which no more of a material can be dissolved, absorbed, or retained by another material.

saturation coefficient. The ratio of the weight of water absorbed by a masonry unit during immersion in cold water to the weight absorbed in boiling water; used as an indication of the resistance of brick to freezing and thawing.

saturation, magnetic. The maximum magnetic field strength in a magnetic material beyond which further magnetization does not occur, and at which point incremental permeability has decreased progressively to approach zero.

saucer wheel. A shallow abrasive wheel of saucerlike shape.

sawdust concrete. A concrete of relatively low strength in which sawdust is incorporated as aggregate; used as a lightweight nailing concrete in construction applications.

sawdust firing. The placement of sawdust in a closed pot containing ware to produce a reducing atmosphere during firing.

saw gummer. A straight or saucer-shaped abrasive wheel used to grind away punch marks formed between the teeth of saws during manufacture; also used as a saw sharpener.

sawtooth crusher. A machine in which material is crushed during passage between sawtooth shafts rotating at different speeds.

scab. (1) A defect in metal sheets and castings for porcelain enameling consisting of a partially detached metal fin joined to the metal surface.
(2) A defect consisting of an undissolved inclusion of sodium sulfate or other similar material in glass.

scalding. The peeling or popping of unfused coatings from the surfaces of ware during the early stages of firing.

scale. (1) The oxide formed on the surface of metal during heating, usually before or during porcelain enameling.
(2) A small fragment of foreign material imbedded in the surface of molded glass articles.
(3) A weighing or measuring device.

scale, fish. A defect in porcelain-enameled surfaces consisting of spontaneously fractured areas of half-moon or fishscale-shaped fragments.

scale, shiner. Very fine shiny scale appearing on thin or overfired porcelain-enamel ground coats.

scaling. (1) The process of forming scale on metals, with or without acid fumes, as a means of cleaning and preparing the surface for subsequent pickling and porcelain enameling.

(2) The removal of rust and other unwanted contaminants from the surfaces of metals.

(3) The peeling or flaking of concrete, usually pavements, under the influence of de-icing agents.

scaling, acid. The process of dipping or sprinkling raw-metal shapes for porcelain enameling with sulfuric or hydrochloric acid prior to annealing to assist in the removal of severe deposits of rust and other surface contaminants on the metal.

scallop. A decorative motif consisting of a continuous series of curves forming an edge or design on a product.

scalp. (1) To remove surface layers of undesired materials from another bulk material.

(2) To remove portions of fine or coarse pit-run aggregate in a preliminary screening operation prior to use of the parent aggregate in concrete.

scalping screen. A screening device used to remove undesirable fine and coarse particles from a material.

scan. To make an examination of an area, product, space, or property as a monitoring activity.

scandium boride. ScB_2; mol. wt. 66.74; m.p. 2249°C (4080°F); sp.gr. 3.67. See borides.

scandium carbide. ScC; mol. wt. 57.10; sp.gr. 3.59; potentially useful as a high-temperature semiconductor.

scandium nitride. ScN; mol. wt. 59.11; m.p. 2700°C (4892°F); sp.gr. 3.6; useful in space applications and as crucible material for preparations of high-purity single crystals of gallium and other compounds.

scandium oxide. Sc_2O_3; mol. wt. 138.2; m.p. >2300°C (4172°F); sp.gr. 3.864; specific heat 0.153 (0–100°C); a network former in glass; also used in high-temperature systems and electronic applications.

scandium phosphide. ScP; mol. wt. 76.13; sp.gr. 3.28; see phosphides.

scandium silicate. (1) Sc_2O_3; mol. wt. 198.26; m.p. 1950°C (3542°F); sp.gr. 3.49. (2) $Sc_2O_3 \cdot 2SiO_2$; mol. wt. 258.32; m.p. 1860°C (3380°F); sp.gr. 3.39.

scanistor. An integrated semiconductor optical-scanning device which converts images into electrical signals.

scanning electron microscope. An electron microscope in which a beam of electrons sweeps over a specimen, measuring the intensity of the secondary electrons generated at the point of impact of the beam on the specimen, and relaying the signal into a cathode-ray display which is scanned in synchronism with the scanning of the specimen.

scar. A mark in a porcelain-enameled surface produced by firing a coating which previously had been scarred or similarly damaged after drying. The term also may be applied to the defect as it appears in the dry state before firing.

scatter coefficient. The rate of the increase in the reflectance of porcelain enamels with thickness at infinitesimal thickness over a black background. See reflectance.

scattering. Dispersing into different directions.

schamotte. A refractory clay or grog which has been calcined for use as a nonplastic material in ceramic body compositions.

scheelite. $CaWO_4$; mol. wt. 288.1; sp.gr. 5.9–6.1; hardness (Mohs) 4.5–5; used as a source of tungsten and in phosphors.

schist. A variety of metamorphic rocks, such as mica, feldspar, horneblende, and quartz, which may readily be split into thin plates or slabs.

scintillation. The multiple discharges or small arcs which originate in the more conductive areas of insulating surfaces, and span less conductive areas; that is, the spark of light produced in some substance by the absorption of an ionizing particle or photon.

scintillation spectroscope. A scintillation counter adapted to measure the energy and intensity of gamma rays from radioactive elements.

scintillator. A material that emits optical photons in response to ionizing radiation used in optical instruments such as spectrometers, scintillation detectors, cameras, counters, etc.

scleroscope. An instrument for determining the relative hardness of materials by measuring the height to which a standard steel ball rebounds from its surface when the ball is dropped from a standard height.

scoop. A shovel-like instrument designed for the movement or transport of loose materials.

scored finish. The grooved appearance of the face surface of a structural clay body as it comes from the die.

scored finish tile. Structural tile having a scored face surface designed to receive and to give increased bond with mortar, plaster, or stucco.

scoring. The process of forming a groove, scratch, notch, or similar indentation on the surface of a material, usually before firing.

scotch block. A rammed refractory gas port in an open-hearth steel furnace.

scotch method of roofing application. A method of applying rectangular asbestos-cement roofing shingles which overlap at the top and one side to form a rectangular or square pattern.

scouring. (1) The process of cleaning and smoothing the surface of bisque-fired ceramic ware with a coarse abrasive in a revolving drum.
(2) The mechanical cleaning or finishing of a hard surface by using an abrasive and low or light pressure.

scouring block. A chemically bonded abrasive block composed of Al_2O_3, SiC, or similar material used in the grinding and polishing of metals and ceramic surfaces.

scove brick. An unfired brick used in the construction of scove kilns, which see.

scove kiln. An updraft kiln constructed of unfired brick having no permanent parts, and which may be fired with gas, oil, coal, or wood.

scoving. The outer layer, usually wet clay, applied to a scove kiln to make the kiln gas tight.

scrap. A reject from a manufacturing operation which is unsuitable for reclaiming or salvage.

scrapings. The overspray of porcelain enamel collected and recovered from a spray booth.

scrapping. The removal of excess body from slip-cast ware before removal of the ware from the mold.

scraps. The excess body removed from slip-cast ware during forming.

scratch. Any marking or tearing of a surface produced during manufacture or handling having the appearance of being caused by a sharp instrument.

scratch coat. A layer of plaster having a scratched surface to improve its bond with a subsequent layer.

scratch hardness, scratch resistance. The resistance of a surface to scratching. See scratch test.

scratch test. A hardness test in which a diamond point or other cutting instrument is drawn across a surface and the length or width of the resulting scratch is compared with a related standard. See Bierbaum scratch hardness.

screed. A straight-edged tool or guide of wood or metal for making the first strike-off of a surface of concrete or plaster while removing any excess of the material and for smoothing the concrete or plaster surface.

screen. (1) A wire mesh or perforated plate mounted on a suitable frame employed to separate coarser parts of a loose, flowing conglomerate material from the finer parts by allowing the passage of the smaller parts while retaining those of the larger.
(2) A silk, wire, or similar material, in mesh or gauze form, through which pigmented inks are forced onto the surface of ware to produce a design.

screen analysis. A technique to determine the particle size or particle-size distribution of powders and the solid constituents of porcelain enamels, glazes, and other slips or slurries by calculating the percentage of solids retained in each of a graduated series of sieves of various sizes.

screening ink. An oil suspension of ceramic pigment used in the silk-screen process to imprint designs on glass, porcelain enamel, and other ceramic surfaces, and which develops its color on firing. Also known as ceramic ink.

screening paste. See screening ink.

screening plate. A metal plate containing openings of specified size used to control the fineness of grinding in dry pans and hammer mills.

screen, scalping. A screening device designed to remove undesirable fine and coarse particles from a material.

screen, shaker. A mechanically vibrating screening device employed to separate materials into desired particle sizes.

screen, vibrating. Wire-mesh screens which are vibrated mechanically, by solenoid or magnetostriction, and which sometimes are heated to increase efficiency.

screw contact. An electrical contact fabricated with an external thread for attachment to a support member.

screw conveyor. A conveyor consisting of a helical screw which rotates on a single shaft in a stationary trough or casing such that granular material may be moved along a horizontal, inclined, or vertical plane.

screw feeder. A device consisting of an augur or rotating helicoid screw employed to transfer pulverized or granular material from one piece of equipment to another.

screw press. A press in which the slide of the press is actuated by a screw mechanism.

scribing. The scoring of a bisque porcelain-enamel coating with a sharp tool, often combined with a brushing operation, as a form of decoration.

scrubber. A machine for cleaning coarse aggregate consisting of a horizontal rotating cylinder containing blades that lift and tumble the aggregate, usually in the presence of water, to remove clay and other soft particles and coatings.

scrub marks. A surface blemish on glass, usually appearing as a series of vertical markings, caused by friction during processing or handling.

scuff, scuffing. Physical damage to the surface of glass or other ceramic ware caused by scratching, gouging, abrasion, or wear.

scum, scumming. (1) A layer of unmelted material floating on the surface of molten glass.

(2) An area of poor gloss on an otherwise bright, glossy surface of porcelain enamel, glass, and glazes.

(3) Clouds appearing around decalcomania which are caused by varnish residues.

(4) A layer of soluble salts or fuel residues which are oxidized on the surface of building brick during the firing operation.

scurf. (1) A hard carbonaceous deposit on the surfaces of retorts, coke ovens, and the like caused by the cracking of gases during use.

(2) To remove scurf by scraping, rubbing, or wiping.

scutch. A steel bricklayer's hammer used for cutting, trimming, and dressing brick.

seal. Any device or system that creates a nonleaking union between two mechanical or process-system elements; a tight closure or joint.

seal, air. A curtain of air blown across the entrance and exit of a dryer or furnace to contain heat within the structure.

seal, airborne. A repair in which a refractory powder is blown and collected in a defective or leaking area of a hot retort to develop a seal against further leakage.

seal, butt. A flat metal washer brazed tightly to a flat metallized ceramic surface.

seal, external. A metal flange or collar surrounding a cylindrical ceramic part.

sealed pores. Pores or small bubbles entrapped in a ceramic body which have no outlet to the exterior of the body, calculated by the formula:

$$Pc = \frac{W_d}{D_a} - \frac{W_d}{D_t}$$

in which Pc is the volume of sealed or closed pores, W_d is the dry weight of the specimen, D_a is the apparent density, and D_t is the true density of the specimen.

sealed porosity. The ratio of the volume of sealed pores to the bulk volume of a ceramic, expressed as percent.

seal, foundation. A slab of concrete placed in the bottom of a wet excavation to seal the bottom and to facilitate subsequent work.

seal, glass-to-metal. A seal in which a glass and metal are fused together at their interfaces, usually in a prescribed shape.

sealing. The process of joining two items of glass, or glass and a metal, by heating an interface to reduce the viscosity of the glass to permit it to flow and bond with the other glass or metal.

sealing compound. (1) A bituminous material for filling and sealing joints and cracks in concrete.

(2) A curing compound for concrete.

sealing glass. A glass with special thermal-expansion and flow characteristics to enable it to bond with another glass or solid.

sealing surface. That portion of the finish of a glass container or other ceramic which makes contact with the sealing gasket or liner.

seal, internal. A seal between a metal and the inner wall of a ceramic component.

seal, pin. Wire positioned and sealed through the inside diameter of a bushing.

seal, ram. A seal in which a metal sleeve is forced to form a thin circumferential line of contact over the sharp edge of a ceramic shape, and then completed by brazing or plating a metal over the joint.

seal, sand. A seal consisting of metal plates attached at the bottom and parallel to the sides of a kiln car and immersed in a trough of sand parallel to the car rails in the kiln to prevent the advent of hot gases to the bottom of the car as it passes through the firing zone.

seal, tapered. A thin, metal sleeve fitted over a thick, tapered, ceramic cylinder so as to form a tight seal.

seam. (1) A mark on the surface of glass or a ceramic resulting from the joint of the matching mold parts.

(2) To grind, slightly, the sharp edges of a piece of glass.

seat. (1) A prepared position on the siege where a pot is placed.

(2) To fit an item to conform with the configuration of another item at the point of contact. See siege.

seat earth. A natural deposit or bed of clay situated beneath a seam or layer of another mineral.

seating block. A fireclay refractory shaped to support a boiler.

seawater magnesia. Magnesia recovered from water from the seas. See magnesia.

second. A marketable product of inferior grade or one which does not conform to the quality of a standard product.

secondary air. Combustion air injected over the flame or fuel bed of a kiln to enhance completeness of combustion.

secondary clay. A clay which has been moved geologically from the site of its formation to another.

secondary crusher. A crushing device used after the initial crushing operation to reduce further the particle size of a material.

secondary expansion. The permanent expansion of fireclay brick during service.

secondary recrystallization. The process by which large grains are nucleated and grow at the expense of a fine-grained but essentially strain-free matrix.

second side. The final or exposed face side of a plate glass to be ground and polished.

sediment analyzer, Woods Hole. A technique for measuring particle sizes of clays based on the measurement of pressure changes resulting from sedimentation in a suspension of the particles in water.

sedimentary clay. A clay which, geologically, has been moved from its point of origin to another.

sedimentation. (1) The process of the deposition or settling of matter suspended in a liquid.
(2) The appearance of free water on the surface of fresh concrete resulting from the settlement of solid particles and the consequent relative movement of water upward.

sedimentation rate. The speed at which particles settle from a liquid suspension.

sedimentation volume. The volume of particles settled from a liquid suspension.

seed. (1) A small gaseous inclusion in glass.
(2) A small single crystal of a semiconductor material used to start the growth of a large crystal.

Seger cone. Any of a series of pyramidal thermometric devices made of materials or mixtures of materials which deform at known temperatures and which are used to indicate the thermal history of ceramic bodies and glazes during the firing operation.

Seger formula. A molecular formula applied to glazes and porcelain enamels in which the oxide constituents are classified in three groups, RO (the alkaline oxides whose sum equal 1), R_2O_3 (the intermediate oxides), and RO_2 (the so-called acidic oxides).

segmental arch. A circular or rounded arch having a curved surface less than a semicircle.

segmental wheel. An abrasive wheel composed of segments of bonded abrasives assembled to form a complete wheel.

segmented belt. A coated abrasive belt made of sections of belt spliced together.

segments. Bonded abrasive sections of various shapes to be assembled to form a continuous or intermittent circular grinding surface.

segregation. (1) The separation of the ingredients of a mixture, such as fine portions from coarser portions.
(2) The separation of coarse aggregates in concrete from the mortar or main mass of the concrete.

seignette-electric. A ferroelectric crystal, such as Rochelle salt, potassium sodium tartrate ($KNaC_4H_4O_6 \cdot 4H_2O$), which displays ferroelectricity, and which is used in ceramic capacitors, transducers, and dielectric amplifiers.

selectivity. The characteristic of a testing system that is a measure of the extent to which an instrument is capable of differentiating between a desired signal and disturbances of other frequencies or phases in electromagnetic testing.

selenium. Se; at. wt. 79.2; m.p. 217°C (420°F); volatilizes at 688°C (1270°F); sp.gr. 4.2–4.8; used in glass as a decolorizer, both in elemental and compound forms; also employed to produce rose and ruby colors in glass, glazes, and porcelain enamels.

selenium ruby. A ruby-red glass containing selenium oxide, cadmium sulfide, arsenic oxide, and carbon, and produced in a reducing atmosphere.

self-cleaning enamels. Porcelain-enamel coatings containing additions of selected materials which, when applied to culinary ovens, will promote oxidation of grease and oven spills continuously during oven use.

self-emulsifier. A penetrating material which emulsifies spontaneously in water, and which may be rinsed from a specimen more readily than if it were removed by dissolving in the rinse water.

self-slip. An archeological term describing the fine layer resulting from the wet-surface smoothing of clay vessels.

self-sustaining discharge. An electrical discharge in which all carriers necessary for the transport of current in the discharge are produced by the discharge itself.

seller. The supplier of a material defined by a purchase order.

selvedge. The formed edge of a ribbon of rolled glass.

semiconducting crystal. A crystal, such as silicon or germanium, which exhibits an electrical conductivity between that of a metal and an insulator.

semiconducting glaze. A ceramic glaze containing metal oxides in sufficient quantities to promote a degree of electrical conductivity to prevent surface discharge or flashover.

semiconductor. Any material having an electrical conductivity between that of a metal and an insulator.

semiconductor device. An electronic instrument in which electronic conduction occurs within a semiconductor.

semiconductor diode. A two-electrode semiconductor or two-terminal device which employs the properties of semiconductors which exhibit rectifying properties at the point of contact.

semiconductor laser. A diode laser in which stimulated

emission of coherent light occurs at a pn junction when electrons and holes are driven into the junction by carrier injection, electron-beam excitation, impact ionization, optical excitation, or other means.

semiconductor, metal oxide. A semiconductor system in which the insulating layer is the oxide of the metal substrate.

semifriable alumina. A hard abrasive grade of recrystallized alumina in which the Al_2O_3 content ranges between 96 and 98%.

semimatte glaze. A ceramic glaze exhibiting only a moderate degree of gloss that is considered to be between high gloss and matte in appearance.

semimuffle furnace. A gas or oil-fired furnace constructed with a partial muffle to prevent the flame from impinging directly on the ware being fired, but in which the products of combustion gases can come in contact with the ware. See muffle.

semipermanent mold. A reusable mold.

semiporcelain. A trade term designating dinnerware having a moderate water absorption of 0.3 to 3.0%.

semisilica fireclay brick. A fireclay brick containing not less than 72% silica.

semivitreous. The degree of vitrification indicated by a moderate water absorption of 0.3 to 3.0% except for wall and floor tile in which the water absorption may range from 3.0 to 7.0%.

semivitreous china. A dinnerware or other ceramic product exhibiting a moderate degree of water absorption.

sensible heat. The heat which raises the temperature of a body in which it comes in contact; the sum of the internal energy of a body or system plus the product of the system's volume multiplied by the pressure exerted on the system by its surroundings.

sensitivity. The least amount of concentration that can be determined by a method.

sensitizing compounds. Metal salts in aqueous or organic solutions which form an invisible film on glass and ceramic surfaces, and which initiate or hasten subsequent surface treatments such as silvering and plating; examples are the chlorides of tin, gold, and palladium, and some salts of aluminum, barium, cadmium, iridium, and silver.

sensor. A generic term for an instrument which measures a value or detects a change in a value such as physical quantity.

sentinel pyrometers. Small cylinders of standardized compositions which melt at predetermined temperatures; used to measure and control thermal treatments of materials in kilns and furnaces.

separated aggregate. Concrete aggregate which has been classified into fine and coarse components.

separating powder. A powder applied to a surface, as in a mold, to facilitate the removal of ware after forming.

separation, chemical. Removal, isolation, or separation of a substance from a sample by methods which involve a knowledge of the chemical properties of the substance as opposed to mechanical or physical properties and techniques.

separation, gravity. The separation of mixtures into layers according to their respective densities, usually in a stream of air, in a liquid suspension, by means of a vibrated sloping table, or other technique.

separation, magnetic. The use of a magnetic separator to remove iron from milled porcelain enamels, glazes, and other slurries.

separation, mechanical. The separation of solid or liquid substances from a sample by settling or gravity separation, centrifugal action, vibratory action, filtration, screening, or other technique involving physical differences in the materials comprising the sample.

separation, physical. See separation, mechanical.

separator. A device employed to separate different kinds and sizes of materials from others.

separator, air. An apparatus which employs a current of air rising in a cone or cylinder to separate one material from another on the basis of particle size or density.

separator, magnetic. An apparatus employing a strong magnetic field to remove magnetic contaminants from bodies, glazes, porcelain enamels, and other slurries.

sepiolite. $3MgO \cdot 4SiO_2 \cdot 5H_2O$; mol. wt. 451.2; sp. gr. 2.0; hardness (Mohs) 2–2.5; a soft, lightweight, absorbent clay.

sequence. An orderly progression of operations to assure optimum utilization of production facilities.

sericite. $K_2O \cdot 3Al_2O_3 \cdot 6SiO_2 \cdot 2H_2O$; mol. wt. 802.4; a fine–grained, potassium mica. See muscovite.

serpentine. $3MgO \cdot 2SiO_2 \cdot 2H_2O$; mol. wt. 277.1; sp. gr. 2.5–2.65; hardness (Mohs) 2.5–4; sometimes used in forsterite refractories.

serrated saddle. A grooved or notched item of kiln furniture to support ceramic whiteware on edge during firing.

service life. (1) The period of time an item may be used economically before breakdown.

(2) The elapsed time until the end-point is reached in an adsorption or other process.

service life, accelerated. Tests in which the failure of an item or product is hastened by the use of conditions more severe than those anticipated in service as a means of estimating normal operating life or use.

service test. A test conducted on a product under simulated or actual conditions of use to determine if the product will perform satisfactorily under conditions of use.

servitor. A workman who shapes the stem and base of goblets and footed stemware.

sessile drop. A method of measuring the surface tension of a liquid on the surface of a material, such as a metal or ceramic body, in which the mass, depth, and shape of the liquid drop are observed.

set. (1) The consistency and flow properties of a porcelain-enamel slip which affect its suspension characteristics, rate of drain, residual thickness, and uniformity of coating.

(2) To place ware in a kiln.

set, false. The premature stiffening or the erratic, abnormal, quick setting of cement in concrete, such as is caused by unstable gypsum in the cement. See setting time.

set, final. The period of time elapsed between the mixing of water with plaster or cement and the point at which the plaster, cement paste, or concrete begins to become rigid.

set, flash. The rapid and permanent hardening of fresh mortar, cement, or cement paste with the evolution of heat.

set, initial. The period of time elapsed between the mixing of water with plaster or cement, and the point at which the plaster, cement paste, or concrete begins to lose plasticity. See setting time, setting rate.

setter. (1) An item of kiln furniture shaped to conform with the under surface of ware and which serves to support the ware in the kiln during firing.

(2) A type of sagger designed to conserve kiln space, the contour of its upper side conforming with the contour of the lower surface of the ware to be fired so that saggers may be stacked or arranged compactly in the kiln.

(3) The operator placing ware in a kiln.

setting. (1) The arrangement or placement of ware in a kiln.

(2) The hardening of plaster or cement.

setting block. Blocks of lead or other nonabsorbent material bedded in a glazing compound on which glass is positioned in a window or other opening.

setting pocket. A technique for hand-placing refractory shapes in a kiln to minimize deformation and the development of stresses in the ware during firing.

setting rate. (1) A comparative term referring to the time required for a glass surface to cool within the limits of the working range.

(2) The elapsed time in which lime, mortar, plaster, or concrete hardens.

setting time. The period of time elapsed between the mixing of water with plaster or cement and certain arbitrary points in the hydration or setting process as determined by the penetration of a standard vicat needle into a sample of the plaster, cement paste, or concrete in a specified period of time. See vicat needle.

setting-up (set-up) agent. An electrolyte, such as $MgCO_3$, K_2CO_3, Na_2O_3,or $MgSO_4$, added to porcelain enamels, glazes, and other slurries to flocculate and increase the suspension properties of clays.

settle mark. A wrinkled surface appearing on glassware as a result of uneven cooling during the forming process.

settlement crack. A crack in the soffit of a concrete beam, or at the top of a concrete wall or column where it joins a slab, the crack resulting from stresses developing in the joint during the continuous placement of the concrete.

settling. The sedimentary process which causes particles of clay, glaze, porcelain enamel, or other materials suspended in water or other liquids to sink to the bottom of a container.

settling tank. A tank or reservoir into which slurries of various components are placed to permit settling of solid materials to be accomplished by gravity.

set-up wheel. An abrasive wheel fabricated by compressing a series of sheets of abrasive-coated fabrics into wheel form.

sewage (wastewater). The spent water of a community which is a mixture of liquid and water-carried waste.

sewer. A pipeline constructed to convey sewage to a disposal area.

sewer brick. A low-absorption, abrasion-resistant brick intended for use in the construction of drainage systems.

sewer pipe. An impervious pipe, sometimes glazed, intended for use in the transport of water and sewage to a disposal area.

sewer, storm. A pipeline intended to convey storm and surface water from an area.

sewer tile. An impervious tile of circular cross-section intended for use in drainage systems.

sgraffito. A decoration used on pottery and other ware with an enameled surface on which a linear drawing is scratched through an unfired engobe or glaze to expose a differently colored body or contrasting fired surface beneath; the item then is refired.

shaded sandblast. The obscuration of a surface by sandblasting, the texture ranging from clear to full obscuration.

shadow wall. A more or less solid structure built on the top of the bridge wall of a glass tank, or suspended from the crown, to limit the flow of heat from the glass-melting zone to the refining zone of the tank.

shaft. (1) The vertical conduit or flue for venting combustion and other gaseous wastes.

(2) The refractory-lined cone-shaped section of a blast furnace or cupola above the hearth and melting zone and extending to the throat.

shaft, feed. Vertical shafts under the fire holes in top-fired kilns for the combustion of fuel and dispersion of heat through the kiln setting.

shaft kiln. An essentially vertical, refractory-lined furnace for heating lump material; the raw material normally is fed into the top of the kiln, passing through hot gases from burners stationed near the bottom, and emerges as a calcined product at the bottom of the kiln.

shaft mixer. A continuous blender consisting of a bladed rotating shaft which mixes and forces materials through an open trough such as in a pug mill, which see.

shaker screen. A mechanically vibrating screening device employed to separate materials into desired particle sizes or to separate larger sizes from the smaller.

shaker table. A slightly tilted vibrating table having a flat, rectangular, and sometimes riffled surface used to separate solid materials according to density and particle size, the larger and heavier particles moving to the bottom edge or the table first.

shale. A thinly stratified or laminated, sedimentary, and consolidated rock with well-marked cleavage, composed of clay, quartz, mica, and other minerals.

shale clay. Finely ground shale, sometimes used as a clay.

shape. (1) The geometrical configuration or visual appearance of a solid body.

(2) The process or act of forming a body to a desired configuration.

shape factor. The ratio of the major dimension of a particle to the minor dimension.

shape standard. A series of refractory units in various sizes and shapes which, because of their extensive or essential use, are stocked by the manufacturer or can be made from stock molds.

shaping block. A wood paddle or block used in the shaping of glass on a blowpipe.

shard. Fired pottery milled to a powder form suitable for use as a replacement for grog or silica to reduce shrinkage without altering the composition of a ceramic body. See grog; sherd.

sharp fire. Combustion with an excess of air and a short flame.

shatterproof glass. Two sheets of glass with a sheet of transparent plastic molded between the sheets under heat and pressure. See safety glass.

shaving. (1) The shaving, scraping, and paring of leather-hard clay from the wall and foot of a pot on a lathe or potter's wheel.

(2) The shaping or removal of excess material from a grinding wheel before firing the wheel.

Shaw kiln. A gas-fired chamber kiln in which a portion of the heat is introduced beneath the floor of the kiln to minimize temperature differences in the firing zone.

shear. (1) The deformation or fracture of a solid under a load which causes one face of the fractured solid to slide against an adjoining face along a parallel plane.

(2) A manual tool or a mechanical device consisting of two opposing sliding blades between which a material is cut.

(3) Scissors.

shear cake. A counterweighted refractory slab used as a gate or door on a small furnace or oven.

Shearer plastometer. An instrument to measure the flow properties of slurries calculated as the time for a specified volume of the slurry to flow through a tube of specified diameter.

shear fire. A thin flame employed to sever the moil from a shaped glass article.

shear mark. A scar appearing in glassware as a result of the cooling action of the cool cutting shear on the hot, but rigid, glass.

shear modulus. The maximum stress per unit of area that a specimen can withstand in shear without breaking.

shear strain. The ratio of the relative lateral displacement between two points lying in parallel planes in a solid to the vertical distance between the points shearing.

shear strength. The maximum shear stress a material can withstand without rupture.

shear stress. The force exerted by the material on one side of a plane surface pushing on the material on the other side of the surface, the force being parallel to the surface. See shear.

sheet glass. A generic term including sheet, plate, rolled, float, and other forms of glass that are of a flat nature.

sheet-steel enamel. A porcelain enamel designed for application to ware fabricated from sheet steel.

sheet, tangle. Pieces of mica which split well in some sections, but tear in others.

shellac bond. A bonding material in which shellac is the major constituent, and which is used in the manufacture of shellac-bonded abrasives.

shellac wheel. A grinding wheel in which the abrasive grains are bonded together with a shellac-type bonding medium.

shelling. (1) The breaking away of a layer of refractory from the roof of an all-basic, open-hearth furnace.
(2) The flaking of glaze from ware due to failure to develop sufficient bond during firing.

shell molding. Forming a rigid, porous, self-supporting refractory mold by sprinkling a mixture of molding sand and thermoplastic material over a preheated metal pattern and then curing in an oven.

shell roof. A thin, curved, platelike roof, usually constructed of concrete.

shells. (1) The outer walls of hollow structural clay tiles and building blocks.
(2) The outer walls of a structure or vessel.

shell wall. A fireclay, refractory wall protecting the metal casing of air preheaters.

shelving. The erosion of the horizontal joints of fireclay refractories in a glass tank.

sherd. Fired pottery in small fragments or ground to powder form for use as a replacement for grog or silica in pottery bodies to reduce shrinkage without altering the composition of the body. Also known as pitchers, and shard.

shield. The material placed around a nuclear reactor, or other source of radiation, to reduce or prevent the escape of radiation or radioactive particles from the reactor.

shield, heat. A layer of substance which provides protection from heat.

shielding glass. A transparent glass containing quantities of the oxides of the heavy elements, such as lead, which absorb high-energy electromagnetic radiation, and which are employed to shield one region of space from ionizing radiation emanating from another, such as in nuclear applications.

shift, phase. In electromagnetic testing, a change in the phase relationship between two alternating quantities of the same frequency.

shiner, shiner scale. A defect characterized by minute, shiny fishscaling occurring on a thin or overfired porcelain-enamel ground coat.

shingle. A thin sheet of building material, such as asphalt shingles or porcelain-enameled steel panels, placed in overlapping rows as a roof covering or siding of a building.

ship and galley tile. A quarry tile with an indented pattern on its face to produce an antislip surface when walked upon.

shipper-receiver difference. The difference between the quantity stated by a shipper as having been shipped and the quantity stated by the receiver as having been received.

shivering. The splintering of fired glazes, porcelain enamels, or other ceramic coatings from a base material due to critical compressive stresses.

shock, resistance to thermal. The ability of a solid material to withstand sudden heating or cooling without cracking or other damage.

shock, thermal. Exposure to sudden heating or cooling.

shoe. An open-ended crucible placed in the opening of a glass-melting pot for heating blowpipes.

shop. (1) A group of workmen engaged in an assigned activity or producing a particular item or end product.
(2) A room, area, or other enclosure in which a particular work is done.
(3) A factory.

shore hardness. The hardness of a material reported as the height of rebound of a metal ball dropped vertically through a glass tube from a specified height to the surface of a specimen.

shorelines. A defect in the surface of porcelain enamels characterized by a series of lines in a pattern similar in appearance to the lines produced on a shore by receding water.

short clay. A nonplastic clay having low-green strength.

shortest arc. The limiting state of an electric arc in which the total arc voltage approaches the sum of the cathode and anode falls.

short finish. An imperfection in plate glass resulting from incomplete polishing.

short fire. An oxidizing flame.

short glass. (1) A fast-setting glass.
(2) A body of low or poor workability.

short glaze. An area on the surface of ware in which insufficient glaze was applied to obtain a desired finish or appearance.

shot. Small spherical particles appearing in fibrous products such as glass or mineral wool which has been attenuated in a fast-moving stream of air or steam.

shot blasting. Blast cleaning or treatment of the surface of ware in which small steel balls are impelled by a blast of compressed air.

shotcrete. A mixture of cement, sand, and water applied through a hose with high-velocity compressed air, and which will adhere tenaciously to a prepared concrete or other surface.

shot, flint. A sharp-edged sand used in sandblasting.

shoulder-angle tile. Small wall-tile shapes used to finish the top and bottom of corner installations.

shovel. (1) A hand tool equipped with a flattened scoop at the end of a handle for moving bulk solid materials.
(2) A mechanical device equipped with a flattened, broad blade or scoop for moving bulk solid materials.

showering. A type of corona discharge of luminous streamers or plasma occurring in an electrical field of a value just below that required for complete breakdown.

shredder. A mechanical device employed to cut or tear clays and other plastic materials into sizes more amenable to subsequent handling and processing.

shrend. The process of making cullet by directing molten glass into a stream of water. See cullet.

shrinkage. (1) The reduction in the dimensions of a body or substance during drying or firing.
(2) Contraction of concrete in the plastic state or after it has become rigid, but before it has developed appreciable strength.

shrinkage, burning. A synonym for firing shrinkage, which see.

shrinkage crack. A fissure resulting from uneven shrinkage of a body.

shrinkage, drying. The reduction in the dimensions of a substance or body during drying. See drying shrinkage.

shrinkage, firing. The reduction in the dimensions of a substance or body during firing. See firing shrinkage.

shrinkage, linear. The reduction in the length of a specimen during drying and firing. See drying shrinkage, firing shrinkage.

shrinkage, plastic. The shrinkage of a concrete paste while the concrete is still plastic, or after the concrete has become rigid, but has not yet developed appreciable strength.

shrinkage rate. The amount of shrinkage of a substance per unit of time.

shrinkage volume. The contraction of a moist body during drying or firing, or both, expressed as the volume percent of the original volume.

shrinkage water. That portion of the water of plasticity of a body which, when removed, contributes to the drying shrinkage of the body.

shrink-mixed concrete. Concrete in which the ingredients are partially mixed and then placed in a truck mixer where mixing is completed while in transit to the site of its use.

shrink wrapping. The process of encasing a product in plastic, and then heating the plastic so that it will shrink to fit tightly.

shuttle kiln. A kiln in which loaded cars are introduced at one end, ejecting cars of fired ware at the opposite end; the process is then reversed in which the ejected cars are unloaded, reloaded, and charged into the kiln at the end of ejection, and the process continued in shuttle-like fashion.

sial. (1) A silica and alumina-rich rock.
(2) A borosilicate glass of high thermal and chemical resistance.

side arch brick. A type of brick having face surfaces inclined toward each other in the shape of a wedge.

side-blown converter. A steel converter in which the air or oxygen blast strikes the molten iron through tuyeres arranged along the refractory wall.

side-construction tile. Tile designed to receive its principal stress at right angles to the axis of the cells.

side-cut brick. A brick which is wire-cut along the side instead of the end.

side-feather brick. A featheredge brick cut along the 9-inch by 3-inch plane.

side-fired furnace. A furnace in which fuel is supplied through ports in the side.

side-grinding. The practice of grinding on the side of an abrasive wheel mounted between flanges.

side lap. The shortest horizontal distance between the exposed side edge of a course of roofing or siding material and the most proximate underlying area of roof deck or side wall not covered by the preceding adjacent course.

side pocket. A refractory-lined chamber at the bottom of a glass tank to catch slag and dust from waste gases before they enter the regenerator.

side-port furnace. A furnace with ports located along the sides through which fuel may be introduced or gases may escape.

siderite. $FeCO_3$; mol. wt. 115.8; sp.gr. 3.83–3.88; hardness (Mohs) 3.5–4; a mineral used as a colorant in ceramic bodies and glazes.

side skew. A brick having one side inclined at an angle other than 90 degrees to the two largest faces; used in the production of circular or curved structures.

siege. The refractory floor of a pot furnace or glass tank.

sienna. A yellowish-brown earth containing hydrated iron oxide; useful as a colorant in slips, bodies, and glazes, particularly celadons.

sieve. A perforated or meshed device through which particles of a material or mixture are passed to separate them from coarser ones; through which soft materials are forced for reduction to particles of finer sizes; or through which liquid is strained.

sieve analysis. The determination of the size distribution of a material on a series of sieves of decreasing size, usually expressed in terms of weight per cent of the sample retained on each sieve.

sieve mesh. Any standardized opening, square in shape, bounded by four meshed wires in a sieve.

sieve shaker. A mechanical device in which a stack of sieves, arranged in progressively reducing mesh or opening sizes, is shaken vigorously so that size fractions of a sample may be collected for analysis or use.

sighting tube. A ceramic tube, inserted in a kiln, through which an optical pyrometer is sighted to obtain a measurement of the temperature of the kiln.

sigma function. A property of a mixture of air and water vapor equal to the difference between the enthalpy and the product of the specific humidity and the enthalpy of water at the thermodynamic wet-bulb temperature.

signal, differentiated. An output signal or measured indication that is proportional to the rate of change of the input signal or indication, particularly in electromagnetic testing.

signal glass. Glasses of various colors used in signal devices.

signal intensity. The electric-field strength of an electromagnetic wave transmitting a signal.

silcrete. A silica-bonded conglomerate of sand and gravel.

silex. (1) A finely ground, pure form of quartz.
(2) A thermal-and physical-shock resistant glass containing approximately 98% of quartz.

silica. SiO_2; mol. wt. 60.06; m.p. 1710°C (3110°F); b.p. 2230°C (4046°F); sp.gr. 2.2–2.6; hardness (Mohs) 7; the most common mineral in the majority of sands; occurs in five crystalline polymorphs-quartz, tridymite, cristobalite, coesite, and stishovite; in cryptocrystalline form as chalcedony, and in amorphous and hydrated forms as opal; used in the manufacture of glass, abrasives, numerous whiteware bodies and glazes, porcelain enamels, refractories, foundry molds, electric and electronic products, carborundum, ferrosilicon concrete and mortars, and other products.

silica brick. A refractory brick usually made from gannister, and containing at least 90% of silica bonded with hydrated lime and fired at a high temperature; characterized by high strength at elevated temperatures, high-thermal conductivity, high-abrasion resistance, and poor resistance to molten basic slags. Used in furnace and kiln arches such as the roofs of open-hearth furnaces, caps of glass tanks, the crowns of copper reverberation furnaces, etc. See gannister.

silica brick, drop-machine. Brick formed by automatically dropping a quantity of prepared mix into a mold from a considerable height so as to fill and compact the mix in the mold before pressing.

silica brick, super-duty. A silica brick in which the total of alumina, titania, and alkalies is significantly lower than normal.

silica cement, silica fireclay. A refractory mortar consisting of a finely ground mixture of quartzite, silica brick, and fireclay in various proportions.

silica flour. Finely ground quartz sand employed as an additive in casting slips.

silica, free. Silica in clay, bodies, glazes, and other ceramic compositions which remains chemically uncombined with other elements.

silica, fused. A glass made by melting silica, quartz, or sand, or by flame hydrolysis of silicon tetrachloride. The former sometimes is identified as fused quartz or quartz glass; the product of either also is known as silica glass.

silica gel. An amorphous, highly absorbent form of silica.

silica glass. A transparent or translucent glass composed almost entirely of high-purity quartz or sand, or by hydrolysis of silicon tetrachloride; also known as fused silica, vitreous silica.

silica modulus. The ratio of SiO_2 to $Al_2O_3+Fe_2O_3$ in hydraulic cement.

silica refractory, fused. A product composed predominantly of fused noncrystalline silica.

silica sand. Sand containing a very high percentage of free silica.

silicate. A compound composed of silicon, oxygen, and one or more metals.

silicate bond. A type of bond consisting essentially of sodium silicate matured by baking at a temperature of approximately 260°C (500°F).

silicate grinding wheel. A grinding wheel in which the abrasive grain is bonded with sodium silicate plus filler materials.

silicate of soda. $Na_2O{\cdot}SiO_2$, with ratios of Na_2O and SiO_2 varying widely, and with varying proportions of water; used as a deflocculant in ceramic bodies, and as a major component in air- and heat-curing cements.

silica, vitreous. A chemically stable, refractory glass made from silica alone.

siliceous. Containing a high percentage of silica.

siliceous fireclay brick. Fireclay brick containing appreciable quantities of uncombined silica, and which usually is low in fluxing constituents.

silicide resistor. A binary compound of silicon, usually with a more electropositive element such as chromium or molybdenum, which is used under conditions where radiation hardness and high-resistance values are required.

silicides. Binary compounds of silicon, usually with a more electropositive element or radical such as chromium, molybdenum, titanium, etc.; used as abrasives, refractories, semiconductors, etc.

silicon. Si; at. wt. 28.06; sp. gr. 2.0–2.49; m.p. 1410°C (2570°F); b.p. 2480°C (4496°F); used to make silicon-containing alloys as an intermediate for silicon-containing compounds, and in transistors and rectifiers.

silicon boride. (1) SiB; mol. wt. 71.34; stable in air to 1370°C (2498°F); a borosilicate glass containing SiB_4 and Si; is stable in air to 1550°C (2822°F). (2) SiB_6; mol. wt. 92.98; m.p. 1950°C (3542°F). See borides.

silicon carbide. SiC; mol. wt. 40.06; sublimes with decomposition at about 2210°C (4010°F); sp. gr. 3.17; hardness (Mohs) 9; coefficient of thermal expansion (25–1400°C) 4.4×10^{-6}; thermal conductivity 90 Btu/h/ft^2/in./°F; the regular grade is black and very tough, but the green grade is more friable; used extensively as an abrasive and refractory because of its high resistance to thermal shock; SiC also is a semiconductor and is used in light-emitting diodes to produce green or yellow light. Also used in kiln furniture, retorts, nozzles, combustion chambers, and nuclear reactors.

silicon carbide composites. (1) $SiC \cdot Si_3N_4$; mol. wt. 180.28; sp.gr. 2.5–2.8. (2) $SiC \cdot Si_2ON_2$; mol. wt. 140.20; sp. gr. ~2.7. (3) SiC·C; mol. wt. 52.06; sp. gr. 2.3–2.8.

silicon carbide, green. A friable form of silicon carbide, which see.

silicon carbide refractories. Refractory products in which silicon carbide is the predominant constituent; characterized by high thermal-shock resistance, wear resistance, and chemical resistance.

silicon diode. A crystalline diode in which silicon is the semiconductor.

silicon dioxide. See silica.

silicone. Any of a number of polymers containing alternate silicon and oxygen atoms whose properties are determined by the organic groups attached to the silicon atoms; the silicones are fluid, resinous, rubbery, water repellant, and stable at high temperatures; employed as a mold-release compound, as a sealant for porous ceramics, and as a coating for glass and other ceramics to improve scratch resistance, chemical durability, and strength.

silicon ester. An organic silicate sometimes used as a ceramic binder.

silicon monoxide. SiO; mol. wt. 44.06; a hard, abrasive, amorphous solid employed as a thin surface film to protect optical parts, mirrors, and the like.

silicon nitride. Si_3N_4; mol. wt. 140.21; sublimes at >1871°C (3400°F); sp. gr. >3.18; hardness (Mohs) >9; exhibits high resistance to thermal shock and chemicals; used as a catalyst support and for stator blades in high-temperature gas turbines. See nitrides.

silicon oxynitride. Si_2ON_2; mol. wt. 100.2; a stable refractory used as plates, crucibles, and tubes for the fusing of salts and nonferrous metals.

silicon tetrachloride. $SiCl_4$; a source of pure silica for use in the production of silica glass.

silicosis. A lung disease caused by inhalation of siliceous particles.

silk-screen printing. A decorating process in which a design is printed on glass, glazes, porcelain enamels, and other surfaces through a tightly stretched silk mesh, woven wire, or similar screen by means of a rubber squeegee, the areas not to be coated being blocked by a suitable resist medium.

sill. The horizontal member of a structure forming the bottom of a furnace door.

sillimanite. $Al_2O_3 \cdot SiO_2$; mol. wt. 162.0; decomposes at 1545°C (2813°F) to form mullite and silica; on further heating 1810°C (3290°F) it forms corundum and glass; sp. gr. 3.23; hardness (Mohs) 6–7; used in special porcelains, refractories, pyrometric tubes, chemical laboratory ware, and patching compounds for furnaces.

sillimanite refractory. A refractory shape in which sillimanite is the predominant ingredient.

silo. A tall, cylindrical structure in which large quantities of powdered or granulated raw materials are stored and dispensed.

silver. Ag; at. wt. 107.88; m.p. 961°C (1760°F); b.p. 2212°C (4012°F); sp. gr. 10.53; employed in precipitated, powdered, fluxed, or paste form as a decoration for pottery, glass, and porcelain-enameled ware, as a soft solder, and as an electrical contact material.

silver-antimony telluride. AgSbTez; an intermetallic p-type thermoelectric alloy.

silver brazing. Brazing in which silver alloys are filler metals.

silver carbonate. Ag_2CO_3; mol. wt. 276.0; m.p. 230°C (464°F); decomposes at 270°C (518°F); used to produce iridescent stains or sheens on glazes.

silver chloride. AgCl; mol. wt. 143; m.p. 455°C (851°F); b.p. 1550°C (2822°F); sp.gr. 6.077; employed in yellow glazes, purple of Cassius, and silver lusters.

silver-copper brazing alloy. An alloy of silver and copper having a liquidus of 779°C (1435°F). See brazing.

silver-copper-tin brazing alloy. An alloy of silver, copper, and tin of good ductility and a melting point of 600°C (1112°F). See brazing.

silvering. A chemical application of a film of silver, either directly or by the reduction of a silver compound, on a glass surface for electrical and light-reflection applications.

silver marking. Gray marks on glazes made by the abrasion of cutlery.

silver nitrate. $AgNO_3$; mol. wt. 170.0; m.p. 212°C (413°F); decomposes at 444°C (833°F); sp.gr. 4.328; used in glass manufacture, as a yellow colorant in glazes, and as a silvering compound for mirrors.

silver oxide. Ag_2O; mol. wt. 232.0; decomposes above 300°C (572°F); sp.gr. 7.14; used as a yellow colorant in glass and glazes, and as a glass polishing material.

silver solder. An alloy of silver, copper, and zinc having a melting point lower than that of silver but above that of lead-tin solders; used in making ceramic-metal seals.

silver sulfide. Ag_2S; mol. wt. 247.8; m.p. 825°C (1517°F); sp.gr. 6.85–7.32.

Singer test of glaze fit. A glaze is placed in an unfired dish, fired to normal maturing temperature, and examined for defects.

single-crystal alumina. Crystals of high-purity alumina, each grain being essentially a single complete crystal, produced by recrystallization from a molten bath. See alumina, single crystal.

single embossing. A process in which a design is worked on a glass surface by a white acid treatment followed by one further treatment so that two visible shades are produced.

single fire. The process of maturing an unfired body and glaze or a multiple coating of porcelain enamel in a single firing operation.

single-roll crusher. A crushing apparatus consisting of a corrugated or toothed rotating cylinder which pinches material against stationary bars or plates.

single-screened ground refractory material. A refractory material that contains its original gradation of particle sizes resulting from crushing, grinding, or both, and from which particles coarser than a specified size have been removed by screening.

single-strength glass. Sheet glass of a thickness between 0.085 and 0.101 inch; used in windows, picture frames, and other applications where great strength is not a major requirement.

single toggle jaw crusher. A mechanical apparatus in which solid materials are crushed by passing between two jaws, one oscillating and the other stationary.

sinkhead. A reservoir of ceramic slip or molten metal placed above a ceramic or metal casting, respectively, to supply additional material as the casting solidifies and shrinks.

sinter, sintering. The bonding of powdered materials by solid-state reactions at temperatures lower than those required for the formation of a liquid phase.

sinterable powder. A powder or compact of powder in which the bonding of adjacent surfaces of the particles may be accomplished by heating but without melting.

sintered alumina. A commonly coarse crystalline, but sometimes microcrystalline, abrasive formed by sintering mixtures relatively high in alumina but usually containing associated minerals such as diaspore and various silicates. See alumina, sintered.

sintered filter. A porous article of sintered material such as glass, silica, or other ceramic employed as a filter medium to separate particulate matter from liquids.

sintered glass. A porous article in which particles of glass of selected or random sizes are compacted and sintered to produce a bonded, but unsealed, item of desired shape and strength sufficient for an intended use, such as aeration, filtration, etc.

sintered pellet. A briquet or compact, usually cylindrical, formed by pressing a powder in a die and then sintering. See sinter.

sintering furnace. Any furnace in which materials are sintered. See sinter.

sintering, pressure. The forming of a ceramic shape by applying pressure and heat simultaneously at temperatures sufficient to produce sintering.

sintern. The process or product obtained by heating a ceramic or a mixture of ceramics to a coherent mass without melting.

sinter point. The temperature at which a clay ceases to be porous.

sinter reaction. The process in which a chemical reaction and sintering occur simultaneously at an elevated temperature.

siporex. A slurry of sand, aluminum powder, and lime or cement cast into molds to produce roofing slabs, wall blocks, and other building materials of high sound and heat insulation.

size. Any of various glutinous materials, varnishes, resins, etc., employed as a surface treatment to render desired properties to the surfaces of glass, ceramics, and molds, usually resistance to abrasion.

size analysis. The determination of the proportion of particles of a particular size or sizes in a granular or powdered sample.

skewback. The course of brick having a beveled or inclined face from which an arch is sprung. See sprung arch.

skewbrick. A brick having one surface beveled or inclined, at an angle other than 90 degrees, to at least two other faces.

skew edge. A brick having one side inclined at an angle other than 90 degrees to the ends.

skid. A movable platform on which materials or ware are placed for handling and moving.

skim coat. A thin finish coating of plaster consisting of a putty of lime and fine white sand.

skim gate. A barrier in a glass-melting tank which traps and prevents slag, scum, and unmelted materials from entering the firing chamber of the tank.

skimmer block, skimmer gate. A refractory gate or wall designed in a glass tank, porcelain-enamel, smelter, or similar furnace to prevent slag and impurities from passing into the feeder channel or smelting chamber.

skimming pocket. An area in a glass-melting tank from which slags and other impurities may be removed from the surface of the molten mass.

skintle. (1) The placement of brick in an irregular pattern so that they are out of alignment with the face.

(2) The placement of brick in a kiln in an oblique position to the courses above and below.

skip hoist. An apparatus employed to raise materials to an elevated level for storage or use.

skittle pot. A small, refractory glass-melting pot.

skiving. The shaving, grinding, or machining of thin layers of excess material in the finishing of spark plugs, insulators, and other ceramic products prior to firing.

skull. The solidified material or dross remaining in a vessel after its contents have been poured.

sky firing. Completing the firing of an updraft bisque kiln by inserting and burning wood slivers in the top of the kiln to increase the draft.

skylight. (1) Flat or appropriately contoured glass installed at an angle greater than 15 degrees from the vertical in a building.

(2) A glazed opening in a roof to admit light.

skylight glass. Plate glass of very poor quality.

slab. A section of concrete laid as a single unjointed unit.

slabbing. (1) The breaking away of a layer of refractory from the roof of a furnace or kiln.

(2) The forming of ware, usually square or rectangular, from sheets of damp, plastic clay, the joints being sealed by a clay slurry.

slab glass. Optical glass obtained by forming or cutting chunk glass into plates or slabs of suitable size for future processing.

slag. (1) The partially fused mixture of spilled batch, overflowed glass, breeze coal, and clay from the floor of a pot furnace or glass tank.

(2) Material formed by the fusion of oxides in a metallurgical process or the fused reaction product between a refractory and a flux.

(3) A nonmetallic by-product of steel blast furnaces which is crushed and sized for use as concrete aggregate.

(4) An electric furnace by-product in the manufacture of phosphate which may be used as a source of alumina in the manufacture of glass.

(5) A pozzolanic material sometimes used in the production of portland cement.

slag, air-cooled blast-furnace. The material resulting from solidification of molten blast-furnace slag under atmospheric conditions.

slag, blast furnace. The nonmetallic product consisting essentially of silicates and aluminosilicates of calcium and other alkaline ingredients which are developed in a molten condition simultaneously with iron in a blast furnace.

slag cement. A hydraulic cement consisting essentially of an intimate and uniform blend of granulated blast-furnace slag and hydrated lime in which the slag constituent is more than a specified minimum percentage.

slagging of refractories. A destructive chemical reaction between refractories and external agencies at high temperatures resulting in the formation of a liquid.

slag line. A horizontal line formed along the refractory wall of a glass, metal, or similar melting tank which is caused by the erosion and corrosion of the refractories at the air-refractory-batch interface.

slag notch. An opening in the hearth to permit the flow of slag from a blast furnace.

slag, phosphate. A by-product of electric phosphate furnaces, processed for use in the manufacture of glass.

slag pocket. A refractory-lined area constructed at the bottom of a melting tank to prevent entry of slag and impurities into a regenerator.

slag sand. Finely crushed slag used in cement and mortars. See slag, (3).

slaking. The disintegration or crumbling of materials when exposed to or saturated with water and air.

slate. A dense fine-grained metamorphic mineral which breaks into thin sheets or slabs; used as a flooring material, roofing material, abrasive, blackboards, etc.

slater's cement. A caulking compound, usually gray in color, used to cover exposed bolt heads, the side and end laps of corrugated roofing, and other areas to prevent penetration of water.

sleek. A fine, scratchlike, smooth-boundaried imperfection in glass usually caused by a foreign particle in the polishing medium during the polishing operation.

sleeper block. The refractory blocks forming the sides of the throat of the submerged passage between the melting and working ends of a glass tank.

sleeper wall. The refractory walls of the submerged passage between the melting and working ends of a glass tank.

sleeve brick. Tube-shaped firebrick used for lining slag vents.

sleeves. Tubular fireclay shapes that encase an immersed metal rod in the valve assembly of a bottom-pouring ladle.

sleeve, wheel. A form of flange used on precision grinding machines where the wheel hole is larger than the machine arbor, and usually designed so that the shell and sleeve are assembled as a unit.

slide conveyor. A trough or chute for the downward movement of materials under gravitational pull.

slide-off transfer. A printed decoration which, when wet, may be slipped from its backing to the surface of ware being decorated, and which subsequently may or may not be fired. See decal.

slide potentiometer. A potentiometer which employs a sliding contact along a length of resistance wire to regulate the voltage in the wire in temperature measuring and control instruments. See potentiometer.

sliding. A porcelain-enamel defect similar to sagging in which patches of the coating slip or slide during drainage to produce a coating of uneven thickness.

sliding-bat kiln. A type of tunnel kiln in which ware is placed on tile or slabs and pushed mechanically or manually through the firing zone.

sliding contact. An electrical or other contact which accomplishes its function while sliding against its mating contact.

slinger process. A forming process in which a wet batch is thrown on a pallet, formed into a column, cut to shape, dried, and fired.

slip. A suspension or slurry of finely divided ceramic materials in a liquid.

slip casting. A forming process in the manufacture of shaped articles in which the material to be cast is ground and mixed to a creamy slurry with water and then poured into plaster molds which rapidly absorb the added water, producing a solid body which has the inside shape of the mold; when the wall thickness of the cast item is attained, the excess slurry is poured from the mold; when the cast item has dried to sufficient strength for safe handling, it is removed from the mold for further processing.

slip clay. A clay having a high percentage of fluxing impurities which fuse at a relatively low temperature to produce a natural glaze; characterized by a fine-grained structure and low-firing shrinkage.

slip coating. A ceramic or mixture, other than a glaze, which is applied and fired on a ceramic body, to develop specific characteristics or properties.

slip form. A sliding form that produces a continuous placement of concrete as the form is moved along either vertically, as in a silo, or horizontally, as for a canal lining.

slip glaze. A glaze consisting primarily of readily fusible clay or silt and other ingredients blended to a creamy consistency in water.

slip house. The room or area in a factory where ceramic slips are prepared and stored for subsequent use.

slip kiln. A structure, consisting of suitable containers, which employs waste heat to dry or reduce the water content of slurries.

slip, mechanical. The fine layer resulting from the wet-surface smoothing of clay vessels.

slip process. A method of preparing a ceramic body in which water is added to dry-blended bodies in a quantity sufficient to produce a fluid suspension for use as such or for subsequent processing.

slip stain. A stain incorporated in a glaze or slip instead of in the body.

slip stone. A small slender abrasive stone used to remove blemishes from the surface of ceramic ware and to sharpen metal tools.

slip trailing. A process of forming a pattern on a clay surface by flowing or squeezing viscous slip through a fine orifice onto the surface of the ware.

slip, vitreous. A slip coating applied and matured on a ceramic body, producing a vitrified or glassy surface.

slipware. Pottery decorated by the application and firing of slips.

sliver. (1) Bundles of noncontinuous or short-length glass fibers that have reached the stage of fabrication into yarn wherein they are parallel, overlapping, and have no twist.

(2) A long, slender piece or splinter.

slop. A homogeneous slurry of glaze ingredients and water applied to ware by dipping, spraying, or brushing.

slope. The incline of a roof expressed as a ratio of the number of inches or millimeters of vertical rise per horizontal foot or meter.

slop weight. The weight of a unit volume of a slop.

slotting wheel. A thin grinding wheel, usually organic bonded, used for cutting grooves or slots in a workpiece.

slow wheel. The practice of perfecting the surface of a handmade article by turning it on a rotating base such as a plate, wood block, or sherd.

sludge. A semisolid waste or collection of settlings from a process.

sludge pan, sludge pit. A container or area in which sludge is collected for subsequent recycling or disposal.

slug. (1) Any nonfibrous glass inclusion in a glass-fiber product.
(2) A geometric shape made by pressing and which is fed to the granulation step of processing.
(3) A small roughly shaped article for subsequent processing.
(4) A length of clay extruded from a pug mill.

slugged bottom. An imperfection in the bottom of a bottle or container in which the glass is heavy, or thick on one side and very light or thin on the opposite side.

slug press. The process of initial compaction of fine powders prior to granulation or subsequent processing.

slum. Fireclay containing a substantial amount of fine coal particles as an impurity.

slump. (1) To drop, sag, or slide down suddenly.
(2) A measure of the fluidity, softness, or wetness of fresh concrete determined by measuring the number of inches in a sample slumps or settles when a conical form is removed from the sample.
(3) A measure of the consistency of a porcelain enamel, glaze, or other slip or slurry, made by spreading a specified volume of slip over a flat plate.

slurry. A mixture or suspension of ground frits, clays, or other ceramic materials in water or other liquid.

slush. A grout made of portland cement, sand, and water mixed to a relatively thin slurry which may be poured, slushed, or spread over a surface area.

slushing. The coating of ware by dipping, shaking, or spinning to obtain a uniform distribution of slip and to remove excess material from the surface of the ware.

smalt. A blue pigment for glass and other ceramics consisting of fused cobalt oxide, sand, and potash.

smear. (1) A material spread over a surface or the process of spreading a material over a surface.
(2) A surface crack on the neck of a glass bottle.

smectite. Montmorillonitic clays characterized by swelling and high cation-exchange properties. See montmorillonite.

smelt. (1) A specific batch or lot of frit.
(2) The process of melting a batch of frit.

smelter. A refractory-lined furnace or tank in which the ingredients of a frit are melted.

smelter, batch. Any smelter into which a batch is charged, melted, and discharged according to a specified periodic cycle.

smelter, continuous. Any smelter into which a batch is charged, melted, and discharged continuously.

smelter drippings. Drippings of molten glassy material from an accumulation of the material on the crown of a smelter.

smelter, rotary. A cylindrical smelter rotating slowly on a horizontal axis to accomplish more efficient mixing of its molten contents.

smithsonite. See zinc carbonate.

smoke. Streaked areas in flat glass appearing as slight discolorations.

smoked. (1) The discoloration of glass or a glaze resulting from a reducing flame.
(2) Glass covered with a smoky film from open-fired lehrs.

smoked glass. Commercial glassware produced in gray or smoky-brown colors, sometimes by chemical additions to the glass and sometimes by exposure to a reducing atmosphere during melting and cooling.

smoked glaze. A smokey or discolored glaze due to over-reduction of ingredients or the entrapment of minute carbon particles in the glaze during firing.

smoking. (1) The slow preheating of a kiln.
(2) A reducing kiln atmosphere.

smoking, water. The first period of firing in which all mechanically held water is removed from a body as the temperature in the kiln is increased.

smoky inclusions. Dispersed metal oxide inclusions in mica which appear in various pastel colors when observed in transmitted light.

smooth-finish tile. Tile and other surfaces which are not altered or marked during manufacture, and which retain the plane surface as formed by the die.

smooth glass. A finely ground glass surface ready for polishing.

smoothing mill. A machine equipped with a fine-grained polishing wheel for the beveling of glass.

smooth rolls. A crusher in which material is passed between a rotating set of smooth rolls.

smother kiln. A kiln into which smoke can be introduced for the blackening of pottery.

snagging. The removal of defects and excess materials

from ware, such as gates, sprues, fins, parting lines, and the like by the use of a grinding wheel.

snagging, automatic. Snagging by use of automatic or semiautomatic grinders where the pressure between the wheel and work, and the traverse over the work, is controlled from a station remote from the wheel.

snake, snaking. (1) The progressive longitudinal cracking in continuous flat-glass operation.
(2) The variation in the width of a sheet during the drawing of sheet glass.

snakeskin glaze. A decorative effect on pottery obtained by using glazes of high-surface tension or very low expansion, causing the glaze to crawl during firing to produce an appearance of snakeskin.

snap. A device for gripping a piece of formed glass for fire polishing and finishing.

snap header. A building brick of half the standard length. See brick, standard.

soak, soaking. (1) Holding a kiln at a constant temperature for a long period of time.
(2) Maintaining a kiln at maximum firing temperature to obtain a desired degree of chemical or physical reaction in a body being fired.
(3) To immerse a material or body in a liquid to obtain thorough wetting.

soaking pit. A conditioning furnace in which molten glass is brought to a uniform temperature for casting.

soap brick. A brick modified so that the width is one-half the standard dimension. See brick, standard.

soapstone. $Mg_3Si_4O_{10}(OH)_2$; mol. wt. 349.22; sp.gr. 2.7–2.8; hardness (Mohs) 1–1.5; generally known in the industry as steatite or massive talc. See talc.

socket. An opening or hollow that forms a holder into which an item is inserted.

soda. Any of the forms of sodium carbonate.

soda ash. Commercial grade of Na_2CO_3; mol. wt. 106.0; decomposes at 852°C (1570°F); sp.gr. 2.53; used as a fluxing component in glass, porcelain enamels, and glazes, and as a neutralizer in the metal treatment of metals for porcelain enamels.

soda-lime glass. Glass containing approximately 72% SiO_2 (sand), 15% Na_2O (soda ash, sodium nitrate, sodium sulfate), and 9% CaO (limestone, dolomite); used for window and plate glass, containers, art objects, light bulbs, and industrial products.

soda nitre. $NaNO_3$; mol. wt. 85.0; m.p. 308°C (588°F); sp.gr. 2.27; hardness (Mohs) 1.5–2.0; employed in glass, porcelain enamels, and glazes as an oxidizing agent and flux.

sodium aluminate. $Na_2O{\cdot}Al_2O_3$; mol. wt. 163.93; m.p. 1650°C (3000°F); employed in porcelain-enamel and glaze slips to improve suspension and working properties, and in the production of milk glass because of its opacifying or obscuration properties.

sodium antimonate. $Na_2O{\cdot}Sb_2O_3$; mol. wt. 394.5; stable to 1427°C (2600°F); used as an opacifier and high-temperature oxidizing agent in porcelain enamels, as a fining and decolorizing agent in glass, and as yellow colorant in glazes.

sodium bicarbonate. $NaHCO_3$; used as a deflocculating agent, as a body wash to improve body-glaze reactions, and as a metal cleaning agent in solutions.

sodium bifluoride. $NaHF_2$; mol. wt. 62; an etchant for glass.

sodium bisulfate. $NaHSO_4$; mol. wt. 120.06; used in the manufacture of brick and magnesia cements, and as a flux to decompose minerals.

sodium borate. $Na_2B_4O_7{\cdot}10H_2O$. See borax.

sodium carbonate. Na_2CO_3; mol. wt. 106.0; decomposes at 852°C (1570°F); sp.gr. 2.53; used as a flux in glass, glazes, and enamels, and as an acid neutralizer in the treatment of metals for porcelain enamels.

sodium carboxymethylcellulose. Employed as a thickener and binder in bodies and glazes.

sodium chloride (common salt). NaCl; mol. wt. 58.5; m.p. 804°C (1480°F); sp.gr. 2.161; used in the production of salt glazes on some types of ceramic ware. See salt glaze.

sodium cyanide. NaCN; mol. wt. 49; m.p. 563°C (1047°F); employed as an addition to improve the performance of neutralizer baths in preparing steels for porcelain enameling.

sodium dichromate. $Na_2Cr_2O_7{\cdot}2H_2O$; mol. wt. 298.06; m.p. 320°C (608°F); decomposes at 400°C (752°F); sp.gr. 2.52; an orange-yellow colorant for glazes and porcelain enamels.

sodium diuranate. $Na_2U_2O_7{\cdot}6H_2O$; mol. wt. 742.4; a yellow-orange pigment used in bodies, glazes, and porcelain enamels; also used in the manufacture of fluorescent uranium glass.

sodium fluoride. NaF; mol. wt. 42; m.p. 993°C (1820°F); sp.gr. 2.76; used as a flux and as a gas or bubble-type opacifier in porcelain enamels.

sodium fluosilicate. Na_2SiF_6; mol. wt. 188.0; decomposes at red heat; sp.gr. 2.7; employed as a flux and opacifier in porcelain enamels, and as an opalizer in glass.

sodium metagermanate. Na_2GeO_3; mol. wt. 166.6; m.p. 1078°C (1978°F); used in special glasses and in electronic devices such as diode rectifiers and transistors.

sodium metasilicate. $Na_2O \cdot SiO_2$; mol. wt. 112.06; m.p. 1089°C (1993°F); employed to clean drawing compounds from metals for porcelain enameling.

sodium metatantalate. $NaTaO_3$; mol. wt. 252.4; m.p. 630°C (1166°F); a ferroelectric material crystallizing in a perovskite structure having a Curie point of 475°C (889°F).

sodium molybdate. Na_2MoO_4; mol. wt. 206.0; m.p. 687°C (1270°F); sp.gr. 3.28; employed as a deflocculant, adherence promoter, and rust inhibitor in porcelain enameling.

sodium niobate. $NaNbO_3$; mol. wt. 161.1; a ferroelectric material having a Curie point of 360°C (680°F).

sodium nitrate. $NaNO_3$; mol. wt. 85.0; m.p. 308°C (588°F); sp. gr 2.267; hardness (Mohs) 1.5–2; employed as an oxidizing agent and flux in glass, glazes, and porcelain enamels.

sodium nitrite. $NaNO_2$; mol. wt. 69.01; m.p. 271°C (519°F); decomposes above 320°C (608°F); sp. gr. 2.15–2.17; employed as a metal cleaner, acid neutralizer, rust inhibitor, and tear-resistant additive in porcelain-enamel slips.

sodium pentaborate. $Na_2B_{10}O_{16} \cdot 10H_2O$; mol. wt. 590.4; sp. gr. 1.72; used as a flux in glass manufacture.

sodium phosphate. (1) A general term for many compounds of sodium and phosphorous.

(2) $Na_2HPO_4 \cdot 12H_2O$; mol. wt. 358.21; m.p. 35°C (95°F); loses $5H_2O$ on exposure to air at ordinary temperatures and $10H_2O$ at 100°C (212°F); employed in the production of opalescent glass, in the purification of clays, as a water conditioner, and as a deflocculant in porcelain enamels and glazes.

sodium pyrophosphate. $Na_4P_2O_7$; mol. wt. 266.04; m.p. 988°C (1812°F); sp. gr. 1.82; employed in aqueous solutions as a metal cleaner for porcelain enamels, as an electrolyte to adjust and control the viscosity and flow characteristics of porcelain enamels and other slips and slurries.

sodium selenite. Na_2SeO_3; mol. wt. 173.0; sp. gr. 3; used as a decolorizer in glass, and to produce rose and ruby colors in glass, porcelain enamels, and glazes.

sodium silicate. $Na_2O \cdot SiO_2$; mol. wt. 122.06; employed in cements, concrete hardeners, mortars, and abrasive wheels primarily as a binder and deflocculating ingredient.

sodium silicofluoride. Na_2SiF_6; mol. wt. 188.0; decomposes at red heat; sp. gr. 2.7; employed as a flux and opacifier in porcelain enamels and to produce opalescence in glass.

sodium stannate. $Na_2SnO_3 \cdot 3H_2O$; mol. wt. 266.76; used as a source of tin oxide as an opacifier in glass, porcelain enamels, and glazes.

sodium sulfate. Na_2SO_4; mol. wt. 142.04; m.p. 888°C (1630°F); sp. gr. 2.67; used in glazes and glass as a source of sodium oxide and as an antiscumming agent.

sodium tannate. A sodium salt of tannic acid used as a deflocculating agent.

sodium tantalate. $NaTaO_3$; mol. wt. 252.4; a ferroelectric material having an ilmenite structure and a Curie point of 475°C (887°F).

sodium uranate. $Na_2O \cdot UO_3$; mol. wt. 348.13; m.p. 1646°C (2950°F); used as a yellow-orange colorant for glass, porcelain enamels, and glazes, and in the production of fluorescence in uranium glasses.

sodium uranyl carbonate. $2Na_2CO_3 \cdot UO_2CO_3$; mol. wt. 542.0; used in the production of fluorescent greenish-yellow glass.

sodium vanadate. Na_3VO_4; mol. wt. 183.95; m.p. 866°C (1592°F); a ferroelectric material having a Curie point of 330°C (626°F).

soft. A term applied to a clay, glaze, porcelain enamel, or glass that is fusible at a relatively low temperature.

softening point, glass. The temperature at which a glass fiber of uniform diameter elongates at a specific rate under its own weight.

soft fire. A flame with a deficiency of air.

soft-fired ware. Clay products fired at a relatively low temperature, resulting in ware of relatively high absorptions and low compressive strengths.

soft glass. (1) A glass having a relatively low softening point or which is easily melted.

(2) A glass which is easily scratched or abraded.

soft mica. Mica which tends to delaminate when bent.

soft-mud brick, soft-mud process. Molded brick formed by machine, or frequently by hand, from wet soft clay bodies containing 20 to 30% of water.

softness. A porcelain-enamel surface of relatively low resistance to abrasion or scratching, or a surface produced by firing at a relatively low temperature.

soft paste. A relatively low-fired china produced from a body containing a glassy frit and a considerable quantity of fluxes.

soft soap. Potash soap used as a parting compound in the making of plaster molds.

soft solder. Solders of alloys of lead and tin which melt at temperatures below 371°C (700°F).

soilability. The relative ease by which dirt and other extraneous matter becomes attached to or builds up on the surface of a material.

soil cement. A compacted mixture of soil, cement, and water used to adjust the engineering properties of the soil.

solar furnace. An image-type furnace in which high temperatures are produced by focusing rays from the sun into a relatively small space.

solarization. A change in the transmission or color of glass when the glass is exposed to sunlight or other strong radiation.

solar-screen. A structure which blocks or diminishes the influence of the rays of the sun.

solder. Alloys of various metals which are applied and fused to the joint between two metal surfaces to unite them without heating the metals being joined to their melting points.

solder, hard. Solders composed of metal alloys which melt at temperatures above 371°C (700°F).

soldering, initial. The time and temperature at which a ceramic or a ceramic coating begins to show evidence of flow.

solder, sealing glass. A sealing glass having a relatively low softening temperature for use as an intermediate bonding material.

solder, silver. An alloy of silver, copper, and zinc having a melting point lower than that of silver but above that of lead-tin solders; used in making ceramic-metal seals.

soldier course. A course of refractory brick set on end in the bottoms of some types of ladles, furnaces, and glass tanks.

sole. The refractory brickwork forming the bed of a coke oven.

solenoid. An assembly consisting of a coil of metal wire wound around a metal core which slides along the coil axis under the influence of a magnetic field; used as an automatic switch.

solid casting. The forming of ceramic ware by introducing a body slip into a porous mold usually consisting of two major sections, one section forming the contour of the outside and the other forming the contour of the inside of the ware, and allowing a solid cast to form between the two mold faces.

solid contact. A monolithic electrical contact member.

solid insulator. Any solid material such as glass, porcelain, or other ceramic used as an electrical insulator.

solid masonry unit. A masonry unit whose net cross-sectional area in every plane parallel to the bearing surface is 75% or more of its gross cross-sectional area measured in the same plane.

solid solution. (1) A homogeneous crystalline phase composed of different mineral groups dissolved in one another either in all proportions or over a limited range of compositions.

(2) A crystal structure in which an atom, molecule, or ion is substituted for another atom, molecule, or ion that is chemically different but of similar size and shape.

solid solution omission. A crystal in which certain atomic positions are unfilled.

solid state. Pertaining to electronic devices that can control electric current without the use of moving parts, heated filaments, or vacuum gaps.

solidus. The portion of a temperature diagram which consists of the curve connecting the temperature at which a solid solution is in equilibrium with its vapor and with the liquid solution, and therefore connecting melting temperatures of solid solutions.

solubility. The amount of a substance that can be dissolved in another substance or solution.

soluble boron in boron carbide. The boron that dissolves from boron carbide by separate reflux digestions with two different acids, 0.1 *M* hydrochloric acid (hydrochloric acid-soluble boron assumed to be boric acid) and 1.6 *M* nitric acid (nitric acid-soluble boron assumed to be boric acid plus free boron).

soluble developer. A developer employed in liquid penetrant inspection which is completely soluble in its carrier, but not a suspension of powder in a liquid, which dries to an absorptive coating.

soluble salts. In ceramic technology, the term usually refers to sulfates, chlorides, and some silicates of lime, soda, potash, and magnesia.

solution ceramics. A metal-salt solution applied to a surface which is converted to a ceramic or glassy coating when a flame is sprayed over the coated surface or the solution is sprayed on a hot surface, or both; exhibits high resistance to thermal shock.

solvent action. The ability of a liquid to dissolve a material.

solvent developer. Any finely divided solid substance suspended in a volatile solvent which, when the solvent dissolves a penetrant to bring it to the surface of a discontinuity, will absorb the penetrant and dry to fix an indication.

solvent remover. A liquid which will remove excess surface penetrant from test specimens or parts by hand-wiping.

sorb, sorption. To take up and hold by either of the processes of absorption or adsorption.

sorel cement. A strong, hard cement formed by the interaction of magnesium chloride and calcined magnesia.

sort, sorting. To classify a product or substance on the basis of some characteristic or property.

soundness. (1) The degree of freedom of a product or substance from a defect or flaw.
(2) The volume stability of portland cement after it has set.

sour. To age a ceramic slurry or clay by storing in a damp environment to improve the plasticity and workability of the material.

source material. Any material, except special nuclear material, which contains 0.05% or more of uranium, thorium, or any combination of the two.

spacer. A device serving to hold two members at a specified or predetermined distance from each other.

spall. A fragment or chip broken from a masonry or ceramic unit by a blow, by the sudden reaction to heat, by prolonged exposure to heat or atmospheres which result in dimensional changes in the unit, or some other severe condition.

spalling, mechanical. Breaking or cracking of refractories in service under the influence of impact or pressure.

spalling of refractories. The chipping, cracking, or breaking of a refractory brick or unit in service which usually results in the detachment of a portion of the brick or unit to expose new surfaces.

spalling of refractories, mechanical. The chipping, cracking, or breaking of a refractory brick or unit in service caused by stresses, impact, pressure, or other forces of mechanical origin.

spalling of refractories, structural. The chipping, cracking, or breaking of a refractory brick or unit in service caused by stresses resulting from changes in the shape or structure of the brick or unit.

spalling of refractories, thermal. The chipping, cracking, or breaking of a refractory brick or unit in service as a result of nonuniform stresses imposed in the unit by differences in temperature.

spalling, spontaneous. A porcelain-enamel defect characterized by chipping or flaking of the coating without apparent cause.

spalling test, panel. A test in which a panel of refractory brick is subjected to a series of heating and cooling cycles, and the loss in weight of the panel due to spalling is taken as an index of the spalling behavior of the refractory.

spalling, thermal shock. Fracture and chipping of porcelain enamel produced by the sudden cooling or quenching of the hot enamel surface by the inadvertent subjection to water or other liquid.

span. The horizontal distance between the supports of an arch.

spandrel glass. Architectural glass which is used as a curtain wall in a nonvision area or in the cladding of a building.

spangles. Magnetic iron fired in a glaze for decorative effects.

spar. An abbreviation for feldspar.

spar, heavy. $BaSO_4$; mol. wt. 233.42; sp. gr. 4.3–4.6; hardness (Mohs) 2.5–3.5; employed in glass as a flux, to reduce seeds, increase toughness, increase brilliance, reduce annealing time, and prevent devitrification.

spark-gap inspection. A technique for the detection of pinholes and cracks in glass-coated iron or steel products in which a high-frequency discharge from a spark generator fanned across the surface of the coating collects to form a spark at the site of a pinhole or fracture.

sparking out. The practice of allowing the work piece and grinding wheel to traverse in relation to each other without additional infeed until all contact between the two ceases.

spark test. See spark-gap inspection.

spathic. The ability or process of being split.

spathic iron. $FeCO_3$; mol. wt. 115.8; sp. gr. 3.83–3.88; hardness (Mohs) 3.5–4; employed as a colorant in ceramic bodies and glazes.

special design, concrete. A concrete pipe design for sizes, loads, and service conditions which are not covered by pipe of standard design.

special nuclear material. Pu^{239}, U^{233}, uranium containing more than the natural abundance of U^{235}, or any material artificially enriched in any of these substances.

special purpose tile. A glazed or unglazed floor or wall tile designed to meet specific appearance or physical requirements not covered by standard tiles, such as size, shape, thickness, decoration, keys or lugs on the backs or sides, electrical properties, high coefficient of friction, or special resistance to staining, frost, alkalies, acids, thermal shock, or impact.

special requirements. The requirements provided to meet a particular need not covered or included under established procedures or specifications.

specific activity. The activity per unit mass of a pure radionuclide.

specification. A precise statement of a set of requirements to be satisfied by a material, product, or service indicating, whenever appropriate, the procedure by means of which it may be determined if the requirements are satisfied. As far as is practical, it is desirable that the requirements be expressed numerically in appropriate units together with their limits.

specific gravity. The ratio of the weight of a unit volume of a substance to that of a standard material under standard conditions of pressure and temperature; the specific gravity of solids and liquids is based on water as the standard.

specific gravity, apparent. The specific gravity of a body based on the volume of solid material plus the volume of sealed pores. See apparent specific gravity.

specific gravity, bulk. The specific gravity of a body based on the volume of solid material with all included pores. See bulk specific gravity.

specific gravity, true. The specific gravity of a body based on its volume excluding all pores.

specific gravity, volume. The specific gravity of a body based on the volume of solid material plus all included pores.

specific heat. The ratio of the amount of heat required to raise a mass of material one degree in temperature without a chemical or phase change to the amount of heat required to raise a mass of reference substance, usually water, one degree in temperature at constant volume or pressure.

specific humidity. The ratio of the mass of water vapor in a system of moist air to the total mass of the system.

specific surface. The surface area per unit weight or volume of a solid substance.

specific volume. The volume of a substance per unit of weight; the reciprocal of the density.

specific weight. The weight of a substance per unit volume.

specified dimensions. The dimensions to which a product or unit must conform.

specimen. An individual unit of a material or product selected for examination, testing, display, or reference.

specimen, job-cured. A specimen of concrete prepared and cured at the site of use and under the same conditions to which the commercial installation is or will be exposed; the specimen may be tested or retained for future reference.

specking. A defect in porcelain-enameled surfaces consisting of small visible specks or spots, frequently dirt, fired on the ware.

speckled ware. A decorative surface finish in which spots of one color appear in a relatively uniform pattern over a surface of another color or shade.

spectrochemical carrier distillation. An emission spectrographic technique in which a carrier material is added to a sample to facilitate the vaporization of the sample or the fractional distillation of the sample.

spectrometer. An instrument equipped with a system of ordered marks or reference standards for measuring the position of spectral lines. See spectroscope.

spectrometry, atomic absorption. The measurement of light absorbed at the wavelength of resonance lines by the unexcited atoms of an element.

spectrophotometer. An instrument which measures the apparent reflection or transmission of visible light as a function of wavelength, particularly in terms of intensity or color.

spectroscope. Any of a number of instruments which are used to resolve, observe, and record the intensity, particularly peak intensity, of spectral lines.

spectroscope, scintillation. A scintillation counter adapted to measure the intensity of gamma rays from radioactive elements or substances.

speed, peripheral. (1) The speed at which any point on the face or rim of a wheel is traveling when the wheel is revolving; expressed as a unit of length per unit of time.

(2) The product of the circumference of a wheel or rotating surface expressed in units of length and the number of revolutions per unit of time.

speed, working. (1) The rate of table traverse during abrasive grinding.

(2) The rate at which ware is rotated during centerless or internal grinding operations.

spent fuel. Nuclear-reactor fuel which is no longer effective.

sp. gr. An abbreviation for specific gravity, which see.

sphene. $CaTiSiO_5$; mol. wt. 169.98; m.p. 1386°C (2525°F); sp. gr. 3.4–5.5; hardness (Mohs) 5–5.5; employed in colorants such as chrome-tin pink.

spider. (1) A defect appearing as a star-shaped fracture in porcelain-enameled ware.

(2) A wheel-like casting consisting of a rim and radial spokes on which felt polishing pads are mounted.

(3) An assembly of radiating tie-rods on the top of a furnace.

(4) A metal unit of two or more radial arms employed to hold a core and disintegrate laminations of clays and bodies in a pug mill.

spigot. (1) The end of a pipe which is overlapped by a portion of the end of an adjoining pipe.

(2) A faucet or device for drawing a liquid from a pipe or container; for example, a water tap.

spindle. A slender rod which turns or on which something else turns.

spinel. A group of minerals of the general formula AB_2O_4, in which A is a divalent metal or mixture of divalent metals such as magnesium, ferrous iron, zinc, manganese, cobalt, calcium, copper, barium, nickel, and strontium, and B is a trivalent metal such as aluminum, ferric iron, and chromium; used in the manufacture of ceramic colors and refractories.

spinneret. A small platinum thimble containing one or more holes through which molten glass is pulled in the making of glass threads or filaments.

spiral conveyor. A conveyor consisting of a screw-type shaft employed to transfer materials on a horizontal, inclined, or vertical plane, and which is based on the principle of the Archimedes screw.

spitout. A glaze defect consisting of aggravated pinholes or craters which are developed during glost firing due to the evolution of gas bubbles from the body or glaze constituents or to vapors in the decorating fire.

split. (1) A brick modified to a thickness of one-half of the usual dimensions.
(2) A glass defect consisting of a crack or check extending from one surface to the other.
(3) To divide a sample into smaller parts.

split feed. A liquid-phase adsorption process in which a powder is added to a solution to be treated in two or more steps.

split mold. A casting mold made in two or more parts to permit the easy removal of ware after casting.

splittings. Trimmed or untrimmed mica produced from block, thins, and splitting blocks to a thickness of less than 0.03 millimeter (0.0012 inch). See mica.

splittings, bookform. Sheets of mica supplied in book form from the same block. See mica.

splittings, loose. Splittings of mica of heterogeneous shapes packed loosely in bulk form. See mica.

splittings, powdered loose. Loose splittings of mica dusted with mica powder. See mica.

splittings, thick. Loose splittings of mica of thicknesses greater than 0.03 millimeters (0.0012 inch), powdered loose splittings of thicknesses greater than 0.025 millimeters (0.001 inch). or bookform splittings greater in thickness than the average permitted for the grade. See mica.

splittings, thin. Splittings of mica having thicknesses less than the minimum average for a grade. See mica.

spluttering. The popping of glaze fragments from ware which fuse to the setters or shelves during firing.

spodumene. $LiAlSi_2O_6$; mol. wt. 186.03; sp.gr. 3.13–3.20; hardness (Mohs) 6.5–7; an ore of lithium having very low thermal expansion employed as a flux, and to improve resistance to thermal shock in glass, porcelain enamels, glazes, and ceramic bodies.

spoil. Bricks which may be placed and removed at the base of a kiln flue to control the draw of the flue, the pressure in the firing chamber, and to maintain the oxidation or reduction characteristics of the kiln atmosphere.

sponging. The removal of surface blemishes from unfired ceramic ware by the use of a damp sponge.

spongy enamel. A defect in fired porcelain enamel characterized by masses of large bubbles occurring in localized areas and having a sponge-like appearance.

spontaneous spalling. Spontaneous fracture, chipping, or flaking of porcelain enamel from ware without apparent external cause. Also known as spontaneous chipping.

spoon proof. A specimen of molten glass taken for analysis and observation from a ladle during various stages of melting and fining.

spot check. A random sampling of a material or observation of a process.

spout. (1) A device through which a material is charged into or discharged from an area.
(2) The part of a glass feeder which carries the orifice, revolving tube, and needle.
(3) The refractory block through which molten glass flows to a forming maching.

spray booth. A chamber, open on one side, in which coatings are applied to ware by means of an atomizing gun; booths usually are equipped with exhaust fans and collectors to collect overspray materials and to prevent dust from entering work areas.

spray dryer. A device in which an atomized suspension of solids in a liquid is dried by direct contact with hot gases or by impingement on a hot surface.

spray gun. A device of gun-like shape designed to deliver an atomized liquid or suspension.

spraying, electrostatic. The technique of spraying an electrically charged material onto the surface of a grounded item to obtain a more uniform coating and to minimize overspray.

spraying, flame. The process of depositing and fusing a ceramic coating on ware by projecting the coating through a flame to preheat the coating particles and surface to be coated, and then fusing the coating in place.

spraying, plasma. The process of depositing and fusing a ceramic coating on ware by projecting the coating through the arc of a plasma gun to preheat the coating particles and the surface to be coated, and then fusing the coating in place; the arc is extremely hot and, therefore, the process is used to apply high-melting materials to refractory surfaces.

spraying, thermal. See spraying, flame; spraying, plasma.

spray nozzle. The discharge opening of a spray gun in which a suspension is atomized.

spray quenching. The rapid cooling of a molten material in a spray of water or other liquid.

spreader. (1) A machine which deposits, distributes, and spreads concrete on a pavement prior to the finishing operation.
(2) A steel or wood spacer put temporarily in a form to keep the walls apart and hold them in alignment until concrete is poured.

spreader block. A refractory block of triangular cross-section employed to divide and distribute coal being charged into a coke oven.

sprigged ware. Pottery that is decorated by the application of a bas-relief ornamentation by hand pressing or by casting in molds.

sprigging. The adding of more wet clay to a body during forming to fill out thin sections or to be shaped into a form of decoration.

spring contact. A contact between surfaces effected by means of a spring device or by means of a spring-like property of one or both materials.

springer. A course of brickwork having an inclined face from which an arch or furnace roof may be sprung.

springing. The breaking of handles from cups, mugs, pots, etc., due to inherent stresses at one or both joints.

springline. (1) The line of contact between the inside surface of a skewback and an arch in a furnace.
(2) The points on an internal surface of the transverse section of a pipe intersected by the line of maximum horizontal dimension.
(3) The mid-height of the internal wall in box sections of furnaces.

sprue. (1) A slug-like material that forms in the discharge channel of a porcelain-enamel frit smelter or glass tank.
(2) The discharge channel of a melting tank or furnace.

sprung arch, sprung roof. A curved structure spanning the working zone in a furnace, and which is supported by abutments at the sides or at the ends of the furnace.

spun glass. An individual filament or a mass of fine threads of attenuated glass, often having a delicate spiral threading or filagree.

spur. A triangular item of kiln furniture used to support glazed ware to prevent the ware from sticking to the shelves of the kiln during firing.

sputtering. The application of porcelain enamel or glaze to ware in droplets to produce a mottled or speckled appearance instead of the usual smooth, uniform surface.

square-cut glass. Optical glass cut into small squares which are separated and designated by weight; used in the production of ground and polished optical units.

squeegee. A rubber-like blade for distributing and rubbing oil suspensions of ceramic pigments over and through silk screens in the decoration of ware.

squeegee oil. A mixture of liquid organic materials employed as the suspension vehicle in screening inks and pastes.

squeegee paste. A suspension of ceramic pigments in oil used in the silkscreen process to imprint designs on glass, porcelain enamel, and other ceramic surfaces, and which develops its color on firing.

stabilizer. An oxide, such as CaO, Al_2O_3, and TiO_2, added to a frit, glaze, or color oxide to stabilize the color during firing.

stability. (1) The resistance of a glass to devitrification.
(2) The chemical and weather resistance of a glass.

stability, physical. The ability of a solid substance to resist changes in its physical characteristics under conditions of service.

stack. Any structure or part of a structure that contains a flue or flues for the discharge of gases, particularly combustion gases from smelters, kilns, and furnaces.

stacker. A device for placing and spacing glass articles properly on a continuous lehr belt for thermal treatment.

stack gas. Furnace and kiln gases that have been exhausted into a stack or a flue.

stain. (1) Color applied to glass by dipping the item in a solution of a color-forming metal salt and then heating the dipped item to a temperature at which the color is formed and absorbed by the glass surface.
(2) Color applied to glassware by subjecting the item to the vapors of a color forming salt at elevated temperatures in a closed furnace.
(3) A ceramic color, usually one of the transition metals in combination with other elements, applied to a body, glaze, or porcelain enamel as an addition to the body, glaze, or porcelain-enamel composition.
(4) An imperfection such as chemical corrosion of the glass or ceramic coating surface.
(5) An unwanted discoloration of the surface of a body or coating.

stainability. The relative easy by which a material is penetrated and discolored by a foreign material.

stained glass. Glass colored by various means, such as by incorporating colorants in the glass batch or by applying and firing a clear colored enamel on the surface of the glass; used in the production of mosaics, church windows, etc.

stain, glaze. Calcined ceramic pigments, usually metal oxides, mixed with a glaze before it is applied to ware;

some serve as pigments, some as precipitates, and some go into solution in the fired glaze.

stains, slip. Calcined ceramic pigments incorporated with a slip instead of as a body ingredient; a means of reducing the quantity of colorant required to produce a desired color.

stamping. A means of decoration or marking of ware by the use of a rubber stamp to apply a stamping ink to the surface of the ware.

stamping ink. A suspension of finely milled ceramic pigment in a suitable medium, usually an oil, which may be applied by means of a stamp, and which develops its color and permanence on firing.

standard. (1) A reference used as the basis for comparison or calibration.
(2) A concept that has been established to serve as a model or rule in the measurement of quantity or quality, or the establishment of a procedure or practice.

standard, acceptance. The acceptance level established to govern quality, quantity, performance, or other property of a material or product.

standard, calibration. Any standard having known parameters which may be used to adjust the sensitivity setting of test instruments at some predetermined level and for periodic adjustment of the instruments.

standard design. A proven or published design for a product.

standard deviation. The standard or allowable deviation of a single determination (sigma) divided by the square root of the group (nu).

standard, primary. A standard calibrated by measurement of the parameters, usually different from the one for which it will be used as a standard.

standard, reference. A reference used as the basis for comparison or calibration.

standard shapes. A series of refractory units in various sizes and shapes which, because of their extensive or essential use, are stocked by the manufacturer or can be made from stock molds.

standard, working. Any specification or standard of quality in current or regular use.

stannic chloride. $SnCl_4$; mol. wt. 260.5; m.p. −33°C (−1°F); b.p. 114°C (237°F); sp. gr. 2.28; used to produce an abrasion-resistant coating on glass and as an electrically conducting film on glass and ceramics.

stannic oxide. SnO_2; mol. wt. 150.7; m.p. 1127°C (2060°F); sp. gr. 6.6–6.9; thermal expansion 3.8×10^{-6}; used as an opacifier in white and colored porcelain enamels, glazes, and alabaster, milk, and opaque glasses.

stannous chloride. $SnCl_2$; mol. wt. 189.61; m.p. 246.8°C (475°F); used as a conductor and resistor coating on glass, porcelain enamels, and ceramics for surface heating.

stannous chromate. $SnCrO_4$; mol. wt. 234.71; used as a colorant in the decoration of porcelain and pottery.

stannous oxide. SnO; mol. wt. 134.7; decomposes with combustion; sp.gr. 6.3; used as an intermediate in preparation of stannous salts in the glass industry.

staple fiber. An individual filament made by attenuating molten glass; the fibers are of relatively short length, generally less than 17 inches.

starch. A group of carbohydrates or polysaccharides of the general composition $(C_6H_{10}O_5)_n$ used as a component in sizes for glass textile yarns.

star dresser. A tool using star-shaped metal cutters which may be rotated for truing and dressing grinding wheels.

star marks. A porcelain-enamel defect characterized by a star-shaped fracture in which lines radiate from a point opposite a firing pin or from impact with a sharp object prior to firing.

starred glaze. A partially devitrified glaze in which star-shaped crystals develop at the surface during firing.

starved glaze. A glaze applied on ware to an insufficient thickness to obtain good coverage.

static balance. The condition which permits a grinding wheel, or other rotating part on a frictionless horizontal arbor to remain at rest in any position.

statistical analysis. The evaluation of data by statistical methods.

statistical bias. A constant or systematic error in test results as may exist between the true value and a test result obtained from one method, between test results obtained from two methods, or between test results from a single method between, for example, different operators or laboratories.

statistical quality control. A means of controlling the quality of a product or process by the use of statistical techniques.

statistics. The drawing of inferences from data on samples obtained under specified conditions by use of the probability theory.

stator blade. A structural member of the stationary part of a motor, dynamo, turbine, or other machine about which a rotor turns.

steadyrest. A supplementary support for pieces being ground on a cylindrical grinder.

steady-state current. The current in a circuit after it has reached equilibrium.

steam. Water in a vapor or gaseous state.

steam curing. The rapid curing of concrete in an atmosphere of steam, either at atmospheric or elevated pressures, as in an autoclave.

steam-rack dryer. A room equipped with steam pipes as the source of heat arranged so as to permit the stacking of pallets of wet greenware for drying.

stearates. Salts or esters of stearic acid used as internal lubricants in the dry pressing of technical ceramics.

stearic acid. $CH_3(CH_2)_{16}COOH$; a colorless wax-like solid used as a lubricant in ceramic products and to promote abrasion resistance in heavy clay items.

steatite. $Mg_3Si_4O_{10}(OH)_2$; approximate mol. wt. 379.22; m.p. above 1300°C (2372°F); sp.gr. 2.7–2.8; hardness (Mohs) 1–1.5; used in the production of low-loss electrical insulators, dinnerware, wall-tile, and as a component in forsterite and cordierite bodies.

steatite porcelain. A vitreous ceramic whiteware for technical application in which steatite (magnesium metasilicate $MgO{\cdot}SiO_2$), is the essential crystalline phase.

steatite talc. Massive talc or its pulverized product having the general formula $3MgO{\cdot}4SiO_2{\cdot}H_2O$. Although the compound term appears in the literature, the use of the terms steatite and talc together actually is a redundancy.

steatite whiteware. Any ceramic whiteware in which magnesium metasilicate, $MgO{\cdot}SiO_2$, is the essential crystalline phase.

steel, cold-rolled. A low-carbon, cold-reduced sheet steel used in porcelain enameling.

steel, decarburized. A special sheet steel having an extremely low carbon content used in the production of porcelain-enameled ware in which the normally called cover coat is applied directly to the metal; that is, without the use of a ground coat.

steel, glass-lined. A steel container lined with a special class of porcelain enamels having high resistance to chemical attack at elevated temperatures and pressures.

steel, high-carbon. Steel containing more than 0.5% of carbon; the carbon tends to cause blistering or similar defects in porcelain enamels.

steel lines. Lines visible in a porcelain-enamel coating which follow the rolling pattern of the steel.

steel, pretensioned. Steel rods, wires, or strands placed in tension in freshly poured concrete; when the concrete has set, the tension is released to place the concrete in compression.

steel, zero-carbon. A decarburized steel suitable for the application of a porcelain enamel cover coat without need for a ground coat.

Steinbuhl yellow. See barium chromate.

stellite. A family of hard temperature- and wear-resistant alloys of cobalt, chromium, and tungsten used as firing racks and tools in porcelain enameling.

stem, pulled. A stem of glassware pulled from the under side of a bowl while the glass is still in the plastic state.

stemware. Glass tableware, such as goblets, compotes, etc., constructed with a slender stem between the bowl and the base.

stencil. A sheet of heavy paper, plastic, or metal in which lettering or designs are cut and through which designs are applied to ware by spraying or brushing; in some instances, a dried but unfired coating is brushed from a previously fired undercoating to produce the design.

stick. A bonded abrasive, stick-like form used for hone-sharpening, for precision honing, and for the dressing of abrasive wheels.

sticking up. The process of joining two ceramic articles, such as a handle to a cup or a knob to a tureen.

sticky. A fat, rich-appearing, plastic concrete mix.

stiff. A concrete mix that is too dry, lacks plasticity, and exhibits low slump characteristics.

stiffening, premature. The erratic and abnormally quick setting of cement in concrete due to the presence of unstable gypsum in the cement.

stiff glazes. Glazes which exhibit little or no run, either after application or during firing.

stiff-mud brick. Brick produced by extruding a stiff but plastic clay, containing approximately 12 to 15% of moisture, through a die.

stiff-mud process. The process of extruding a stiff but plastic clay through a die.

still. An apparatus consisting of a vessel in which a liquid is heated and vaporized, and then cooled in a tower or chamber in which the vapor is condensed and collected.

stilliards. Racks used for storage or for transporting clayware from one point to another prior to firing.

stilt. A tripod-like setter with sharp points at the end of each arm on which glazed ware is placed and fired.

stilt marks. Marks left on the bottom of a glazed item caused by its sticking to the stilt following the firing operation; these marks are stoned or ground off before shipment of the item.

sting-out. Hot air and flame exhausted through openings in furnaces, kilns, and glass tanks due to the existence of positive pressures in the firing zones.

stinkers. Soured storage barrels used for the aging of clays and slips to obtain improved qualitites of the ware.

stippled finish. (1) Spattered or pebbly textured porcelain-enameled or glazed finishes produced by distributing and firing droplets of different colored enamels or colored glazes over the surface of the ware.

(2) A stippled or mottled effect produced on the surface of glass by treatment with a mixture of acid and an inert substance, resulting in a variable penetration of acid over the surface of the glass.

stoichiometric. Having the precise weight relationship of elements as demanded by the chemical formula, and by which the quantities of reactants and products of a chemical reaction are determined.

stockpile. A reserve of materials or products accumulated for future use or shipment.

stoke. A unit of kinematic viscosity equal to the kinematic viscosity of a fluid having a dynamic viscosity of one poise and a density of one gram per cubic centimeter.

stoker. A mecahnical device for feeding coal or other solid fuel to a furnace.

Stoke's law. The frictional force on a sphere moving through a fluid at constant velocity is equal to 6π times the product of the velocity, the fluid viscosity, and the radius of the sphere; large particles suspended in a liquid settle more rapidly than smaller particles.

stone. (1) A defect consisting of a crystalline inclusion in glass.

(2) To rub a concrete surface with a carborundum stone.

(3) To remove blemishes from fired ware by means of a fine-grained rubbing stone.

stone china. An opaque, nonporous dinnerware, made from a clay that will vitrify; the ware may be glazed or unglazed.

stone, Cornish. Partly decomposed granite consisting of quartz, feldspar, and fluorine-bearing minerals; used as a flux in pottery.

stone, crushed. The product obtained by crushing rocks, boulders, and cobblestone, substantially all faces of which result from the crushing operation.

stone, rubbing. A shaped abrasive used in the removal of surface imperfections in porcelain enamel and glazes.

stoneware. A vitreous or semivitreous ceramic ware of fine texture and high-chemical resistance made primarily from nonrefractory fireclay, for laboratory, industrial, and some domestic uses, such as tanks, sinks, and chemical containers.

stoneware, chemical. Vitreous ceramic whiteware used as containers for the storage, transporting, or reacting of chemicals.

stoneware clay. A semirefractory plastic clay which will fire to a dense, vitrified body of high strength which may or may not be glazed.

stoning. The removal of imperfections and undesired portions of porcelain enameled ware and glazed ceramics by means of an abrasive rubbing stone.

stopper, stopper head. (1) A refractory shape, usually made of clay and graphite, which is employed as a movable valve-head seating in a nozzle brick, the assembly forming a valve for molten metal in a bottom-pouring ladle.

(2) A movable refractory controlling the flow of molten glass from a tank.

(3) A refractory or clay plug at the discharge channel of a porcelain-enamel smelter or glass tank.

stopping. The filling of holes and cracks in bisque ware with clay mixtures prior to glazing.

storm anchor. A corrosion-resistant metal fastener with a flat base and a shank which fastens the concealed lower corner of each asbestos-cement shingle to the exposed edge of an adjacent shingle.

storm sewer. A pipeline designed to carry storm or surface water from an area.

storm water. The collection of run-off water during or following rainfall.

stove clay. A seldom used synonym for fireclay.

straight brick. A rectangular brick, 13½ inches or less in length, in which the thickness is less than the width.

straight throat. The passage between the melting and refining zones of a glass melter which is located at the same level as the bottom of the melter.

straight wheel. A grinding wheel having sides or faces that are straight and parallel, with each side at right angles to the arbor hole.

strain. Elastic deformation due to stress.

strain disk. A disk of glass having a calibrated amount of birefringence at a specified location, and used as a comparative measure of the degree of stress or the degree of annealing of glassware.

strainer core. A porous refractory employed to remove slag and sand inclusions during the pouring of cast iron.

strainlines. A defect in finished porcelain-enameled ware appearing as a line or series of lines in a strain pattern, and having the appearance of cracks healed by fusion.

strain point. The temperature which corresponds to a

specific rate of elongation (when measured by ASTM Method C336) or a specific rate of midpoint deflection of a glass beam (when measured by ASTM Method C598); at the strain point of glass, internal stresses are substantially relieved in a matter of hours.

strand. Glass fibers twisted or laid together in thread or yarn form.

strand count. The thickness of a strand of glass filaments reported as the number of specified lengths per unit of weight.

stratification. The formation of layers in a body during pugging or other process.

strength. (1) The ability of a material or product to resist force.
(2) A term indicating the relative thickness of sheet glass.

strength, bond. (1) The adhesive strength of a mortar joint or the strength of a wall produced by the type of bond used.
(2) The ability of a heterogeneous product to resist stress loading.

strength, compressive. The maximum compressive stress which a material is capable of developing, based on the original area of the cross section.

strength, crushing. The property of a material to resist physical breakdown when subjected to a slowly increasing and continually applied crushing force. See crushing strength.

strength, dielectric. The voltage gradient at which dielectric breakdown of the insulating material occurs under specific conditions of test.

strength, double. Sheet glass with a thickness between 0.115 and 0.143 inch.

strength, dry. The ability of a dried, but unfired, body to withstand force without damage.

strength, fatigue. The maximum stress a body can withstand over a specific number of test conditions without failure.

strength, film. The relative resistance of a dried, but unfired, porcelain-enamel or glaze coating to mechanical damage.

strength, flexural. The ability of a solid material to withstand a flexural or transverse load.

strength, green. The resistance of a formed, but unfired, ceramic body to mechanical damage, particularly impact and transverse loads.

strength, hydrostatic. The property of a pipe, under specified conditions, to withstand internal pressure of specified magnitude.

strength, impact. The property of a material to resist damage when subjected to sharp blows or shock loading.

strength, magnetic field. The measured intensity of a magnetic field at a point, usually expressed in oersteds.

strength, shear. The maximum shear stress a material can withstand without rupture.

strength, single. Sheet glass with thickness between 0.085 and 0.101 inch.

strength, tensile. The ability of a material to withstand tension without fracture. See tensile strength.

strength, transverse. The ability of a material to withstand a flexural or transverse load. See modulus of rupture.

strength, ultimate. (1) The maximum strength of a material or product, usually reported in terms of weight per unit of length at the instant of failure.
(2) The maximum load supported by a concrete pipe as determined by three-edge testing procedures and reported as newtons per linear meter per millimeter of inside diameter or as pounds per square inch of inside diameter or horizontal span.

strength, yield. The unit of stress corresponding to a specific amount of permanent deformation in a solid.

stress. (1) Any condition of tension or compression existing within a glass, particularly due to incomplete annealing, temperature gradient, or inhomogeneity.
(2) A mutual force of action between bodies in contact with each other caused by external forces, such as tension or shear, the intensity of the force usually being reported in terms of weight per unit of area.
(3) An applied force or system of forces which tend to strain or deform a body.

stress crack. An internal or external crack in a solid body resulting from tensile, compressive, or shear forces.

stress, grinding. The residual stress, tensile or compressive, or a combination of both, generated in a workpiece during a grinding operation.

stress, impact. The stress imposed on a solid body by a sharp, suddenly applied force; reported as force per unit of area.

stress, internal. The stress existing in a solid body which is independent of external forces; for example, the stress remaining in glass induced by a particular heat treatment.

stress, thermal. Stress induced by temperature changes in a body which is unable to expand or contract freely.

stress, unit. The measure of stress or load per unit area of a substance.

stress, yield. The minimum stress at which creep will occur in a solid body. See creep.

stretcher. A brick laid flat in a wall with its length parallel to the face of the wall. Also known as stretcher bond.

stria. (1) A cord of low intensity, generally of major interest in optical glass, but also of concern in other glasses in which uniformity of the glass is important.
(2) Occurring or produced in layers.

striking. The development of opacity or color in porcelain enamels and glasses during cooling, reheating, or special thermal treatment.

string. (1) An imperfection in glass consisting of a straight or curled line, usually resulting from the slow solution of a large grain of sand or other substance.
(2) A thread of porcelain enamel drawn from a molten smelter batch for observation as a means of estimating the degree or completeness of the smelting operation.

string dryer. An intermittent tunnel-type dryer of high humidity used in the treatment of building brick.

stripping yard. The area in which plate glass is removed from the polishing table following the grinding and polishing operation.

strontium aluminate. (1) $SrO \cdot Al_2O_3$; mol. wt. 205.54; m.p. 2010°C (3650°F). (2) $SrO \cdot 2Al_2O_3$; mol. wt. 307.48; m.p. 1771°C (3220°F); sp. gr. 3.03.

strontium aluminum silicate. $SrO \cdot Al_2O_3 \cdot 2SiO_2$; mol. wt. 325.36; m.p. 1660°C (3020°F); sp. gr. 3.12; hardness (Mohs) 5–7.

strontium arsenate. $3SrO \cdot As_2O_5$; mol. wt. 540.72; m.p. 1637°C (2980°F); sp. gr. 4.60.

strontium boride. SrB_6; mol. wt. 152.55; m.p. 2235°C (4055°F); sp. gr. 3.42; a potential material for use in energy sources when using the radioisotope, for high-temperature insulation, for nuclear absorption control rods, and as control additives.

strontium carbide. SrC_2; mol. wt. 111.63; m.p. about 2000°C (3632°F); sp. gr. 3.2. See carbides.

strontium carbonate. $SrCO_3$; mol. wt. 147.6; decomposes at 1100° to 1340°C (2012° to 2444°F); sp. gr. 3.62; used in television tubes and iridescent glasses, ceramic ferrites, and ceramic bodies and glazes.

strontium fluoride. SrF_2; mol. wt. 125.63; m.p. 1190°C (2174°F); sp. gr. 2.4; used as single-crystal components in lasers.

strontium nitride. Sr_3N_2; mol. wt. 290.91; m.p. 1483°C (2700°F). See nitrides.

strontium oxide. SrO; mol. wt. 103.63; m.p. 2430°C (4406°F); sp. gr. 4.7; converts to the hydroxide in water; used as colorant in glass.

strontium phosphate. $3SrO \cdot P_2O_5$; mol. wt. 452.86; m.p. 1766°C (3210°F); sp. gr. 4.53.

strontium scandate. $SrO \cdot Sc_2O_3$; mol. wt. 241.8; sp. gr. 4.59.

strontium silicate. (1) $SrO \cdot SiO_2$; mol. wt. 163.69; m.p. 1580°C (2876°F); sp. gr. 3.65. (2) $2SrO \cdot SiO_2$; mol. wt. 267.26; m.p. $>$1705°C (3100°F); sp. gr. 3.84; hardness (Mohs) 5–7.

strontium stannate. $SrO \cdot SnO_2$; mol. wt. 254.33; m.p. $>$1400°C (2552°F); used in titanate bodies to reduce the Curie temperature.

strontium sulfate. $SrSO_4$; mol. wt. 183.7; m.p. 1605°C (2921°F); sp. gr. 3.71–3.94; hardness (Mohs) 3–3.5; used to impart iridescence on the surfaces of glass and pottery glazes, and as a fining agent in the production of crystal glasses.

strontium sulfide. SrS; mol. wt. 119.69; m.p. $>$1999°C (3630°F); sp.gr. 3.67.

strontium titanate. $SrO \cdot TiO_2$; mol. wt. 183.53; m.p. 2080°C (3776°F); theoretical sp.gr. 5.11; a dielectric material used in electronics and electrical insulation, and in low-melting glazes.

strontium uranate. $SrO \cdot UO_3$; mol. wt. 389.77; m.p. 1799°C (3270°F).

strontium zirconate. $SrO \cdot ZrO_2$; mol. wt. 226.85; m.p. 2700°C (4892°F); sp.gr. 5.48; used in dielectric compositions to reduce the Curie temperature.

structural clay facing tile. Tile designed for use in interior and exterior unplastered walls, partitions, and columns.

structural clay tile. Hollow burned clay masonry building units with parallel cells or cores, or both, used as facing tile, partition tile, load-bearing tile, fireproofing tile, header tile, and furring tile.

structural facing unit. A structural or building unit designed for use in areas where one or more faces will be exposed in the finished wall and for which specifications include color, finish, and other properties influencing appearance.

structural glass. (1) Opaque or colored glass, frequently ground and polished, used for structural purposes, particularly in windows.
(2) Glass block, usually hollow and often with patterned faces, used for structural purposes such as in walls, partitions, and windows.

structural products. Building material units which may be load-bearing (loads in addition to their own weight) or nonload bearing (only their own weight).

structural spalling of refractories. Spalling of a refractory unit resulting from stresses caused by differential changes in the structure of the unit.

structure. (1) The proportion and arrangement or spacing of abrasives in a grinding wheel.

(2) The arrangement and interrelation of the parts of an object.

(3) The state of agglomeration of particles in carbon black.

structure number. The number, generally from 0 to 15, designating the spacing of abrasive grains relative to their grit size in a grinding wheel.

stub. The portion of a grinding wheel remaining after it has been worn down to the discarding diameter.

stucco. A mixture of portland cement, sand, and a small percentage of lime blended into a smooth, plaster-like consistency which is applied to exterior walls and other surfaces of a building or structure.

subbase. A compacted layer of material placed on the subgrade to support the base on which a concrete pavement is constructed.

subgrade. The foundation on which a concrete pavement is constructed.

sublots. Subdivisions of a lot or shipment of a material.

submarine throat, submerged throat. A throat with the level below the bottom of a glass melter.

submerged wall. A refractory wall submerged below the level of molten glass in a glass-melting tank forming the throat between the melting and refining chambers of the tank.

subsidence. The deformation of a material under load.

substance. The thickness of sheet or rolled glass expressed as weight per unit of area.

substrate. A surface upon which a coating or film has been applied.

subsurface discontinuity. A defect which does not extend through the surface of the item in which it exists.

sucking. The sucking or drawing of vaporized lead and other glaze constituents into a porous refractory.

suction. The absorption of liquids into the pores of a concrete surface.

suction process. Any process in which molten glass is gathered into a mold by vacuum.

suction rate. The weight of water absorbed by a partially immersed brick in one minute, usually expressed as a unit of weight per minute.

sulphation. A powder, stain, or scum forming on the surface of a glaze, during or after firing, caused by sulphur compounds emanating from the body or present in the furnace atmosphere.

sulphide. A compound in which one or more sulphur atoms is attached to a nonoxygen atom such as carbon or a metal. Refractory sulphides have received scant consideration for technical applications, but now are of interest in nuclear fuels and direct-energy conversion, particularly the sulphides of plutonium, thorium, and uranium. The sulphides, in general, are prone to chemical and physical instabilities. Melting points range from 1100°C(2012°F) to approximately 2428°C(4400°F), although some decompose above 593°C(1100°F).

sulphoaluminate cement. A hydraulic cement consisting of a mixture of gypsum and high-alumina cement.

sulphonated oils. Sulphuric acid-treated animal and mineral oils used as wetting agents and defoaming agents in glazes, porcelain enamels, and other slips and slurries.

sulphur. S; at. wt. 32.06; m.p. 119.3°C (248°F); b.p. 444.6°C (831°F); changes to beta form at 94.5°C (203°F); sp.gr. 2.06; index of refraction 1.957; used as a colorant in glass to produce yellows and ambers, and with cadmium sulphide is used in selenium ruby glass.

sulphuric acid. H_2SO_4; mol. wt. 98.08; boils from 210° to 338°C (410° to 640°F); sp.gr. 1.8; used in the pickling of steel for porcelain enameling, and occasionally as a mill addition for acid-resisting porcelain enamels to counteract the alkaline nature of the coating.

sulphur-impregnated abrasive. A bonded abrasive product in which all connected pores are filled with sulphur.

sulphuring. The scumming or staining of a glaze caused by sulphur compounds in the atmosphere during and after firing.

sump throat. The submerged passage between the melter and refiner of a glass tank situated at a level below the bottom of the melter.

sun-dried brick. Large, roughly molded clay brick of varying sizes, frequently made with additions of damp straw, which are dried in the sun.

sun-pumped laser. A continuous-wave laser in which the energy of the sun is focused on the laser crystal.

supercooled liquid. A liquid cooled below its freezing point without solidification or crystallization.

superduty fireclay brick. A fireclay refractory having a pyrometric cone equivalent not less than Cone 33, not more than one per cent linear shrinkage in the reheat test, 1598°C (2910°F), and not more than four per cent weight loss in the panel-spalling test. See spalling test, panel.

superduty silica brick. Silica brick having a total alumina, titania, and alkali content significantly lower than normal.

superstructure. The parts of a glass tank above the sidewall tank blocks.

supply voltage. The potential voltage available from a power source of electric current.

surface. The outer layer of a substance.

surface area. (1) The measured extent of an area covered by a surface, excluding thickness.
(2) The total exposed area of the surface of a pulverized solid usually expressed as some unit of area per gram.

surface-area distribution. The distribution of surface area in accordance with some parameter such as pores of different size or diameter.

surface clay. An unconsolidated, unstratified clay occurring on the surface of the earth.

surface combustion. The combustion of fuel gases or mixtures of gases and air by impingement on or through a heated refractory.

surface density. The quantitative distribution of a substance on a surface expressed as per unit of surface area.

surface finish. The character of a solid surface in terms of roughness and irregularities after final treatment.

surface, gassy. An imperfection characterized by poor gloss and fuzzy texture on a porcelain-enamel surface.

surface grinding. The abrading or grinding of a plane surface.

surface mark. A relatively long, narrow, shallow groove, cut, or other abrasion in the surface of a solid.

surface oxides. Oxygen-containing compounds and complexes formed on the surface of a substance or object.

surface polishing. The polishing of plate glass and other surfaces to remove imperfections.

surface reaction. A chemical or physical reaction taking place only on the surface of an item.

surface resistance. The electrical resistance of an insulating product, usually measured between the opposite sides of a square on the surface of the insulator.

surface sealing. That portion of the finish of a glass container which makes contact with the sealing gasket or liner.

surface, specific. The surface area per unit of weight or volume of a solid substance.

surface tension. Cohesive forces that attract molecules of a liquid to each other, tending to minimize the surface area and cause the surface to act somewhat like a plastic film.

surface treatment. Any treatment of the surface of a material to render it receptive to subsequent coating or to develop a desired property such as resistance to abrasion or weathering.

sursulfatec cement. A type of cement composed of slag (70%) and calcium sulfate (30%); a cement developed and used in France.

suspended arch. An arch in a furnace in which the brick shapes are suspended from overhead supports.

suspender. See suspension agent.

suspension. A system in which denser particles, usually solid, are distributed throughout a less dense liquid or gas.

suspension agent. A chemical compound such as an inorganic salt which is added to a porcelain-enamel or glaze slip to promote suspension of the solid particles in the liquid medium.

swab test. A low-voltage test in which an electrical discharge is fanned across a porcelain-enameled surface to detect a discontinuity in the coating by means of a spark concentrating in the discontinuity.

swarf. A mixture of grinding chips and fine particles of an abrasive and bond resulting from a grinding operation.

sweat, sweating. The process of placing a heated and expanded metal ring or collar around a ceramic part and allowing it to shrink around the ceramic to produce a tight, adherent seal or joint.

sweet. An easily workable glass.

swelling clay. A clay which will absorb large quantities of water.

swing-field magnetization. A magnetic field induced in two different directions to detect defects in a part which are oriented in different directions in the part.

swing-frame grinder. A grinding machine suspended above the work piece by a chain at its center of gravity so that it may be turned and swung in any direction for the in-place grinding of work too heavy for manual handling.

swing press. A screw press, often hand operated, used to form special shapes in small quantities.

switch. A mechanical device for opening and closing an electric circuit.

sworl. Marks formed on the bottom of a pot by a cutting or grinding wheel.

syenite. An igneous rock composed chiefly of an alkali feldspar containing quartz, feldspathoids, mica, or hornblende in minor quantities.

synthetic graphite. A crystalline graphitic material made by processing carbon at high temperature and pressure.

synthetic magnesite. Magnesite made by a chemical process using seawater or other saline solutions as a source material.

synthetic quartz. A quartz crystal grown at high temperature and pressure around a seed of quartz which is suspended in a solution containing natural quartz crystals.

synthetic test solution. A solution of two or more components prepared under specified conditions for use in the evaluation of adsorbents.

systematic sampling. The taking of samples from a batch or manufacturing operation at fixed time intervals or in fixed quantities, or both.

T

Taber abrader. An instrument for measuring the resistance of surfaces to abrasion consisting of loaded abrasive wheels rotating on the surface being tested.

table. The platform of a grinding machine supporting work being ground.

table, round. A type of table supporting plate glass during grinding and polishing.

table, shaker. A slightly tilted, vibrating table having a flat, rectangular, and sometimes riffled surface used to separate solid materials according to density and particle size.

table, traverse. A reciprocating platform of a grinding machine supporting work being ground.

tableware. Plates, cups, saucers, and related items employed on the dining table in the serving of food.

tabular alumina. Alpha Al_2O_3; mol. wt. 101.9; m.p. 2040°C (3704°F); sp.gr. 3.4–4.0; hardness (Mohs) 9; used in refractories, electroceramics, high-quality porcelains and other ceramics, and abrasive products. See alumina, tabular.

tabular crystal. A flat crystal with parallel faces.

taconite. A low-grade iron ore containing hematite, magnetite, and fine-grained silica.

tailings. (1) Screened particles of a material that are too coarse or too fine for an intended use.

(2) The undesirable residue from a magnetic separation.

(3) Worthless residue from a mining, milling, or similar process.

take-out. A mechanical device for removing a finished glass article from a glass-forming machine.

talc. $3MgO{\cdot}4SiO_2{\cdot}H_2O$; mol. wt. 379.2; m.p. above 1400°C (2552°F); sp.gr. 2.6–2.8; hardness (Mohs) 1–1.5; employed in wall tile, refractories, electroceramics, dinnerware, and other ceramic bodies; particularly valuable in improving thermal-shock resistance.

Talwalker-Parmellee plasticity index. The ratio of total deformation of a clay at fracture to the average stress beyond the proportional limit.

tamped pipe. Concrete pipe formed by tamping dry, no-slump concrete into rotating, vertical molds.

tamping. The forming of articles by the repeated pounding of dampened bodies into molds.

tank. (1) A refractory-lined, glass-melting unit.

(2) A container in which ceramic slips and slurries are stored for subsequent use.

tank block. A refractory block used to line the melting zone of a glass tank.

tank, continuous glass. A glass-melting tank into which batch is charged continuously at a rate essentially equivalent to the amount continuously withdrawn.

tank, day. A periodic glass-melting unit in which a day's supply of glass is charged, melted, and withdrawn in the production of handmade glass.

tank furnace. A furnace containing a receptacle or tank in which glass is melted.

tank glass. Glass melted in a large tank as distinct from a pot.

tank, glass. See tank furnace.

tank, pressure. An air-tight container from which porcelain enamel, glaze, and other slips are delivered to a processing system, such as a spraying operation, by means of air pressure.

tank, settling. A tank or reservoir into which slurries of various components are placed to permit settling of solid materials to be accomplished by gravity.

tannic acid, tannin. A water-soluble, organic powder obtained from nutgalls, tree bark, and other plants; used as a deflocculant and binder in slips and slurries.

tantalum aluminide. $TaAl_3$; mol. wt. 262.41.

tantalum beryllide. (1) $TaBe_{12}$; mol. wt. 291.9; m.p. 1850°C (3360°F); sp.gr. 4.18. (2) $TaBe_{17}$; mol. wt. 519.4; m.p. 1990°C (3610°F); sp.gr. 5.05. (3) $TaBe_2$

mol. wt. 199.90; sp.gr. 8.95. (4) $TaBe_3$; mol. wt. 209.10; sp.gr. 8.18–8.23. See beryllides.

tantalum boride. (1) TaB_2; mol. wt. 203.14; m.p. 3200°C (5792°F); sp.gr. 12.5. (2) TaB; mol. wt. 192.31; m.p. 2400°C (4352°F); sp.gr. 14.3. (3) Ta_3B_4; mol. wt. 587.78; m.p. 2650°C (4802°F); sp.gr. 13.6. (4) Ta_2B; mol. wt. 373.82; m.p. 1899°C (3450°F). (5) Ta_3B_2; mol. wt. 566.14; m.p. 2038°C (3700°F). See borides.

tantalum capacitor. An electrolytic capacitor employing tantalum in some form as the anode.

tantalum carbide. (1) TaC; mol. wt. 193.4; m.p. 3875°C (7007°F); sp.gr. 14.5; hardness (Mohs)+9. (2) Ta_2C; mol. wt. 374.8; m.p. 3400°C (6152°F); used in cutting tools and dies.

tantalum nitride. (1) TaN; mol. wt. 195.4; m.p. 3360°C (6080°F). (2) Ta_2N; mol. wt. 376.8; loses nitrogen at 1900°C (3452°F). (3) Ti_3NS; mol. wt. 614.2; m.p. 3400°C (6152°F); sp.gr. 14.1.

tantalum nitride resistor. A thin-film resistor with a deposit of tantalum nitride on a substrate such as sapphire.

tantalum oxide. Ta_2O_5; mol. wt. 442.8; m.p. 1800°C (3272°F); sp.gr. 7.6; used in optical glass and ferroelectric components.

tantalum phosphide. TaP; mol. wt. 212.53; m.p. 1660°C (3020°F); sp.gr. 11.1. See phosphides.

tantalum silicide. (1) $TaSi_2$; mol. wt. 237.6; m.p. about 2400°C (4352°F); sp.gr. 9.14. (2) Ta_2Si; mol. wt. 391.06; m.p. 2499°C (4530°F); sp.gr. 13.54. (3) Ta_5Si; mol. wt. 935.56; m.p. 2510°C (4550°F); sp.gr. 12.86. (4) Ta_5Si_3; mol. wt. 991.68; m.p. ~2499°C (4530°F); sp.gr. 13.06.

tantalum sulfide. (1) TaS; mol. wt. 213.56; sp.gr. 9.20. (2) TaS_2; mol. wt. 245.62.

tap. (1) To drain molten vitreous compositions from a smelter through an opening in the smelter floor.

(2) To remove excess slag from the bottom of a pot furnace.

(3) To center an inverted pot on a wheel for trimming and decoration.

tap density. The apparent density of a powder or granulated material resulting when the receptacle containing the material is vibrated or tapped manually under standard or specified conditions.

tapered seal. A thin metal sleeve fitted over a thick, tapered ceramic cylinder so as to form a tight seal.

tapered wheel. A flat-faced grinding wheel tapered with the greater thickness at the hub.

tapestry brick. A brick having a rough, unscored, textured surface.

tap hole. A hole in the bottom of a smelter or ladle through which a molten batch is drained.

tap-hole clay. A damp, plastic, refractory clay formed into a wad and used to seal the tap hole of a smelter or melting furnace.

tapping. The removal of the tap-hole plug to drain a smelter or furnace of its molten charge.

tar-bearing basic ramming mix. A tar-bearing basic-refractory mixture which is rammed into place to form a monolithic structure in the heat zone of a furnace. See pitch.

tar-bearing basic refractory. A refractory shape composed of basic refractory grains to which tar has been added during manufacture. See pitch.

tare weight. The combined weight of an empty container and its accessories.

tarnish. The dulling discoloration, or staining of a surface by exposure to air or reactive atmospheres.

tarras cement. A volcanic tuff having pozzolanic properties; used as a hydraulic cement. Also known as trass.

tea-dust glaze. An opaque, iron oxide-bearing stoneware glaze of greenish color.

teapot ladle. A type of ladle containing a refractory dam under which molten metal flows; designed to prevent slag from reaching the ladle spout.

tear. A crack or a torn section in glass caused by sticking to hot metal.

tearing. A pattern of healed cracks in porcelain enamel in which the undercoat or metal may be observed.

teaser. A workman supervising the charging, temperature control, and operations of a glass-melting tank or furnace.

teeming. The pouring of a molten batch from a pot or ladle into molds.

teeming, bottom. The filling of ingots or molds in which the molten batch enters the molds from the bottom.

teeming, uphill. The process of discharging molten steel from a ladle through refractory tubes into molds in such a manner that the steel is introduced at the bottom of the mold instead of at the top.

tellurium. Te; at. wt. 127.6; m.p. 452°C (845°F); b.p. 1390°C (2534°F); sp.gr. 62.4; used as a yellow, green and blue colorant in glass and glazes.

temper. (1) To moisten and mix clay, plaster, mortar, etc., to proper consistency for use.

(2) The degree of residual stress in annealed glass.

(3) To strengthen, harden, or toughen glass by controlled heat treatment.

temperature. The thermal state of a body in terms of its ability to transfer heat to other bodies.

temperature, absolute. The temperature measured on the absolute scale in which 0° abs. is −273.15°C (−459.67°F), the scale units being equal in magnitude to the Centigrade scale.

temperature, annealing. The temperature at which the viscosity of glass is 10^{13} poises and is essentially free of stress.

temperature deformation. The minimum temperature at which a body begins to deform.

temperature, firing. (1) The peak temperature reached during the firing of ware.

(2) The degree of sensible heat attained by ware during the maturing of the coating. See sensible heat.

temperature gradient. The degree or measured rate of temperature change between two points of reference in a substance or in an area.

temperature-gradient furnace. A slender laboratory furnace of relatively small cross-section in which a controlled temperature gradient is maintained along its length.

temperature, ignition. The lowest temperature at which combustion will occur spontaneously when a substance is heated in air under specified conditions.

temperature, liquidus. The temperature at which the liquid and the crystalline or solid phases of a substance are in equilibrium.

temperature, maturing. The temperature at which a body, glaze, porcelain enamel, or other ceramic must be fired for some period of time to attain a desired degree of maturation.

temperature, melting. The minimum temperature at which a substance changes from the solid to the liquid state.

temperature, oxidizing. The temperature at which an oxidizing reaction will occur, such as the oxidation of carbonaceous products in clays and other minerals.

temperature, recrystallization. The temperature at which a substance will change from the liquid to a crystalline phase during cooling.

temperature, refining. The temperature just above the melting point of glass and vitreous compositions at which the molten batch is sufficiently fluid to permit the solution or escape of gaseous inclusions.

temperature, softening. The temperature at which a substance such as a body, glaze, or porcelain enamel begins to exhibit a tendency to flow.

temperature, transformation. The temperature at which a change occurs in a phase of a material during heating or cooling.

temperature, yield. The minimum temperature at which permanent deformation will occur in a solid body under specified conditions of stress.

tempered glass. Glass that has been cooled from near its softening point to room temperature under rigorous control to increase its mechanical strength and thermal endurance by the formation of a compressive layer at its surface.

tempered safety glass. A glass that has been tempered, so that it will break into granular instead of jagged fragments as a result of particular stress patterns created in the glass, by a rigidly controlled heat treatment. See tempered glass.

tempering. (1) The treatment of clays, ceramic bodies, plaster, mortar, and similar materials with water or steam to obtain desired working and forming characteristics.

(2) The heat-treatment of glass and metals to develop improved resistance to mechanical and thermal damage.

tempering pan. A mechanical, pan-type mixer in which clays and bodies are blended with water to working consistencies.

tempering water. The water or moisture added to a body or clay mix to develop desired working properties.

template. (1) A guide pattern used in the shaping of ware during manufacture.

(2) A pattern through which porcelain may be sprayed on ware or through which previously applied and dried porcelain enamel may be removed by brushing to produce a desired design.

temporary wicket. Temporary closure of refractory or insulating block in a furnace or kiln, such as at the ends of checker-chambers.

tendon. A tensioned steel bar or strand of wires anchored in concrete to induce compressive stress in the concrete when set.

tenmoku. A lustrous, iron bearing, black, stoneware glaze which blends to a red-dust color on thinner parts, on firing.

tensile specimen. A bar of a material of specified dimensions used to measure the resistance of the material to fracture in tension.

tensile strength. The maximum stress a material subjected to a pulling or stretching load can withstand without breaking, calculated as the load in pounds per square inch or kilograms per square centimeter, reported for the cross-sectional area of the specimen at the point of fracture.

tensile stress. The stress developed in a specimen under a pulling load.

tension. A force which tends to lengthen a solid, such as by pulling.

tension, surface. The cohesive force, acting on the surface of a liquid which tends to minimize the surface area.

terbium boride. (1) TbB_4; mol. wt. 202.48; sp.gr. 6.55. (2) TbB_6; mol. wt. 224.12; sp.gr. 5.39. See borides.

terbium carbide. (1) Tb_3C; mol. wt. 489.6; sp.gr. 8.88. (2) Tb_2C; mol. wt. 354.4; sp.gr. 8.33. (3) TbC_2; mol. wt. 183.2; sp.gr. 7.17. See carbides.

terbium nitride. TbN; mol. wt. 173.21; sp.gr. 9.57. See nitrides.

terbium oxide. (1) Tb_2O_3; mol. wt. 366.4; m.p. 2387°C (4328°F); used in electronic ceramics. (2) Tb_4O_7; mol. wt. 748.88.

terbium silicide. $TbSi_2$; mol. wt. 215.32. See silicides.

terra alba. Pure white, uncalcined gypsum; $CaSO_4 \cdot 2H_2O$.

terra cotta. (1) An unglazed, low-fired, ornamental earthenware such as tile, roofing, vases, statuettes, building block, and primitive pottery.

(2) A hard-fired, glazed or unglazed, clay building unit, generally larger than facing tile or brick; used for ornamental purposes in architectural applications.

terra di siena. A ferric oxide pigment used in glazes.

terra rosa. A variety of hematite sometimes used as a red pigment in glazes.

terra sigillata. A fine-textured, glossy, embossed, red pottery.

terrazzo. A mosaic-type floor obtained by embedding special aggregate, such as marble or granite chips, in concrete, followed by grinding and polishing to a smooth surface after the concrete has hardened.

tertiary air. Preheated air added to the waste-gas flue of a furnace or kiln being fired under reducing conditions.

tessara. A small rectangular ceramic tile or glass used in a mosaic design.

tessha. A more metallic and broken version of tenmoku, which see.

test batch. A sample of concrete taken from a production mix for testing.

test certificate. A document certifying the validity of a performed test.

test coil. A section of a coil assembly that excites or detects the magnetic field in a material in a comparative system.

test cones. Pyramidal-like shapes composed of selected oxide mixtures which deform at known temperatures and which are used to indicate the thermal history of fired ware; commonly known as pyrometric cones.

test cylinder. A cylinder of concrete used as a test specimen.

test, eddy-current. A nondestructive test method in which a change in the induced eddy current flow in a specimen indicates certain properties or defects in the specimen, such as bubbles, inclusions, fractures, etc.

test, electromagnetic. A nondestructive test method which employs electromagnetic energy to yield information relative to the qualities of the specimens being tested.

test, nondestructive. Any test which does not damage the material or product being tested.

test, performance. A test of the ability of a product to meet performance or service requirements.

tetrasodium pyrophosphate. $Na_4P_2O_7$; employed as a suspension and dispersing agent in porcelain enamels and ceramic glazes.

texture. (1) The visual and tactile characteristics of a surface.

(2) The relationship between shapes and sizes of pores and grains in a refractory product.

textured brick. A brick treated to alter its surface appearance from that produced by the die, such as by scratching or scoring.

thallium oxide. Tl_2O; mol. wt. 424.3; m.p. 300°C (572°F); b.p. 1865°C (3389°F); used to increase the index of refraction of optical glass.

theoretical air. The amount of air theoretically required for complete combustion.

theoretical density. The density of a material calculated from the number of atoms per unit cell and the measurement of the lattice parameters.

therm. A unit of heat equal to 100,000 British thermal units.

thermal analysis. The analysis of the properties of a material which are heat-related.

thermal analysis, differential. The study of the thermal reactions, including rate, nature, and intensity, occurring in a substance or body during continuous heating and cooling, particularly to and from elevated temperatures.

thermal barrier. An insulating material which will prevent or deter the transfer of heat or cold from one body or area to another.

thermal black. A relatively coarse carbon black made by the thermatomic process for use as a pigment.

thermal capacity. (1) The amount of heat a body will absorb, expressed as Btu per degree of temperature per unit of mass.

(2) The quantity of heat required to raise the temperature of a body or substance one degree. See British thermal unit.

thermal conductivity. The rate of heat flow through a body per unit of area per unit of time per unit of temperature in a direction perpendicular to the surface.

thermal diffusivity. The rate of temperature change in a body or substance from one point or area to another.

thermal efficiency. The ratio of inlet versus output temperature to the maximum temperature range possible.

thermal emissivity. The ratio of heat radiated by a body to that of a perfect black body at the same temperature.

thermal endurance. The ability of glass or other body to resist thermal shock or to withstand deterioration during exposure to high temperatures.

thermal expansion. The reversible or permanent change in the dimensions of a body when heated.

thermal expansion coefficient. The fractional change in the length or volume of a material per degree of temperature change.

thermal glass. A low-expansion glass in which boron oxide is substituted for calcium oxide in ordinary soda-lime glass, and which may be heated and cooled rapidly without breaking.

thermal-gradient furnace. A tubular furnace of small cross-section in which a controlled temperature gradient is maintained along its length.

thermal incompatibility. A condition in which part of an aggregate in concrete exhibits a different coefficient of expansion or other thermal property from the other constituents, resulting in damage or distress to the concrete when hardened, particularly crumbling.

thermal-insulating cement. A dry cementitious composition containing additions of substances of low-thermal conductivity which, when blended with water, form a mixture which may be placed or applied as a covering to provide a thermal barrier.

thermal resistance. The resistance of a body to the flow of heat, calculated as the temperature difference between the opposite faces of the body divided by the rate of heat flow.

thermal shock. Exposure of a body or coating to sudden and severe changes in temperature.

thermal-shock failure. The fracture or crazing of a porcelain enamel, glass or glaze when subjected to sudden cooling from an elevated temperature, as by the application of cold liquids.

thermal-shock resistance. The ability to withstand sudden changes in temperature without fracture.

thermal-shock test. A test in which a body, glass, glaze, or porcelain enamel is subjected to selected conditions of sudden temperature change to determine their thermal endurance properties.

thermal spalling. The breaking or cracking of refractories sufficient to expose new surfaces caused by sudden or nonuniform temperature changes which create irresistable stresses in the unit.

thermal spraying. The spraying of droplets of molten powders on a substrate by means of a heated applicator.

thermal strength. The physical strength of a solid at an elevated temperature.

thermal stress. Stress induced by temperature changes in a body unable to expand or contract freely.

thermionic. The flow of an electrically charged particle or ion emitted by a conducting material at high temperatures.

thermister. An electrical resistor whose resistance varies sharply with changes in temperature.

thermit reaction. An exothermic reaction in which a metal oxide is reduced when heated with finely divided aluminum. See exothermic reaction.

thermocouple. A temperature-measuring device consisting of two dissimilar conductors joined together at their ends which generate thermoelectric voltage when heated; the voltage, being proportional to the temperature difference between the junctions, is calibrated to indicate temperature.

thermodynamics. The study of the relationships between the properties of matter influenced by changes in temperature, and the conversion of energy from one form to another; the conversion of heat into work and vice versa.

thermoelectricity. Electricity produced in a circuit consisting of two different metals whose junctions are at different temperatures; used in thermocouples for the measurement and control of temperatures.

thermogravimetric analysis, differential. A technique of thermal analysis in which the rate at which a material changes weight on heating is plotted against temperature.

thermometer. An instrument that measures temperature.

thermometer, maximum. An instrument which registers the maximum temperature attained during a period of time.

thermometer, minimum. An instrument which registers the minimum temperature attained during a period of time.

thermonuclear reactions. Nuclear transformations involving the light atoms, as in the hydrogen bomb.

thermoplastic. The property of softening when heated and hardening when cooled without change in properties.

thermoplastic decoration. A process of applying colors dispersed in a thermoplastic medium through a hot screen, the design freezing in place on contact with the cold surface of ware being decorated.

thermoscope. A device for estimating temperature changes of a body based on corresponding changes in the volume of the body.

thermosetting. The property of a body or material to solidify when heated, and then cannot be remelted without destroying its original characteristics.

thickening. Increasing the viscosity of a slip. See viscosity.

thick film. A resistor or other circuit component with a resistance film over one-thousandth of an inch in thickness. See resistor.

thickness. (1) The vertical depth of a coating.
(2) The dimension of a product, such as tile, measured at right angles to the wall, floor, or other surface to which it is applied.

thickness gauge. Any device designed to measure the thickness of a coating, sheet, or object.

thickness gauge, magnetic. An instrument to measure the thickness of porcelain enamel in which the magnetic force required to lift a magnet from the coating surface is calibrated to indicate the distance between the coating surface and the coating-metal interface.

thief, sample. A device of suitable design employed to remove a sample from a batch or lot of a material for subsequent evaluation or analysis.

thimble. (1) An L-shaped refractory device used to stir pot-made optical glass.
(2) A conical refractory item of kiln furniture with a projection at its bottom on which ware is supported during the decorative fire.

thin film. A film a few molecules thick deposited on a glass, ceramic, or other semiconductor substrate to form a capacitor, resistor, or other circuit component.

thin section. A material which is ground and polished to a thickness of about 0.03 millimeter for examination of its optical properties by a polarizing microscope.

thiokol. A series of polysulfide rubbers highly resistant to oils and solvents; used as tank linings, tubing, gaskets, and other applications where chemical and weather resistance is required.

third party. A person or organization other than the principals involved in a dispute.

thixotropy. The property of a suspension to liquefy when agitated and to thicken or solidify on standing.

thoria. See thorium oxide.

thorianite. Radioactive ThO_2; mol. wt. 264.1; m.p. 3300°C (5972°F); sp.gr. 9.7–9.8; hardness (Mohs) 7; a mineral, sometimes containing uranium and rare earth metals.

thorium aluminide. Th_2Al; mol. wt. 491.27; sp.gr. 9.67. See aluminides.

thorium beryllide. $ThBe_{13}$; mol. wt. 351.75; sp.gr. 4.10; hardness (Knoop) 1170–1340. See beryllides.

thorium bismuthide. Th_2Bi; mol. wt. 673.3; m.p. 1799°C (3270°F).

thorium boride. (1) ThB_4; mol. wt. 275.4; m.p. above 2500°C (4532°F); sp.gr. 8.5. (2) ThB_6; mol. wt. 297.0; m.p. 2200°C (3992°F); sp.gr. 7.1.

thorium carbide. (1) ThC; mol. wt. 244.1; m.p. 2625°C (4757°F); sp.gr. 10.65. (2) ThC_2; mol. wt. 256.1; m.p. 2655°C (4811°F); sp.gr. 9.6. Both forms are used in nuclear fuels.

thorium dioxide. See thorium oxide.

thorium fluoride. ThF_4; mol. wt. 308.1; m.p. 1111°C (2032°F); used in ceramics for high-temperature applications.

thorium nitride. (1) ThN; mol. wt. 246.1; m.p. 2630°C (4766°F). (2) Th_2N_3; mol. wt. 506.3.

thorium oxalate. $Th(C_2O_4)_2 \cdot 2H_2O$; mol. wt. 426.1; decomposes to ThO_2 above 300°–400°C (572°–752°F); used as a source of ThO_2 in ceramics.

thorium oxide. ThO_2; mol. wt. 264.15; m.p. 3300°C (5972°F); sp.gr. 9.7; hardness (Mohs) 7; used in high-temperature crucibles, cermets, incandescent gas mantles, nonsilica optical glass, cathodes and coatings in electron tubes, and in nuclear fuels.

thorium phosphide. (1) Th_3P_4; mol. wt. 820.57; sp.gr. 8.59. (2) ThP; mol. wt. 263.18; sp.gr. 7.0 See phosphides.

thorium reactor. A nuclear reactor in which thorium encases enriched uranium ore to give breeder operation.

thorium selenide. ThSe; mol. wt. 311.35; m.p. 1882°C (3420°F). See intermetallic compound.

thorium silicate. $ThO_2 \cdot SiO_2$; mol. wt. 324.21; m.p. 1979°C (3595°F); sp.gr. 5.3; hardness (Mohs) 5–7.

thorium silicide. (1) ThSi; mol. wt. 260.21; sp.gr. 8.99–9.03. (2) $ThSi_2$; mol. wt. 288.27; m.p. 1660°C (3020°F); sp.gr. 7.79–8.23. (3) Th_3Si_2; mol. wt. 752.57; sp.gr. 9.80–9.81. See silicides.

thorium sulfide. (1) ThS; mol. wt. 264.21; m.p. ~ 2471°C (4400°F); sp.gr. 9.56. (2) ThS_2; mol. wt. 296.27; m.p. ~ 1904°C (3460°F); sp.gr. 7.36. (3) Th_2S_3; mol. wt. 560.48; m.p. ~ 1949°C (3540°F);

sp.gr. 7.87. (4) Th_4S_7; mol. wt. 1153.02. Used in the manufacture of crucibles. See sulfides.

thread grinding. The cutting of threads on a part by the use of a bonded abrasive tool.

three-edge bearing test. A technique for applying load to a concrete pipe in testing its external load-crushing strength, the load being applied at the center of a specimen resting on two outside points. See load-crushing strength.

threshold concentration. The minimum concentration at which a substance can be detected by odor or taste.

threshold level. A value above or below which a specimen is rejected.

threshold odor test. The evaluation of the odor level in a fluid by dilution with an odor-free liquid.

throat. (1) The submerged passage between the melting and refining chambers of a glass-melting tank.
(2) The constricted area between the port and firing chamber of an open-hearth furnace.

throat, straight. The passage between the melting and refining chambers of a glass-melting tank which is located at the same level as the bottom of the tank.

throat, submarine. The passage between the melting and refining chambers of a glass-melting tank placed at a level below the bottom of the tank.

throat, submerged. See throat, submarine.

throat, sump. See throat, submarine.

throwing. The throwing of a prepared pottery body on a revolving potter's wheel and shaping by hand.

throwing marks. Grooves and ridges on the surface of a shape formed by hand-throwing.

thulium boride. (1) TmB_4; mol. wt. 212.68; sp.gr. 7.09. (2) TmB_6; mol. wt. 234.32; sp.gr. 5.59. See borides.

thulium carbide. (1) Tm_3C; mol. wt. 520.2; sp.gr. 9.90. (2) Tm_2C_3; mol. wt. 374.8. (3) TmC_2; mol. wt. 193.4; sp.gr. 8.17. See carbides.

thulium nitride. TmN; mol. wt. 183.41; sp.gr. 10.84. See nitrides.

thulium oxide. Tm_2O_3; mol. wt. 386.8; sp.gr. 8.7; used as a radiation source in x-ray equipment after irradiation in a nuclear reactor.

thulium silicide. $TmSi_2$; mol. wt. 225.52. See silicides.

thulium sulfide. Tm_5S_7; mol. wt. 1071.42. See sulfides.

thwacking. The final shaping of clay roofing tile by pounding it on a wooden form of a desired shape with a wooden paddle.

tied concrete column. A column of concrete reinforced by internal longitudinal bars bound by horizontal ties for stability.

tiering. The pointing of roofing tile with mortar or cement.

tiger eye. A decorative, glass-like formation in an aventurine glaze. See aventurine.

tiger skin. A type of salt glaze characterized by crawling and beading of the glaze to produce the appearance of tiger or leopard skin.

tile. (1) A relatively thin piece of fired clay, concrete, stone, or other material used in functional and ornamental applications on walls, floors, roofs, etc.
(2) A hollow or concave earthenware or concrete product used for drainage and other purposes.

tile, antistatic. Floor tile containing carbon or other electrically conducting substances to dissipate electrostatic charges and to minimize sparking; used in areas where combustible and explosive atmospheres may be present.

tile, combed finish. Tile with a scored back to increase its bond with mortar or plaster.

tile, cove. Tile with a flanged skirting for use in junctions with floors and other corners.

tile, double-shell. Tile with double faces separated by short webs. See web.

tile, drain. Unglazed earthenware or concrete pipe used for drainage purposes.

tile, encaustic. Tile in which a pattern is inlaid with clays of a color other than that of the body.

tile, end construction. Tile designed to receive compressive stress parallel to the axes of the cells.

tile, exposed finish. Tile or hollow clay building block whose surfaces are to be left exposed or painted after installation.

tile, extra-duty glazed. Floor tile with a glaze more abrasion-resistant than conventional glazes.

tile, facing. Tile whose faces are to be exposed in interior or exterior masonry construction.

tile, faïence. A glazed or unglazed, plastic-formed, earthenware tile having a hand-crafted, decorative appearance.

tile, finish. A glazed tile employed in wall construction with the glazed surface exposed.

tile, fire-proofing. Tile used to protect structural members from fire.

tile, floor. Tile used in floor and roof construction.

tile, furring. Nonload-bearing tile used as an unexposed lining in the inside of exterior walls of buildings and other construction; sometimes grooved to receive plaster coatings.

tile, gable. Roofing tile of the same length but half the width of standard tile; used to complete alternate courses of the gable of a tile roof.

tile, garden. Molded tile used as stepping stones in gardens or patios.

tile, glazed. Tile having a fused, impervious facial finish.

tile, glazed interior. A glazed nonvitreous tile suitable for interior use, and which is not required to be resistant to severe conditions such as impact, abrasion, weathering, and the like.

tile, header. Structural tile designed to provide recesses for header units in masonry-faced walls.

tile, hip. A V-shaped roofing tile used as the juncture at the peak of two roof faces.

tile, horizontal-cell. A hollow, burned-clay, masonry unit designed so that the cells are positioned in a horizontal plane when incorporated in a wall.

tile, interlocking. A roofing tile with edges designed to interlock with an adjacent tile in roof and other installations.

tile, load-bearing. Tile designed to support superimposed loads in masonry construction.

tile, mosaic. Glazed or unglazed porcelain or natural clay tile shaped by plastic forming or dust pressing to facial dimensions less than 6-inches square and thicknesses of ¼ to ⅜ inches, frequently mounted on a backing, usually paper, to facilitate setting.

tile, natural-clay. A tile of distinctive, slightly textured appearance formed from dense clay bodies by the plastic or dust-pressing methods.

tile, natural-finish. Glazed or unglazed tile of the natural color of the fired clay body.

tile, nonload bearing. Tile which will support no superimposed loads in masonry construction.

tile, nonlustrous. Tile coated with a matte or nonglossy glaze.

tile, ornamental. A tile designed and decorated primarily for decorative purposes which may or may not be used in functional applications.

tile, pan. An S-shaped roofing tile laid so that the down curve of one tile overlaps the up curve of an adjacent tile.

tile, partition. A structural tile used in the construction of nonloadbearing interior walls and partitions.

tile, porcelain. An impervious, dense, fine-grained, smooth mosaic tile with a sharply formed face, usually made by dust pressing; colors may be granular or of the clear, luminous type.

tile, quarry. An extruded, unglazed floor tile of natural clay or shale, usually 39 cm^2 (6 $in.^2$) and 13 to 19 mm. (½ to ¾ in.) thick; highly resistant to abrasion and many liquids, and is used as flooring.

tile, roofing. Any of several designs of large, natural clay tile with overlapping or interlocking edges.

tile, rough-finish. Tile having a roughened back surface to obtain increased bond with mortar, plaster, or other substances.

tile, salt-glazed. Facing tile with a lustrous glaze formed by the reaction of salt vapors with the silicates of the tile body, the salt being introduced into the kiln near the end of the firing operation.

tile, scored-finish. Tile which has been grooved, scratched, or notched to improve its bond with mortar, plaster, or stucco.

tile, sewer. An impervious, hollow, clay tile of circular cross section used in sewage-movement systems.

tile, ship-and-galley. A quarry tile with an indented facial pattern to prevent slipping of persons walking on the tile, particularly when the tile is wet. See tile, quarry.

tile, side-construction. A structural clay building unit designed to withstand stress at right angles to the axes of the cells.

tile, shoulder-angle. Small wall-tile shapes used to finish the top and bottom of corner installations.

tile, smooth-finish. A smooth-surfaced unglazed tile with the face left as removed from the die.

tile, solar screen. A type of construction in which tile is placed so as to block or diminish the intensity of the rays of the sun.

tile, special purpose. Glazed or unglazed tile of a design, composition, or appearance to meet a specific functional or ornamental requirement.

tile, structural clay. Hollow, burned-clay masonry units with parallel cells or cores, or both.

tile, structural-clay facing. Structural-clay tile designed for use on interior and exterior unplastered walls, partitions, or columns.

tile, under-ridge. A type of roofing tile used under ridge tile at the juncture of the sloping segments.

tile, unglazed. A hard, dense tile of a homogeneous

whiteware body used on floors and walls; the color, texture, and properties are determined by the composition, production method, and firing temperature.

tile, valley. A V-shaped or appropriately curved roofing tile used in the valley or junction at the bottom of two sloping roof segments.

tile, wall. (1) A thin, flat glazed tile used primarily as the exposed surface in interior wall construction.
(2) A hollow concrete or fired-clay block used in the construction of walls.

tile, wind-ridge. A type of ceramic or concrete roofing tile designed for use at the ridge of a pitched roof.

tilt up. A method of building construction in which wall panels are precast in a horizontal position, usually on the building floor, and then tilted into vertical position when the concrete has hardened.

time of set. The time required for freshly mixed concrete to attain initial set or a specified degree of hardness.

tin ash. A mixture of tin oxide and lead oxide used as an opacifier in glazes.

tin enamel. A white porcelain enamel or glaze opacified by tin oxide added to the slip at the mill.

tin-glazed ware. Pottery coated with a tin-enamel type of coating such as that on majolica or Delft ware. See majolica, Delft.

tin luster. An iridescent luster produced by the reduction of tin oxide in a glaze.

tin oxide. SnO_2; mol. wt. 150.7; m.p. 1150°C (2102°F); sp.gr. 6.6–6.9; hardness (Mohs) 6–7; used as an opacifier in porcelain enamels, glazes, and glass, and as a constituent of pink, maroon, purple, yellow, ruby, and gold colors for glass and glazes.

tinsel. Thin platelets of glass used to produce a glittering effect in glazes and glass.

tin sulfide. SnS; mol. wt. 150.76. See sulfides.

tin-titanium. $SnTi_3$; mol. wt. 263.00; m.p. 1666°C (3030°F). See intermetallic compound.

tin-vanadium yellow. A ceramic colorant composed of 80 to 90% of tin oxide and 10 to 20% of vanadium oxide.

tin-zirconium. (1) $SnZr_4$; mol. wt. 483.58; m.p. 1587°C (2890°F). (2) Sn_2Zr_3; mol. wt. 511.06; m.p. 1932°C (3510°F). See intermetallic compound.

tired clay. Clay that has lost its strength by being overworked.

tit. An imperfection consisting of a protrusion on a glass article.

titanate ceramics. Electroceramic compositions consisting of titanium in combination with barium, boron, beryllium, niobium, tin, zirconium, etc., because of their high-dielectric constant; used in capacitors, transducers, etc.

titanates. Salts of titanic acid, H_2TiO_3.

titania. See titanium dioxide, TiO_2.

titania porcelain. A vitreous, white, technical porcelain in which titanium dioxide is the essential crystalline phase.

titania whiteware. Ceramic whiteware in which titanium dioxide is the essential crystalline phase.

titanite. A calcium silicon titanate, $CaO{\cdot}SiO_2{\cdot}TiO_2$; used to produce a crystalline effect or appearance in glazes; sp.gr. 3.4–3.55; hardness (Mohs) 5–5.5.

titanium aluminide. TiAl; mol. wt. 75.07; m.p. 1640°C (2660°F); sp.gr. 4.00. See aluminides.

titanium beryllide. (1) TiBe; mol. wt. 57.30; sp.gr. 4.17. (2) $TiBe_2$; mol. wt. 66.50; m.p. ~1427°C (2600°F); sp.gr. 3.23. (3) $TiBe_{12}$; mol. wt. 158.5; m.p. <1538°C (2800°F); sp.gr. 2.29–2.30. See beryllides.

titanium boride. TiB; mol. wt. 58.92; m.p. 2060°C (3740°F); sp.gr. 5.26; hardness (Mohs) 9+; used as a refractory, high-temperature electrical conductor, and cermet. (2) TiB; mol. wt. 107.02; m.p. 2204°C (4000°F). (3) Ti_2B_5; mol. wt. 150.30; m.p. 2093°C (3800°F). (4) TiB_2; see titanium diboride.

titanium carbide. TiC; mol. wt. 60.1; m.p. 3140°C (5684°F); sp.gr. 4.93; hardness (Mohs) 9+; used in wear-resistant cutting tools, bearings, cermets, arc-melting electrodes, refractories, and high-temperature conductors; characterized by high thermal-shock resistance.

titanium carbide composite. TiC + metal, mol. wt. variable; m.p. approx. 1454°C (2650°F); sp.gr. 5.5–5.68.

titanium carbide, nickel bonded. A high-temperature composition of high strength used in turbine blades, high-temperature structures, tool bits, high-temperature bearings and seals, and other wear-resistant applications.

titanium diboride. TiB_2; mol. wt. 69.5; m.p. 2930°C (5306°F); sp.gr. 4.52; used in refractory, wear-resistant products, bearings and bearing liners, cutting tools, jet nozzles and venturi, crucibles, arc and electrolytic electrodes, resistance elements, high-temperature electrical conductors, contact points, hard-faced welding-rod coatings, metallurgical addition agents, and similar high-temperature applications.

titanium dioxide. TiO_2; mol. wt. 80.0; m.p. 1560°C (2840°F); sp.gr. 3.8; available as rutile, anatase, and brookite; used as an opacifier in porcelain enamels, glazes, and glass, as a component in various dielectrics, and as a constituent in welding-rod coatings.

titanium fluoride. TiF; mol. wt. 123.9; m.p. 284°C (543°F); sp.gr. 2.8; used as a flux in the production of rubies and sapphire abrasives.

titanium hydride. TiH_2; mol. wt. 49.9; used as a solder in bonding glass with metals.

titanium niobate. $TiO_2 \cdot Nb_2O_5$; mol. wt. 346.3; m.p. 1483°C (2700°F).

titanium nitride. TiN; mol. wt. 62.11; m.p. 2930°C (5306°F); sp.gr. 5.29; used in refractories, cermets, and semiconductors.

titanium oxide. (1) TiO; mol. wt. 64.10; m.p. 1749°C (3180°F); sp.gr. 4.93. (2) Ti_2O_3; mol. wt. 144.2; m.p. 2077°C (3770°F); sp.gr. 4.6. (3) Ti_3O_5; mol. wt. 224.3; sp.gr. 4.24. (4) TiO_2; see titanium dioxide. (5) TiO_3; see titanium trioxide.

titanium phosphide. (1) TiP; mol. wt. 79.13; m.p. 1579°C (2875°F); sp.gr. 4.27. (2) Ti_3P; mol. wt. 175.33; sp.gr. 4.64. See phosphides.

titanium silicide. (1) Ti_5Si_3; mol. wt. 324.68; m.p. 2120°C (3848°F); sp.gr. 4.2; used in high-temperature applications where thermal shock is not a factor. (2) TiSi; mol. wt. 76.16; m.p. 1760°C (3200°F); sp.gr. 4.34. (3) $TiSi_2$; mol. wt. 104.22; m.p. ~1499°C (2730°F); sp.gr. 4.15. See silicides.

titanium sulfide. (1) TiS; mol. wt. 80.16; m.p.~2049°C (3720°F); sp.gr. 4.46. (2) Ti_2S_3; mol. wt. 192.38; sp.gr. 3.71. (3) TiS_2; mol. wt. 112.22; sp.gr. 3.28. (4) TiS_3; mol. wt. 144.28; unstable above 593°C (1100°F); sp.gr. 3.25. See sulfides.

titanium tetrachloride. $TiCl_4$; used to produce iridescence in glass.

titanium trioxide. TiO_3; mol. wt. 95.9; used in ivory-colored ceramics, dental porcelain, and dental cements.

titanium-zinc. Ti_3Zn; mol. wt. 209.68; m.p. 1666°C (3030°F); sp.gr. 6.0. See intermetallic compound.

toggle mechanism. A knee-shaped joint consisting of two bars fastened together at one end; when pressure is placed on the joint to straighten it, opposite pressures are transmitted to the open ends.

toggle press. A mechanical press in which the slide is actuated by a toggle mechanism.

tolerance. Permissible variations in specified dimensions or other values.

tolerance interval. An interval computed to include a stated percentage of the number of items from a sample which is compared with a stated probability in a statistical analysis.

tolerance limit. The statistics defining a tolerance interval. See tolerance interval.

tolerance, water. The amount of water a substance can absorb without impairment of its effectiveness or usefulness.

tongue. The male end of a pipe which is overlapped by the end of an adjoining pipe.

tongue tile. The projecting partition between the streams of gas and the port of a glass melting tank.

tool, ceramic. A cutting and machining tool of strong, temperature resistant, ceramic composition used in shaping, forming, and finishing operations.

tool, machine. A hard, fracture resistant attachment to a machining apparatus used to cut, drill, shape, grind, or polish a solid product.

tool tips. Ceramic tips bonded to cutting and machining tools.

tooth. A coarse-grain structure, causing roughness in clay.

toothing. A projection of bricks in a building wall to permit future extensions of the wall.

topaz. $Al_2(FOH)_3SiO_4$; sp.gr. 3.4–3.6; hardness (Mohs) 8; used as a substitute for, or in conjunction with, kyanite in the production of mullite-type, high-alumina refractories.

top-fired kiln. A kiln in which the fuel is introduced into the firing zone through apertures in the kiln roof.

top-hat kiln. A kiln in which the firing zone, placed immediately above the ware on a refractory base, is lowered to surround the ware to be fired. Also known as an envelope kiln.

top lap. The shortest distance between the lower edge of an overlapping roofing shingle and the upper edge of the lapped unit below.

top, manhole. The concrete slab or conical top employed to reduce the diameter of the manhole riser to that of the desired manhole access hole.

topping. (1) A thin layer of high-quality, high-strength concrete applied as a finish to a concrete slab.

(2) A dry mixture of cement and fine aggregate scattered over a concrete slab before final finishing to produce a wear-resistant surface.

top pouring. The direct transfer of molten steel from a ladle into ingot molds, usually by means of refractory nozzles.

torr. A unit of pressure equal to 1/760 of an atmosphere. See atmosphere (3).

torsional modulus. The ratio of torsional rigidity of a bar to its length.

torsional viscosimeter. An instrument designed to estimate the viscosity and thixotropy of a slurry which consists of an outer cylinder containing the slurry to be tested and an inner cylinder supported in the slurry by a wire twisted one complete turn; when released, the overswing of the inner cylinder is taken as an indication of the viscosity of the slurry; the thixotropy of the slip is a comparison of the degree of overswing within a specified time interval. See thixotropy, viscosity.

tortoise-shell finish. A type of decorative finish for pottery and other earthenware resembling the shell of a tortoise, produced by sprinkling and firing colored metal oxides over a dampish unfired glaze surface.

total air. The total quantity of air supplied in the combustion of a fuel.

total equivalent boron content. The sum of the individual equivalent boron values in a neutron cross section. See boron equivalent.

total porosity. The ratio of total void spaces in a body to its bulk volume.

total pressure. The gross load applied to a surface.

total solids. The sum of the suspended and dissolved solids in a slip.

touf (pisée). An adobe-type wall construction of rammed straw-tempered sun-dried clay.

tough alumina. See alumina, tough.

toughened glass. A glass highly resistant to mechanical and thermal shock produced by rapid and rigid control of its cooling rate from near its softening point to room temperature to produce residual internal stresses which remain after the glass has cooled; used in windows, doors, and other installations where breakage may be dangerous.

towing. The smoothing of the outer edges of dried ware with sandpaper, scrapers, cloth wheels, or similar items.

tows. Short, broken, matted fibers of any material.

toxic material. A material which is harmful to the human body.

trace. An extremely small, but detectable, quantity of a constituent or impurity in a substance.

trace element. An extremely small quantity of an element, frequently nonessential, in a substance.

tragacanth. A mucilaginous gum from Asian shrubs used as a binder in glazes and porcelain enamels. Also known as gum tragacanth.

trailing. A method of decorating leather-hard pottery in which a pattern of thick, adherent slip is squeezed through a small orifice onto the pottery surface.

tramp iron. Unwanted metal, such as a nail, bolt, filing, or screw which finds its way into bulk material or a batch.

transducer. A device which converts electrical energy in one system to a mechanical energy in another system.

transfer car. A car equipped with a set of rails upon which loaded cars from a drier or kiln may be moved for transfer from one set of tracks to another set.

transfer decoration. See transfer, slide-off.

transfer glass. Optical glass cooled to room temperature in the pot in which it was melted.

transfer, heat. The movement of heat within a body or from one body to another.

transfer ladle. A refractory-lined ladle used to transfer molten pig iron from the blast furnace to the next processing operation.

transfer printing. A type of decoration in which patterns embossed on paper in color from engravings or lithographs are transferred to bisque, glazed, porcelain-enameled, or other ware.

transfer ring. A raised ring around the outside circumference of a parison used as a gripping aid in the transfer of the parison to the blow mold.

transfer, slide off. A printed design on a specially prepared paper which, when wetted, is transferred from the paper to the surface of the glass, glaze, or porcelain-enamel surface being decorated, and then is permanently fired in place.

transfer track. A set of rotating rail tracks by which kiln or furnace cars may be transferred from one set of tracks to another set.

transfer zone, mass. The region in which the concentration of adsorbate in a fluid decreases from influent concentration to the lowest detectable concentration.

transformation temperature. The temperature at which a change occurs in a phase of a material during heating or cooling.

transformer. An electrical component which transfers electric energy from one or more alternating-current circuits to one or more other circuits by magnetic induction.

transistor. An electronic device consisting of a small block of a semiconducting material to which three or more electrical contacts are made, usually with two rectifying contacts being spaced in close proximity, and one nonrectifying contact, for use as an amplifier, detector, switch, or similar application.

transit-mixed concrete. Concrete mixed in a truck mixer en route from the proportioning plant to the job site.

translucency. The property of a material to admit and diffuse light so that objects beyond cannot be clearly distinguished.

translucent glass. A glass transmitting light with varying degrees of diffusion and which impedes or obscures vision to the degree that objects seen through it cannot be seen distinctly.

transmutation. The transformation of a nuclide into a nuclide of a different element.

transmutation color. A color, such as in a porcelain enamel, glaze, or glass, which may be changed by the intentional or accidental introduction of another colorant or impurity into the batch, or by melting a composition in a crucible in which a composition of a different color previously had been melted.

transmutation glaze. A flambé or flow glaze containing copper to produce a variegated appearance. Also known as rouge flambé.

transparent coating. A clear colorless or tinted porcelain enamel, glaze, or other coating through which the base material or intermediate coating may be seen.

transuranium elements. Radioactive elements having atomic numbers greater than that of uranium, 92.

transverse-arch kiln. A chamber kiln in which the arch of the roof is set at right-angles to the length of the kiln.

transverse strength. The maximum bending stress per unit of area which a specimen can withstand without breaking. See modulus of rupture.

trap. A device to prevent the passage of selected substances, such as dust, sulfur, water, gas, etc., while permitting the passage of other substances.

trass. Also known as tarras. A light-colored, powdered volcanic ash resembling pozzolana in composition, and which is used in hydraulic cements. See pozzolan.

traverse. To travel or move across, over, or through.

traverse table. A reciprocating platform on a grinding machine supporting the ware being ground.

traverse, wheel. The prescribed rate at which a grinding wheel moves across a work piece.

treading clay. A clay kneaded by the human heel; a process used in primitive-type potteries.

treadle bar. The foot pedal operating a potter's wheel.

treatment. A material incorporated in a grinding wheel during manufacture to improve its grinding action and to minimize its tendency to fill with grinding residues.

tremie. A large metal tube with a hopper at the top and a valve arrangement at the bottom used in the placement of concrete under water.

tremie seal. A foundation seal placed under water by a tremie, usually in an area enclosed by sheet piling. See tremie.

tremolite. $2CaO \cdot 5MgO \cdot 8SiO_2 \cdot H_2O$; sp.gr. 3.0–3.3; hardness (Mohs) 5–6; a fibrous talc used as a substitute for asbestos in acid-resisting applications.

trial mix. A preliminary batch of concrete mixed to determine the optimum proportions of ingredients to produce concrete having specified properties.

trials. Small samples which are withdrawn from a kiln during firing for use as a guide to temperature and atmospheric conditions therein.

triangle bar. Alloy bars of triangular cross section upon which porcelain-enameled ware is placed for firing.

trickle. To drip slowly or to flow in a gentle intermittent stream.

tridymite. A high-temperature polymorph of SiO_2; mol. wt. 60.1; sp.gr. 2.28–2.3; hardness (Mohs) 7; used in ceramic bodies to improve thermal shock resistance and to minimize crazing.

Trief process. A process for making concrete in which a slurry of wet-ground slag is mixed with portland cement and aggregate.

trim. To remove edges and excess material and shape leather-hard bodies by means of a wheel.

trimmed block. Dressed or crude mica which has been split into prescribed thicknesses, and has been side trimmed to remove irregularities, imperfections, and contaminants.

trimmer. (1) Tile of various sizes and shapes, such as bases, caps, corners, moldings, etc., employed to complete tile installations.

(2) A workman employed to remove fins, edges, and other irregularities from ware.

trimming. The process of removing edges, fins, mold marks, and other irregularities from ware.

trimorphous. The property of crystallizing in three different forms.

triple-cavity mold. A mold containing three cavities for the simultaneous forming of three glass articles.

triple-cavity process. A process in which three gobs of glass are accepted and formed in a mold simultaneously.

triple-gob process. The triple-cavity process of glass forming.

tripoli. Rotten stone; a porous, siliceous, sedimentary rock used as an abrasive and polishing powder.

trisodium phosphate. Na_3PO_4; a cleaning compound and water softener.

tristimulus colorimeter. An instrument that measures color by determining the intensities of three different colors, white, blue, and yellow, although other colors may be used.

triuranium octoxide. U_3O_8; mol. wt. 842.42; decomposes at 1450°C (2642°F); sp. gr. 8.39; natural uranium oxide used in nuclear applications.

trivet. A stainless steel or heat-resistant alloy formed into a shape suitable to support porcelain-enameled ware during firing.

trommel. A tilted, revolving, cylindrical screen used to separate coarsely crushed materials.

Tropenas converter. A refractory-lined converter in which the air blast strikes the molten batch through tuyeres.

trough. A channel through which materials flow from one point to another.

trowel. A flat, rectangular, or triangular shaped hand tool used to apply, spread, and shape concrete, mortar, and plaster.

truck chamber kiln. A chamber-type kiln through which ware is pushed on refractory platforms or bats.

truck-mixed concrete. Concrete mixed in a truck-mounted mixer.

truck mixer. A rotating mixer mounted on a motor truck in which concrete is mixed en route from the proportioning plant to the job site.

true, truing. Trimming the outside surface of a grinding wheel so that its rotation will be concentric with the axis.

true density. The weight of a unit volume of a substance excluding its pore volume and interparticle voids when measured under standard or specified conditions.

true porosity. The ratio of the total volume of open and closed pores and interparticle voids to the bulk volume of a material.

true specific gravity. The ratio of the density of a material to the density of water at 4°C (60°F) that has a volume equal to the true solid volume at standard conditions of pressure and temperature.

true volume. The volume of a solid material, neglecting pore volume.

T-stake. A T-shaped steel form used to shape metal bowls and the like to be processed into porcelain-enameled artware.

tube furnace. A furnace in which the fuel combustion takes place in a design of heat-resisting alloy tubes to prevent combustion gases from contact with ware being fired.

tube mill. A revolving cylinder containing grinding media in which the material to be ground is introduced at one end and removed, after grinding, at the opposite end.

tube, revolving. A hollow cylinder, concentric with the needle of a feeder, revolving in molten glass, the feeder delivering gobs of glass to a forming unit.

tubing, lens-fronted. Graduated glass tubing designed for the containment of liquids for use in temperature, pressure, and similar instruments of measurement, but modified so as to magnify the liquid column for easy reading.

tub, tempering. A vertical pan-type mixer in which materials are blended with water and then fed directly into a vertical pug mill, the same shaft serving the mixer and pug mill.

tuckstone. Shaped refractory blocks placed on top of flux blocks in a glass tank to protect the flux blocks against combustion gases, and to serve as a seal between the flux blocks and the side and end walls of the glass tank.

tuck wall. A course of tuckstones or a wall in a glass tank.

tuff. A volcanic ash.

tuille. A vertically operated damper or counterweighted door in a glass tank to control the flow of molten glass or to protect a newly set pot.

tumbler. A drinking glass without a foot or stem.

tumbling. A surface-finishing operation in which small articles are loosely rotated in a barrel with abrasives or polishing compounds to remove burrs, protrusions, and surface imperfections, and to produce polished surfaces.

tungstates. Compound mixtures of tungsten oxide and the oxides of other metals in proportions equivalent to the general formula $WO_3{\cdot}Me_2O_3$, and having melting points usually in the range of 1483°–1705°C (2700°–3100°F).

tungsten aluminide. WAl_2; mol. wt. 237.94; m.p. 1649°C (3000°F). See aluminides.

tungsten beryllide. (1) WBe_2; mol. wt. 202.40; sp.gr. ~10.2. (2) WBe_{12}; mol. wt. 294.4; sp.gr. ~4.4. (3) WBe_{13}; mol. wt. 303.6. (4) WBe_{22}; mol . wt. 386.4; sp.gr. 3.23. See beryllides.

tungsten boride. (1) WB; mol. wt. 194.8; m.p. 2860°C (5180°F); sp.gr. 16.0. (2) WB_2; mol. wt. 205.6; m.p. 2900°C (5252°F). (3) W_2B; mol. wt. 378.8; dissociates at 1900°C (3452°F); sp.gr. 16.7. (4) W_2B_5; mol. wt. 422.0; m.p. 2200°C (3992°F); sp.gr. 17.2. Used as a refractory for furnaces and chemical equipment.

tungsten carbide. (1) W_2C; mol. wt. 380.0; m.p. 2855°C (5171°F); sp.gr. 17.2. (2) WC; mol. wt. 196.0; m.p. 2780°C (5036°F); sp.gr. 15.5; hardness 2100 (K100). Used in tools, dies, cermets, and wear-resistant parts, and as an abrasive.

tungsten carbide, cemented. See tungsten carbide composite.

tungsten carbide composite. WC·Co; mol. wt. 254.94; m.p. 1454°C (2650°F); sp.gr. 11.0–15.2; hardness 80–92 Rockwell A; used in machine tools and abrasives. Also known as cemented carbide.

tungsten nitride. (1) WN; mol. wt. 198.01; m.p. ~593°C (1100°F); sp.gr. 15.94. (2) WN_2; mol. wt. 212.02. (3) W_2N; mol. wt. 382.01; m.p. 798°–871°C (1470°–1600°F); sp.gr. >16.0. See nitrides.

tungsten oxide. WO_3; mol. wt. 232.0; m.p. 1473°C (2683°F); sp.gr. 7.2; used as a yellow colorant in ceramics.

tungsten phosphide. (1) WP; mol. wt. 215.03; decomposes at 1449°C (2640°F); sp.gr. 12.40. (2) WP_2; mol. wt. 246.06; sp.gr. 9.17. See phosphides.

tungsten silicide. (1) WSi_2; mol. wt. 240.1; m.p. 2050°C (3722°F); sp.gr. 9.3. (2) W_3Si_2; mol. wt. 608.12; decomposes at 2321°C (4210°F); sp.gr. 12.21.

tungsten sulfide. WS_2; mol. wt. 248.12; m.p. 1126°C (2060°F). See sulfides.

tungsten-zirconium. W_2Zr; mol. wt. 459.22; m.p. 2149°C (3900°F). See intermetallic compounds.

tunnel dryer. A tunnel-shaped, continuous dryer through which loaded cars are moved.

tunnel kiln. A tunnel-shaped, continuous kiln or furnace consisting of preheating, firing, and cooling zones through which ware is transported on cars.

tunnel updraft kiln. A tunnel kiln in which air and combustion gases are caused to move upward through a kiln setting to the exhaust flues.

turbidometer. A device which measures the loss in intensity of a light beam passing through a slip or suspension of particles as a means of determining the concentration of solids in the suspension. See Klein turbidometer, Wagner turbidometer.

turbine. A type of engine in which a central drive shaft equipped with curved vanes is spun at high speed by water, steam, or gas pressure to convert kinetic energy to mechanical power.

turbine blade. A bucket, paddle, or blade composed of a strong, high-temperature and thermal-shock resistant ceramic, cermet, or alloy used as a vane on the drive shaft of a turbine.

turbine, gas. An air-breathing internal combustion engine consisting essentially of an air compressor, combustion chamber, and turbine wheel; used more for propulsion than power generation.

turning. The shaping of an article on a lathe or potter's wheel.

turning, rough. The rapid and efficient removal of excess stock from a workpiece by a grinding or milling machine.

tuscan red. An iron oxide pigment used in glazes.

tuyere. An opening through the walls of a blast furnace or forge through which air is forced to facilitate combustion.

tuyere brick. A refractory shape containing one or more holes or passages through which air is introduced into a furnace.

tweel. A refractory door or damper in a glass tank to control the flow of molten glass or to protect a newly set pot.

tweel block, tuile block. A refractory block used in the production of a counterweighted door of a glass furnace. See tweel.

twin-plate polishing. A process in which both faces of sheet glass are ground and polished simultaneously.

two-way slab. A slab of concrete supported by beams along the four edges or sides with steel-bar reinforcement parallel to the two faces of the slab.

U

ulexite. $NaCaB_5O_9 \cdot 8H_2O$; sp.gr. 1.96; hardness (Mohs) 1–2.5; a mineral used in glazes as a flux and glass former.

ultimate analysis. The chemical composition of a material calculated on the basis of its constituent oxides.

ultimate strength. (1) The maximum strength of a material or product, usually reported in terms of weight per unit of length at the instant of failure.

(2) The maximum load supported by a concrete pipe as determined by three-edge testing procedures and reported as newtons per linear meter per millimeter of inside diameter or horizontal span or as pounds per square inch of inside diameter or horizontal span.

ultrahigh purity. A grade of a reagent material of extreme purity; that is, containing only ultratrace levels of impurities designated in the region below one microgram per gram.

ultrasonic. A term used to designate frequencies above the audiofrequency range.

ultrasonic cleaning. The use of ultrasonic vibrations as an auxiliary force in the cleaning bath to remove soil from sheet metals being prepared for porcelain enameling.

ultrasonic devices. Instruments employed to generate ultrasonic conditions or to evaluate materials or products under such conditions.

ultrasonics. The science dealing with phenomena occurring in ultrasonic ranges; that is, in frequencies above the range of human hearing.

ultratrace. A term employed to designate an extremely small chemical component, usually an impurity, in a material in the region below one microgram per gram.

ultraviolet-absorbing glass. Glass containing appropriate quantities of elements such as cerium, chromium, cobalt, copper, iron, lead, manganese, neodymium, nickel, sulfur, titanium, uranium, or vanadium which absorb ultraviolet rays without appreciable effect on the transmission of visible light rays.

ultraviolet transmitting glass. Glass containing no appreciable quantities of ultraviolet-absorbing elements and which will transmit ultraviolet rays in high concentrations without impedance. See ultraviolet-absorbing glass.

umber. (1) An earth consisting chiefly of a hydrated iron oxide, and sometimes an oxide of manganese, employed as a brown or reddish-brown colorant in bodies and glazes.

(2) A characteristic brown or reddish-brown color.

umpire. A person or laboratory of recognized capability selected to resolve or arbitrate a difference such as, for instance, a shipper–receiver difference.

unbalanced capacitance. The difference in capacitance of the two insulated conductors to the shield, expressed as the percentage of the capacitance between the conductors, or per cent unbalance. See capacitance.

unburned brick. Brick manufactured by processes which do not require kiln firing to develop the strength of the finished product; for example, adobe brick, chemically bonded refractories, etc.

unburned refractory. Refractory shapes which are installed and placed in service without prior burning.

uncombined water. Water added to a body or slip to produce plasticity, or workability, and which is removed by evaporation during drying or during the early stages of firing. Also known as mechanical water.

unctuous. Rich in organic matter and easily workable, as a clay; having an oily appearance or soapy feel.

undercar temperature. The temperature in the area or segment of a kiln beneath a kiln car as it transports ware through the kiln.

underclay. A layer of argillite or clay stone immediately underlying a bed of coal; sometimes used as a component in ovenware bodies.

undercloak. The section of a tile roof consisting of an intermediate layer of a material between the tile and the supporting laths.

undercutting. (1) To cut away a material so as to leave an overhanging portion.

(2) Faulty cutting of flat glass resulting in an edge that is oblique to the surface of the glass.

underdrain. A type of asbestos-cement pipe having a multiplicity of perforations along its length intended for use in surface or subsurface draining of fields, streets, and similar applications.

underfire, underfired. Incomplete fusion of porcelain enamels or glazes resulting in failure to form a smooth, glassy surface.

underglaze, underglaze decoration. A finely milled ceramic color, decoration, or other coating applied directly to the unfired or bisque-fired surface of ceramic ware and subsequently covered with a transparent glaze, and then fired concurrently with the glaze.

underloading. (1) Insufficient charging of a ball mill for the proper grinding of materials.

(2) Reducing the charge of a ball mill to obtain faster grinding.

(3) Insufficient loading of a furnace or kiln to obtain faster furnace comeback and, as a result, faster firing.

under-load refractoriness. A measure of the resistance of a refractory to the combined effect of heat and loading, often expressed as the temperature of shear or 10% deformation when heated under 25 or 50 pounds per square inch.

under-ridge tile. Roofing tile used under the tile forming the ridge in the construction of the top of tile roofs.

undersanded. The condition of concrete in which it appears to contain insufficient additions of fine aggregate.

undersize. In the production of concrete, the aggregate materials smaller than the specified minimum screen size.

unfired brick. Brick manufactured by using materials and procedures which do not require kiln firing to develop the strength required of the finished product. For example, adobe brick, and chemically bonded refractories. Also known as unburned brick.

unglazed tile. A hard, dense tile, employed on floors

and walls, composed of a whiteware body which is of homogeneous composition throughout, and which derives its color and texture from the materials from which the body is made. The color and characteristics of the body are determined by the raw materials used in the body, the method of manufacture, and the thermal treatment.

UNH. Uranyl nitrate (hexahydrate). $UO_2(NO_3)_2 \cdot 6H_2O$; mol. wt. 502.82; m.p. 60.2°C (140°F); b.p. 118°C (246°F); sp.gr. 2.807; used in ceramic glazes and porcelain enamels as a red, yellow, or orange colorant.

uniform flow. A flow of constant velocity or volume of gas, liquid, or a powdered or granulated solid.

unit cost. The total cost of producing a unit of a product.

unit mold. A one-piece mold in which ware is cast.

unit stress. The measure of stress or load per unit of area of a substance.

universal grinding machine. A machine on which cylindrical, internal, or face grinding may be done as required.

unsoundness. A term employed in the portland cement industry to indicate expansion after the cement has set, frequently causing cracking or crumbling of the cement or concrete.

updraft. A kiln in which the movement of combustion gases proceeds upward from the firebox through the kiln setting to the exhaust flues.

updraw. The process of continuous drawing of glass of various cross sections in a vertical plane from an orifice to produce glass rod or tubing.

uphill teeming. The process of discharging molten steel from a ladle through refractory tubes into molds in such a manner that the steel is introduced at the bottom of the mold instead of at the top.

upright. An item of kiln furniture employed to support ware during firing.

uptake. A refractory-lined passage in an open-hearth furnace to conduct hot air and gaseous fuels into the combustion area.

urania. The oxides of uranium, but usually UO_2; used to produce red, yellow, and orange pigments. See uranium oxide.

uraninite. A mineral, probably uranium dioxide in its original form, but altered by radioactive decay, usually containing more or less uranium trioxide, lead, radium, and helium; sp.gr. 7.5–9.7; hardness (Mohs) 5.5–6. See uranium dioxide.

uranium. (1) A white lustrous, radioactive, metallic element.

(2) In the ceramic context a term referring to the oxide or other compound of uranium.

uranium aluminide. UAl_2; mol. wt. 292.11; m.p. 1587°C (2890°F); sp.gr. 8.26.

uranium antimonide. (1) U_3Sb_4; mol. wt. 1201.59; m.p. 1693°C (3080°F). (2) USb; mol. wt. 359.94; m.p. 1849°C (3360°F). (3) U_4Sb_3; mol. wt. 1317.99; m.p. 1799°C (3270°F).

uranium barium oxide. Barium diuranate, BaU_2O_7; mol. wt. 725.64; an orange to yellow powder employed as a ceramic colorant, particularly for porcelains.

uranium beryllide. (1) UBe_{13}; mol. wt. 357.77; m.p. 1849°C (3630°F); sp.gr. 4.37; hardness 1400 Vickers. (2) UB_4; mol. wt. 281.45; m.p. 2482°C (4500°F); sp.gr. 9.38; hardness 2500 Vickers. (3) UB_{12}; mol. wt. 368.01; m.p. 2232°C (4050°F); sp.gr. 5.86.

uranium boride. (1) UB_2; mol. wt. 259.81; m.p. 2371°C (4300°F); sp.gr. 12.73; hardness 1400 Vickers. (2) UB_4; mol. wt. 281.45; m.p. 2482°C (4500°F); sp.gr. 9.38; hardness 2500 Vickers. (3) UB_{12}; mol. wt. 368.01; m.p. 2232°C (4050°F); sp.gr. 5.86.

uranium carbide. (1) One of two carbides, uranium monocarbide (UC) and uranium dicarbide (UC_2) which are used chiefly as nuclear fuels. See uranium monocarbide and uranium dicarbide. (2) U_2C_3; mol. wt. 512.34; m.p. 1777°C (3230°F); sp.gr. 12.88.

uranium content equivalent. A concentration of U^{236} that will provide a fast-neutron cross-section equivalent to that of a specific impurity element.

uranium, depleted. Uranium having a smaller percentage of U^{235} than the 0.7% found in natural uranium.

uranium dicarbide. UC_2; mol. wt. 262.14; m.p. about 2370°C (4298°F); sp.gr. 11.28; decomposes in water, acids, and alkalies, is very pyrophoric, and ignites in air at 400°C (750°F); used as a nuclear fuel.

uranium dioxide. UO_2; mol. wt. 270.14; m.p. 3000°C ± 200°C (5432°F ± 390°F); sp.gr. 10.9; spontaneously flammable; used to pack nuclear fuel rods, and as a coloring agent in glazes; use now confined to nuclear applications.

uranium, enriched. Uranium containing concentrations of U^{235} greater than its natural value of 0.711 weight %. Enriched uranium may range from $>$0.711 to $>$93% of U^{235}.

uranium monocarbide. UC; mol. wt. 250.14; m.p. 2550°C (4622°F); sp.gr. 13.63; used in the preparation of nuclear fuel rods.

uranium, natural. Uranium with an isotopic composition as it appears in nature, 0.711%, which has not been altered.

uranium (uranyl) nitrate. $UO_2(NO_3)_2 \cdot 6H_2O$; mol. wt.

502.8; m.p. 60.2°C (140°F); sp.gr. 2.807; used as a red-yellow and orange pigment in glass and clayware. Also known by the acronym UNH.

uranium nitride. (1) UN; mol. wt. 252.18; m.p. 2905°C (5260°F); sp.gr. 14.32. (2) UN_2; mol. wt. 266.19; unstable above 704°–787°C (1300°–1450°F); sp.gr. 11.73. (3) U_2N_3; mol. wt. 518.37; decomposes above 704°C (1300°F); sp.gr. 11.24.

uranium, normal. Uranium containing the same weight percentage of U^{235}as it occurs in nature. It may be obtained by blending uranium of different isotopic compositions or by processing in a different factory. Loosely, it means natural uranium.

uranium oxide. Uranium forms several oxides, including U_2O_3; UO_2; UO_3; U_3O_8; and $UO_4 \cdot xH_2O$; used in nuclear applications and as a green and yellow ceramic colorant.

uranium-oxygen ratio. The mole ratio of uranium to oxygen in a sample of uranium oxide.

uranium peroxide, uranium tetroxide. $UO_4 \cdot xH_2O$; decomposes at 115°C (240°F); sp.gr. 2.5 at 15°C (59°F); used as a red, orange, or yellow colorant.

uranium phosphate. $UO_2 \cdot P_2O_5$; mol. wt. 412.23; m.p. 1549°C (2820°F).

uranium phosphide. (1)UP; mol. wt. 269.2; sp.gr. 9.68. (2) U_3P_4; mol. wt. 838.63; sp.gr. 9.83.

uranium selenide. (1) USe; mol. wt. 317.37; m.p. 1849°–1999°C (3360°–3630°F); sp.gr. 11.3. (2) U_2Se_3; mol. wt. 713.94; m.p. 1571°C (2860°F).

uranium silicide. (1) USi; mol. wt. 266.23; m.p. 1598°C (2910°F); sp.gr. 10.40; hardness 745 Knoop. (2) USi_2; mol. wt. 294.29; m.p. 1142°–1649°C (2090°–3000°F); sp.gr. 8.15–9.25; hardness 700 Knoop. (3) US_3; mol. wt. 322.35; m.p. 1649°C (3000°F); sp.gr. 8.15; hardness 445 Knoop. (4) U_3Si; mol. wt. 742.57; m.p. ~ 959°C (1760°F); sp.gr. 15.58. (5) U_3Si_2; mol. wt. 770.63; m.p. 1666°C (3030°F); sp.gr. 12.20; hardness 796 Knoop.

uranium sulfide. (1) US; mol. wt. 270.23; m.p. ~ 2371°C (4300°F); sp.gr. 10.87. (2) US_2; mol. wt. 302.29; m.p. ~ 1849°C (3660°F); sp.gr. 7.52. (3) U_2S_3; mol. wt. 572.52; m.p. 1927°C (2500°F); sp.gr. 8.78.

uranium telluride. UTe; mol. wt. 365.67; m.p. 1549°–1649°C (2820°–3000°F); sp.gr. 8.8.

uranium trioxide. UO_3; mol. wt. 286.14; sp.gr. 8.34; decomposes when heated; a radioactive red or yellow powder employed as an orange coloring agent in ceramics.

uranium yellow. Sodium diuranate, $Na_2U_2O_7 \cdot 6H_2O$; mol. wt. 742.39; a yellow-orange solid used as a yellow pigment in glazes and enamels, and in the manufacture of fluorescent uranium glass.

uranous oxide. U_2O_3; mol. wt. 524.34; the least important of the uranium oxides. See uranous-uranic oxide.

uranous-uranic oxide. U_3O_8; mol. wt. 842.42; decomposes at 1450°C (2642°F); sp.gr. 8.39; naturally occurring uranium oxide used in nuclear applications.

uranyl uranate. U_3O_8; mol. wt. 842.51; decomposes above 1300°C (2372°F); sp.gr. 8.39.

urea. $CO(NH_2)_2$; mol. wt. 60.06; m.p. 132.7°C (270°F); decomposes before boiling; sp.gr. 1.335; used as a binder in porcelain enamels to minimize or eliminate tearing.

U-type furnace. A continuous furnace shaped in the form of a ''U'' or hairpin in which porcelain enamels, during firing, enter and leave the furnace at adjacent stations; the firing zone usually is located in the ''U'' portion of the furnace.

uvarovite. $3CaO \cdot Cr_2O_3 \cdot 3SiO_2$; mol. wt. 500.44; sp.gr. 3.5–4.3; hardness (Mohs) 6.5–7.5; a variety of garnet colored green by the presence of chromium; and used as a coloring agent; also used as an abrasive.

uviol glass. A glass which is highly transparent to ultraviolet radiation.

V

vacancy. A defect in a solid-state crystal consisting of an unoccupied lattice position in the crystal structure.

vacuum. (1) A space devoid of matter.

(2) A space in which air or gas is contained at a reduced pressure, usually below normal atmospheric pressure.

vacuum and blow process. The process employed in the manufacture of bottles in which glass is gathered by vacuum and subsequently blown.

vacuum bottle. A bottle or other container equipped with an evacuated liner to prevent the influx of heat or cold into the contents of the container from the surrounding environment.

vacuum casting. The forming of ceramic ware by introducing a slip into a permeable mold and hastening the removal of water from the slip by the application of a vacuum to the outer walls of the mold to produce a rigid or semirigid article.

vacuum chamber. The section of an auger extrusion machine through which plastic clay is kneaded under a vacuum to remove air and gases from the mass.

vacuum concrete. A fast-curing concrete of improved durability, high strength, increased surface hardness, and improved resistance to crushing obtained by subjecting freshly poured concrete to a vacuum to remove entrapped air and the excess water not required for setting of the concrete.

vacuum degassing. The removal of gases from materials and bodies by subjection to a vacuum at elevated temperatures.

vacuum deposition. The deposition or condensation of a vaporized coating of a material on the cold surface of another in a partial vacuum.

vacuum diffusion. The diffusion of selected impurities into a semiconducting material in a vacuum to induce desired properties.

vacuum drying. The technique of speeding the removal of moisture from a material or body by means of a vacuum applied in conjunction with a conventional drying system.

vacuumed clay. Clay which has been subjected to vacuum treatment to remove air bubbles to increase its density and improve its green strength in ceramic bodies.

vacuum firing. Firing of ware in an evacuated furnace to reduce the porosity and to prevent oxidation of the body.

vacuum furnace. A furnace or heating device constructed so as to permit the firing chamber to operate under a vacuum.

vacuum mat. A combination screen and textile filter placed over freshly poured concrete and through which by application of a vacuum, air and water are sucked out to produce a dense concrete.

vacuum mixer. A machine in which the clay or body is moistened and deaired concurrently as it enters the mixing chamber to be blended to forming consistency.

vacuum pug mixer. A pug mill consisting of a trough with a longitudinal shaft on which blades are mounted for the pugging of clay mixtures, the trough being situated in a vacuum chamber to permit concurrent deairing of the mixture.

vacuum pump. A device for exhausting air or other gases from an enclosed space.

valence. The property of an atom or group of atoms which determines the number of other atoms or groups with which it will unite chemically.

valve, air relief. A small automatic valve placed at a high point in a pipeline to exhaust air or other entrapped gases from the line.

vanadium. In the ceramic context, a term for an oxide of vanadium; V_2O_5; V_2O_4; and V_2O_3.

vanadium aluminide. V_5Al_6; mol. wt. 470.56; m.p. 1671°C (3040°F).

vanadium beryllide. (1) VBe_2; mol. wt. 69.36; m.p. >1649°C (3000°F); sp.gr. 3.80. (2) VBe_{12}; mol. wt. 161.36; sp.gr. 2.35.

vanadium boride. (1) VB; mol. wt. 61.8; m.p. 2100°C (3812°F); oxidizes between 1000°–1100°C (1830°–2010°F); sp.gr. 5.1; hardness (Mohs) 8–9; electrical resistivity 16 microhm/cm. (2) VB_2; mol. wt. 72.6; m.p. 2400°C (4350°F); sp.gr. 5.5. (3) VB_3; mol. wt. 83.4; m.p. 2300°C (4170°F); sp.gr. 5.5. (4) V_3B_2; mol. wt. 174.52; m.p. 2066°C (3750°F). (5) V_3B_4; mol. wt. 196.16; m.p. 2271°C (4130°F).

vanadium carbide. (1) VC; mol. wt. 62.95; m.p. 2830°C (5126°F); sp.gr. 5.77; used as a component in ceramic cutting tool compositions. (2) V_2C; mol. wt. 113.92; m.p. ~ 2166°C (3930°F); sp.gr. 5.75.

vanadium dioxide. VO_2; mol. wt. 82.95; m.p. 1541°–1637°C (2813°–2980°F); sp.gr. 4.65.

vanadium nitride. (1) VN; mol. wt. 64.96; m.p. 2050°C (3270°F); sp.gr. 5.63. (2) V_2N; mol. wt. 115.93; sp.gr. 5.99. See nitrides.

vanadium oxide. VO; mol. wt. 66.96; m.p. 2049°C (3720°F).

vanadium pentoxide. V_2O_5; mol. wt. 181.9; m.p. 690°C (1275°F); sp.gr. 3.36; used as a red, green, pink, or yellow colorant and flux in glasses, porcelain enamels, and glazes; inhibits ultraviolet transmission in glass; also a glass former.

vanadium phosphide. (1) VP; mol. wt. 81.99; m.p. 1315°C (2400°F); sp.gr. 5.0. (2) V_3P; mol. wt. 183.91.

vanadium silicide. (1) VSi_2; mol. wt. 107.08; m.p. ~ 1699°C (3090°F); sp.gr. 5.10. (2) V_3Si; mol. wt. 180.94; m.p. 1732°–2049°C (3150°–3720°F); sp.gr. 5.74. (3) V_5Si_3; mol. wt. 338.98; m.p. 2149°C (3900°F); sp.gr. 5.27.

vanadium sulfate. $VOSO_4 \cdot 2H_2O$; mol. wt. 199.0; used as a green and blue pigment in glass, porcelain enamels, and glazes.

vanadium sulfide. (1) VS; mol. wt. 83.02; m.p. ~ 1899°C (3450°F); sp.gr. 4.89. (2) V_2S_3; mol. wt. 198.1; m.p. 1927°C (3500°F); sp.gr. 4.70. (3) V_2S_5; mol. wt. 262.22; sp.gr. 3.00.

vanadium tetroxide. V_2O_4; mol. wt. 165.9; m.p. 1967°C (3570°F); sp.gr. 4.339; used in refractory compositions fusing at temperatures above 1540°C (2805°F); forms low-porosity body with BeO; refractories tend to be unstable in air.

vanadium trioxide. V_2O_3; mol. wt. 149.9; m.p. 1970°C (3580°F); sp.gr. 4.84; used in glazes as a flux, as a yellow pigment in combination with tin or zirconium oxides, and as a blue pigment in combination with zirconium silicate.

vanadium zirconium. V_2Zr; mol. wt. 193.14; m.p. 1499°C (2730°F).

van der Waals adsorption. The binding of an adsorbate to the surface of a solid by forces having energy levels approximating those of condensation.

van der Waals equation. A thermodynamic equation relating the pressure, volume, and absolute temperature of a gas with reference to the finite size of the molecules and the attractive force between them; calculated by the formula:

$$P = [RT/(v-b)]-(a/v^2)$$

in which P is the pressure, v is the volume per mole, T is the absolute temperature, R is the gas constant, and a and b are constants.

vane feeder. A device consisting of a rotating horizontal shaft equipped with blades to feed ground clay from the bottom of a hopper or bin to a mixer or other receptacle.

vapor. A gas which is at a temperature below its critical temperature and which can be liquified by the application of appropriate pressure without reduction in temperature.

vapor barrier. A layer of plastic or other impervious sheeting placed under concrete floors and on walls to prevent the passage of air and moisture through the concrete.

vapor deposition. See vacuum deposition.

vapor glaze. A glaze composed of lead, sodium, and boric oxides which volatilize from a melt during firing, but will condense and reliquify on ceramic surfaces on cooling.

vaporization. The conversion of a liquid or solid to its vapor state, particularly by heating.

vaporization, heat of. The heat absorbed per unit mass of a material at its boiling point which completely converts the material to a gas at the same temperature; that is, the amount of heat required to evaporate 1 mole of liquid at constant temperature and pressure.

vapor pressure. The pressure exerted by the vapor of a solid or liquid when in equilibrium with its solid or liquid form.

variance. A measure of the dispersion of a series of results around their average; it is the sum of the squares of the individual deviations from the average of the results divided by the number of results minus one.

varister. A resistor whose resistance varies automatically in proportion to the current passing through it; its resistance drops with an increase in applied voltage.

varsol. An aliphatic petroleum solvent used to clean silk screens.

varved clay. A natural clay deposited in layers or in a sequence of layers, one coarse and the other fine or silty.

v-cuts. V-shaped edge cuts with an included angle of 120 degrees or less in mica sheets which are used as electrical insulators in special applications.

v-drain. A defect evidenced by a second flowing of a porcelain-enamel slip on ware which occurs after it appears that draining has been completed; a double-draining type of phenomenon.

vebe apparatus. A vibrating slump-testing device employed to evaluate the consistency of concrete.

Vegard's law. The linear relationship between the lattice parameters and the composition of solid solutions, expressed as atomic percentage.

vegetable inclusions. A misnomer which describes inclusions of dispersed metal oxides in electrical insulation; these appear as areas of pastel colors in transmitted light.

vegetable oils. Hydrogenated oils of peanuts, soybeans, coconut, and the like which are employed in the sizings for glass-textile yarns as lubricants to improve the resistance of the fibers to abrasion.

vehicle. A fluid in which a material is dissolved or held in suspension to facilitate a subsequent operation, such as an oil or varnish in graining pastes or printing inks.

Vello process. A continuous drawing process for the production of glass tubing or cane in which the molten glass is fed downward to the draw through an annular orifice.

vellum glaze. A semimatte glaze having a satin-like appearance due to the presence of minute crystals of zinc silicate, zinc titanate, or lead titanate in the fired glaze surface.

velocity. The rate of change in the position or displacement of a body expressed as a unit of length or distance per unit of time; the rate or distance traveled by a substance per unit of time.

velvet finish. A surface finish on glass produced by two white acid treatments; a tinge of color is embossed on the surface during the process to obtain complete obscuration.

veneer. (1) A single wythe of masonry not structurally bonded but applied for facing purposes.

(2) The decorative surface of an asbestos-shingle or sheet which usually is pigmented or granular to provide color to the areas in which the shingle or sheet is installed.

veneer, adhesion-type. Thin sections of ceramic held in place by the adhesion of mortar to the unit and its backing without the use of metal or other anchors.

Venetian red. A pure red ceramic pigment composed of 15 to 40% of a high-grade ferric oxide and 60 to 80% of calcium sulfate.

vent. (1) An opening to permit the discharge or release of pressure from enclosed areas such as a pressure tank, steam boiler, etc.

(2) An opening to permit passage or escape of liquids, gases, vapors, fumes, heat, etc., from an area such as a pickling room, furnace room, the interior of a furnace, etc.

ventilating fan. An electrically or mechanically operated device to remove contaminated spent air and to introduce fresh or cooling air into a desired area.

ventilation. The process of replacing air in an area with fresh air.

ventilator. A device that exhausts and replaces stale, contaminated, or other air from an area with circulating fresh air.

verge. The edge of a sloping roof projecting over a gable, the point where roofing tile are edge bedded for improved appearance and to deflect rain water back onto the roof for drainage.

verification tests, quality. Tests performed within a system to verify and maintain a desired level of quality in a process or product.

vermiculite. A group of micaceous minerals of the general formula $(MgFe)_3(SiAlFe)_4O_{10}\cdot 4H_2O$; when heated, they exfoliate from 16 to 20 times their original size, and are used as ingredients in lightweight concrete, acoustic and fireproof plaster, asbestos tile, acoustic tile, and refractory insulators for their insulating values and light weight.

Verneuil method of crystal growth. A process in which a powder, such as corundum, is dropped through an oxyhydrogen flame so that it falls in a molten state onto a crystal seed of the same material, the mass then growing to form jewels and bearings for watches and other delicate instruments.

vertical retort. A vertical refractory chamber lined with silicon-carbide brick in which zinc is smelted.

vesicular. Having a cellular structure; a term applied to fireclays which have become bloated by overfiring.

vesicular structure. A body containing a conglomeration of small, spherical cavities, usually filled with air or a gas.

vestibule. The area at the entrance of a dryer tunnel where cars of greenware may be stored.

vibrating ball mill. A ball mill in which conventional milling is combined with a vibratory or bouncing action of the mill to obtain more efficient and rapid grinding.

vibrating feeder. An electrical or mechanical device employed to impart a vibrating or jarring action on a hopper or bin to prevent packing of its contents, and to increase and control the rate of flow of the material from the hopper or bin.

vibrating pebble mill. See vibrating ball mill.

vibrating screens. Wire-mesh screens which are vibrated by any of a variety of means to increase efficiency and minimize clogging; sometimes the screens may be heated to obviate the influence of moisture on the material being screened.

vibration. (1) Rapid periodic motion in alternately opposite directions.

(2) The act of rendering fresh concrete into a quasiliquid state by subjecting the mass, in forms, to high-frequency vibratory impulses to consolidate the concrete in the forms.

vibration limit. The setting or hardening point of concrete, as determined by a penetration needle test, beyond which the concrete no longer can be made plastic by vibration. See Vicat needle.

vibratory pressing. A process for forming refractory and other ceramic shapes in which the ground particles of the material being formed are packed closely together by rapid impact-type vibrations of the top and bottom dies.

vibrocast pipe. Concrete pipe made by placing fresh concrete in a stationary vertical mold or form and then subjecting the unit to internal or external vibratory forces.

vibroenergy mill. A ball mill designed to vibrate on both horizontal and vertical planes.

Vicat needle. An instrument to evaluate the consistency and setting time of cement by measuring the depth to which a special or standardized needle under a standard load will penetrate into the cement.

vice, vise. A clamping device consisting of two jaws which may be operated by a screw or lever and which is designed to hold a work piece in position.

Vickers hardness. A measure of the hardness of glasses, glazes, and other surfaces in which a diamond pyramid indenter with a 136° angle between opposite faces is forced into the surface of a test specimen under variable loads; the hardness is reported as the Vickers number, the surface area of the indentation in square millimeters divided into the pressure in kilograms.

viscid. Having thick, syrupy, and adhesive qualities.

viscosimeter. An instrument designed to measure the flow resistance of fluids.

viscosimeter, Brookfield. An instrument measuring the viscosity of a slurry in which the resistance of an electrically operated cylinder to rotation is determined.

viscosimeter, torsion. An instrument designed to estimate the viscosity and thixotropy of a slurry which consists of a fixed outer cylinder containing the slurry to be tested and an inner cylinder supported in the slurry by a wire twisted one complete turn; when released, the overswing of the inner cylinder is taken as an indication of the viscosity of the slurry; the thixotropy of the slip is a comparison of the degree of overswing with a specified time interval.

viscosity. The resistance of a fluid to free flow expressed in poises or dyne-seconds per cm^2, the resistance resulting from internal friction in the liquid due to molecular attraction or the combined effects of cohesion and adhesion.

visible penetrant. A liquid of low-surface tension containing a fluorescent chemical or a dye of intense color, usually red, which is employed as a visual indication of the porosity, the existence of cracks and other discontinuities, and other surface and subsurface characteristics of a surface.

vitreous, vitrified. The state of being glassy in brilliance, brittleness, and composition, the degree of vitrification being evidenced by low-water absorption, generally 0.3% or less. In floor and wall tile, low-voltage electrical porcelains, and products of similar compositions and usage, however, bodies up to 3.0% water absorption are considered to be vitreous.

vitreous china. Any glazed or unglazed vitreous ceramic product, such as dinnerware, sanitary ware, and artware which is not used for technical purposes.

vitreous clay pipe. A clay pipe fired in a kiln to induce vitrification and which then is glazed to assure watertightness for use in drainage applications.

vitreous enamel. A substantially vitreous or glassy inorganic coating bonded to a metal base by fusion at a temperature above 425°C (800°F) for protective purposes. More widely known as porcelain enamel.

vitreous luster. A glassy appearance.

vitreous sanitary ware. Vitreous products such as sinks, lavatories, toilet bowls, urinals, and similar items used for sanitary or hygenic purposes.

vitreous silica. A transparent or translucent glass consisting almost entirely of silica, and which exhibits low thermal expansion, high resistance to thermal shock, and high resistance to chemical attack. Also known as silica glass, fused silica.

vitreous slip. A ceramic material or mixture of materials which will produce a vitrified surface when applied and fired on a ceramic body.

vitrification. The progressive reduction in the porosity of a body as a result of heat treatment and fusion during which a glassy or noncrystalline material is formed.

vitrification clay. A clay which will tend to vitrify on heating to elevated temperatures, but usually without deformation until its vitrification temperature is reached.

vitrification range. The temperature interval between the temperature at which a body or substance first begins to fuse and the temperature at which the body begins to deform due to melting.

vitrified bond. The bond created by the fusion of ceramic materials.

vitrified wheel. A grinding wheel in which the abrasive ingredients are strongly bonded or held together by means of a glassy or porcelanic bond.

vitriol. A sulfate of any of various metals such as copper, iron, zinc, etc.

vitroceramic. A glass containing nucleating agents which may be formed in the conventional glass-forming manner and then devitrified by heat treatment to produce a body of crystalline rather than an amorphous structure.

void. An unfilled space enclosed within an apparently solid body.

volatiles. Materials which vaporize during firing.

volatility. The tendency of a material to vaporize at a given temperature and pressure.

volt. A unit of electric force equal to the difference in electric potential between two points on a conducting wire carrying a constant current of one ampere when the power dissipated between the two points is one watt.

voltage. The value of an electromotive force or difference in potential expressed in volts.

voltage, dielectric breakdown (breakdown voltage). The difference in potential at which failure occurs in an electrical insulating material located between two electrodes.

voltage, fritting. The voltage at which an electric breakdown occurs between two mating contacts separated by an insulating film, when the field strength is excessive.

voltage, supply. The potential available at the source of an electric current.

voltaic current. The electric current produced by chemical action as in a battery composed of a primary cell or cells.

voltameter. An instrument for measuring a current or potential.

volt-ampere. The unit of power equal to the product of one volt and one ampere, the equivalent of one watt.

voltmeter. An instrument, such as a galvanometer, calibrated in volts, for the direct measurement of differences in electric potential.

volume. The space occupied by a substance.

volume, apparent. The volume of a body, including its sealed pores. See apparent volume.

volume, bulk. The volume of a solid material, including the volume of its open and sealed pores. See bulk volume.

volume change. The change in the volume of hardened concrete resulting from expansion and contraction due to wetting and drying or to variations in temperature.

volume, pore. The total combined volume of open and closed pores contained per unit volume or weight of a solid.

volume, pore distribution. The relative distribution of the total pore volume among pores of different sizes or diameters incorporated in a substance.

volume, sedimentation. The volume occupied by solid particles that have settled from a liquid suspension.

volume shrinkage. The contraction of a moist body during the drying or the firing process, or both, expressed as the volume ''percent'' of the original volume. See drying shrinkage, firing shrinkage.

volumetric analysis. Quantitative analysis in which accurately titrated volumes of standard chemical solutions are employed to estimate the amount of a particular constituent present in solution.

volumetric glassware. Glassware that is marked with graduations for volumetric measuring.

volume, true. The volume of solid material, excluding the volume of open and sealed pores in the material.

v-value. The reciprocal of the light-dispersive power of a material. See nu value.

vycor. A nearly pure silica glass made from a sodium borosilicate glass in which the acid-soluble phases have been removed by an appropriate acid treatment.

wad. (1) A hand-shaped rope of stiff clay mud placed around a pottery body or plaster-of-paris mold to hold the body in place on a potter's wheel during the shaving and trimming process.

(2) A strip of low-grade fireclay separating saggers and for leveling the supports and shelves in a kiln.

wafer. A small slice of a semiconductor, such as barium titanate, on which matrices of microcircuits can be fabricated, or which can be cut into dice for the fabrication of capacitors, transistors, diodes, resistors, and other devices.

Wagner turbidometer. An apparatus for the determination of the particle size of powdered substances based on the turbidity of a suspension of the powder in a suitable medium at specified levels and settling times. See turbidometer.

wagon, pot. A cart or wagon employed to transport a pot arch to a pot furnace.

waler. A horizontal reinforcement to prevent the forms for newly poured concrete from bulging. Sometimes spelled whaler.

wall anchor. A steel strap attached to a joist and built into brickwork as a reinforcement.

wallboard. Panels of gypsum plasterboard, asbestos-cement sheet, and similar products used in the surfacing of walls and ceilings.

wall, bridge. The section of a glass-melting furnace forming a bridge or a separation between the melting and the refining zones.

wall, farren. A hollow wall, four inches in thickness; common in house construction.

wall, flash. A refractory wall in a kiln placed so as to prevent flames from impinging on the ware during firing.

wall, gable. (1) The wall of the charging end of a glass-melting furnace.

(2) A wall crowned by a gable.

wall, jamb. (1) The side wall of a furnace or kiln between the flux block and crown, but not including the ends.

(2) The refractory wall between the pillars of a

glass-melting pot furnace and in front of or surrounding the front of a pot.

wall, monkey. The section between the front and back walls of an open-hearth furnace and the port sidewalls.

wall, panel. A nonloadbearing wall.

wall pipe. The structural element of concrete, or concrete and steel, between the inside and the outside surfaces of a concrete pipe.

wall, ring. The refractory wall of the unit delivering hot air to the tuyeres of a blast furnace.

wall, shadow. A structure on the top of the bridge, or suspended from the crown, designed to limit the flow of heat from the glass-melting section to the refining section of a glass tank.

wall, shell. A refractory fireclay wall protecting the metal casing of air preheaters.

wall, sleeper. See wall, submerged.

wall, submerged. A refractory wall submerged below the level of molten glass in a glass-melting tank forming the throat between the melting and refining chambers of the tank.

wall tile. (1) A thin, flat, glazed tile used primarily as the exposed surface in interior wall construction.
(2) A hollow concrete or fired-clay block used in the construction of walls.

ware. A manufactured article or product.

ware clay. An infrequently used synonym for ball clay, which see.

warming in. The reheating of glass for further working or for the development of opacity.

warping. To turn, bend, twist, or bow out of shape.

warping joint. A joint in a pavement which permits movement of concrete slabs so as to minimize uncontrolled cracking.

wash. A thin slurry of powdered clay, talc, alumina, or other substance applied as a coating on the face of a mold before casting to prevent the cast item from sticking to the mold.

washability. The relative ease with which a porcelain enamel, glaze, or other surface can be cleaned by washing with conventional soap and water.

washbanding. The application of a thin, brush coating of color over a glaze as a decoration.

washboard. An unintended and undesirable wavy or rippled glass, glaze, or porcelain-enameled surface.

washed clay. Purified clay of low silica and grit content obtained by making a thin slurry with water and allowing the impurities to be removed by settling.

washing off. The process of removing decal papers from glaze and porcelain-enamel surfaces before firing.

wash, kiln. A slurry of refractory materials applied to kiln linings, kiln shelves, kiln furniture, etc. as a means of protection against combustion gases, volatile gases, and glaze drippings during firing.

wash, mill. Water introduced into a newly emptied mill to clean the lining and balls before the next mill charge, and sometimes to recover valuable residues remaining in the mill.

wash, mold. A suspension or emulsion used to coat the cavity of a mold to facilitate the release of ware from the mold after it has been formed.

wash water. Water carried on a truck mixer or agitator for use in washing the mixer drum after a batch of concrete has been discharged.

waste. Any material remaining, after completion of an operation or process, which is no longer useful or of value.

waste heat. Sensible heat emanating from a combustion or other heating system which may be exhausted into the atmosphere or may be put to some subsequent use, such as in a dryer or in the heating of a factory area.

waste-heat dryer. A dryer which utilizes heat retrieved from a kiln, furnace, or other source.

waste mold. A mold, such as a plaster mold, into which concrete is poured and allowed to harden and then is destroyed as the cast item is removed.

waster. A defective refractory product which is broken up or crushed for use as grog. See grog.

water absorption. A measure of the amount of water a body or substance will suck up, or assimilate, under standardized conditions, expressed as a percentage of the dry weight of the body or substance.

water, adsorbed. Water held on the surface of a substance by molecular forces.

water-cement ratio. (1) The ratio of the weight of water to the weight of dry cement in a concrete or mortar batch exclusive of the water in or absorbed by the aggregate.
(2) The number of gallons of water per sack of cement in a batch of concrete, exclusive of the water in or absorbed by the aggregate.

water, combined. Water which is contained as part of the chemical composition of a substance which cannot be removed by ordinary drying. Also known as water of constitution, water of crystallization, water of hydration.

water content. The quantity of liquid water present in a substance or mixture of substances.

water, deionized. Water that has been purified of salts and is the equivalent of distilled water.

water expansion. The increase in the dimensions of a body resulting from the absorption of or reaction with water.

water-extractable material. Substances which may be removed from a body or material by solution in and washing with water.

waterfall process. The process of applying a porcelain enamel, glaze, or other coating by pouring or flowing the coating over an object and allowing it to drain to the desired thickness, the thickness being controlled by the viscosity and flow properties of the slip as well as the angle of drain.

water finishing. The process of producing a smooth surface on ceramic greenware by washing and wiping the ware carefully with a soft damp sponge, chamois, or cloth.

water-floc test. A test in which the tendency of a hydraulic cement to resist flocculating when aged in a substantial volume of water is taken as a measure of the durability of the cement.

water, free. See water, mechanical.

water gain. The appearance of a free water film on the surface of concrete due to the migration of water from the interior to the surface during the settling of the solid particles.

water glass. $Na_2O \cdot xSiO_2$; a class of water-soluble powders used in aqueous solutions as a binder and deflocculant in ceramic bodies and glazes; sometimes used as a fluxing ingredient in porcelain enamels and glazes; also used as a concrete hardener.

water, heavy. Water enriched with deuterium; used as a moderator in nuclear reactors.

water, hydration. See water, combined.

water, hygroscopic. (1) Water introduced into ceramic bodies from the atmosphere which can be removed by simple drying.

(2) Chemically combined water which can be removed from a body only by heating it to elevated temperatures.

water, interlayer. Water naturally contained between the layers of multilayer minerals which can be removed by drying.

water, lattice. Water contained as an integral part of a mineral structure and which requires higher than ordinary drying temperatures for removal.

water lines. (1) Visible lines where water condensed on the surface of an unfired porcelain enamel and washed the coating in streaks or lines from the surface of the coated article before drying was completed.

(2) The separation of water from a freshly applied porcelain enamel during draining, resulting in a streaked appearance.

(3) Lines where the movement of water in an unfired porcelain enamel has produced a concentration of salts and color, blisters, or depressions.

water marks. (1) A shallow depression in a porcelain enamel or glaze caused by the presence of an inadvertent drop of water on the unfired coating.

(2) A discoloration in pottery glazes caused by the leaching of soluble salts from the glaze by a drop of water accidentally splashed on the unfired glaze surface.

water, mechanical. Water added to a ceramic body to produce plasticity and workability and which can be removed by drying, or during the early stages of firing. Sometimes known as free water.

water migration. The flow of water through the interstices of a body, such as the flow from the interior to the surface of a body during drying.

water of constitution. See water, combined.

water of plasticity. Water contained in a body which contributes to its working and forming properties; reported as a percentage by weight of the dry body.

water, pore. Water contained in the pores of a body which contribute to its tempering during forming and to the open-pore structure of a fired body.

water-proof concrete. A concrete in which a waterproofing admixture has been incorporated or to which an impervious coating has been applied to its surface to decrease its permeability.

waterproofer, integral. A material or mixture of materials added to concrete to reduce capillary flow of water through the concrete.

waterproofing agent. A substance incorporated as an integral component, or a coating applied to the surface of concrete, brick, or other structure to render it impervious to penetration by water.

waterproofing, membrane. The application of alternate layers of a bituminous material, frequently hot, such as an asphalt-impregnated felt, or a plastic sheet, to a concrete surface below ground level to render the concrete impervious to penetration by water.

water reducer. An admixture that reduces the amount of water required per batch of concrete without deleterious influence on the workability and slump characteristics of the concrete.

water repellent. Any hydrophobic materials, such as waxes, silicones, soaps, and the like, used to render a surface resistant to wetting by water, but not completely waterproof. See hydrophobic.

water retentivity. The property exhibited by concrete, mortars, and plaster which prevents the loss of water to adjoining masonry units of high-suction characteristics, or which prevents bleeding when in contact with impervious units.

water, shrinkage. The portion of the water of plasticity of a body which contributes to the shrinkage of the body when the water is removed during drying.

water smoking. The removal of mechanically held water in a body during the early stages of firing.

water spot. A shallow depression in a porcelain enamel or glaze caused by an accidental drop of water splashed on the surface of the coating before firing.

water, storm. The collection of run-off water during or following rainfall.

water streak. A striped or washed-out pattern occurring in an unfired porcelain enamel or glaze; this may be due to a thin and uneven coating thickness resulting from poor draining characteristics of the coating during the dipping and sometimes the spraying operations, the slip usually being too thin.

water-struck brick. Brick formed in wetted molds to minimize sticking during removal from the molds.

water, tempering. The water added to and contained in a ceramic body to control its consistency and workability during forming.

water tolerance. The amount of water a body can assimilate before its workability is impaired.

water, uncombined. See water, tempering.

water, wash. See wash water.

wauk. A plastic clay body which has been beaten and rolled into the approximate shape of the mold into which it is to be formed.

wave. An imperfection of sinuous or wavy appearance in glass or other surface, usually due to uneven thicknesses or striations.

wax. Any of a group of substances composed of hydrocarbons, fatty acids, esters, and alcohols that usually are solid at room or slightly higher temperatures; used in the vacuum impregnation of insulations and coatings for ceramic capacitors and other electronic components, as a binder to hold ceramic parts to steel plates for attachment to magnetic chucks for mechanical grinding, and as a binder for dry-pressed bodies and glaze suspensions.

wax emulsion. A colloidal suspension of wax in a solvent used as a binder, lubricant, and suspension agent in ceramic bodies and glazes.

wax resist. A coating of wax applied to the surface of ware to prevent or inhibit glazes, slips, colors, or etching agents from adhering to specific areas in a decoration process.

weakened-plane joint. A groove formed in freshly poured concrete, or sawed in hardened concrete, to form a line of weakness along which the concrete will crack during drying or use rather than to crack in a random pattern.

wear. To impair or reduce a surface under the physical conditions of use.

wear, mechanical. The deterioration of a surface as a result of mechanical action.

weathering. (1) Deterioration of a surface during exposure to atmospheric conditions such as wetting, drying, sunlight, freezing, thawing, and changes in temperature.

(2) The aging of clay by exposure to the weather to disintegrate the clay and improve its plasticity.

weatherometer. An instrument designed to simulate the conditions encountered in weathering, such as rainfall, sunlight, etc.; used as an accelerated evaluation of the resistance of materials and finishes to weathering.

web. The partitions dividing hollow structural tile into cells.

Webb effect. The increase in the volume of a pottery slip during deflocculation. See deflocculate.

wedge brick. A brick with its two main faces meeting in an acute angle.

wedged bottom. An imperfection in the bottom of a glass bottle characterized by very thick glass on one side and very thin glass on the other side.

wedge stilt. A tripod-like item of kiln furniture with cone-shaped points at the end of each of the arms; employed as a setter in the glost firing of ware.

wedging. The process of homogenizing moist clay by kneading and hand working.

weep hole. An opening in mortar joints and concrete structures positioned to permit the passage of accumulated moisture through the structure to a point where it may be evaporated or drained away.

weight, application. The weight of a porcelain-enamel coating applied per unit of metal area, usually expressed as dry weight unless specifically noted as wet weight; cover coats generally are reported in grams per square foot on one side of a one-square-foot test panel, and ground coats in ounces or grams per square foot on both sides of the test panel.

weight, atomic. The relative weight of an element based on a scale in which carbon-12 is assigned a weight value of 12.

weight, dipping. The weight of porcelain-enamel slip retained per unit of area on ware coated by dipping.

weight, equivalent. The weight of an element or compound which will combine with or replace 1.008 parts by weight of hydrogen, 8.00 parts of oxygen, or the equivalent weight of any other element or compound.

weight, gross. The total weight of a container or vehicle, including its contents.

weight, molecular. The sum of the atomic weights of all atoms contained in a molecule.

weight, net. The weight of the contents of a container or vehicle.

weight, slop. The weight of the solid substances in a slip or slurry per unit volume of the slurry.

weight, tare. The weight of an empty container or vehicle.

weld. A fused joint formed between two components, or the process of forming such a joint.

weld blisters. An imperfection consisting of broken or unbroken bubbles in porcelain enamels caused by the evolution of gases along the line of a weld in the base metal during the firing operation.

welding glass. A special colored and tempered glass designed to protect the eyes of a welder from harmful radiations and flying sparks during welding operations.

weld marks. A groove or fissure formed at the junction of two or more parts when the intended fusion of the parts is incomplete.

well. A reservoir constructed in a melting furnace for the collection of molten materials such as metal, glass, slag, etc.

well-hole pipe. A refractory pipe or tube directing the flame upward from the well of a melting furnace.

Westlake process. An automatic glass-forming process simulating the procedures of a hand shop in which the molten glass is gathered by vacuum, and formed by blowing into molds.

wet. A term describing the consistency of fresh concrete.

wet-grinding. (1) The milling of ceramic bodies, glazes, and porcelain enamels in a liquid medium, usually water.

(2) The application of a liquid coolant to a work piece and also the grinding wheel during abrasive grinding.

wet milling. The grinding of porcelain-enamel frits with selected mill additions and water in a ball mill to form a slip suitable for application to metal by dipping, spraying, or other technique.

wet pan. An apparatus consisting of heavy mullers revolving on the bottom of a slotted revolving pan in which wet or damp materials are mixed and ground.

wet pressing. The process of forming plastic ceramic bodies in dies by the mechanical or manual application of pressure.

wet process. (1) The process of preparing ceramic bodies in which the ingredients are blended with sufficient liquid, usually water, to form a slurry for casting.

(2) The process in which the ingredients of portland cement are charged into the cement kiln in the form of a slurry.

wet-process porcelain enameling. The technique of applying porcelain enamel to metal in slip form, usually by dipping or spraying, followed by drying and firing to a smooth, impervious, glassy finish.

wet-rubbing test. A test of the resistance of porcelain enamel, glaze, and other surfaces to abrasion by rubbing with damp selected abrasives.

wetting, wettability. The ability of a liquid film to spread over and adhere to the surface of a solid.

wetting agent. A substance, such as a soap, detergent, or other surface-active material, which will reduce the surface tension of water or other liquid and cause the liquid to spread over or penetrate the surface of another material more easily.

wetting off. The use of a fine jet of water to sever a hand-blown glass article from a blowpipe.

wet ware. Ware placed into the glost kiln without prior drying.

whaler. See waler.

wheel, abrasive. A grinding wheel composed and formed of an abrasive and a suitable binder.

wheel, composite. A bonded-abrasive wheel containing a mixture of abrasives of different specifications.

wheel, concentric. A bonded-abrasive wheel consisting of two or more concentric sections, each section being made of abrasives of different specifications.

wheel, cone. A relatively small bonded-abrasive wheel shaped in the form of a cone.

wheel, cup. A bonded-abrasive wheel in the shape of a cup or bowl.

wheel, cut-off. A thin, bonded-abrasive grinding disk used to cut, slice, or groove a solid material.

wheel, cylinder. A bonded-abrasive grinder in the shape of a cylinder on which work is ground on its side instead of the peripheral surface.

wheel, diamond. A bonded-abrasive grinding wheel containing sized commercial diamonds as the abrasive.

wheel, dish. A dish-shaped, bonded-abrasive grinding wheel.

wheel, disk. A bonded-abrasive grinding disk mounted on a suitable backing, sometimes flexible, for use in grinding and polishing operations.

wheel, feed. A wheel regulating the speed and pressure of a grinding wheel on the surface of a part being processed on a centerless grinder.

wheel, grinding. A bonded-abrasive wheel or disk-shaped article mounted on a mechanically actuated axis for use in the grinding and polishing of solid surfaces.

wheel head. The outer or upper surface of a grinding wheel.

wheel, kick. A foot-operated device which actuates a potter's wheel.

wheel, magnesite. An abrasive grinding wheel in which the bonding medium is a magnesium oxychloride cement.

wheel, mounted. An abrasive wheel mounted on an axle of a mechanically operated grinding and polishing machine.

wheel, parting. A thin abrasive disk mounted on the spindle of a mechanically operated grinder for use in cutting, slicing, and grooving solid materials.

wheel, polishing. A fine-grained abrasive disk or wheel used for mechanical polishing.

wheel, recessed. An abrasive grinding wheel with a contoured recess on one or both sides.

wheel, regulating. A wheel on a centerless grinder which regulates the speed and pressure of an abrasive wheel on a work piece during the grinding operation.

wheel, resinoid grinding. A grinding wheel in which the abrasive grains are bonded with a thermosetting resin.

wheel, rubber. A grinding wheel in which the abrasive grains are bonded with rubber.

wheel, saucer. A shallow, saucer-shaped abrasive grinding wheel.

wheel, segmented. A type of grinding wheel composed and assembled of segments, instead of being of composite construction.

wheel, set-up. An abrasive wheel fabricated by compressing a series of sheets of abrasive-coated fabrics into wheel form.

wheel, shellac. A grinding wheel in which the abrasive grains are bonded together with shellac or a shellac-base adhesive.

wheel sleeve. A flange used as an adaptor for grinding wheels when the hole in the wheel is larger than the arbor.

wheel, slotting. An abrasive grinding wheel of thin cross section employed to produce slots on a work piece.

wheel, slow. A relatively slow-turning potter's wheel or similar device used in finishing the surface of hand-made pottery.

wheel, straight. A grinding wheel having sides or faces that are straight and parallel, and each side at right angles to the arbor hole.

wheel, tapered. An abrasive grinding wheel with slightly tapered sides, the hub being thicker than the circumference.

wheel traverse. The rate at which a grinding wheel moves across a work piece.

wheel, vitrified. A grinding wheel in which the abrasive grains are bonded by a vitrified ceramic composition.

whelp. A refractory of standard cross-sectional dimensions, but of substantially greater length; that is, for example, a brick approximately 3⅕ inches thick, 4 inches wide, and more than 8 inches in length.

whirler. (1) Flat ceramic ware in which the bottom sagged during the firing of the item.
(2) A rotating plaster mold in which bone china sometimes is cast in order to obtain a uniform thickness.
(3) A faulty plate, platter, saucer, or dish which will not rest firmly on its foot.

white acid. A mixture of hydrofluoric acid and ammonium bifluoride employed in the etching of glass surfaces.

white-acid embossed. A fully obscured glass surface obtained by a single acid treatment.

white alumina. A recrystallized alumina abrasive.

white cement. A pure white portland cement made from raw materials of extremely low-iron content or from materials which have been sintered in a reducing atmosphere.

white clay. A high-quality kaolin which fires to a white color.

white, dryer. A superficial white discoloration on the surface of ceramic ware due to the migration of soluble salts from the interior to the surface during drying.

white feldspar. A milky white or colorless variety of sodium feldspar.

white flint glass. A colorless glass having high light-dispersing qualities; used in optical instruments.

white graniteware. A term designating an exceptionally strong white earthenware body.

white-hard clay. A clay from which the water of plasticity has evaporated at its surface.

white, kiln. A white scum formed on ceramic ware, particularly structural clay products, due to the migration of soluble salts from the interior of the ware during drying or to the reaction of surface ingredients with harmful impurities in the combustion atmosphere during firing.

white lead. Basic lead carbonate of variable compositions, usually $2PbCO_3 \cdot Pb(OH)_2$; mol. wt. 776.0; decomposes at 400°C(725°F); sp.gr. 6.14; used as a fluxing constituent in glazes, porcelain enamels, and glass.

white portland cement. Finely milled white cement made from pure calcite limestone and a white-burning clay.

white spot. An imperfection in a colored glaze due to the separation of the pigment in the glaze as the glaze flows and heals when fired over a prior defect.

White's test. A microscopic technique for identifying the presence of free CaO in dolomitic refractories and portland cement by observing the presence of elongated needles when the material is wetted by a mixture of nitrobenzene, phenol, and water.

whiteware. A general term for a ceramic body which fires to a white or ivory color.

whiteware, alumina. Any whiteware product containing substantial amounts of alumina, Al_2O_3, as an essential ingredient. See alumina.

whiteware, cordierite. Any whiteware product containing substantial quantities of cordierite, $2MgO \cdot 2Al_2O_3 \cdot 5SiO_2$, as an essential ingredient. See cordierite.

whiteware, forsterite. Any whiteware product containing substantial quantities of forsterite, $2MgO \cdot SiO_2$, as an essential ingredient. See forsterite.

whiteware, mullite. Any whiteware product containing substantial quantities of mullite, $3Al_2O_3 \cdot 2SiO_2$, as an essential ingredient. See mullite.

whiteware, steatite. Any whiteware product containing substantial quantities of magnesium metasilicate, $MgO \cdot SiO_2$, as an essential ingredient.

whiteware, titania. Any whiteware product containing substantial quantities of titania, TiO_2, as an essential ingredient. See titanium dioxide.

whiteware, zircon. Any whiteware product containing substantial amounts of zircon, $ZrO_2 \cdot SiO_2$, as an essential ingredient. See zircon.

whiteware, zirconia. Any whiteware product containing substantial quantities of zirconia, ZrO_2, as an essential ingredient. See zirconia.

whiting. Natural $CaCO_3$; mol. wt. 100.1; sp.gr. 2.7; used in earthenware and vitreous sanitaryware bodies, glazes, glasses, and porcelain enamels as a refractory or neutral component.

wicket. A temporary refractory closure or door in a furnace or kiln such as may be placed, for example, near the ends of the checker chamber for charging and removing ware from the kiln.

wilkenite. A type of bentonite used in a variety of ceramic bodies and refractories, and as a suspension agent in porcelain enamels and glazes. See bentonite.

willemite. $2ZnO \cdot SiO_2$; mol. wt. 222.8; sp.gr. 3.3; hardness (Mohs) 5.5; used in the production of crystalline glazes.

Williamson kiln. A type of cross-fired tunnel kiln in which both direct-fire and muffle segments are incorporated.

Williamson's blue. A series of iron-bearing blue pigments.

willow blue. A diluted cobalt-blue colorant.

Winchester bottle. A 2½ liter, straight-sided glass bottle.

wind. A term sometimes used to describe air bubbles in ware.

winding. A wire wound in the shape of a coil or spiral, frequently around a ceramic core.

window dip. The dipping of ware sideways into a slip or glaze.

window glass. A continuously drawn soda-lime glass produced in sheet form; used primarily in the construction of windows.

wind-ridge tile. A type of ceramic or concrete roofing tile designed for use at the ridge of a pitched roof.

winning. The process of extracting a raw material from some source, such as an ore or reclaimed product, and converting the material to a useful product.

wire, brass. Brass wire of different and suitable diameters used in the cutting of pugged clay columns.

wire cloth. A fabric of wire mesh or screen woven in squares or rectangles, such as is used in sieves and screens.

wire-cut brick. Building and refractory brick units cut from extruded columns by means of a taut wire.

wired safety glass. A glass with an embedded network of wire which resists shattering when broken.

wire glass. See wired safety glass.

wire glass, polished. Wire glass which has been ground and polished on one or, usually, both sides.

witherite. $BaCO_3$; mol. wt. 197.4; m.p. 1360°C (2480°F); sp.gr. 4.27–4.35; hardness (Mohs) 3–3.75;

used extensively in various optical, plate, and tableware glasses, pottery bodies, and as a flux in glazes and porcelain enamels; also used in structural clay bodies to prevent efflorescence.

wollastonite. $CaO \cdot SiO_2$; mol. wt. 116.1; m.p. (incongruent) 1544°C (2811°F); sp.gr. 2.8–2.9; hardness (Mohs) 4.5–5.5; used in refractories, cements, dielectric bodies and glazes, mineral wool, wall board, and various whiteware bodies, and as a flux in welding-rod coatings.

wonderstone. A synonym sometimes used for pyrophyllite, which see.

Wood's glass. Glass having a high-transmission factor for ultraviolet radiation, and which is almost opaque to visible light.

Woods Hole sediment analyzer. A technique for determining the particle size distribution in a clay suspension based on changes in pressure resulting from the settling of the clay particles.

Wood's process. A technique employed in the production of glass tubing and rod in which molten glass is drawn from an orifice; the tubing is formed by drawing rod around a refractory cone.

wool drag. The smearing of color during the application of background color on pottery and other ceramic ware.

wool, glass. A fleecy mass of glass fibers used as insulation, filters, etc.

wool, mineral. A slag wool similar to glass wool made by attenuating molten slag in a fast-moving stream of air or steam; used as insulation, fireproofing, filters, etc.

wool, rock. A mass of intertwined fibers formed by blowing air or steam through a stream of molten rock or slag; used for thermal and acoustic insulation, fireproofing, filters, etc.

work. (1) Something produced by the exertion of effort.

(2) A piece or item being operated on in the process of manufacture, such as grinding, polishing, or other process.

workability. (1) The property of being workable.

(2) The combination of properties which contribute to the ease with which concrete, mortar, ramming mixes, and plastic masses can be mixed, handled, transported, and placed with a minimum of effort or loss of homogeneity.

workability agent. An admixture employed in concrete, mortar, and other plastic mixes to improve their workability.

workability index. A measure of the consistency and molding properties of plastic masses, particularly refractories.

work board. A long narrow board on which green ware is placed for drying and for transport from one location to another.

working end. The end compartment or section of a glass-melting tank from which molten glass is taken for forming.

working molds. Plaster-of-paris molds in which ceramic bodies are shaped by casting, jiggering, or roller forming.

working range. The temperature range in which glass may be shaped into ware, the upper temperature being such that the glass is sufficiently fluid or plastic for forming, and the lower temperature being such that the formed ware will retain its shape as formed.

working standard. Any specification or standard of quality in current or regular use.

work speed. (1) The rate of table traverse during abrasive grinding.

(2) The rate at which work is rotated during centerless and internal grinding operations.

wreathing. A slightly raised crescent on the inside wall of slip-cast ware.

writing, magnetic. A nonrelevant indication caused when a magnetic part comes in contact with another ferromagnet.

wye. Any solid item fabricated in the shape of a ''y.''

wythe. (1) Each continuous vertical section of masonry one unit in thickness.

(2) The thickness of the masonry units separating flues in a chimney.

X

xanthates. Various salts of xanthic acid employed as flotation agents in the beneficiation of minerals for ceramic and other uses.

xenocryst. A crystalline material which is not in equilibrium with other minerals in a rock.

xenolith. A rock inclusion which is unrelated to the rock in which it occurs.

xenothermal deposit. A mineral which has been formed at low pressure and high temperature, usually at shallow to moderate depths.

xenotime. YPO_4; mol. wt. 183.9. sp.gr. 4.4–5.1; hardness (Mohs) 4–5; a lustrous, vitreous, natural phosphate ranging from white to brown in color.

xerography. A dry method of photocopying in which an image is projected through a camera lens onto a smooth electrostatically charged metallic plate which has been coated with selenium or other photoconductive material; a pigmented powder containing a developing resin is cascaded or dispersed over the plate, adhering to the plate in thicknesses proportional to the patterned charges remaining on the plate after exposure, the intensity of the pattern varying with the darkness and shading of the image; the image then is transferred electrostatically to paper or other surface placed in contact with the plate, and fixed by heat or other means to provide a positive reproduction of the subject.

xeroradiography. A xerographic technique in which x-rays are employed to project an image onto a photosensitive plate.

x-ray. A high-frequency electromagnetic ray of extremely short wave length, 0.06–120Å, emitted as a result of a sudden change in the velocity of an electric charge such as occurs when cathode rays strike a solid, usually metallic, in a vacuum tube.

x-ray absorption. The absorption of energy from an x-ray beam by a medium through which the beam is passing.

x-ray analysis. Determination of the atomic distribution, structure, chemical analysis, and anisotropic behavior of crystalline materials by means of x-rays.

x-ray crystallography. The study of the structure, identity, texture, and behavior of crystals by x-ray techniques.

x-ray diffraction. Scattering of an x-ray beam into many beams at definite angles to the original beam in the orderly pattern characteristic of different atoms; used as an analysis of crystal structure.

x-ray photograph. A shadow produced on photosensitive film by exposure to x-rays.

x-ray protective glass. A glass containing a high percentage of lead oxide and, occasionally, barium oxide, which exhibits a high degree of opacity to x-rays. Sometimes known as document glass.

x-ray spectroscope. An instrument for comparing the color intensity between corresponding parts of different spectra, or between parts of the same spectrum.

x-ray spectrum. Patterns of emission from matter bombarded by high-velocity electrons.

x-ray tube. A vacuum tube containing a metal target on which x-rays are produced by impact of electrons.

x-unit (Xu). A unit used to express the wave length of x-rays or gamma rays.

Y

yard, laying. The area in which rough glass is laid in plaster on tables for grinding and polishing.

yard, stripping. The area in which glass is removed from the polishing table following the grinding and polishing operation.

yarn. A continuous strand of glass, ceramic, asbestos, or other fiber.

yellow, antimony. Lead antimonate, $Pb_3(SbO_4)_2$; mol. wt. 993.0; an orange-yellow powder used as a colorant in overglazes, porcelain enamels, and glass.

yellow ochre. $FeO(OH){\cdot}nH_2O$; used as a yellow pigment. Also known as limonite.

yellow ware. A buff or yellow semivitreous or earthenware body which sometimes is coated with a clear, colorless glaze.

yield. (1) The measure of stress at which a permanent change will occur in the shape of a solid body without causing the body to fracture.

(2) The number of cubic feet of concrete produced per sack of cement, calculated as total volume per batch divided by the number of sacks per batch.

yield point. (1) The minimum unit of stress at which continuous flow will occur in a clay-water or similar mass when subjected to some force such as tension, compression, torsion, or shear.

(2) The minimum unit of stress at which a solid material will deform without an increase in the applied load.

yield strength. The unit of stress corresponding to a specific amount of permanent deformation of a solid.

yield stress. The minimum stress at which creep will occur in a solid body. See creep.

yield temperature. The minimum temperature at which permanent deformation will occur in a solid body under specified conditions of stress.

yoke. A C-shaped solid or laminated piece of soft magnetic material around which a coil to carry magnetizing current has been wound.

yoke magnetization. A longitudinal magnetic field induced in a material, or in an area of a material, by means of an external yoke-shaped electromagnet.

Young's modulus. The ratio of tensile stress to elongation within the elastic limit of a solid body; also known as the modulus of elasticity, which see.

ytterbium boride. (1) YbB_3; mol. wt. 206.06; sp.gr. 6.74. (2) YbB_4; mol. wt. 216.88; sp.gr. 7.31. (3) YbB_6; mol. wt. 238.52; sp.gr. 5.57; hardness 3800 Vickers. See borides.

ytterbium carbide. (1) Yb_3C; mol. wt. 532.8; sp.gr. 10.26. (2) YbC_2; mol. wt. 197.6; sp.gr. 8.10. See carbides.

ytterbium-iron garnet. $Yb_3Fe_2(FeO_4)_3$; mol. wt. 737.9; used as a resonator at microwave frequencies.

ytterbium nitride. YbN; mol. wt. 187.61; sp.gr. 11.33. See nitrides.

ytterbium oxide. Yb_2O_3; mol. wt. 394.1; m.p. 2346°C (4255°F); sp.gr. 9.18; used as a component in electrically conducting ceramics, glass ceramics, special refractories, phosphors, etc.

ytterbium selenide. YbSe; mol. wt. 252.8; m.p. ~ 1521°C (2770°F).

ytterbium silicate. (1) $Yb_2O_3{\cdot}SiO_2$; mol. wt. 455.26; m.p. 1979°C (3595°F); hardness (Mohs) 5–7. (2) $2Yb_2O_3{\cdot}3SiO_2$; mol. wt. 970.58; m.p. 1949°C (3540°F); hardness (Mohs) 5–7. (3) $Yb_2O_3{\cdot}2SiO_2$; mol. wt. 515.32; m.p. 1777°C (3230°F); hardness (Mohs) 5–7.

ytterbium silicide. $YbSi_2$; mol. wt. 229.72. See silicides.

ytterbium sulfide. (1) YbS; mol. wt. 205.66; sp.gr. 6.75. (2) YbS_2; mol. wt. 237.72. (3) Yb_2S_3; mol. wt. 443.38; sp.gr. 6.04. (4) Yb_3S_4; mol. wt. 649.04; sp.gr. 6.74. See sulfides.

ytterbium telluride. YbTe; mol. wt. 301.1; m.p. 1738°C (3160°F). See intermetallic compounds.

yttrium aluminate. (1) $Y_2O_3{\cdot}Al_2O_3$; mol. wt. 327.74; sp.gr. 5.50. (2) $2Y_2O_3{\cdot}Al_2O_3$; mol. wt. 553.54; m.p. 2838°C (3700°F). (3) $3Y_2O_3{\cdot}5Al_2O_3$; mol. wt. 1187.1; m.p. 1982°C (3600°F).

yttrium beryllide. YBe_{13}; mol. wt. 208.5; sp.gr. 2.56. See beryllides.

yttrium boride. (1) YB_2; mol. wt. 110.54; sp.gr. 2.91. (2) YB_3; mol. wt. 121.36; sp.gr. 3.97. (3) YB_4; mol. wt. 132.18; sp.gr. 4.36. (4) YB_6; mol. wt. 153.82; m.p. 2299°C (4170°F); sp.gr. 3.70; hardness 3260 Vickers. See borides.

yttrium carbide. (1) Y_3C; mol. wt. 278.7; sp.gr. 5.41. (2) Y_2C_3; mol. wt. 213.8; m.p. 1799°C (3270°F); hardness 900 Vickers. (3) YC_2; mol. wt. 112.9; m.p. 2299°C (4170°F); sp.gr. 4.33; hardness 700 Vickers. (4) YC; mol. wt. 100.9; m.p. 1949°C (3540°F). See carbides.

yttrium carbonate. $Y_2(CO_3)_3{\cdot}3H_2O$; mol. wt. 411.9; used as a phosphor in refractory gas mantles.

yttrium ferrite. $3Y_2O_3$; mol. wt. 1475.8; m.p. 1560°C (2840°F); sp.gr. 5.17. See intermetallic compound.

yttrium nitride. YN; mol. wt. 102.91; m.p. ~ 2671°C (4840°F); sp.gr. 5.90. See nitrides.

yttrium oxide. Y_2O_3; mol. wt. 225.8; sp.gr. 4.84; m.p. 2410°C (4370°F); used to make red phosphors for television tubes, in the production of microwave filters, and, with ZrO_2, in the manufacture of special high-temperature refractories.

yttrium phosphide. YP; mol. wt. 119.93; sp.gr. 4.32. See phosphides.

yttrium silicate. (1) $Y_2O_3{\cdot}SiO_2$; mol. wt. 285.86; m.p. 1979°C (3595°F); sp.gr. 4.49; hardness (Mohs) 5–7. (2) $2\,Y_2O{\cdot}3SiO_2$; mol. wt. 631.78; m.p. 1949°C (3540°F); sp.gr. 4.39; hardness (Mohs) 5–7. (3) $Y_2O{\cdot}2SiO_2$; mol. wt. 345.92; m.p. 1777°C (3230°F); sp.gr. 4.06; hardness (Mohs) 5–7.

yttrium silicide. (1) YSi_2; mol. wt. 145.02; m.p. 1521°C (2770°F); sp.gr. 4.35. (2) Y_3Si_5; mol. wt. 407.0. See silicides.

yttrium sulfide. (1) YS; mol. wt. 120.96; m.p. 2038°C (3700°F); sp.gr. 4.95. (2) Y_5S_7; mol. wt. 668.92; m.p. 1521°C (2770°F); sp.gr. 4.18. (3) Y_2S_3; mol. wt. 273.98; m.p. 1598°C (2910°F); sp.gr. 3.87. (4) YS_2; mol. wt. 153.02; m.p. 1660°C (3020°F); sp.gr. 4.35. See sulfides.

yttrium telluride. Y_2Te_3; mol. wt. 560.3; m.p. 1943°C (3530°F). See intermetallic compound.

Z

Zacharaison rules of glass formation. (1) An oxygen atom is linked to no more than two glass-forming atoms;
(2) the coordination number of glass-forming atoms is small;
(3) the oxygen atoms share corners with each other, but not faces or edges; and
(4) the polyhedra are linked in a three-dimensional network.

zaffre. An impure form of cobalt oxide used in the production of smalt.

Zahn cup. An orifice-type viscometer in which the time required for a measured quantity of glaze or porcelain-enamel slip to flow through an opening of specified size is taken as a measure of the slip viscosity.

zebra roof. A roof used in basic open-hearth furnaces consisting of silica and chrome-magnesite refractories arranged in alternate rings, resulting in a striped appearance.

Zeeman effect. The splitting of the spectroscopic lines of a source of radiation when subjected to a moderately intense magnetic field.

Zener current. The current through an insulator when placed in an electric field of sufficient intensity to excite an electron directly from the valence to the conduction band.

Zener voltage. The field required to excite the Zener current, usually of the order of 10^7 volts per centimeter. See Zener current.

zeolite. A class of hydrous aluminum silicates of the approximate composition $Na_2O \cdot Al_2O_3 \cdot xSiO_2 \cdot yH_2O$; used in ion-exchange reactions.

zeolite process. A water-softening process involving the cationic exchange of the sodium in zeolite for the calcium and magnesium in hard water. See cationic.

zero, absolute. The lowest possible temperature at which particles whose motions constitute heat would be at rest; approximately −273.1°C (−459.7°F).

zero-carbon steel. Sheet steel of extremely low-carbon content on which porcelain enamel cover coats usually may be applied and fired without the need of a ground coat.

zero-point energy. The kinetic energy remaining in a substance at absolute zero.

zeta potential. The difference in the electrical potential at the interface between the solid and liquid phases of a colloidal solution in which a sharp potential fall occurs in a one-ion thick layer of the solid phase, and a gradual rise or fall extending some distance into the liquid phase is observed.

zig-zag kiln. A type of kiln in which the dividing walls are staggered in a manner so as to force the heat to flow through the kiln in a zig-zag pattern.

zinc aluminate. $ZnO \cdot Al_2O_3$; mol. wt. 183.3; m.p. 1950°C (3542°F); sp.gr. 4.58; a spinel used as a refractory lining in containers for the melting of selected metals.

zinc ammonium chloride. $xZnCl_2 \cdot 2NH_4Cl$; sp.gr. 1.8; used as a soldering flux in glass-to-metal and ceramic-to-metal seals.

zinc antimonide. ZnSb; mol. wt. 187.2; used as a p-type leg in thermionic-power generators; unstable at approximately 1000°C (1832°F). See thermionic.

zinc blende. Natural ZnS; mol. wt. 97.4; sp.gr. 3.9–4.1; hardness (Mohs) 3.5–4.0.

zinc borate. Various compounds of ZnO and B_2O_3 used as fluxes in ceramic compositions.

zinc carbonate. $ZnCO_3$; mol. wt. 125.4; loses CO_2 at 300°C (572°F); sp.gr. 4.42–4.45; used in Bristol and other glazes. Also known as smithsonite.

zinc cement. A quick-hardening cement composed of zinc oxide made into a paste by the use of a zinc chloride solution.

zinc chloride. $ZnCl_2$; mol. wt. 136.3; m.p. 290°C (554°F); b.p. 732°C (1348°F); sp.gr. 2.91; used in special cements, glass-etching compositions, dental cements, etc.

zinc chromate. Various compounds of ZnO and Cr_2O_3 used as a yellow ceramic colorant.

zinc crown glass. An optical glass containing substantial amounts of zinc oxide as an auxiliary flux. See optical crown glass.

zinc ferrite. $ZnO \cdot Fe_2O_3$; mol. wt. 241.1; a spinel melting at about 1590°C (2894°F); sp.gr. 5.33.

zinc flash. A colored surface produced on brick by the introduction of zinc into the fireboxes of the kiln at the end of the firing operation, the zinc vapors depositing on the surface of the brick to form various shadings ranging from yellow to green.

zinc fluoride. ZnF_2; mol. wt. 103.4; m.p. 872°C (1600°F); sp.gr. 4.84; used as a gaseous opacifier and flux in porcelain enamels and glazes.

zinc fluosilicate. $ZnSiF_6 \cdot 6H_2O$; mol. wt. 315.5; decomposes on heating; sp.gr. 2.1; used as a concrete hardener.

zinc glass. Glass of the ordinary soda-lime type in which part of the calcium oxide is replaced by zinc oxide.

zincite. ZnO; mol. wt. 81.4; sp.gr. 5.4–5.7; hardness (Mohs) 4–4.5; used in the production of zinc oxide of high purity.

zinc molybdate. $ZnMoO_4$; mol. wt. 225.4 m.p. about 900°C (1652°F); occasionally used as an adherence promoting agent in white porcelain enamels.

zinc niobate. (1) $ZnO{\cdot}Nb_2O_5$; mol. wt. 347.58; m.p. 1398°C (2550°F). (2) $3ZnO{\cdot}Nb_2O_5$; mol. wt. 510.34; m.p. 1305°C (2385°F).

zinc nitride. Zn_3N; mol. wt. 210.2; a semiconducting material used in electronic applications.

zinc oxide. ZnO; mol. wt. 81.4; sublimes at 1800°C (3272°F); sp.gr. 5.6–5.8; used as an opacifier and fluxing ingredient in glass, glazes, porcelain enamels, magnetic ferrites, dental cements, and special piezoelectric compositions.

zinc phosphate. $Zn_3(PO_4)_2{\cdot}4H_2O$; mol. wt. 458.2; m.p. 900°C (1652F); sp.gr. 3.08; used in dental cements and in the production of phosphors.

zinc selenide. $ZnSe$; mol. wt. 144.6; m.p. above 1100°C (2012°F); sp.gr. 5.33; used in infrared optical windows.

zinc silicate. (1) $2ZnO{\cdot}SiO_2$; mol. wt. 222.8; m.p. 1509°C (2748°F); sp.gr. 3.3; hardness (Mohs) 5.5. (2) $ZnO{\cdot}SiO_2$; mol. wt. 141.48; m.p. 1510°C (2750°F); sp.gr. 4.1; hardness (Mohs) 5–7.

zinc silicofluoride. $ZnSiF_6{\cdot}6H_2O$; mol. wt. 315.5; decomposes on heating; sp.gr. 2.1; employed as a hardener for concrete.

zinc sulfide. ZnS; mol. wt. 97.4; sp.gr. 3.9–4.1; m.p. 1020°C (1868°F); sublimes at 1180°C (2156°F); hardness (Mohs) 3.5–4.0; used in white and opaque glasses, in x-ray and television tubes, phosphors, and similar products.

zinc telluride. $ZnTe$; mol. wt. 192.9; m.p. 1238°C (2260°F); sp.gr. 5.54.

zinc titanate. (1) $ZnO{\cdot}TiO_2$; mol. wt. 161.5; m.p. above 1500°C (2822°F); a dielectric material. (2) $2Zn{\cdot}TiO_2$; mol. wt. 226.9; m.p. above 1500°C (2822°F); used in dielectric applications.

zinc zirconium silicate. $ZnO{\cdot}ZrO_2{\cdot}SiO_2$; mol. wt. 264.5; m.p. 2080°C (3776°F); used primarily as an ingredient in ceramic glazes.

zircon. $ZrO_2{\cdot}SiO_2$; mol. wt. 182.9; softening temperature 850°–950°C (1562°–1742°F); m.p. 2250°C (4082°F); sp.gr. 4.68; hardness (Mohs) 7.5; used in porcelain enamels and glazes as an opacifier and to improve color stability and crazing resistance; also used in refractories, abrasives, grinding wheels; precision molds for the casting of alloys, electrically resisting cements, and in conventional electrical and technical porcelains.

zirconal phosphate. $ZrO_2{\cdot}P_2O_7$; mol. wt. 388.4; a possible semirefractory capable of withstanding temperatures up to 1600°C (2912°F).

zircon flour. Finely milled zirconium silicate used as a mill wash, which see.

zirconia. ZrO_2; mol. wt. 123.2; m.p. about 2700°C (4892°F); sp.gr. 5.73; hardness (Mohs) 6.5; used as an opacifier in porcelain enamels and glazes, as an abrasive in polishing and grinding compounds, as setter plates for the firing of ceramics, ferrites, and titanates, as wind-tunnel liners, as a refractory structural material in nuclear applications, as a highly corrosion-resistant ceramic, and as a refractory for high-temperature use.

zirconia brick. A refractory brick composed of substantial amounts of zirconia; used to line metallurgical furnaces for resistance to basic slags.

zirconia refractories. A refractory composed essentially of zirconium oxide.

zirconia whiteware. Any whiteware product containing substantial amounts of zirconia, ZrO_2, as an essential ingredient.

zirconium aluminate. $ZrO_2Al_2O_3$; mol. wt. 224.9; used as a component in high-temperature refractories.

zirconium aluminide. (1) $ZrAl$; mol. wt. 118.19; m.p. 1637°C (2980°F). (2) Zr_4Al_3; mol. wt. 445.79; m.p. 1532°C (2790°F); sp.gr. 5.30.

zirconium antimonide. Zr_2Sb; mol. wt. 304.21; m.p. 1899°C (3450°F).

zirconium beryllide. (1) $ZrBe_{13}$; mol. wt. 210.8; m.p. 1930°C (3506°F); sp.gr. 2.72. (2) Zr_2Be_{17}; mol. wt. 338.8; m.p. 1980°C (3596°F); sp.gr. 3.08. Both are intermetallic compounds with good strength at elevated temperatures; used as moderators in nuclear reactors.

zirconium boride. (1) ZrB; mol. wt. 102.04; m.p. 798°–1248°C (1470°–2280°F); sp.gr. 6.7. (2) ZrB_2; see zirconium diboride. (3) ZrB_{12}; mol. wt. 221.06; m.p. 2566°C (4650°F); sp.gr. 3.63. See borides.

zirconium-boron composites. (1) $ZrB_2{\cdot}B$; mol. wt. 123.68; m.p. ~ 2983°C (5400°F); sp.gr. 5.2–5.4. (2) $ZrB_2{\cdot}MoSiO_2$; mol. wt. 268.92; m.p. >2371°C (4300°F); sp.gr. 4.87–5.50.

zirconium carbide. ZrC; mol. wt. 103.2; m.p. 3540°C (6404°F); sp.gr. 6.44; hardness (Mohs) 8–9; employed as an abrasive, refractory, incandescent filament, and cutting tool. See carbides.

zirconium diboride. ZrB_2; mol. wt. 112.9; m.p. 3040°C (6404°F); sp.gr. 6.1; used in cutting tools, metal-casting refractory molds, refractory pouring spouts, rocket nozzles, combustion chamber liners, thermocouple tubes, and other high-temperature products. See borides.

zirconium dioxide. ZrO_2; mol. wt. 123.2; m.p. about 2700°C (4892°F); sp.gr. 5.73; hardness (Mohs) 6.5; used as an opacifier in porcelain enamels and glazes, as an abrasive in grinding and polishing compounds, as setter plates for the firing of ceramic ware, ferrites, and titanates, as wind-tunnel liners, as a refractory structural material in nuclear applications, as a corrosion-resistant ceramic, and as a refractory component for high-temperature applications.

zirconium dioxide porcelain. A porcelain in which zirconium dioxide, ZrO_2, is a major component.

zirconium dioxide refractory. A refractory of low-thermal conductivity in which zirconium dioxide, ZrO_2, is a major ingredient; the stabilized ZrO_2 refractories are used at temperatures above 2200°C (3992°F).

zirconium germanide. (1) Zr_2Ge_3; mol. wt. 400.24; m.p. 1549°C (2820°F). (2) Zr_2Ge; mol. wt. 255.04; m.p. 1910°–2277°C (3470°–4130°F). (3) Zr_3Ge; mol. wt. 346.26; m.p. 1587°C (2980°F).

zirconium hydroxide. $Zr(OH)_4$; mol. wt. 159.2; decomposes to ZrO_2 at 550°C (1022°F); sp.gr. 3.25; used in glass manufacture.

zirconium napthenate. An amber-colored, transparent, heavy liquid sometimes used in porcelain enamels and glazes.

zirconium nitride. ZrN; mol. wt. 105.2; m.p. 2930°C (5306°F); sp.gr. 7.32; a brass-colored powder used in refractories, crucibles, and cermets. See nitrides.

zirconium oxide. ZrO_2; mol. wt. 123.2; m.p. about 2700°C (4892°F); sp.gr. 5.73; hardness (Mohs) 6.5; used as an opacifier in porcelain enamels and glazes, as setter plates and in bat washes for the firing of ceramic ware, ferrites, and titanates, as a refractory structural material in nuclear applications, as a corrosion-resistant ceramic, and as a refractory for high-temperature use.

zirconium oxide porcelain. A porcelain in which zirconium oxide, ZrO_2,is a major component.

zirconium oxide refractory. A refractory in which zirconium oxide, ZrO_2,is a major constituent, and which is characterized by low-thermal expansion.

zirconium phosphate. $ZrO_2{\cdot}P_2O_7$; mol. wt. 388.4; used in the production of glass containers and in refractories withstanding temperatures up to 1600°C (2912°F).

zirconium phosphide. ZrP; mol. wt. 122.25; sp.gr. 5.43–5.57. See phosphides.

zirconium silicate. $ZrO_2{\cdot}SiO_2$; mol. wt. 182.9; softening temperature 850°–950°C (1562°–1742°F); m.p. 2250°C (4082°F); sp.gr. 4.68; hardness (Mohs) 7.5; used as a component in refractories, porcelain enamels, abrasives and grinding wheels, precision molds for the casting of alloys, electrically resisting cements, and electrical and technical porcelains.

zirconium silicide. (1) Zr_4Si; mol. wt. 392.94; m.p. 1631°C (2970°F). (2) Zr_2Si; mol. wt. 210.5; m.p. 2166°C (3930°F); sp.gr. 5.99; hardness (Vickers) 1180–1280. (3) Zr_5Si_3; mol. wt. 540.28; m.p. 2249°C (4040°F); sp.gr. 5.90; hardness (Vickers) 1280–1390. (4) Zr_3Si_2; mol. wt. 329.78; m.p. 2210°C (4010°F). (5) Zr_4Si_3; mol. wt. 449.06; m.p. 2227°C (4040°F). (6) Zr_6Si_5; mol. wt. 687.62; m.p. 2249°C (4080°F). (7) ZrSi; mol. wt. 119.28; m.p. 2121 °C (3850°F); sp.gr. 5.56; hardness (Vickers) 1020–1180. (8) $ZrSi_2$; mol. wt. 147.34; m.p. 1604°C (2920°F); sp.gr. 4.88; hardness (Vickers) 830–1060. See silicides.

zirconium spinel. A spinel having a melting point of 1710°C (3110°F).

zirconium sulfide. (1) ZrS; mol. wt. 123.28; m.p. 2099°C (3810°F); sp.gr. 4.56 (2) Zr_2S_3; mol. wt. 278.62; sp.gr. 4.29. (3) ZrS_2; mol. wt. 155.34; m.p. 1549°C (2820°F); sp.gr. 3.82. (4) ZrS_3; mol. wt. 187.40; unstable above 871°C (1600°F); sp.gr. 3.82. See sulfides.

zircon porcelain. A vitreous ceramic whiteware used in technical applications, crucibles, combustion boats, thermocouple tubes, etc., and in which zircon, ZrO_2, is an essential component.

zircon refractory. Any refractory product composed substantially of zircon, $ZrO_2{\cdot}SiO_2$.

zircon sand. A natural sand containing considerable amounts of zirconia, ZrO_2, titania, TiO_2, and related materials.

zircon whiteware. Any ceramic whiteware containing zircon, $ZrO_2{\cdot}SiO_2$, as an essential ingredient.

zirkite. A mineral source of zirconium dioxide, ZrO_2, used in refractories and low-expansion bodies of high thermal-shock resistance.

Zisman apparatus. An instrument for measuring contact-potential differences between solid-solid and solid-liquid interfaces.

zone control. A system of independent heating and temperature controls for each zone of a furnace or kiln.

zone melting. A method of separating or purifying a substance by fusion in which a series of molten zones traverse a rod or charge of a metal or other substance.

zone refining. A procedure for purifying materials in which a narrow molten zone is moved slowly along the length of a specimen in such a manner that impurities are retained in and moved along with the molten material to the end of the specimen where they are collected and removed by severing; the process may be repeated until the desired degree of purity is attained.

Appendices

Table of the Chemical Elements

Element	*Symbol*	*Atomic number*	*Atomic weight*
actinium	Ac	89	227(?)
aluminum	Al	13	26.98
americum	Am	95	243.13
antimony	Sb	51	121.75
argon	Ar	18	39.95
arsenic	As	33	74.92
astatine	At	85	210(?)
barium	Ba	56	137.34
berkelium	Bk	97	248(?)
beryllium	Be	4	9.01
bismuth	Bi	83	208.98
boron	B	5	10.81
bromine	Br	35	79.91
cadmium	Cd	48	112.40
calcium	Ca	20	40.08
californium	Cf	98	251(?)
carbon	C	6	12.01
cerium	Ce	58	140.12
cesium	Cs	53	132.91
chlorine	Cl	17	35.45
chromium	Cr	24	52.00
cobalt	Co	27	58.93
copper	Cu	29	63.55
curium	Cm	96	247(?)
dysprosium	Dy	66	162.50
einsteinium	Es	99	252(?)
erbium	Er	68	167.28
europium	Eu	63	151.96
fermium	Fm	100	257(?)
fluorine	F	9	19.00
francium	Fr	87	223(?)
gadolinium	Gd	64	157.25
gallium	Ga	31	69.72
germanium	Ge	32	72.59
gold	Au	79	196.97
hafnium	Hf	72	178.49
helium	He	2	4.00
holmium	Ho	67	164.93
hydrogen	H	1	1.01
indium	In	49	114.82
iodine	I	53	126.90
iridium	Ir	77	192.20
iron	Fe	26	55.85
krypton	Kr	36	83.80
lanthanum	La	57	138.91
lawrencium	Lr	103	256(?)
lead	Pb	82	207.19
lithium	Li	3	6.94
lutetium	Lu	71	174.97
magnesium	Mg	12	24.31
manganese	Mn	25	54.94
mendelevium	Md	101	258(?)
mercury	Hg	80	200.59
molybdenum	Mo	42	95.94
neodymium	Nd	60	144.24
neon	Ne	10	20.18
neptunium	Np	93	237.00
nickel	Ni	28	58.71
niobium	Nb	41	92.91
nitrogen	N	7	14.01
nobelium	No	102	255(?)
osmium	Os	76	190.20
oxygen	O	8	16.0
palladium	Pd	46	106.40
phosphorus	P	15	30.97
platinum	Pt	78	195.09
plutonium	Pu	94	239.05
polonium	Po	84	210.05
potassium	K	19	39.10
praseodymium	Pr	59	140.91
promethium	Pm	61	145(?)
protactinium	Pa	91	231.10
radium	Ra	88	226.00
radon	Rn	86	222.00
rhenium	Re	75	186.20
rhodium	Rh	45	102.91
rubidium	Rb	37	85.47
ruthenium	Ru	44	101.07
samarium	Sm	62	150.35
scandium	Sc	21	44.96
selenium	Se	34	78.96
silicon	Si	14	28.09
silver	Ag	47	107.87
sodium	Na	11	22.99
strontium	Sr	38	87.62
sulfur	S	16	32.06
tantalum	Ta	73	180.95
technetium	Tc	43	97(?)
tellurium	Te	52	127.60
terbium	Tb	65	158.92
thallium	Tl	81	204.37
thorium	Th	90	232.04
thulium	Tm	69	168.93
tin	Sn	50	118.69
titanium	Ti	22	47.90
tungsten	W	74	183.85
uranium	U	92	238.03
vanadium	V	23	50.94
xenon	Xe	54	131.30
ytterbium	Yb	70	173.04
yttrium	Y	39	88.90
zinc	Zn	30	65.37
zirconium	Zr	40	91.22

Temperature-Conversion Table*

C	0	10	20	30	40	50	60	70	80	90
	F	F	F	F	F	F	F	F	F	F
−200	−328	−346	−364	−382	−400	−418	−436	−454		
−100	−148	−166	−184	−202	−220	−238	−256	−274	−292	−310
− 0	+ 32	+ 14	− 4	− 22	− 40	− 58	− 76	− 94	−112	−130
0	32	50	68	86	104	122	140	158	176	194
100	212	230	248	266	284	302	320	338	356	374
200	392	410	428	446	464	482	500	518	536	554
300	572	590	608	626	644	662	680	698	716	734
400	752	770	788	806	824	842	860	878	896	914
500	932	950	968	986	1004	1022	1040	1058	1076	1094
600	1112	1130	1148	1166	1184	1202	1220	1238	1256	1274
700	1292	1310	1328	1346	1364	1382	1400	1418	1436	1454
800	1472	1490	1508	1526	1544	1562	1580	1598	1616	1634
900	1652	1670	1688	1706	1724	1742	1760	1778	1796	1814
1000	1832	1850	1868	1886	1904	1922	1940	1958	1976	1994
1100	2012	2030	2048	2066	2084	2102	2120	2138	2156	2174
1200	2192	2210	2228	2246	2264	2282	2300	2318	2336	2354
1300	2372	2390	2408	2426	2444	2462	2480	2498	2516	2534
1400	2552	2570	2588	2606	2624	2642	2660	2678	2696	2714
1500	2732	2750	2768	2786	2804	2822	2840	2858	2876	2894
1600	2912	2930	2948	2966	2984	3002	3020	3038	3056	3074
1700	3092	3110	3128	3146	3164	3182	3200	3218	3236	3254
1800	3272	3290	3308	3326	3344	3362	3380	3398	3416	3434
1900	3452	3470	3488	3506	3524	3542	3560	3578	3596	3614
2000	3632	3650	3668	3686	3704	3722	3740	3758	3776	3794
2100	3812	3830	3848	3866	3884	3902	3920	3938	3956	3974
2200	3992	4010	4028	4046	4064	4082	4100	4118	4136	4154
2300	4172	4190	4208	4226	4244	4262	4280	4298	4316	4334
2400	4352	4370	4388	4406	4424	4442	4460	4478	4496	4514
2500	4532	4550	4568	4586	4604	4622	4640	4658	4676	4694
2600	4712	4730	4748	4766	4784	4802	4820	4838	4856	4874
2700	4892	4910	4928	4946	4964	4982	5000	5018	5036	5054
2800	5072	5090	5108	5126	5144	5162	5180	5198	5216	5234
2900	5252	5270	5288	5306	5324	5342	5360	5378	5396	5414
3000	5432	5450	5468	5486	5504	5522	5540	5558	5576	5594
3100	5612	5630	5648	5666	5684	5702	5720	5738	5756	5774
3200	5792	5810	5828	5846	5864	5882	5900	5918	5936	5954
3300	5972	5990	6008	6026	6044	6062	6080	6098	6116	6134
3400	6152	6170	6188	6206	6224	6242	6260	6278	6296	6314
3500	6332	6350	6368	6386	6404	6422	6440	6458	6476	6494
3600	6512	6530	6548	6566	6584	6602	6620	6638	6656	6674
3700	6692	6710	6728	6746	6764	6782	6800	6818	6836	6854
3800	6872	6890	6908	6926	6944	6962	6980	6998	7016	7034
3900	7052	7070	7088	7106	7124	7142	7160	7178	7196	7214

°C	°F
1	1.8
2	3.6
3	5.4
4	7.2
5	9.0
6	10.8
7	12.6
8	14.4
9	16.2
10	18.0

°F	°C
1	0.56
2	1.11
3	1.67
4	2.22
5	2.78
6	3.33
7	3.89
8	4.44
9	5.00
10	5.56
11	6.11
12	6.67
13	7.22
14	7.78
15	8.33
16	8.89
17	9.44
18	10.00

*Dr. L. Waldo, *Metallurgical and Chemical Engineering*, March, 1910.

Examples: 1648°C = 2984°F + 14.4°F = 2998°F
3267°F = 1790°C + 7.22°C = 1797°C

Weights and Measures

Metric System

Weight

1000	grams	= 1 kilogram	(kg.)
100	grams	= 1 hektogram	(hg.)
10	grams	= 1 decagram	(dg.)
1	gram	= 1 gram	(gm.)
0.1	gram	= 1 decigram	(dg.)
0.01	gram	= 1 centigram	(cg.)
0.001	gram	= 1 milligram	(mg.)

Length

1000	meters	= 1 kilometer	(km.)
100	meters	= 1 hektometer	(hm.)
10	meters	= 1 dekameter	(dm.)
1	meter	= 1 meter	(m.)
0.1	meter	= 1 decimeter	(dm.)
0.01	meter	= 1 centimeter	(cm.)
0.001	meter	= 1 millimeter	(mm.)

Capacity (liquid)

1000	liters	= 1 kiloliter	(kl.)
100	liters	= 1 hektoliter	(hl.)
10	liters	= 1 dekaliter	(dl.)
1	liter	= 1 liter	(l.)
0.1	liter	= 1 deciliter	(dl.)
0.01	liter	= 1 centiliter	(cl.)
0.001	liter	= 1 cubic centimeter	(cc.)

Weights and Measures

United States System

Weight
(Avoirdupois)

1 long ton	= 2240 pounds (lb.)		
1 short ton	= 2000 pounds		
1 pound	= 16 ounces (oz.)	= 7000 grains	
1 ounce	= 437.5 grains (gr.)		
1 grain			

(Troy)

1 pound (lb.)	= 12 ounces
1 ounce (oz.)	= 20 pennyweights
1 pennyweight (dwt.)	= 24 grains
1 grain	
1 avoirdupois pound	= 1.21528 troy pounds
1 avoirdupois pound	= 14.583 troy ounces

Length

1 inch (in.)		
1 foot (ft.)	= 12 inches	
1 yard (yd.)	= 36 inches	
1 rod (rd.)	= 198 inches	= 16.5 feet
1 chain (ch.)	= 792 inches	= 66 feet
1 mile (mi.)	= 5280 feet	= 1760 yards

Capacity (liquid)

1 fluid ounce (fl. oz.)	
1 pint (pt.)	= 16 fluid ounces
1 quart (qt.)	= 2 pints
1 gallon (gal.)	= 4 quarts

Weights and Measures

Comparison of the Metric and U.S. Systems of Measures and Weights

Weight Relations

(Avoirdupois)

I		II	
1 gram	=	0.35274	ounce (avoir.)
1 kilogram	=	35.274	ounces
1 kilogram	=	2.2046	pounds
1 metric ton	=	2204.62	pounds
1 metric ton	=	1.10231	short tons
1 metric ton	=	0.984206	long ton

(Troy)

I		II	
1 gram	=	0.032151	ounce (troy)
1 kilogram	=	32.151	ounces
1 kilogram	=	2.6792	pounds

(Avoirdupois)

I		II	
1 ounce (avoir.)	=	28.3495	grams
1 pound	=	453.5924	grams
1 short ton	=	907.185	kilograms
1 long ton	=	1016.647	kilograms

(Troy)

I		II	
1 ounce (troy)	=	31.1035	grams
1 pound	=	373.2418	grams

To convert units of one system into those of the other, multiply the number of units by the equivalent opposite that unit in column II.

For example: To convert 3.5 kilograms into avoirdupois ounces, multiply 3.5 by 35.274 ($3.5 \times 35.274 = 123.46$ ounces).

Length

I		II		I		II	
1 millimeter	=	0.03937	inch	1 inch	=	2.5400	centimeters
1 centimeter	=	0.3937	inch	1 foot	=	30.480	centimeters
1 meter	=	39.37	inches	1 foot	=	0.3048	meter
1 meter	=	3.2808	feet	1 yard	=	91.440	centimeters
1 meter	=	1.09361	yards	1 yard	=	0.9144	meter

To convert units of one system into those of the other, multiply the number of units by the equivalent opposite that unit in column II.

For example: To convert 8.12 centimeters into inches, multiply 8.12 by 0.3937 ($8.12 \times 0.3937 = 3.197$ inches).

Comparison of the Metric and U.S. Systems of Measures and Weights (continued)

Capacity (liquid measure)

I		II	
1 liter	=	33.815	fluid ounces
1 liter	=	2.1134	pints
1 liter	=	1.0567	quarts
1 liter	=	0.26418	gallon
1 gallon	=	3.7853	liters
1 quart	=	0.9463	liter
1 pint	=	0.47317	liter
1 ounce	=	29.5729	cc.

To convert units of one system into those of the other, multiply the number of units by the equivalent opposite that unit in column II.

Volume

1 cubic centimeter	=	0.0616234	cubic inch
1 cubic meter	=	35.3145	cubic feet
1 cubic meter	=	1.30794	cubic yards
1 cubic inch	=	16.38716	cubic centimeters
1 cubic foot	=	0.028317	cubic meter
1 cubic yard	=	0.76456	cubic meter

Area

1 square millimeter	=	0.00155	square inch
1 square centimeter	=	0.15501	square inch
1 square meter	=	1550.1	square inches
1 square meter	=	10.7643	square feet
1 square meter	=	1.19603	square yards
1 square foot	=	929.034	square centimeters
1 square yard	=	0.836131	square meter

Conversion Table*
for
Volumes and Weights

To Convert from	Multiply by							
	To Cu. In.	To Cu. Ft.	To Cu. Yd.	To Fl. Oz.	To Pint	To Quart	To Gallon	To Grains
Cu. in	1.00000	$.0_35787$	$.0_42143$	.554112	.034632	.017316	.004329	252.891
Cu. ft.	1728.00	1.00000	.037037	957.505	59.8442	29.9221	7.48052	436996
Cu. yd	46656.0	27.0000	1.00000	25852.6	1615.79	807.896	201.974	117990_3
Fl. oz.	1.80469	.001044	$.0_43868$	1.00000	.062500	.031250	.007813	456.390
Pint	28.8750	.016710	$.0_36189$	16.0000	1.00000	.500000	.125000	7302.23
Quart	57.7500	.033420	.001238	32.0000	2.00000	1.00000	.250000	1460.45
Gallon	231.000	.133681	.004951	128.000	8.00000	4.00000	1.00000	58417.9
Grain	.003954	$.0_52288$	$.0_78475$	.002191	$.0_31369$	$.0_46850$	$.0_41712$	1.00000
Oz., troy	1.89805	.001098	$.0_44068$	1.05173	.065733	.032867	.008217	480.000
Oz., av	1.72999	.001001	$.0_43708$	.958608	.059913	.029957	.007489	437.500
Lb., troy	22.7766	.013181	$.0_34882$	12.6208	.788800	.394400	.098600	5760.00
Lb., av	27.6799	.016018	$.0_35933$	15.3378	.958611	.479306	.119826	7000.00
Cc. or gram	.061024	$.0_43531$	$.0_51308$	.033814	.002113	.001057	$.0_32642$	15.4323
Liter or kg	61.0237	.035315	.001308	33.8140	2.11337	1.05669	.264172	15432.3
Cu. M	61023.7	35.3146	1.30795	33814.0	2113.37	1056.69	264.172	154320_3

To Convert from	Multiply by						
	To Oz. Troy	To Oz. Av.	To Lb. Troy	To Lb. Av.	To Cc. or G.	To Ltr. or Kg.	To Cu. M.
Cu. in	.526857	.578037	.043905	.036127	16.3871	.016387	$.0_41639$
Cu. ft.	910.408	998.848	75.8674	62.4280	28316.9	28.3169	.028317
Cu. yd	24581.0	26968.9	2048.42	1685.56	764556	764.556	.764556
Fl. oz.	.950813	1.04318	.079234	.065199	29.5736	.029573	$.0_42957$
Pint	15.2130	16.6908	1.26775	1.04318	473.177	.473177	$.0_34732$
Quart	30.4260	33.3816	2.53550	2.08635	946.354	.946354	$.0_39463$
Gallon	121.704	133.527	10.1420	8.34541	3785.42	3.78542	.003785
Grain	.002083	.002286	$.0_31736$	$.0_31428$	.064799	$.0_46479$	$.0_76479$
Oz., troy	1.00000	1.09714	.083333	.068571	31.1035	.031104	$.0_43110$
Oz., av	.911457	1.00000	.075955	.062500	28.3495	.028350	$.0_42835$
Lb., troy	12.0000	13.1657	1.00000	.822857	373.242	.373242	$.0_33732$
Lb., av	14.5833	16.0000	1.21528	1.00000	453.593	.453593	$.0_34536$
Cc. or gram	.032151	.035274	.002679	.002205	1.00000	.001000	.000001
Liter or kg	32.1507	35.2739	2.67923	2.20462	1000.00	1.00000	.001000
Cu. M	32150.7	35273.9	2679.23	2204.62	1000000	1000.00	1.00000

*Olsen, Van Nostrand's Chemical Annual, 1926.

Note: The small subnumeral following a zero indicates that the zero is to be taken that number of times; for example, 0.0_31428 is equivalent to 0.0001428.

End Points of Orton Pyrometric Cones

Cone No.	End point large cone, °C	End point small cone, °C	Cone No.	End point large cone, °C	End point small cone, °C
022	585		10	1285	1330†
021	602	643†	11	1294	1336†
020	625	666†	12	1306	1355†
019	668	723†	13	1321	1349†
018	696	752†	14	1388	1398‡
017	727	784†	15	1424	1430‡
016	764	825†	16	1455	1491‡
015	790	843†	17	1477	1512‡
014	834	—	18	1500	1522‡
013	869	—	19	1520	1541‡
012	866	—	20	1542	1564‡
011	886	—	23	1586	1605§
010	887	919†	26	1589	1621§
09	915	955†	27	1614	1640§
08	945	983†	28	1614	1646§
07	973	1008†	29	1624	1659§
06	991	1023†	30	1636	1665§
05	1031	1062†	31	1661	1683§
04	1050	1098†	31½		1699§
03	1086	1131†	32	1706	1717§
02	1101	1148†	32½	1718	1724§
01	1117	1178†	33	1732	1743§
1	1136	1179†	34	1757	1763§
2	1142	1179†	35	1784	1785§
3	1152	1197†	36	1798	1804§
4	1168	1209†	37	—	1820§
5	1177	1221†	38	—	1850¶
6	1201	1255†	39	—	1865¶
7	1215	1264†	40	—	1885¶
8	1236	1300†	41	—	1970¶
9	1260	1317†	42	—	2015¶

*National Bureau of Standards, 1956.

†Small uncalcined cones, heated in air at 60°C/hr. rise.

‡P.C.E. cones heated to 1200°C in 1 hr., and then at 300°C/hr. rate.

§P.C.E. cones heated to 1400°C in 1 hr., and then at 150°C/hr. rate.

¶P.C.E. cones heated at 600°C/hr in combustion furnace (1926). All large cones heated at a rate of 60°C/hr.

Some Factors for Calculating Properties of Glass Composition

Composition	THERMAL EXPANSION				Heat conductivity	Density		Tensile strength W.&S. Kg. per sq. mm.	Crushing strength W.&S. Kg. per sq. mm.	Elasticity	Hardness	Specific gravity
	M.&H. × 10^7	W.&S. × 10^7	E.&T. × 10^7	F.&P. × 10^7	P.&F.	W.&S.	Baillie			C.&T.	Auerbach	H.&C.
SiO_2	0.8	0.8	0.15		0.0220	2.3	2.24	0.09	1.23	40	+ 3.32	0.001913
Al_2O_3	5.0	5.0	0.42		0.0220	4.1	2.75	0.05	1.0	120	+10.1	
B_2O_3	0.1	0.1	−1.98		0.0160	1.9	3.00	0.065	0.90		+ 0.75	0.002272
Na_2O	10.0	10.0	12.96	12.5	0.0160	2.6	3.20	0.02	0.52	110	− 2.65	0.002674
K_2O	8.5	8.5	11.7		0.0010	2.8	3.20	0.01	0.05		+ 3.9	0.001860
PbO	4.2	3.0	3.18		0.0080	9.6	10.30	0.025	0.48		+ 1.45	0.000512
ZnO	2.1	1.8	2.1	1.85	0.0160	5.9	5.94	0.15	0.6		+ 7.1	0.001245
CaO	5.0	5.0	4.89		0.0320	3.3	4.30	0.20	0.20	240	− 6.3	0.001903
MgO	0.1	0.1	1.35			3.8	3.25	0.01	1.1	300		0.002439
BaO	3.0	3.0	4.2	5.7	0.0110	7.0	7.20	0.05	0.65		+ 1.95	0.000673
As_2O_5	2.0	2.0				4.1	2.90	0.03	1.00			0.001276
P_2O_5	2.0	2.0			0.0160	2.55		0.075	0.76		+ 1.32	0.001903
Sb_2O_5	3.6											
SnO_2	2.0											
TiO_2	4.1											
ZrO_2	2.1											
Na_3AlF_6	7.4											
NaF	7.4											
AlF_3	4.4											
CaF_2	2.5											
Cr_2O_3	5.1											
CoO	4.4											
CuO	2.2											
Fe_2O_3	4.0											
NiO	4.0											
MnO_2	2.2											

M.&H.—Mayer and Havas: Sprechsaal, *42* 497 (1909); Sprechsaal, *44* 188, 207, 220 (1911).
W.&S.—Winkelmann and Schott: *Ann. Phys. Chem.*, *51* 730, 735 (1894).
E.&T.—English and Turner; *Jour. Soc. Glass Tech.*, *4* 115 (1920); *Jour. Soc. Glass Tech.* *5* 121 (1921); *Jour. A. Cer. S.*, *12* 760 (1929).
F.&P.—Fetterolf and Parmelee: *Jour. A. Cer. S.*, *12* 214 (1929).
P.&F.—Paalborn and Focke: *Jena Glass*, p. 212.
—Baillie: *Jour. Soc. Chem. Ind.*, *40* 141 (1921).
C.&T.—Clark and Turner: *Jour. Soc. Glass Tech.*, *3* 260 (1919).
H.&.C.—Hodkin and Cousen: *Glass Technology*, D. Van Nostrand, 1925.
—Auerbach: *Glass Technology*, D. Van Nostrand, 1925.

Ceramic Colors*†

Color	Fired under oxidizing conditions	Fired under reducing conditions
White	Aluminum oxide	
	Antimony oxide	
	Arsenic oxide	
	Calcium borate	
	Cerium compounds	
	Magnesium carbonate	
	Magnesium oxide	
	Silver	
	Tin oxide	
	Titanium dioxide	
	Zinc oxide	
	Zirconium dioxide	
Black	Chrome ore + pyrolusite + cobalt oxide	Bismuth salts
	Chromium	Carbides
	Cobalt	Carbon
	Iridium compounds	Iridium sesquioxide
	Iron oxides	Lead salts
	Manganese oxides	Molybdenum compounds
	Nickel oxides	Nickel oxide
	Pyrolusite	Sulfides
	Uranium oxide + copper oxide	Uranium oxide
Gray	Antimony gray	Antimony (metallic)
	Iridium oxide	Carbon compounds
	Osmium oxide	Chromium
	Palladium oxide	Cobalt
	Platinum salts	Copper
	Rhodium oxide	Manganese
	Ruthenium oxide	Molybdenum compounds
		Nickel compounds
		Stannous oxide
		Uranium
		Vanadium salts
Silver	Palladium	
	Platinum	
	Silver (metallic)	
Red	Bismuth uranate	Copper salts
	Cadmium sulfide + cadmium selenide + barytes	
	Iron oxide, iron salts	
	Lead chromate (basic)	
	Lead uranate	
	Manganese oxide	
	Manganese pink	
	Neodymium salts	
	Purple of Cassius	
	Sodium diuranate	
Pink	Chrome-tin combinations	

Color	Fired under oxidizing conditions	Fired under reducing conditions
Orange	Bismuth uranate Cadmium sulfide and selenide Chrome iron ore Iron oxide + chromates Iron titanate Lead chromate (basic) Lead uranate Manganese oxide + titanates Manganese tungstate Sodium diuranate Uranium titanate	
Gold	Metallic gold	
Yellow	Barium chromate Bismuth salts Cadmium sulfide Ceric oxide Gold salts Iron oxide (with litharge) Lead chromate Lead oxide + antimony oxide Manganese dioxide Molybdenum salts Nickel oxide Praseodymium salts Pyrolusite Silver salts Sodium diuranate Vanadium stannate Zinc chromate	Praseodymium salts Vanadium stannate
Green (sea)	Cobalt antimonate Copper compounds (lead glaze) Nickel-zinc oxides	Chromic oxide Cobalt titanate
Green (leaf)	Chromic oxide Cobalt titanate Copper salts Nickel oxide + zinc oxide Praseodymium salts	Cobalt titanate Praseodymium salts
Blue (ice)	Copper compounds Nickel oxide + zinc oxide	Titanium dioxide (rutile)
Blue (deep)	Cobalt compounds Neodymium compounds	Titanium dioxide (rutile) Vanadium compounds
Violet	Nickel oxide Pyrolusite Purple of Cassius	Copper (colloidal metal) Titanium dioxide (rutile)

*Wolf, Josef: Welche Grundstoffe und ihre Verbindungen werden als Farbemittel in der Tonwaren-, Glas-, und Emailerzeugung Verwendet, *Sprechsaal 70* 48, 49, 50, 601, 612, 625 (1937).

†Singer, Felix and Singer, Sonja S.: *Industrial Ceramics,* Chemical Publishing Co., Inc., 1963.

Weight and Approximate Thickness of Sheet Steel*
United States Standard Gauge for Sheet and Plate Steel

Number of gauge	*Weight per square foot in pounds avoirdupois*	*Equivalent thickness of steel in decimal parts of an inch*	*Approximate thickness in fractions of an inch*
0000000	20.0	.4900	1/2
000000	18.75	.4594	15/32
00000	17.50	.4288	7/16
0000	16.25	.3981	13/32
000	15.00	.3675	3/8
00	13.75	.3369	11/32
0	12.50	.3063	5/16
1	11.25	.2757	9/32
2	10.625	.2604	17/64
3	10.0	.2451	1/4
4	9.375	.2298	15/64
5	8.75	.2145	7/32
6	8.125	.1991	13/64
7	7.50	.1838	3/16
8	6.875	.1685	11/64
9	6.25	.1532	5/32
10	5.625	.1379	9/64
11	5.0	.1225	1/8
12	4.375	.1072	7/64
13	3.75	.0919	3/32
14	3.125	.0766	5/64
15	2.1825	.0689	9/128
16	2.50	.0613	1/16
17	2.25	.0551	9/160
18	2.00	.0490	1/20
19	1.75	.0429	7/160
20	1.50	.0368	3/80
21	1.375	.0337	11/320
22	1.25	.0306	1/32
23	1.125	.0276	9/320
24	1.00	.0245	1/40
25	.875	.0214	7/320
26	.75	.0184	3/160
27	.6875	.0169	11/640
28	.625	.0153	1/64
29	.5625	.0138	9/640
30	.50	.0123	1/80

*Andrews: *Enamels,* The Twin City Printing Co., 1935.

Bibliography

Bibliography

The collection of words and terms for inclusion in a dictionary for an industry so large, so diverse, so complex, and so widespread as the ceramic industry has been a monumental task. The selection of these words and terms literally has involved a review of more than a hundred-odd textbooks, professional papers appearing in technical and trade publications, corporate literature, advertisements, technical dictionaries, and personal correspondence. The selection, to a considerable degree, was discretionary and was predicated primarily on the frequency with which the terms appeared in the literature, as well as on the importance or potential importance they were assigned in the industry. Certainly, it was an effort to bring information from many sources into a single collection.

Most of the definitions are original and are based on the experiences of the author, not only in the United States but in an impressive number of countries on all continents as well. Others are composites of several published definitions, while still others are reproduced as they have been defined in the literature. It should be noted that information on many raw materials is scant; these materials, however, have been or are being investigated more or less as laboratory curiosities and are included for the benefit of researchers and others who might find them of interest.

The sources of information most frequently consulted, and to whom the author is deeply indebted and most appreciative, are as follows:

The American Ceramic Society Bulletin

The American Ceramic Society Journal

The American Heritage Dictionary of the English Language, American Heritage Publishing Co., Inc., and Houghton Mifflin Company, 1969.

American Society for Testing and Materials, Designation Number:

B 542 Electrical contacts
C 11 Gypsum
C 43 Structural clay products
C 51 Lime
C 71 Refractories
C 119 Natural Building Stones
C 125 Concrete and concrete aggregates
C 162 Glass
C 219 Hydraulic cement
C 242 Whitewares
C 286 Porcelain enamel and ceramic-metal systems
C 460 Asbestos cement
C 709 Manufactured carbon and graphite
C 822 Concrete pipe
D 1711 Electrical inspection
D 2652 Activated carbon
E 268 Electromagnetic testing
E 269 Magnetic particle inspection
E 270 Liquid penetrant inspection
F 109 Surface imperfections on ceramics

Anderson, Harriette. *Kiln-Fired Glass.* Radnor, Pa: Chilton Book Company, 1970.

Andrews, A. I. *Ceramic Tests and Calculations.* New York: John Wiley & Sons, Inc., 1928.

Andrews, A. I. *Enamels.* Lewiston, Me: The Twin City Printing Co., 1933.

Barna, Gordon L. Kaiser Refractories. Personal communication.

Belger, Joseph H. Lingl Corporation. Personal communication.

Bryant, E. E. *Porcelain Enameling Operations.* Cleveland, Ohio: Enamelist Publishing Company, 1964.

Budnikov, P. P. *The Technology of Ceramics and Refractories.* Cambridge, Mass.: MIT Press, 1964.

Ceramic Industry Magazine. "Annual Raw Materials Issue." January 1978.

Chesters, J. H. *Refractories: Production and Properties.* The Iron and Steel Institute, 1973.

Dodd, A. E. *Dictionary of Ceramics.* Totowa, N.J.: Littlefield, Adams & Co., 1967.

Encyclopedia Americana. New York: Americana Corporation. 1976.

Encyclopedia Britannica. Chicago: University of Chicago, 1982.

Encyclopedia International. New York: Golier, Incorporated, 1963.

Hamer, Frank. *The Potter's Dictionary of Materials and Techniques.* London, England: Pitman Publishing, 1975.

Kepple, John B. American Glass Research, Inc. Personal communication.

Kerr, John M. Babcock & Wilcox Co. Personal communication.

Kingery, W. D. *Introduction to Ceramics.* New York: John Wiley & Sons, Inc., 1960.

Lapedes, Daniel., ed. *Dictionary of Scientific and Technical Terms.* 2ed. New York: McGraw-Hill Book Company, 1978.

Leach, Bernard. *A Potter's Book.* Central Islip, N.Y.: Transatlantic Arts, Inc., 1969.

Lee, P. William. *Ceramics.* New York: Reinhold Publishing Corp., 1961.

Lynch, J. F., Ruderer, C. G., and Duckworth, W. H. *Engineering Properties of Selected Ceramic Materials.* Columbus, Ohio: The American Ceramic Society, 1966.

McGrath, Raymond, Frost, A. C., and Beckett, H. E. *Glossary of Terms Connected with Glass and Glazing.* London, England: The Architectural Press, 1961.

McMillan, P. W. *Glass Ceramics.* New York: Academic Press, 1969.

Morey, George W. *Properties of Glass.* New York: The Guinn Co., 1947.

Nelson, Glenn C. *Ceramics, A Potter's Handbook.* New York: Holt Rinehart and Winston, Inc., 1971.

Norton, F. H. *Ceramics for the Artist Potter.* Reading, MA: Addison-Wesley Publishing Co., Inc., 1956.

Norton, F. H. *Elements of Ceramics.* Reading, MA: Addison-Wesley Press, Inc., 1952.

Norton, F. H. *Fine Ceramics.* New York: McGraw-Hill Book Company, 1970.

Onoda, George Y., Jr. and Hench, Larry L. *Ceramic Processing Before Firing.* New York: John Wiley & Sons, 1978.

Parmelee, Cullen W. *Ceramic Glazes.* New York: Industrial Publications, 1948.

Perkins, Walter W. Norton Co. Personal communication.

Popper, P. *Special Ceramics 1962.* New York: Academic Press, 1963.

Rado, Paul. *An Introduction to the Technology of Pottery.* Elmsford, N.Y.: Pergamon Press, 1969.

Rose, Arthur and Rose, Elizabeth. *The Condensed Chemical Dictionary.* New York: Reinhold Publishing Corporation, 1956.

Roy, Vincent A. *Ceramics.* New York: McGraw-Hill Book Company, Inc., 1959.

Ryshkewitch, Eugene. *Oxide Ceramics.* New York: Academic Press, 1960.

Salmang, Hermann and Francis, Marcus. *Ceramics, Physical and Chemical Fundamentals.* Woburn, MA: Butterworths & Co., 1961.

Shand, E. B. *Glass Engineering Handbook.* New York: McGraw-Hill Book Company, Inc., 1958.

Singer, Felix and Singer, Sonja A. *Industrial Ceramics.* New York: Chemical Publishing Company, Inc., 1963.

Stewart, G. H. *Science of Ceramics.* New York: Academic Press, 1962.

Van Schoick, Emily C. *Ceramic Glossary.* Columbus, OH: The American Ceramic Society, 1964.

Waddell, Joseph J. *Practical Quality Control for Concrete.* New York: McGraw-Hill Book Co., Inc., 1962.

Webster's New Twentieth Century Dictionary. 2ed., Unabridged. New York: Simon and Schuster, 1979.

World Book Encyclopedia. San Francisco, CA: Field Enterprises Educational Corporation, 1969.

Encyclopedia Americana. New York: Americana Corporation. 1976.
Encyclopedia Britannica. Chicago: University of Chicago, 1982.
Encyclopedia International. New York: Golier, Incorporated, 1963.
Hamer, Frank. *The Potter's Dictionary of Materials and Techniques*. London, England: Pitman Publishing, 1975.
Kepple, John B. American Glass Research, Inc. Personal communication.
Kerr, John M. Babcock & Wilcox Co. Personal communication.
Kingery, W. D. *Introduction to Ceramics*. New York: John Wiley & Sons, Inc., 1960.
Lapedes, Daniel., ed. *Dictionary of Scientific and Technical Terms*. 2ed. New York: McGraw-Hill Book Company, 1978.
Leach, Bernard. *A Potter's Book*. Central Islip, N.Y.: Transatlantic Arts, Inc., 1969.
Lee, P. William. *Ceramics*. New York: Reinhold Publishing Corp., 1961.
Lynch, J. F., Ruderer, C. G., and Duckworth, W. H. *Engineering Properties of Selected Ceramic Materials*. Columbus, Ohio: The American Ceramic Society, 1966.
McGrath, Raymond, Frost, A. C., and Beckett, H. E. *Glossary of Terms Connected with Glass and Glazing*. London, England: The Architectural Press, 1961.
McMillan, P. W. *Glass Ceramics*. New York: Academic Press, 1969.
Morey, George W. *Properties of Glass*. New York: The Guinn Co., 1947.
Nelson, Glenn C. *Ceramics, A Potter's Handbook*. New York: Holt Rinehart and Winston, Inc., 1971.
Norton, F. H. *Ceramics for the Artist Potter*. Reading, MA: Addison-Wesley Publishing Co., Inc., 1956.
Norton, F. H. *Elements of Ceramics*. Reading, MA: Addison-Wesley Press, Inc., 1952.
Norton, F. H. *Fine Ceramics*. New York: McGraw-Hill Book Company, 1970.
Onoda, George Y., Jr. and Hench, Larry L. *Ceramic Processing Before Firing*. New York: John Wiley & Sons, 1978.
Parmelee, Cullen W. *Ceramic Glazes*. New York: Industrial Publications, 1948.
Perkins, Walter W. Norton Co. Personal communication.
Popper, P. *Special Ceramics 1962*. New York: Academic Press, 1963.
Rado, Paul. *An Introduction to the Technology of Pottery*. Elmsford, N.Y.: Pergamon Press, 1969.
Rose, Arthur and Rose, Elizabeth. *The Condensed Chemical Dictionary*. New York: Reinhold Publishing Corporation, 1956.
Roy, Vincent A. *Ceramics*. New York: McGraw-Hill Book Company, Inc., 1959.
Ryshkewitch, Eugene. *Oxide Ceramics*. New York: Academic Press, 1960.
Salmang, Hermann and Francis, Marcus. *Ceramics, Physical and Chemical Fundamentals*. Woburn, MA: Butterworths & Co., 1961.
Shand, E. B. *Glass Engineering Handbook*. New York: McGraw-Hill Book Company, Inc., 1958.
Singer, Felix and Singer, Sonja A. *Industrial Ceramics*. New York: Chemical Publishing Company, Inc., 1963.
Stewart, G. H. *Science of Ceramics*. New York: Academic Press, 1962.
Van Schoick, Emily C. *Ceramic Glossary*. Columbus, OH: The American Ceramic Society, 1964.
Waddell, Joseph J. *Practical Quality Control for Concrete*. New York: McGraw-Hill Book Co., Inc., 1962.
Webster's New Twentieth Century Dictionary. 2ed., Unabridged. New York: Simon and Schuster, 1979.
World Book Encyclopedia. San Francisco, CA: Field Enterprises Educational Corporation, 1969.